The Great Rift

The Great Rift

LITERACY, NUMERACY, *and the* RELIGION–SCIENCE DIVIDE

MICHAEL E. HOBART

HARVARD UNIVERSITY PRESS
Cambridge, Massachusetts
London, England
2018

Printed in the United States of America

First printing

Library of Congress Cataloging-in-Publication Data

Names: Hobart, Michael E., 1944– author.

Title: The great rift : literacy, numeracy, and the religion-science divide / Michael E. Hobart.

Description: Cambridge, Massachusetts : Harvard University Press, 2018. | Includes bibliographical references and index.

Identifiers: LCCN 2017045244 | ISBN 9780674983632 (hardcover : alk. paper)

Subjects: LCSH: Religion and science—History. | Numeration—History. | Mathematics—History. | Mathematics, Medieval. | Science, Renaissance. | Signs and symbols—History.

Classification: LCC BL265.M3 H63 2018 | DDC 201/.65—dc23 LC record available at https://lccn.loc.gov/2017045244

For Sally, Abigail, and Julia

L'amor che move il sole e l'altre stelle.

—Dante, *Commedia*

Contents

Preface and Acknowledgments

This book uses the history of information technology—in particular, the shift from alphabetic literacy to modern numeracy—to narrate and explain the origins of the contemporary rift between science and religion. Its own origins date to several decades ago when, as a young PhD candidate, I interviewed for a job. I was then approaching the end of my dissertation on the French philosopher Nicolas Malebranche ("Science and Religion in the Thought of Nicolas Malebranche"), and I had prepared for the interview with some thoughts about how it could be turned into a book. (It was subsequently published by a university press.) But the interviewer showed no interest whatsoever in that prospect. Instead he asked what I planned to do after finishing with Malebranche. Translation? What was to be the title of my second book? I hadn't a clue. I had been working on the dissertation forever, it seemed, and just wanted the damned thing done. So I could only stammer, "Well . . . , I think I'd like to look further into the relation between knowing and believing, between science and religion, perhaps among other early modern intellectuals." Whereupon he looked at me and said, emphatically and with a dripping dose of snarkiness, "But I *know* what I *believe*." I replied he was luckier than most, which did draw a laugh—but not a job (that came later and elsewhere).

Ever since then, that exchange has echoed down the hallways, in the classrooms, and throughout countless discussions with students and colleagues and friends. In particular, I have been struck over the years by the curiosity and intensity with which students address matters pertaining to science and religion and the divide between them—often wondering how or whether these separate worlds of thought might accord with one another, what these expansive and often quarreling categories might portend for their own lives.

And in their existential concerns I saw an ongoing mirror of my own, which have remained at the forefront of my reflections throughout a lengthy teaching career. While some colleagues and contemporaries, like my interviewer, have no doubt resolved to their satisfaction the doubts of their youth, many others among us—even as we approach the midnight of our life's day—still grapple with these fundaments of our being, reveling ever more in questions, less in answers.

Since those interview days, too, the pathway of my scholarship has followed some surprising twists and turns from that ill-fated, diffident response to my interlocutor, and I know now what I didn't then: where I was going. As a scholar, beginning with my study of the religion-science divide in the philosophy of Malebranche, I explored several seemingly different historical and philosophical topics with an eye toward engaging tensions and polarities in our thinking, fault lines that crisscross our intellect. Through several studies I came to appreciate that coursing through these topics and carving one of the most central fault lines in our modern age has been what I am calling here the analytical temper, the marked propensity for reverse engineering the phenomena of nature—and, by some accounts, mutatis mutandis, of culture and history. Historically, the human capacity for disassembling and reassembling features of our thought, experience, and consciousness lies deep within the mists of our past, most likely appearing with our evolutionary emergence as a symbolic species. But this capacity took a transformative turn with the rise of modern science in the sixteenth through eighteenth centuries, when reverse engineering, in symbiosis with modern, abstract mathematics, shaped the attitudes, practices, and theoretical engagements that typify the analytical temper today.

These earlier studies eventually led me to embark with my close colleague and friend Zachary Schiffman on a historical examination of information technology—*Information Ages: Literacy, Numeracy, and the Computer Revolution*—as a means to understanding such transformations as that just mentioned. We had both become convinced that, properly understood, an information technology stands as a humanly constructed screen between the knowing mind and the world outside, highly shaping and conditioning (though not determining) our understanding of nature and history alike. Explicating this, we believed, would put our own "information age" into some historical perspective. We corroborated our conviction by sketching the creation and development of three such technologies in order to articulate the roles played by literacy (especially the invention of the alphabet and its ties to a classifying mindset), by numeracy (especially the development of abstract, mathematical symbols in the formation of modern science and its contributions to an "analytical vision" of scientific activity), and by the computer revolution (especially its refinement and furthering of the abstraction begun with modern numeracy and its subsequent "digital" union of symbol and electronic circuitry).

Given the broad scope of our extended, historical essay, Zach and I were not able to delve as deeply into some of its chapters as we might have wished, not able to examine the implications of the transitions we outlined. This I now do in the following pages, looking more directly and attentively into the transition from literacy to numeracy. I do so with a particular eye to discovering the Renaissance preconditions that made possible the emergence of modern, relational numeracy as an information technology and from there the modern, abstract mathematics that then settled into the core of scientific thinking, where it remains still. The payoff of this current historical narrative and analysis lies with recognizing and unearthing the momentous consequences that the shift from literacy to numeracy augured for the divide between religion and science, our Great Rift of today. This book thus marks a return to that early exchange with my interviewer and to what I trust is a more engaging and thoroughgoing response: the culmination of a lifetime of teaching, professional scholarship, and personal reflection on this most important of topics.

If nature no longer "abhors a vacuum"—modern science having overturned this Aristotelian dictum—historical writing still does, for it is not created in vacuo. Over the years, the many scholars cited in my notes, as well as countless other people (far too many to acknowledge adequately), have come to populate my own intellectual plenum as friends and critics alike, some more so than others. From the beginning of this project, Zach Schiffman has been steadfastly my staunchest supporter and my most severe critic, as well as an always available whetstone for my own wits. Having read and critiqued every chapter in each of its several drafts, his thought has continued to merge with mine in many ways since the days of our joint authorship, so much so that it's sometimes hard to tell where his leaves off and mine begins, or vice versa. Ours has been a salutary and exceedingly rare collaboration, and only at his insistence that this was truly my "baby" and not his did I agree to his absence on the title page.

Along with Zach, our mutual friend Terrence Millar of the University of Wisconsin at Madison has left a major imprint on the book. A formidable logician and mathematician, for several years Terry has probed and dissected my treatment of the book's mathematical topics with the scalpels of his trade (and with snifters of fine cognac), prodding me always toward greater logical and literary clarity, especially when I was convinced I already had it (though he less so). In a different vein, the three anonymous (to me) readers assigned by Harvard University Press to evaluate the manuscript produced an exceptionally high-quality, prepublication scholarly review, well surpassing normal expectations. The readers not only cheered on the project in numerous ways but also supplied page upon page of valuable and detailed suggestions for making my ideas clearer and more relevant. Collectively, their commentary generated a great deal of intellectual stimulation and reflection, especially on points of disagreement, and it has improved the book far beyond what I had imagined in the initial draft. I also wish to extend a particular note of gratitude and appreciation to my editor at the press, Jeff Dean. From his reading of my initial proposal through the final hard copy held in hand, I have been keenly aware of my good fortune in having his guidance and support throughout the entire process, and in having directly benefited from the experience, talents, and practices of a first-rate editor: prompt and professional correspondence, clear and concise remarks, sage and timely advice, unending and totally appropriate encouragement. Few authors have been so well served.

Among the long queue of those who have read and commented on all or part of the manuscript over the years, or otherwise had a hand in its preparation, pride of place goes to my late father-in-law, Stuart Rockwood Sheedy. Although Stu died before seeing the draft completed (in full possession of his faculties at age ninety-five), he had served for over a quarter century as my exemplar of an "educated" audience. His training in analytical philosophy, his expertise in computer technology, and his lifetime of voracious nonfiction reading—all molded by a retentive memory and critical acumen—made him into an ideal general reader, a moribund species in our culture. More recently, Jerry LeClaire has also assumed this mantle, critically engaging each chapter with his background in science as a Harvard-trained ophthalmologist and with his widely ranging curiosity in the humanities. Others, academic and nonacademic alike, who have read and commented on the entire manuscript or various sections of it include Leonard Helfgott, Donald Kaplan, Pat Martin, John Miller, George Panichas, Jay Saxton, and Daniel Simonson, each of whom has contributed in a unique way to the final product. In Italy, Susanna Cimmino of Florence's Museo Galileo graciously helped locate and obtain numerous photocopies of art works and manuscripts. Finally, although the list of those with whom I have conversed over the years about various topics covered in this book would take another book, I would be remiss in failing to mention the Alta Lake fishing crew of professionals, engineers, and technicians—Jim, John, Mitch, Phil, Gary, and Dick. Amid the panorama of Washington's Columbia River basin, we have spent many a delightful hour in argument about Aristotle, about causality, and about the place of science and technology in our lives, arguments interwoven with others about how best to catch and cook trout. The contributions of all these critics and friends allow me to shoulder, gratefully and sanguinely, if not always gracefully, any remaining burden of error.

In a different category altogether, my wife, Sally Sheedy, deserves special notice. From an undergraduate training in fine arts, from a decade of computer programming in Manhattan companies and the garment industry, and from her current position as a systems librarian in our local community college, she has accrued a capacious set of skills and talents in visual design and computer technology, in library research and database management, and in negotiating the world of business- and techno-speak, which have all proved invaluable throughout the research and writing of this book. Her boss sometimes calls her "goddess" or "wizard," and I can only echo that and add more, for she has magically (at least to my technologically myopic eye) produced or had a deft hand in all the diagrams and images on which I have relied, adding her technical creativity to that of the stellar art production people at the press.

But vastly beyond her technological wizardry and contributions to the book, Sally has forged with me for over a quarter century a shared partnership in the rearing of our twin daughters, Abigail and Julia, who are now happily launched from the nest as fledglings and flying on their own. For many authors, I suspect, writing a book is somewhat like rearing children: before you have kids, you can't imagine life with them, but once they arrive, you can't imagine life without them. And so too with a book project once it takes hold. Early on it became something of a family endeavor, with everyone adding to it in countless ways, large and small, everything from tolerating my late-night writing habits and frequent distractedness to sharing in the rush of discovery and pushing back the shadows from my darker moments. Sally and the kids: dedicating this work to the three women of my life, to my family, is but a pittance for the sustenance they have all given me as an author, and for the sheer laughter and delight and joy . . . and love . . . they continually bestow.

Introduction

The Rift between Science and Religion

> The questions we ask, as well as the answers we are willing to accept, reflect our temper of mind.
>
> —Arthur Zajonc, *Catching the Light* (1993)

Approaching the Divide

Few among us would dispute that we live in an age marked by a deep schism between science and religion. Questions about where we've come from (creationism versus evolution), where we're going (global warming versus Armageddon), and the myriad social and cultural issues between reveal how our concerns and arguments are shaped by our deeper attitudes surrounding these far-reaching realms of our experience. Ultimate convictions about the universe, about our earthly abode, about our history, about society, about well-being and happiness, and about human thought itself shape our public discussions and guide our private affairs. By the time we reach adulthood, these attitudes have become stamped into our psyches, not permanently imprinted, perhaps, but lying instinctively beneath the surface of our thought as fallback positions and justifications for our opinions. Generally we carry on with a hefty and healthy dose of blind disregard of these convictions in our quotidian activities until confronted by controversy or crisis, when they are called on as though a court of appeal. And once subject to the appellate court of thought, the fault line between science and religion reveals its full, and for some foreboding, breadth and depth. This book narrates the story of how such a rift came to exist.

In presuming the temerity of commenting on the chasm between science and religion, we are immediately struck by the difficulties in articulating the issues and in the lack of consensus on even how to proceed, on how to determine which perspectives might yield fruitful rather than barren results. There is no one science, of course, but many sciences; likewise with religion. Each term has a long history of its own, and each historically has cast wide definitional nets around its catch.[1] So much so that defining them in our own day has become quite problematic.[2] Because definitions themselves grow out of disciplines and perspectives, we confront not simply a stark, terminological demarcation between 'science' and 'religion' but an even wider gulf between different ways of thinking about them and the encounter between the two. Our current state of affairs in this matter may be illustrated by brief mention of three prominent, public intellectuals who pose and articulate the issues, each from within his own discipline and interests: Huston Smith, Steven Weinberg, and Steven Jay Gould.

Student of world religions and a specialist in what he terms "the Big Picture" or "worldview," Huston Smith portrays science as a subordinate of religion. We currently face a "spiritual crisis," he observes, because we have abandoned this fundamental assumption. The crisis itself is the offspring of a science-based culture and the "tunnel vision" of "scientism" and its core belief, the "metaphysically sloppy" philosophy of "naturalism." In recent years, scientism in turn has generated its opposite, a postmodern "anything goes" of "New Age" religious sensibilities.[3] Both worldviews run contrary to the "conceptual spine" that bridges differences between the major, traditional religions.[4] That spine embodies the "Great Divide" between "*this-world* and an *Other-world*," which Smith sees as the heart or essence of all religions. In his depiction, the "Other-world" of religious meaning and spiritual reality surrounds the "this-world" of scientific explanation, and we get ourselves into problems (to wit, a spiritual crisis) when we either fail to look reasonably beyond our immediate warren, remaining trapped in science and scientism's tunnel vision, or, at the first prompting, take flight into the never-never land of a New Age, spiritual Milky Way.[5]

Physicist, Nobel laureate, and acknowledged atheist Steven Weinberg looks at the issue from an opposing vista, one based on scientific explanations, where he has spent his career extending our understanding to the reaches of the cosmos and its origins and to the depths of subatomic particles.[6] In recent years he has defended science against a number of attacks in the so-called culture wars between the sciences and the humanities. He does favor a "dialogue between science and religion, but not a constructive dialogue.

One of the great achievements of science has been if not to make it impossible for intelligent people to be religious, then at least to make it possible for them not to be religious." And, he emphasizes, "we should not retreat from this accomplishment." Weinberg acknowledges that science traditionally has answered the "how" questions, religion the "why" questions, and that even when scientists come to realize their dreams of a "final theory," the "why" question will remain standing. Why does one theory work, and not another? But he insists that religion cannot answer the "why" question either. Religion may claim that the universe is "governed" by one sort of God "rather than some other sort of God and it may offer evidence for the belief, but it cannot explain why this should be so."[7] Whatever the explanatory limitations possessed by science, it seems they are owned by religion as well. No one knows the "why" of things. But within the realm of explanation itself, there is no contest: science trumps religion repeatedly.[8]

Steven Jay Gould, noted paleontologist and essayist, seeks a balance between the perspectives of Weinberg and Smith, one that grants equal legitimacy to the scientific and religious (including humanistic, ethical, and value-laden) dimensions of human experience. With a voice of *politique* reasonableness, Gould has identified each of these separate categories, science and religion, as a "magisterium" (from the Latin *magister*, or "teacher") and described them as representing a "domain of authority in teaching," wherein each "form of teaching holds the appropriate tools for meaningful discourse and resolution." Science holds court in the "empirical realm: what is the universe made of (fact) and why does it work this way (theory)." Religion presides over "questions of ultimate meaning and moral value." Science and religion thus each circumscribe their own terrain, or intellectual space, wherein their respective authorities can and do prevail. Further, the domains do not overlap—"NOMA" Gould terms them, "non-overlapping magisteria"—and both are equally vital for a full life. Gould notes that there is "nothing original" in this argument.[9] And, to be sure, the metaphor of the medieval magisterium yields a comforting image. We can till our fields of quantum mechanics and astrophysics, evolution and DNA, during the day, then retreat to the manor house, crawl into bed at night, say our prayers, and sleep the sleep of the untroubled. Many do. Still, others may experience unrest in seeking such comfort, perhaps because they lack the "God gene" (an updated version of what John Calvin called the "dreadful" doctrine of election) or simply entertain doubts about the elevated vantage point or magisterium from which Gould himself defines the subjects.[10]

The disagreement among these exemplary individuals highlights the thorniness of the issues involved in trying to define science and religion and their relations to one another. This impasse has led me to adopt a different approach to the problem. Instead of seeking yet another set of definitions, I shall examine the deep substructures of thought that underlie the uncoupling of the two. Of course, the causes of this uncoupling are, as Sigmund Freud might say, "overdetermined," but recent research points to a novel, albeit counterintuitive starting point for an investigation into the current breach between science and religion: the effect of shifts in the technology of information.

In the last several decades numerous studies have heightened our awareness of our own era as an "information age," or, more accurately, as I have argued elsewhere with my colleague Zachary Schiffman, one of several "information ages."[11] Scholars, scientists, and intellectuals of various stripes have vastly enlarged our understanding of how the mind creates and processes information, of how it generates microcosmic, abstract symbols, which capture and shape the "givens" (data) of experience, and then combines and structures these symbols into macrocosmic, meaningful expressions of well-nigh limitless permutations. "Symbols furnish the substrate—information carries the meaning," in the salutary expression of physicist Hans Christian von Baeyer.[12] Yet despite the far-reaching insights of contemporary information studies, no one has yet exploited them in their implications for comprehending the relation between science and religion. This I shall do in the chapters that follow. My central thesis may be baldly and succinctly stated: the shift between two distinct information technologies—literacy and numeracy—resides at the source of how science and religion went their separate ways, producing the Great Rift between them.

Information Technology and the Two Tempers

As the impasse over definitions suggests, the rift between science and religion is, to adopt the memorable distinction of Edmund Burke (1729–1797), "discernible," not "definable."[13] And, as recollection of Burke also suggests, it is profoundly historical in nature, bound to context and evolving over time.

In our history, *Information Ages,* Schiffman and I argued that processes of abstraction are inherent in the use of language, but with the advent of writing these abstractions took on systematic and sustained form, giving rise to information itself. The first information age was thus born of literacy, and it culminated with the advent of the alphabet.[14] For the Greeks in particular, the subsequent evolution of alphabetic literacy produced an information system based on phonetic letter symbols as the means of encoding information, on definitions as the instruments of information storage and the stuff of thought, and on the classification of words and things as the mode of knowledge issuing from it. Under Aristotle's nurture, these embryos developed into a mature classifying temper, an enduring accomplishment that dominated intellectual activity for some two millennia.

In the Middle Ages, science and religion alike found their expressions via the information technology of literacy and the classifying temper inherited from the Greeks. Words captured and communicated knowledge about the natural world and beliefs about the world beyond, often conterminously in the same phrases. They tendered at once the precision of scientific knowing and the poetry of religious conviction. Conflicts between science and religion were conducted within the "mind-forg'd manacles" of conventional languages and common sense, universes of words and meanings.[15] The same technology underlay common practices, sensibilities, and debates about the descriptive and metaphorical use of words, about definitions and their semantic content, about abstractions and logical rules, and even about argument. Though few in numbers, natural philosophers, theologians, and other people of learning in the medieval world shared a magisterium, to borrow Gould's term, the "appropriate tools for meaningful discourse and resolution." Their magisterium, however, included both science and religion.

During the Renaissance, classificatory thinking faced a crisis brought on by the recovery of ancient texts, by the discovery of new worlds, and, most prominently, by the fifteenth-century invention of the printing press and the mind-boggling proliferation of books and other printed material. There was simply too much information for the existing technology of literacy to absorb, whether through mnemonic devices and note taking, "commonplace" compilations, list making, cross-referencing, branching diagrams, indexing, or other means of information management. An information overload overwhelmed the age's categories and classifying techniques.[16] Sidling in from the byways of thought and supplying a response to the crisis was a new information technology—modern, relational numeracy—which, especially in the world of scientific investigation, would rival and in many ways supplant alphabetic literacy. Already under way and growing incrementally by the time of Gutenberg's press, this new technology provided the basic symbols and operations of counting and measuring, symbols and operations that encode, store, and manipulate bits of information.

Occurring mostly outside the predominantly philosophical mathematics taught in the universities, and its "inventory of (ideal) mathematical entities," practical innovations in all areas of the quadrivium—the traditional mathematical disciplines of arithmetic, music, geometry, and astronomy—produced several burgeoning proficiencies that laid the foundation of the new and increasingly abstract means of processing information.[17] First, in arithmetic the arrival of Hindu-Arabic numerals promoted the practice and awareness of a new counting system from the thirteenth century forward, employed increasingly by merchants, bankers, and accountants, as well as mathematicians and natural philosophers, craftsmen and artisans, musicians and artists. The new system featured a simplified notation, a place-value or columnar arrangement of numerals, and the symbolic representation of zero, all of which would eventually contribute to seeing numbers as relations, rather than merely as collections of things or objects. Second, in the world of music the rise of polyphony was accompanied by a newly invented and abstract, musical notation, which allowed composers and musicians to capture and manage as information the elusive ephemera of rhythm and pitch. Standardized units of sound (notes) and silence (rests) facilitated the first measurement of time as an independent, symbolized reality, while the new Hindu-Arabic numerals expressed descriptively the fluid, irrational proportionalities and harmonies that made up the dynamics of musical tone.

Third, beyond arithmetic and music, in the visual arts the discovery of linear perspective spawned innovations in geometry and gave new expression and shape to visual information. Perspective grids, spatial proportionalities tied to changing viewpoints, one-to-one mappings between objects and representations (two-dimensional drawings or three-dimensional sculptures)—all these novel techniques tendered alternatives to definitions as a means of seeing objects in the world. And fourth, in astronomy new technologies brought heavenly motion and time down to earth, paving the way for merging the terrestrial and celestial branches of knowledge, physics and astronomy. With the technological mastery of time through calendars and clocks, time itself became conceptually uncoupled from motion, only to be rejoined with it through a mathematical formula in Galileo's law of free fall, an upshot of reverse engineering phenomena with modern analysis. By Galileo's lifetime (1564–1642) these proficiencies had mushroomed and were cohering into the information technology of relational numeracy, which in turn kindled and enabled fresh readings of ancient mathematics manuscripts as they were recovered and translated in the sixteenth century. Henceforth, as Galileo wrote in a passage often cited, the book of nature came to be understood increasingly as "written in the language of mathematics."[18]

In the ancient world, alphabetic literacy had given rise to the higher, abstract reaches of Greek philosophy, literature, and classificatory science. During the Renaissance, modern relational numeracy would catalyze an extraordinary, pupa-to-butterfly metamorphosis of premodern mathematics into a highly abstract and relational world. In the far-reaching, incisive words of the mathematics historian Jagjit Singh, early mathematical thought was transformed from "thing-mathematics," in which numbers and other symbols stood for concrete collections of things or objects embedded in the perceptual world, into "relation-mathematics," in which mathematics lost such tethers and entered a more rarefied universe, one constructed with the "mutual relations of abstract symbols."[19] With numbers depicted, in effect, as adjectives modifying separable things, the arenas of arithmetic and geometry remained incommensurable thought processes, classified as separate disciplines or genera in Aristotle's view. In the premodern world, mathematics thus remained subordinated to the classifying temper. But once mathematicians started considering their symbols, including numbers, as designating abstract and empty relations separate from material or even spiritual realities, the worlds of space and number, geometry and arithmetic (eventually algebra), were united, most convincingly in the analytical geometries of René Descartes (1596–1650) and Pierre de Fermat (1601–1665).[20]

In a pattern comparable to that of writing, modern numeracy too elicits data (the sensory-supplied "givens") from the flux of experience and informs the mind. Both information technologies rest on the "substrate" of symbols, letters in the case of alphabetic literacy, abstract numbers (or, more technically, numerals) in the case of numeracy. Both groups of symbols are controlled by rule-governed relations and syntactical operations as they transform data into information and information into knowledge. And, with both numeracy and literacy, the higher-level achievements of thought, whether in mathematics or philosophy (or other forms of literature), remain grounded in their respective foundational technologies. Moreover, the dividing line between the basic symbols of counting and higher mathematics is as wide and fuzzy as that existing between fundamental writing and Shakespeare.

Although comparable to literacy as an information technology, modern numeracy departed from it, introducing a new and different, abstract screen between the mind and the outside world, one that deeply conditioned how nature came to be seen and understood.[21] And it required thinking about relations rather than things, leading to the eventual characterization of thought itself as primarily relational, rather than classificatory. The gradual creation of this new information technology eventually emptied the long-standing philosophical, allegorical, and religious significance from thing-mathematics and its expressions. Mathematics became "disenchanted" from its medieval animism.[22] Relational numeracy would elicit a newer and ever-expanding range of analytical possibilities in mathematics in much the same way that alphabetic literacy had brought forth the open-ended, classifying potential of conventional speech.

With its symbols, then, the new information technology midwifed a new type of analysis, which today we identify as "reverse engineering," and which differed markedly from previous forms of analysis. In ancient times, the works of Aristotle (384–322 BCE), Galen of Pergamon (129–ca. 216), and, to a lesser extent, Euclid (fl. 300 BCE) had explained or shown analysis as an intellectual process of breaking down complex ideas into their simpler components and resolving them into fundamental elements, entities, principles, or axioms.[23] These were then logically synthesized, reconnected in step-by-step deductions back to the original, more complex ideas, thereby "proving" them. This mode of deductive analysis had served mainly to rationalize already existing knowledge, not to create new knowledge, except in the sense that by demonstrating the logical relations between premises in a syllogism, one might "discover" new and hidden relations between them. For ancient and medieval thinkers alike, deductive analysis secured the demonstrations of science, and an entire practice and theory of scientific knowing revolved around the system of classifying whose scaffolding such demonstrations erected.

By contrast, the modern "analytic art" was a tool of creation and discovery, not of philosophical, classifying rationalization.[24] 'Taking apart' or 'resolving' came to mean reverse engineering how things or ideas worked. For us, reverse engineering occurs when we take apart a machine or electronic device or natural phenomenon, figure out how its components go together, and then reassemble the gadget or process. Industrial manufacturing secrets are often acquired in this manner. Psychologically we gain immeasurable confidence in "knowing" how something works when we can reverse engineer it. The French mathematician-cryptographer François Viète (1540–1603) relied on this same sense of reverse engineering in analyzing how algebraic equations worked: how the various operations performed on one side of an equals sign affected operations on the other, how the procedures and functions of algebra could be generalized, and how unknowns could be discovered and problems solved. In this context Viète became the first to assign letters of the alphabet systematically as generalizations of numbers, thereby inventing modern symbolic algebra, whose ambitious goal in his mind was "to solve every problem" (*nullum non problema solvere*).[25]

Not more than a decade after Viète's book appeared, Galileo Galilei too was busy using mathematics as an instrument of creation and discovery, only now in the orbit of physics. He did not employ Viète's symbolic or notational novelties and in fact never used algebra (or even an equals sign), instead relying heavily on the geometrical techniques of Euclid and Archimedes. Nonetheless, with the recently arrived Hindu-Arabic numerals, geometric grids, musical notation, abstract time, and other components of the new information technology in mind, Galileo transformed Euclidean geometry, especially the theory of proportion, into the new form of analysis as reverse engineering. His analyses of motion and matter entailed breaking down complex phenomena, such as bodies in free fall or the motion of pendulums or parabolic motion, into constituent components, and then reassembling the components into functional relations, what we now term formulas. In Galileo's hands reverse engineering became an innovation whose momentous consequences we shall be exploring in some detail. Suffice for now that by the early 1600s, a new sort of mental activity was flowering: algebra, geometry, and physics were being revolutionized.

Within the next century the reverse engineering of mathematical analysis would steadily expand its scope of operations, far transcending its origins in mathematics and working its way into the practices of experimental discovery as well. The analytical arts became the pulsebeat of modern science, and "number, the language of science," was its new, underlying information technology.[26] Along with its development, much of the machine vocabulary we now associate with engineering would blossom as natural philosophers increasingly employed mechanical terms, such as 'watch,' 'clock,' 'gear,' 'pulley,' 'function,' 'operation,' 'mechanism,' and a host of comparable expressions, both as descriptions of natural processes and as metaphors. A mechanistic worldview grew out of analytical thinking about nature.

Moreover, once in place, the corrosive nature of mathematical analysis and its highly abstract and functional thinking about natural processes would over the years expunge any lingering, inhering, or intrinsic religious beliefs and content from scientific claims. A cautionary note here: science and religion would continue to be intertwined in the minds of many, if not most natural philosophers, religious thinkers, and intellectuals flourishing in the kingdom of letters until well into our own times. In fact, many reflective minds initially saw the new science as a way of bolstering religious belief because of its demonstrations of God's handiwork in nature. One could, of course, continue to be at the same time a scientist and a believer; many still are. But any knitting together of the two would be henceforth extrinsic, not internal to the activities themselves. As ways of engaging one's mind with the outside world, science and religion became and would remain separate. A new mind-set, the analytical temper, was in the making. And from within this new, scientific mind-set, religion—not so much conquered as ignored—would gradually and inexorably, like the grin of the Cheshire cat, disappear.

Galileo Astride the Divide

Details in historical accounts assume their significance either by revealing or by contributing to a historical trend, development, or state of affairs. With Galileo's career and accomplishments we find both. His scientific investigations and practices reveal his place among several figures of his era who began to shape the new numeracy into a new, relational mathematics for the sake of studying motion and other natural phenomena.[27] At the same time, this son of Tuscany also stood somewhat taller than his peers as one of the earliest practitioners of the new, analytical method he largely created, delving a little deeper into natural processes, seeing a little further into the future of investigations. For this reason we often count him among the guiding geniuses who contributed to inventing modern science and the Scientific Revolution.[28] His was a singular contribution. Before Galileo, numerous natural philosophers had taken steps, some small, others larger, toward the creation of modern, scientific analysis. With Galileo antecedent factors reached a critical mass. His use of the new numeracy, especially the functional nature of its abstractions, led him to introduce measured time as both a discrete and a continuous phenomenon into a program of experiments, which produced the laws of free fall, pendulums, and projectiles, among other contributions to the analysis of matter in motion. Moreover, with novel and experimentally accessible means of mensuration, he forged a new connection between the reverse engineering of modern analysis and the ancient, formal structure of Greek mathematics. In, of, and beyond his times, Galileo reworked the conventions of the new information technology into the analytical temper.

In addition to his role in the Scientific Revolution, another well-known and revealing image also accompanies Galileo's passionate and sometimes turbulent life—that of his prostration before the Roman Catholic Congregation of the Holy Office, or Inquisition. At the time, Galileo's abasement symbolized his full recantation of belief in the veracity of the Copernican hypothesis that the earth orbits the sun, a hypothesis the Church had earlier declared "as altogether contrary to the Holy Scripture." Since then the entire Galileo "affair" has come to symbolize a singular, eponymous moment in the relation between science and religion, and as such has been the subject of endless scrutiny and debate.[29]

An arresting and flamboyant historical figure, Galileo was supremely self-assured in his abilities and opinions, not to say brash, caustic, and merciless to his intellectual foes in the courtly world of his surroundings. He wielded pointed bon mots like stilettos between the ribs. He was smart. And he was good theater. The drama of his undoing has often been interpreted as the consequence of his larger-than-life personality, the rise and fall of a Renaissance courtier. Whether or not pride always "goeth before destruction, and a haughty spirit before a fall," as Proverbs 16:18 tells us, Galileo certainly displayed an abundance of pride and a haughty spirit (not to mention extensive self-promotion). And he certainly fell. Myth and fact often parade arm in arm throughout the debates surrounding the events of his life, but notwithstanding biases and embellishments one conclusion remains firm: the arena of conflict was Galileo's disobedience to the authority of the Roman Catholic Church on a matter of natural philosophy, or science. And in that arena, science and religion revealed for the first time their modern separation.

The High Middle Ages (ca. 1050–1350 CE) could boast an integrated, idealized harmony of science and religion, with both subject matters grounded in a common way of thinking based on literacy, common sense, and the compulsion to classify. But as relational mathematics became increasingly refined and sophisticated, emerging from a new, abstract, and symbolic information technology, Europeans began rethinking their science. The underlying thesis of the present narrative highlights the practical developments in information technology and early modern science to help explain the changes that eventually would spur the separation of science from religion. Nonetheless, while science took the lead in the separation, so to speak, these developments occurred in a cultural and social context in which religion thoroughly permeated one's perceptions and "belief was the default option," in the noteworthy expression of philosopher-historian Charles Taylor. The ensuing centuries have hewed today's "secular age," in which the prevailing "background" has shifted and the "presumption of unbelief" has become dominant. Religion is but one among many options available to us; we choose to believe or not to believe, or which religion to follow. Yet in the transition to our secular world, as Taylor and other contemporary scholars have also rightly stressed, religion was not simply the passive stage across which marched scientific discovery and technological progress, the nineteenth-century drama of modernity. It supplied actors in starring roles as well.[30]

Premodern theologians and savants of all varieties were continually rethinking their religion, revisiting the central existential questions pertaining to the relation between man and God.[31] Within the relatively flexible authority of the Western Church, central dogmas might be firmly in place, framed in what one scholar has termed the "Aristotelian Amalgam," a "scholastic" fusion of Aristotle's natural philosophy and Christian doctrine crafted by the "Angelic Doctor," Thomas Aquinas (1225–1274).[32] But they were forever undergoing adjustments, shifts in emphasis, extrapolations from earlier positions, and logical extensions, not to mention outright challenges from nominalists, mystics, pietists, and eventually Protestants.[33] Within this context of activity, religious thought in these centuries simultaneously stimulated a turn toward examining nature and, ironically, led to the revival of skepticism, separate trends that would likewise contribute to the eventual separation of science from religion.

For some, Renaissance humanism and philosophy—including the recovery of long-lost texts, the revival of Platonism, the Hermetic tradition, and similar intellectual stimuli—renewed strains of Augustinian thought or evinced an "evangelical turning" to the world or revitalized excitement in astrology, magic, numerology, alchemy, the occult, and mysticism, keeping alive various forms of sacred animism.[34] For others (Johannes Kepler stands out here), the new mathematics itself opened the way to a deeper, more subtle appreciation of the divine author of nature's book. Yet further, others refashioned traditional religious subjects, such as the story in Genesis of Adam's fall, in ways that commended new approaches and even an urgency to wresting secrets from nature.[35] The nub of these widely ranging religious sensibilities, interests, thoughts, and activities was a greater curiosity and attention given to investigating the natural world, its patterned manifestations of God's creativity, and the place of humankind within it.

Moreover, by the sixteenth century religious protesters were directly challenging well-established theological doctrines—of free will, of sin, of grace and salvation, of the Eucharist, and of the institutional nature of Christendom. Beginning with Martin Luther (1483–1546) and rapidly spreading, Protestants confronted the authority of Rome, often with martial violence as the Reformation gathered the secular support of kings and princes—not activities for the weak-kneed. No medieval Innocent III would stem this tide, though many popes and monarchs tried. The struggles between Protestant and Catholic reformers underlay an intensely tumultuous intellectual and cultural context, within which skepticism about older ways of thinking became increasingly pronounced, an ironic consequence of contending religious sects. For the inventor of the essay as a literary form, Michel de Montaigne (1533–1592), skepticism became a new norm; at the dawn of a new era, poet John Donne (1572–1631) summarized incisively: the "new philosophy calls all in doubt."[36] In these circumstances the breakthrough into a new, analytical temper at the heart of science tendered for some an attractive, optimistic, and more secure way of approaching thought in general, including a fundamental rethinking of science's relation to religion.

In the latter vein, and beyond his role in revealing and furthering the emerging fault line between them, Galileo left for future generations to ponder the first modern version of NOMA—those non-overlapping magisteria of science and religion that Steven Jay Gould has advanced. In Galileo's terminology, "sensory experience and necessary demonstrations," voiced in the language of mathematics, tendered the magisterium of science. Cast in "narrations" created by the conventional language of words, things, and meanings, the "dictation of the Holy Spirit" as revealed by God in the "sacred words of Scripture" authored the magisterium of religion.[37] As Gould would reiterate several centuries later, Galileo also insisted that the two did not overlap. Scientists would correct the shortcomings of scientific claims, while trained theologians would unfold the layers of religious interpretation as insight, inspiration, and wisdom dictated. Different cobblers should stick to their own lasts, he believed. Yet even as he laid out his own *politique* position on the matter, Galileo's analytical temper had already taken hold. He could not resist the temptation of meddling, and interpreted passages of scripture according to his own, scientific magisterium. More pointedly, in opposition to the Catholic Church's "narration" of the integration of science and religion, he exercised his own highly developed rhetorical skills to proclaim and promote a narrative of his own, the separation of scientific analysis and religious belief.

C. S. Lewis, Oxford don, literary scholar, and Christian apologist, once wrote that the medieval image of the cosmos and humankind's place within it has long been "discarded."[38] Lewis was referring primarily to literary themes in the Middle Ages that betokened an animated universe, one populated by the varied and frequently mysterious forces—natural and supernatural, divine and satanic—governing people's lives. These literary themes created an image out of the materials provided by the information technology of alphabetic literacy, a world of words and things. That image has since been supplanted by another, at whose visionary and intellectual center resides modern analysis, with its own foundation of modern numeracy as an information technology. Once gaining a toehold in European thought and culture, modern numeracy and analysis burrowed into the core of scientific thinking about nature and from there generated their modern progeny, the bifurcation of science and religion.[39]

In the following chapters we shall explore this remarkable set of occurrences, tracking a narrative pathway through the intellectually lush topography of medieval and Renaissance Europe, a seemingly daunting, even at times bewildering trek. Before we commence, therefore, a map of where we are headed is in order. In Part I we shall begin by surveying some relatively familiar territory, that of the Middle Ages, but with an eye to discerning its underlying information technology—alphabetic literacy—and the influence this technology displayed in framing the presumptive harmony of science and religion. From there the story proceeds in Part II to the birth of a modern, relational numeracy, a new information technology emerging against the backdrop of ancient and medieval mathematics, as well as against the predominant, classifying temper. The focus here is on the practical, mixed mathematics of the Renaissance in arithmetic (business and the Hindu-Arabic counting system), music (polyphony and abstract musical notation), geometry (linear perspective in art), and astronomy (time reckoning with calendars and clocks). In all these areas the qualitative, philosophical, and allegorical modes of thinking about mathematics reigning in the universities gave way to thinking with empty and abstract information symbols, which catalyzed a metamorphosis from the earlier "thing-mathematics" of classifying into the "relation-mathematics" of modern scientific investigation. Finally, in Part III we shall see how this new information technology coalesced in Galileo's scientific work, especially in his analyses of free fall, pendulums, and projectiles, and in his efforts to mathematize matter. With the information symbols of the new mathematics, he pioneered the underlying, analytical temper of modern science and its techniques of reverse engineering. In so doing he separated science's "demonstrations" from the "narrations" of religion, first exposing in his trial the Great Rift between them.

PART I

A Prayer and a Theory: The Classifying Temper

Most High, all-powerful, all-good Lord,
All praise is Yours, all glory, all honour and all blessings.
To you alone, Most High, do they belong,
and no mortal lips are worthy to pronounce Your Name.

Praised be You my Lord with all Your creatures,
especially Sir Brother Sun,
Who is the day through whom You give us light.
And he is beautiful and radiant with great splendour,
Of You Most High, he bears the likeness.

Praised be You, my Lord, through Sister Moon and the stars,
In the heavens you have made them bright, precious and fair.

Praised be You, my Lord, through Brothers Wind and Air,
And fair and stormy, all weather's moods,
by which You cherish all that You have made.

Praised be You my Lord through Sister Water,
So useful, humble, precious and pure.

Praised be You my Lord through Brother Fire,
through whom You light the night
and he is beautiful and playful and robust and strong.

Praised be You my Lord through our Sister, Mother Earth,
who sustains and governs us,
producing varied fruits with coloured flowers and herbs.
Praise be You my Lord through those who grant pardon
for love of You and bear sickness and trial.
Blessed are those who endure in peace,
By You Most High, they will be crowned.

Praised be You, my Lord through Sister Death,
from whom no-one living can escape.
Woe to those who die in mortal sin!
Blessed are they She finds doing Your Will.
No second death can do them harm.
Praise and bless my Lord and give Him thanks,
And serve Him with great humility.

—Saint Francis of Assisi, "Canticle of the Sun" (ca. 1224)

Religio and *Scientia*

ONE OF THE MOST STRIKING images in the annals of hagiography comes to us from a confrontation that occurred on April 16 in the year 1210. Tradition has it that Giovanni di Pietro di Bernardone (nicknamed Francesco, ca. 1181 / 1182–1226), then approximately twenty-nine years of age and wearing a ragged brown cloak, appeared in his innocence, humility, and self-imposed poverty before one of the most politically savvy and potent, worldly-wise personages in Europe, Pope Innocent III (ca. 1160 / 1161–1216). In several previous audiences Francesco (Francis to the English-speaking world) had beseeched His Holiness for permission to found a new religious order in accordance with his recently revealed vocation. Imitating the twelve apostles, he and his eleven ragtag disciples would follow in the footsteps of Jesus as portrayed in the Gospel, the simplest and purest of rules: abandon all possessions, preach repentance, and announce the kingdom of God. Unlike monks in the existing monastic orders, with their cloisters and estates, they would not be withdrawn from the world, but as "lesser brothers" ("Friars Minor") in the fullness of their individual and collective poverty they would serve God in it—in the cities, in the villages, on the roads, among the poorest and most needy of souls.

Tradition also has it that Innocent earlier had rejected Francis's petition, thinking his group just another of the various heretical associations common in northern Italy and southern France. In his first dozen years after donning the papal tiara, Innocent had energetically sought to extirpate heresies (most notably launching the Albigensian Crusade to eliminate the Cathars in southern France) as part of his expansion of papal influence throughout Europe.[1] That influence included wielding the papal swords of interdict and excommunication in the making and unmaking of kings. But now he faced

Francis, Innocent confronting innocence. The supplicant's humble persistence, plus perhaps some sage advice from a cardinal or two and a dream in which Innocent saw Francis supporting the Basilica of Saint John Lateran, the pope's cathedral and therefore the center of Christendom, apparently changed the pontiff's mind. On this auspicious day, he granted approval orally for the formation of the Order of Friars Minor (*Ordo Fratrum Minorum*), known subsequently as Franciscans.

The Franciscan friars not only would hold dear the flotsam and jetsam of humanity but, through the "childlike" character of their founder (the adjective comes from the *Catholic Encyclopedia*), would embrace all of nature as the living manifestation of the loving God. Francis himself regularly conversed with animals, preached to birds, protected bees, attended plants, and even communed with the earth as he gathered stones for restoring ruined churches in his hometown of Assisi. He brought animals to the celebration of Jesus's birth, thereby creating the first crèche or nativity scene, complete with stable, live ox, and ass. On that occasion, wrote Saint Bonaventure, Francis "stood before the manger, bathed in tears and overflowing with joy" while he chanted the Holy Gospel with such divine favor that even the hay from the crèche became blessed. Preserved by the people, it later miraculously cured cattle of diseases and warded off pestilence.[2] Accounts of Francis's deeds multiplied even before his death (1226) and subsequent canonization a scant two years later, contributing to a host of legends and stories. Many of these were compiled in the *Little Flowers of St. Francis* (*Fioretti di San Francesco*), a popular account of the saint's life dating from the late fourteenth century.[3] One legend relates that on his deathbed Francis thanked his donkey for carrying him and helping him throughout his life, whereupon his donkey wept. Apocryphal or not, such stories indicate that in the fullness of his animism, Francis, called by some (for example, G. K. Chesterton) the greatest of Christian saints for his Christlike innocence and redemption, became the patron saint of nature, of all creatures great and small.[4]

At first blush it might seem that a man preaching about God to birds and animals bespeaks the promiscuous jostling of New Age sensibilities, to recall the earlier-cited remark of Huston Smith. Yet a little reflection can bring us to appreciate the expanse separating our modern view from that of the Middle Ages. Whereas we generally detach our scientific understanding of the universe from its ultimate creation and purpose, embracing the natural world as Francis did meant apprehending immediately a divine creation pulsating with its Creator. Francis certainly could not have imagined it otherwise, could not have detected the slightest fault line between science and religion or even dreamed of science and religion (or secular and sacred) as separate categories.[5] For him God and nature were at once alive and real, equally present both in the drama of human sin and redemption and in the ordinary course of natural events. Of course, sin itself prevented humans from immediately or fully appreciating this multitextured, divine reality. It was equally part of redemption to journey within oneself and without, in the world, to an other-world unity with the transcendent deity. The hardships and sacrifice required by such a journey could not be made without divine aid any more than was Francis's vocation simply a young man's reasonable career choice. No hint here either of New Age enthusiasms or of scientism's tunnel vision, which have preoccupied Smith and others.

Although his could not be described as an ordinary career, in his earlier years Francis nonetheless fit in well as an ordinary man of his time.[6] A bourgeois son of a cloth merchant, witty and pleasure seeking, Francis delighted in fine clothes, served in the local militia, cavorted with friends, and sang the songs of troubadours. After his vocational transformation, he continued the last practice, but rather than tales of chivalry and courtly love, bawdy or stately, he sang of the richness and fullness of God as displayed in the natural world surrounding him, hymns of praise and joy. His most famous song was the "Canticle to the Sun" ("Laudes Creaturarum," also translated more directly as "Praise to the Creatures"). Composed in an Umbrian dialect near the end of his life, the canticle stands among the earliest literary compositions in Italian. It manifests and exemplifies Francis's simple joy and lightness of being, immersed in the world of his brothers and sisters: earth, air, fire, water, sun, moon, . . . and death.

His brothers and sisters revealed how Francis typified his own historical age in another way as well, for embedded in the poetry of Francis's song lay an entire, regnant cosmology. From Aristotle forward, the four elements of earth, air, fire, and water had been commonly propounded by natural philosophers as the fundamental ingredients or essences that blended together to create all the diverse phenomena in the sublunary, natural world we inhabit and experience. To them was added a fifth essence, a "quintessence" that accounted for the superlunary heavens—the fixed stars, as well as the "wandering stars" (the Greek *planētēs astēr,* from which our 'planet'), the latter including "Brother Sun" and "Sister Moon." In short, the universe that Francis inhabited with his brothers and sisters could boast an intellectual antiquity of the highest scientific caliber, as well as divine animation and widespread adherence.

The harmony of God and nature that Francis embodied in prayerful life and song was sought and conveyed, no less prayerfully, in the concepts and distinctions of philosophers and schoolmen of the High Middle Ages.[7] Earlier, the basic religion-science lexicon of the period had been laid down by Saint Augustine (354–430), who followed Church fathers Tertullian (ca. 155–ca. 240) and Lactantius (ca. 250–ca. 325) in designating the "true religion" (*religio vera*) as "rightly directed worship." From then until well into the Renaissance, writes historian Peter Harrison, the term 'religion' did not just mean a body of doctrine, practices, or beliefs, as typically rendered in our times; it also carried with it centrally an "inner disposition," personal qualities associated with moral virtues. Augustine had also characterized *scientia* ('science,' the Latin translation via Cicero of Aristotle's *epistēmē,* from which our own 'epistemology') as "knowledge," the "rational cognition of temporal things," and paired it with "wisdom," the "intellectual cognition of eternal things."[8] By the time of Thomas Aquinas it signified the organized knowledge of natural things issuing from the elements, including the principles of organization and demonstration themselves.[9]

But in a roughly parallel fashion to *religio, scientia* too had come to encompass a personal and moral, internal realm, one of intellectual powers that could be guided to perfection by the intellectual virtues. Organizing the innermost topics of their meaning-filled universe, as well as the phenomena surrounding them, scholastics such as Aquinas classified the intellectual virtues as "understanding," "science," and "wisdom," distinctions drawn from Aristotle. The understanding meant grasp of first principles, science the derivation of truths from those principles ("knowledge by demonstration"), and wisdom the comprehension of the highest causes, including the first cause, God.[10] As virtues, scientific practices cultivated habits of mind. Study hard and think geometry, said Aquinas, and the "same specific habit of science" will increase—you'll become a better geometer.[11] So too with other sciences. And with practice might come perfection, or at least virtuous improvement of mental acuity. Extending the organizational hierarchy, these natural, intellectual virtues in turn were infused and interwoven with the supernatural gifts of the spirit, the virtues of faith, hope, and charity.[12]

Thus did *religio* and *scientia* alike reveal divinely grounded personal qualities of life, as well as bodies of knowledge, for Aquinas and other scholastics, a contemplative and introspective ideal, with religion manifesting inner, spiritual actions of devotion and prayer, and science hallowing moral excellence through intellectual virtues.[13] Both were habit forming. Like the churches Francis devoutly restored, stone by divine stone, the habitual practices of worship and learning led, with God's help, to greater, even beatific edifices of knowledge and belief. A moral and spiritual unity, an ideal harmony, underlay the practices and content of both religion and science. Just as the divinely inspired Francis praised the God he saw everywhere in nature, so too did the summatory intellect of Aquinas organize for Christians all that could be known about divine wisdom and God's natural creation. "Everything . . . is from God," he declared. Theology reigned as the "queen" of our learning, for it produced the closest approximation to divine wisdom that humans could hope to attain. Philosophy, the "love of wisdom," served as theology's "handmaiden," for it tendered the reasoning through which one could perhaps realize some intellectual knowledge of God, severely limited though it may be. And *scientia* provided the understanding of nature that revealed divine presence in the pattern of events, as well as in the principles and virtues through which events were comprehended.[14]

The tools of philosophy and science utilized by Aquinas and other scholastics were forged largely in the furnace of the formidable Aristotle, student of Plato and teacher of Alexander the Great. "The Philosopher," as he was invoked reverentially (Dante tagged him "the master of those who know"), had brought the information technology of alphabetic literacy to its philosophical and scientific apogee through a fully developed, classificatory way of thinking. With a vast array of manuscripts covering nearly every subject discussed at the time, from the principles of logic to the details of acting, there was little his far-reaching intellect did not embrace. Not only was he expansive with his vision and views, he was equally formidable with his positions and arguments. His thought was essential: one could argue for him or against him, but not without him.[15]

Except . . . , that is just what the small number of thinkers had to do for much of the early Middle Ages. Aristotle's thought traveled a long odyssey before arriving in Western Christendom during the thirteenth century. Before then it had been available only in dribs and drabs, mostly through some of his logical works translated by the late Roman philosopher Boethius (ca. 480–524). After Boethius's imprisonment (during which he composed *The Consolation of Philosophy*) and subsequent execution, further translations of works from Greek to Latin virtually ceased for over six hundred years as western Europe underwent a centuries-long period of social transformation, political fragmentation, and cultural consolidation.[16] Recovering from the ninth- and tenth-century folk migrations and marauding raids of the Vikings, Magyars, and Saracens, Europe witnessed the gradual emergence of its towns and cities. Spearheaded by an agricultural revolution and dramatic population growth, urban life was secured by the formation of craft and commercial guilds, legal systems, banking practices, cathedral schools, new religious orders, and other institutions. Chief among the last was a creation "unique to world civilization," the university, whose expanding curriculum and voracious appetite for new learning spurred the recovery of philosophical and scientific works from antiquity, many of which had been preserved in Arabic translations made by Islamic scholars.[17] During the "great age of translation" into Latin, the entire extant corpus of Aristotelian philosophy towered among the recovered works.

Medieval universities were singularly situated to receive the works of Aristotle, as well as other Greek and Arabic authors. Often emerging from earlier monastic and cathedral schools, they had grown up in response to the changing demographic conditions of the eleventh and twelfth centuries in western Europe.[18] Urban life demanded trained functionaries—folks who could read and write, plus ones who could keep accounts, however crudely. More people needed serving: monarchs, towns, and merchants needed scribes and bookkeepers; churches needed priests; disputants needed law courts and legal minds; the ailing always needed doctors. In the *studium generale* ("general studies," as universities were initially called), students prepared themselves for various occupations by studying the seven liberal arts (sometimes called the seven liberal sciences), taught by "faculties of arts." The arts included the trivium of logic, rhetoric, and grammar, plus the quadrivium of arithmetic, music, geometry, and astronomy. Other faculties in theology, law, and medicine taught graduate students, as we would call them, readying them in these disciplines for more specialized and advanced professional careers and vocations. By the thirteenth century universities had become legal corporations, with a host of the various "rights, privileges, and immunities" pertaining thereto still commonly noted on diplomas.[19] Of professors, those staffing the arts were by far the most numerous, instructing upwards of 80 percent of all matriculating students.[20] They were also, collectively, the most intellectually and institutionally independent. While faculties of medicine and theology absorbed much from the new translations, the arts faculties and their students paid particular, critical attention to the newly arrived learning, sometimes in disregard of its religious implications.

In seeming tandem with the rise of universities, translating activity flourished in the twelfth and thirteenth centuries, the years as well of the Christian reconquest of Islamic Iberia and those following the Norman expulsion of Muslims from Sicily.[21] Especially in Toledo, during these years there existed an intermingling of Islamic, Christian, and even Jewish cultures, whose inbred, historical animosities were occasionally superseded by mercantile and intellectual interests. In the year 1085 Alfonso VI of Castile (1040–1109) recaptured Toledo from the Muslims, with its extensive library intact. A flood of translations followed, often from Arabic into Spanish by Muslims and then from Spanish into Latin by Christians; in some cases thought traveled from Arabic to Hebrew to Latin. The European hunger for Greek learning led some to learn Arabic and eliminate the middle steps. Most notable among the "Toledo school of translators" was Gerard of Cremona (ca. 1114–1187), who journeyed from his Italian home to Toledo and lived there for many years, mastering Arabic and translating scientific and philosophical works directly into Latin. His over seventy translations included Ptolemy's *Almagest,* Euclid's *Elements,* the *Algebra* of Al-Khwârîzmî, Arabic commentaries on Aristotle, and all of Aristotle's scientific works. In Sicily, the Norman count Roger I (ca. 1031–1101) expelled the Muslims in 1091. During subsequent years, because of the long-standing presence of Greeks and Greek manuscripts in Sicily, Latin translators there were able to work directly from the originals. Driven by the "poverty of the Latins," scholars from both regions directed a steady stream of manuscripts, derived from Arabic translations and Greek originals, into the new universities.[22]

An intellectual context accompanied these translations. Arabic philosophers, most notably Avicenna (Ibn Sina, ca. 980–ca. 1037) and Averroes (Ibn Rushd, 1126–1198), had earlier grappled extensively with the issue of how

Islamic theology accorded with Hellenistic, especially Aristotelian, science and philosophy.[23] Thus, when Aristotle's works became available to the schoolmen of Western universities, a centuries-old scrim of commentary and discussion fronted the introduction. At first, this pagan philosopher was not well received; many of his doctrines (for example, the eternity of the universe; various limitations on God's absolute power) ran afoul of received Christian theology. Such doctrines were among those called to task when the Church intervened in academic matters with the "Condemnation of 1277," striking down numerous, mostly Aristotelian-based propositions.[24] But despite fierce rivalries and turf battles between philosophers and theologians, struggles not unlike those in modern university settings, the faculties of arts and theology steadily worked out their accommodations with one another, and after midcentury Aristotle's logic, natural philosophy, and metaphysics were increasingly incorporated into university curricula.[25] By then, too, Thomas Aquinas was crafting his extensive amalgamation of Aristotelian philosophy and Christian theology. A word of caution: modern scholars carefully, and rightly, distinguish between the actual works of Aristotle and those of his Muslim and Christian followers in the Middle Ages, who often sought to restrict, reject, or redefine various of his positions in service of their theological interests.[26] "Faith seeking understanding" (*fides quaerens intellectum*), those interests, the interweaving of scientific knowledge and religious belief, held sway as scholastics built their own "Church Intellectual," the heavenly city of the thirteenth-century philosophers.[27]

Still, despite the presumptive harmony of God and nature, of religion and science (or of revelation and reason, as it was most frequently identified at the time), not all was warm and cozy among the new dons. Nor did they lack scholarly vigor, variety, talent, or the courage to follow their thoughts away from the harmonious ideal.[28] One need only sample the *Sic et Non* of Peter Abelard (1079–1142) to realize sheer intellectual brilliance charting its own course, or either of Aquinas's *Summae* to see the jaw-dropping persistence and subtlety of his intellectual architecture, or the works of John Duns Scotus (1265–1308) or, especially, William of Ockham (ca. 1288–ca. 1348) to witness medieval deductive analysis at its finest. Even a limited foray into these and lesser authors discloses, as well, abundant conflicts existing and persisting among the tiny percentage of the population who spent time reflecting on intellectual, natural, or spiritual matters. Theirs was a kingdom of words, contested terrain where arguments over how specific terms accorded with specific phenomena—material, psychological, and spiritual—tended to center on whether the categories were correct and being applied correctly.[29] Categories were often extended, divided, subdivided, restricted, and redefined; so too were the logical rules governing them. Much ink was spilled over nouns, and still more over adjectives as their qualifiers. We can witness both the classifying temper in operation and the intellectual vigor of the age with a brief excursion into one of its central disputes, over whether theology should be understood as a science.

The question was simple, the answers complex. Should *scientia,* Augustine's "rational cognition of temporal things," extend to theology itself? Basing his arguments on Aristotle, Aquinas responded in the affirmative, revealing his own reliance on the information technology of literacy. The natural reasoning of philosophy, the "perfect intellectual operation in man," began by abstracting "material images" or "phantasms" from sensation, he wrote, launching his explanation of how the sensory-laden data of experience were converted into information. From phantasms the mind extracted further abstractions, which corralled information into the definitions of words and created knowledge, the material of thought. And reaching into the higher realms of thought, the compulsion to classify subject matter—arranging subjects hierarchically into genera or categories (disciplines) of knowledge—outweighed all other forms of philosophical and scientific activity.[30] Each scientific genus was separate and determined by the subject matter of its own investigation. Some rested on higher forms of abstraction (for example, music was a dependent offspring of arithmetic), but all were based on appropriately applying the most general, abstract principles of thought and argument to a designated area of study. From these general principles, the light of philosophy could show us some things about God—for example, that he exists, that he is infinite, and that he embodies the metaphysical attributes of perfect being. Science clearly reached into theology. Moreover, the order and direction of study were equally clear for Aquinas: "Faith presupposes natural knowledge."[31] One started with nature, then worked his or her way up to God.

But words, their definitions, and the general principles of reasoning were all ultimately based on the limiting human and temporal experience of sensations, and therefore they could not encage God, who transcended this mortal coil. Philosophy sufficed only up to a point; for salvation "there should be a knowledge revealed by God, besides the philosophical sciences investigated by human reason."[32] But then a problem arose: How could one use the same language to speak of "knowledge revealed" and the knowledge of the "philosophical sciences"? The term 'knowledge' was the same after all. Patiently and carefully, Aquinas answered. Language was used in different ways: "No name belongs to God in the same sense that it belongs to creatures." When speaking about God, his mysteries and miracles and powers, we use language metaphorically; we talk about God with analogies. "Wise" may well refer to humans who act with prudence; it applies to God only "according to proportion." And because we know philosophically that God is infinite, we can extend "wise" analogously to God in calling him infinitely wise.[33] Thus, as a category, "knowledge" had two overlapping subcategories, in effect science and religion, each with its own appropriate use of language. In this manner reason could go part of the way in establishing God and true religion (especially to reasonable but still unbelieving people, those whom Aquinas addressed with the *Summa Contra Gentiles*), but it could not go all the way to accounting for the mystery of God (for example, the Trinity) or God's actual, revealed presence in the world (as with the Incarnation). Those truths must be accepted on faith.[34]

Over a half century later, one of Aquinas's most penetrating critics, William of Ockham, would have little of this argument. Theology had nothing to do with science or human reason, he believed, but rested exclusively on faith in God as revealed in scripture, a position subsequently given the name of "fideism." Like Aquinas, Ockham too began with a basic Aristotelian account of how we know something. Through our senses we perceive things, give them names to define them, and cluster the names into categories. But whereas Aristotle and Aquinas both believed that the categories mirrored reality in the world outside the mind, general words matching general things, Ockham contended that all our categories—our general ideas, classifications, subgenera, and so on—existed only in our minds as a "mental language" whose function is to "convey information" about the world, essentially for our convenience. Otherwise stated, our definitions ultimately signify only the individual things we perceive in the world.[35] From this starting point he derived what eventually became known as "Ockham's razor," a parsimonious principle that one should not "multiply entities beyond necessity." And from this keep-it-simple principle it followed that "universals," all those general categories referring to general entities, were only useful fictions, not actual realities. A universal concept was merely the mental sign that indicated thinking about several things at once, and the terms of science could ultimately be reduced to statements about singular substances or qualities.[36]

Of course Ockham did not deny that God, the greatest of the universal entities, actually exists, only that human science and reason could prove or demonstrate the fact. To do so would require a "simple cognition of the divine nature in itself—the kind of cognition had by one who sees God. But we cannot have this kind of cognition in our present state."[37] Through the screen of faith we may perceive God in all the nooks and crannies of his creation, as had Francis (Ockham too was a Franciscan friar), but our reason per se gives us little purchase on explaining divine matters philosophically. Nor, in Ockham's view, can we talk intelligibly about God by using analogical extensions from our own, limited terms and meanings, a use of language prized by Aquinas. For Ockham terms possessed either a univocal or an equivocal sense, and this posed a dilemma. In the first instance, if we use words univocally (literally, with "one voice"), then their core meanings always remain the same, always based on human perceptions. To characterize God with the human purposes, descriptions, or meanings we articulate in language, therefore, meant endowing the divine with human traits, limiting God to our own frail and diseased comprehension. And God wouldn't be God if he, or she, were like us. In the second instance, if terms are equivocal, signifying one thing when applied to humans and another when applied to God, then we are—not to put too fine an edge on it—uttering ambiguities or vacuities or just sheer nonsense in speaking of the divine.[38] As for the moral virtues or intellectual habits pertaining to science, which Aquinas had also placed under the light of natural philosophy, these were provided humans only by the free gift or grace of God, equally accepted on faith, not discovered through reason. (This was a point of view that would have far-reaching implications for theology, many of which were later adopted by Protestants.) In sum, while reason and argument could be used to explicate and characterize many of the tenets of faith, they could not establish them. Theology stood separate from science. Thus did the "More than Subtle Doctor" refute the "Angelic Doctor."[39]

Viewed in hindsight, one might think that such scholastic disputes as that just sketched reveal cracks and fissures within the classifying temper, structural weaknesses of such magnitude that they would eventually collapse the medieval edifice. Some scholars think this way, and even go so far as to claim that "a major turning point in the history of Western Civilization" arose from these and comparable conflicts in the late thirteenth and fourteenth centuries. They have a point, these scholars, but a smallish one.[40] For behind all the jockeying over what would be included in the categories of reason and revelation, science and religion, both "realists," following Aquinas, and the "nominalists" in Ockham's wake labored under the same, classifying impulses when it came to depicting how we gain knowledge of the world and how it relates to theology.[41] They were more bound together by language and literacy than separated by the thoughts devised with them. Whether of reason or revelation, all human knowledge depended on words, their classifying possibilities, and the logical tools with which they were organized.

Thus none of these disputes widened into our modern rift between science and religion: a Reformation, yes; a Scientific Revolution, no. That rift surfaced only centuries later, and not primarily from a quarrel over categories, concepts and distinctions, nouns and adjectives. Rather, a deeper divide would be exposed, one based on a new set of intellectual practices, a new means of abstracting and processing information that came to mediate the mind and the outside world. On the one hand were those habits of mind that had endured for nearly two millennia, the penchant for classifying and its underlying information technology of alphabetic literacy. On the other hand were the mental habits that eventually fed emerging scientific styles, procedures, language, and methods—habits having their origins in a different information technology, the cornerstone of which would be relational numeracy. In later chapters we shall follow the appearance of this new theater of thought, the new habits and the birth of the analytical temper they sired. For now it suffices to recognize that the new technology would find alternative building blocks to those of literacy—in new and different types of symbols and the rules for their manipulation, in new and different ways of capturing sensory-laden information (the givens of sense data), in new and different layers of abstraction, and in a new and different, critical distance from the natural world these abstractions afforded natural philosophers.

* * *

The breakthrough into the new information technology would be carved initially in bas-relief against two dominant and intertwining themes of the medieval magisterium, themes uniting *scientia* and *religio:* "words and things" and "demonstrable common sense." To appreciate the practical innovations forthcoming in the Renaissance and leading up to relational numeracy, the analytical temper, and the rift between science and religion, we shall need to spend some time with these medieval themes in Chapters 1 and 2, traversing a known landscape with a fresh look at the substratum of information technology in the period's intellectual life. The patterns of thought evoked by these two themes were deposited deep within the layers of the past and had become commonplace by the thirteenth century. The phrase 'words and things' denotes the mind's creating and processing of information, which in turn furnished the technological underpinnings of the science of 'demonstrable common sense.'

Historically, the invention of writing long preceded the creation of ancient, predominantly Aristotelian, science, and in Chapter 1 we shall see how the alphabet emerged as an information technology and how it came to be situated at the core of the medieval magisterium that embraced both science and religion. We should bear in mind that, before the modern era, not only were religion and science thoroughly intertwined, but philosophy and science were likewise virtually indistinguishable from one another. The terms "natural philosophy" and "natural philosopher" expressed this close association, and their currency continued until well into the nineteenth century before giving way to our modern lexical employment of 'science' and 'scientist,' terms coupled historically with a far different array of scientific practices and content. Of a cloth during the Middle Ages, science, religion, and philosophy were all based on the technology of alphabetic literacy and the classifying temper it propagated. Relying heavily on Aristotle, scholastics explained in various ways how words mirrored, or were abstracted from, things of the natural world. Alternatively put, words conformed, corresponded, or "adequated" to things, while words and things together shared some sort of structured or patterned taxonomical arrangement.[42] The details of these arrangements provoked many, often contested differences of opinion and arguments, even while they shared a profound and common assumption: words capture reality.

The same information technology generated what we shall be referring to as 'demonstrable common sense,' which typified the premodern natural sciences, topics we shall explore in further detail in Chapter 2. Among others, Aquinas articulated the idea for scholastics with his version of the "Peripatetic axiom," handed down from the Philosopher himself: "Nothing is in the intellect that was not first in the senses."[43] This meant taking in ordinary, commonsense perceptions and eliciting from them carefully defined categories. In turn the categories allowed one to account for the "causes" of natural phenomena under investigation, understood from a particular point of view, while further philosophical reflection on the categories yielded the underlying principles organizing both phenomena and the mind's tools for comprehending them. Notwithstanding the many permutations of these themes, for medieval schoolmen and intellectuals, 'words and things' and 'demonstrable common sense' together explicated the procedures of classifying through which nature and God alike were known insofar as humanly possible. They embodied the classifying temper.

The procedures, in turn, reached into two additional, subordinate areas that we moderns generally consider outside the classifying mode: analysis and mathematics. It jars our sensibilities to see these two pillars of modern rationality as dependent on being classified in an overall system of knowledge. For us they have become tools for scientific discovery, whereas in the premodern world they explained and organized what was already held in "common." Thus, premodern deductive analysis, here designated by its closely allied medieval expression, "demonstration" (*demonstratio*), focused on both establishing deductively the primary elements of common objects and then classifying objects according to the logical and causal connections between them. Similarly, premodern mathematics can be called "thing-mathematics" because numbers referred to collections of discrete things or objects, because geometrical figures and definitions likewise referred to fixed spaces as continuous things, and because, as different things, the universes of number and space (arithmetic and geometry) remained classified as distinct and incommensurable objects or categories, topics we shall take up in Part II.

In surveying the classifying temper that prevailed in the Middle Ages, its origins in alphabetic literacy and the thought of Aristotle, and its procedures for producing knowledge of the divinely infused natural world, we must be ever aware of the anachronistic tendency to make of our own projected image a cartoon of scholastic thinking as we hasten to modernity and the first signs of the fault lines leading to modern analysis and the estrangement of science and religion. Together these distinct, but inseparable categories, articulated through the terms of *scientia* and *religio,* reason and revelation, were thoroughly interwoven in the medieval cultural and intellectual milieu. They stand as witness to the predominance of the classifying temper—the deeply ingrained penchant for organizing the information of our sensory-laden experience into groups, types, sorts, and classes—as a means of making sense of the world around us. They also testify to a profound belief in the wisdom and reality of a God who created and sustained and bathed the universe with his presence, a divinely animated universe in which the smallest sparrow displayed the mighty splendor of its Creator. They stand as witness to the universe of Francis.

CHAPTER 1

A World of Words and Things

> The reason of anything . . . is signified by its definition.
>
> —Thomas Aquinas, *Disputed Questions on Truth* (1256–1259)

ON DECEMBER 29 IN THE YEAR 1170, according to legend, John of Salisbury witnessed the murder of Thomas à Becket in Canterbury Cathedral by the knights of King Henry II. (Some said John too suffered wounds at the assassins' hands.) Though the eyewitness portion of the legend is no doubt spurious, he was certainly at Canterbury that ill-fated day and quite probably close by. As Becket's secretary and sometime adviser, John was intimately involved in the ongoing struggles with Henry (1133–1189) over ecclesiastical prerogatives and their incomes. Nonetheless, whatever his precise location at that tumultuous moment, he was not likely thinking about one of his life's major concerns: grammar and language. Student of Peter Abelard and Gilbert of Poitiers, among others, John (ca. 1120–1180) rose from humble origins through the force of his own intellect, talents, and charisma to become a courtier, papal secretary, aide to two archbishops of Canterbury, and eventually bishop of Chartres. Along the way he produced the *Policraticus,* a manuscript born of his experience, reflections, and wide-ranging scholarship. One of the central medieval treatises on political theory, it covered topics ranging from a justification of tyrannicide to the propriety of off-color dinner-party jokes. A good observer—like Machiavelli in a different age—John, however, remained too much the man of virtue, too committed to principled stances, and too devoted to an authentic life of moderation for a totally successful career as a courtier.[1] Apparently his heart was not always in it, for, he confessed, "by dint of snatching moments like a thief" whenever possible, he took refuge from the cares of politics and "the burdens of administrative responsibilities" in his own consolation of philosophy, the world of books and letters. There he found the calm to complete his other major manuscript, the *Metalogicon* (ca. 1159), a work as typical of the times as of the man.[2]

On the surface, John's *Metalogicon* presented a program of education, defending somewhat narrowly the medieval arts of the trivium (logic, grammar, and rhetoric). The trivium comprised the first three subjects of the seven liberal arts, and it combined with the quadrivium (arithmetic, music, geometry, and astronomy) to form the foundation of a liberal education at cathedral schools and later at universities. But John's treatise covered much more than curriculum development. Relying heavily on the partially available logic of Aristotle, a broad reading of the classical authors at his disposal, a moderate skepticism, and a "humanistic" approach to topics, he devoted much of his attention to the herein-called words and things, how the mind processes information through conventional language to create and retain knowledge. In his reckoning, phonetic symbols formed words, and words secured the impressions of common sense, providing the intermediary between the knowing mind and the external world. Grammar, he observed, operates as the "basis and root of scientific knowledge," for it treats "the nature of letters, syllables, and words." Even more fundamental, "letters" are "written symbols" that "represent sounds" and "stand for things, which they conduct into the mind through the windows of the eyes."[3]

With his account John of Salisbury effectively gave medieval voice to what has become a modern subject, the information technology of alphabetic literacy, though obviously not so termed in John's day.[4] And with allowances made for the expanse of centuries between us, John could very easily have written the following pages, for many of his insights and observations seem surprisingly familiar, surprisingly modern. All the same, lest we forget, John lived in a different world from ours. His was a world in which he and other medieval thinkers were enveloped by a classifying mentality, a world in which philosophical reflection and contemplation led from the sensory "givens" (literally, "data," the plural of the Latin *datum,* or "thing given") to knowledge, and from knowledge to wisdom, the "highest power of a spiritual nature."[5] It was the divinely animated world of Saint Francis, articulated through the sparse intellectual inheritance of ancient writings. We are separated from the "discarded image" of that world not only by the intervening years and the subsequent flood of information and discovery but also by the transformations of perspective the years have wrought, including the creation of the modern analytical temper and the rift between science and religion, the subject of our present narrative.[6]

Further separating us from John's world is our modern sense of the historical past, often termed historical consciousness, an awareness completely alien to medieval perspectives. John and his confreres lived figuratively at the ahistorical crossroads of a timeless time and an endless eternity (Augustine tied both to the Incarnation of God and to the Parousia, Christ's Second Coming).[7] They viewed their mortal coil predominantly as a temporal and temporary pilgrimage to a higher, everlasting plane of existence, wherein one's "condition of living will be transformed into a better one."[8] For them words and definitions and logical demonstrations were simply a part of the natural, divinely created hierarchy of things, the building blocks of a vertically constructed "Church intellectual" (recalling Henry Adams's matchless expression). Using these tools of literacy to achieve "understanding" made manifest the most exalted of human endeavors, through which one ascended from the city of man to the city of God, to the extent possible "with [God's] grace both preparing the way and providing assistance."[9]

By contrast, we recognize these informational and linguistic building blocks as the products of a drawn-out, multifaceted development over time. Thus, although generally following John's rendering of words and things from literacy to definitions to logical demonstrations, we shall be exploring these topics from a historical perspective, albeit one quite compressed given space limitations. As we shall see, the long trajectory leading up to the premodern, classificatory science of John's era began with the early, abstract symbols of writing, which first created information and laid the foundation for the later evolution of thought.[10]

Early developments in literacy reached their apogee with the phonetic alphabet, the greatest of Greek inventions and bequeathals to Western culture, and subsequently with the philosophical thinking of Aristotle, "from whom, as from a fountain," John wrote, "all have drunk." Definitions lay at the core of Aristotle's thinking, and his medieval acolytes took up those "most efficacious instrument(s)" in explicating a dizzying, dog-chasing-tail account of definitions defining definitions in order to understand the "understanding" itself.[11] (Parsing their prose was not a task for the impatient or the unsubtle or the faint of heart, a caution applying to denizens of the twenty-first century, as well as the thirteenth.) In their ascent from grammar and definitions to the abstract study of formal reasoning in syllogisms, scholastics capped their lofty intellectual cathedral with the logic of demonstrations, whose principal codification was likewise inherited from Aristotle's *Organon* or logical works.[12] In sum, throughout their enterprise John and his fellow scholars relied thoroughly on the historically accrued technology of literacy in order to construct their fusion of *scientia* and *religio.*

Supervening this entire process both historically and hierarchically was *logos,* a capacious concept and inveterate piece of metaphysics that stood variously for the essence of a thing, the definition of a thing, the logic coupling words and things, an argument from reason, and even order in the entire universe. Ancient and medieval natural philosophers sought understanding of nature's patterns through the world's *logos,* its rational order binding intelligibility and being, thought and reality. Scholastic theologians went further, extracting from *logos* another, even more critical layer of awareness. In the opinion of Thomas Aquinas, for example, words displayed two dimensions: exterior and interior. The former correlated with things in the outside world, and the latter with ideas, thereby opening the gateway to the universe of the soul. In this way, human words were intrinsically drawn to the "Word [*logos*] made flesh" (from the Gospel of John), the divine incarnate. *Logos* therefore united words and things, and in so doing it supplied the infrastructure for the harmony of science and religion, a harmony conceived and conveyed through the written word, a harmony of words and the Word.

* * *

We don't know, of course, but we can surmise. Humans likely began the "symbolic species" phase of their evolution with storytelling.[13] More so than the sounds uttered by other mammals (again, as far as we know), human speech and the words composing it evolved to allow for a far greater precision in the symbolic reach of their imagery. Words evoked pictures in the minds of those who heard them, capturing detailed features of things seen, sounds heard, odors smelled, flavors tasted, and objects touched. As bardic traditions suggest, people early on recognized patterns in speech, which they could then elicit and hold in memory, recalling and repeating when desired, varying communication in and through the formulas of rhythm and content, structure and meaning. Long before they wrote, humans abstracted in vocal articulation and composition the sensory-laden ephemera of their momentary experience. They told tales.[14]

From its earliest appearance in the fourth millennium before the Common Era, the invention of writing took the tale-telling words and patterns of speech to another level of abstraction by creating, processing, and storing information. This was accomplished through symbols, abstract marks of one sort that stand for things of another. The term 'abstraction' itself derives from the Latin verb *abstrahere* ("to pull," "drag," or "draw away from"), and historically it has come to connote a mental activity that fixes the flux of ephemeral and momentary experience through two closely intertwining movements. These involve (1) drawing away from experience, such that we are no longer immersed in it and can observe it from some distance or perspective; and (2) pulling or dragging out of experience some indwelling form or pattern or, in the earlier but still charming and useful expression of philosopher José Ortega y Gasset, a "mental extract." This twofold movement of abstraction allows us to look at the products of our own creation, to investigate them with a critical eye toward shaping or forming them into something new, and most significantly when the products are linguistic, to classify systematically similar terms and objects. Writing and information thus elicited further the classifying potential inherent in human speech.[15]

Virtually all students of early writing agree that it began as pictures ("speech put in storage"), and that at some point, under the pressure of early accounting and record-keeping needs, it became more systematic, although the exact journey from pictures to scripts remains quite murky and disputed given the paucity of evidence.[16] Fairly soon after the earliest pictograms, writing began to model itself on spoken language, with phoneticization making its appearance around 3000 BCE. During this phase of development, symbols came to represent sounds, as well as objects, providing a sign for each syllable used in a language (hence the expression "syllabic writing"). Scripts became increasingly stylized. In Mesopotamia, for example, early proto-cuneiform writing gave way to the development of a more systematic and stylized cuneiform tied more directly to vocal utterances. Early Egyptian hieroglyphs likewise began adopting phonetic signs during the third millennium BCE. And in China the first pictograms were subsequently augmented by simple and compound "ideograms" (characters that suggest meaning but do not depict anything concrete) and "semantic-phonetic compounds," a blend of abstract pictures and sounds that still composes some 97 percent of modern Chinese characters.[17]

Pictographic, ideographic, or phonetic, a common feature of all these early systems lay with their burdensome reliance on the reader's ability to interpret the symbol correctly within its syntactic (structural) and semantic (referential) context. For a modern example in English, the picture or symbol of an eye might be interpreted as the physical eye itself, or as the act of seeing, or as the phonogrammic, first-person "I", or in many other ways (for example, as God's "eye") depending on circumstances that require the reader's comprehension. For millennia manifold nuances in both writing and reading left the task of interpreting scripts within the province of specialists.

As it evolved, writing steadily and gradually made manifest speech's classifying potential. Symbols stood for names; names stood for objects. Grouping objects under common names and representing them with common symbols became progressively expanded and more conventional. With the growing complexity of scripts, scribes captured greater amounts of information, and with more precision. Accompanying this acquisition was the intrinsic urge to organize into categories the information writing had created, associating things with common features under the same word. Sumerians, for example, combined the properties of kings under the rubric of "kingship" rather than simply naming individual kings.[18] Both the urge and the potential for classifying expanded as writing increasingly became, in Voltaire's elegant phrase, the "picture of the voice."[19]

Although numerous writing systems were devised throughout the ancient world, the alphabet in all likelihood was invented just once. It began emerging sometime around 1500 BCE in the crossroads of the Eastern Mediterranean, where commercial and cultural exchanges brought with them various, often incompatible, scripts and glyphs, creating a "symbolic stew."[20] (See Figure 1.1.) Such circumstances furthered the already-existing tendency to simplify found in syllabic writing. Modeled on speech, syllabic writing provided a sign for each syllable used in a language, from which issued the practical difficulty of remembering signs for hundreds of different spoken syllables. A "principle of economy" operated to reduce them to a more manageable number, streamlining the script in the process.[21] But even a reduced number of signs kept the burden of interpreting them on the reader, a Gordian knot cut only with the breakthrough into the alphabet.[22] In effect, the

An Evolving Alphabet

Egyptian Hieroglyphs	Proto-Sinaitic	Phoenician & Hebrew	Archaic Greek	Classical Greek	Latin
				A	A
				B	B
				E	E
				K	K
				M	M
				N	N
				O	O
				P	R
				Σ	S
				T	T

FIGURE 1.1. Evolution of the alphabet.

alphabet reduced the basic unit of pronunciation, the syllable, to its constitutive components, individual "letters," which, as John of Salisbury noted, were the "written symbols" that "represent sounds."

Preceded by the highly compressed syllabaries of West Semitic scripts, wherein each sign stood for a set of consonants, the appearance of a mature, fully developed alphabet occurred first with the Greeks, whose earliest inscriptions date from ca. 730 BCE. Using Phoenician signs that had no equivalent sounds in Greek, the children of Homer adapted the syllabic form of writing to the pronunciation of their own language by systematically assigning separate symbols to each consonant and vowel. This was revolutionary, overturning the previous relationship between writing and language. Rather than being a mnemonic device for recalling words in a language one already knows, alphabetic script became a means of transcribing the sounds of language in general, regardless of whether one knew the meaning. Because most languages limit the number of sounds to no more than about forty, the alphabet can, with slight modifications, be used to transcribe virtually any language, a feat it accomplishes by visually representing the two classes of sound formed by the human voice, vowels (those sounds produced by blowing air through the vocal chords and causing them to vibrate) and consonants (sounds derived from stopping, restricting, or shaping the flow of air, chiefly with the tongue and lips).[23] Grammar, recorded our medieval John, "imitates" nature and at "nature's bidding [has] limited the number of elementary vowel sounds to five." Add in the consonants, he continued, and names are "stamped on all substances."[24]

With this revolutionary accomplishment, the alphabet brought to its apogee the ubiquitous urge toward greater classification and abstraction found in earlier forms of writing. Then and still, it supplied a multilayered distancing of information from the immediacy of our sensory-laden surroundings. At the first level, correlating sounds and letters directly means that the words formed are unfettered from context; they no longer rely on the reader to fill in the gaps simply in order to ascertain what word he or she is reading. The correlation between word and object can be made independently of the reader; words have become objective. (Now, if you don't know a word, you can look it up in a dictionary, or on a Rosetta stone, but in order to do so you must already have recognized the word as a specific word.) In a second layer of abstraction, alphabetic literacy increases the distance between the mind and the objects it observes, between the knower and known, which it does by allowing the mind to examine and rework its own, intermediary products from a critical vantage point. While true of all writing, with the alphabet one can focus greater attention on the words themselves, the things to which they refer, and the associations between the two. Instead of trying to comprehend how the pictographic and phonetic symbols make up words, and which ones are being formed, the mind can directly engage and interpret the words and things of the world.

In the critical reworking of its own products through the technology of the alphabet, the mind shifts into a third layer of abstraction, the process of defining, which both limits and clarifies names and their corresponding objects. Through definitions, the nouns and adjectives of conventional language (at least Western, object-oriented languages) allow us to create terms or concepts in order to delineate or determine the impressions we take in from our manifold sensations.[25] They provide us that "mental extract," to recall Ortega's felicitous phrase. Thus evolving from early placeholders, this mental extract or concept permits a new focus of attention on the properties included within it, and it serves as a boundary separating some features of sensation from others. By eliciting these mental extracts, we channel, as it were, our sensations, along with our intuitions, memory, and imagination, into definitions and their corresponding things. We convert raw data into information.

* * *

Alphabetic literacy, then, both sowed and cultivated the development of a mature classifying temper. As much as any classical author, Aristotle nurtured these developments to fruition. Before his lifetime there had been a centuries-long process of alphabetization and emerging literacy in the Hellenic world whose causes, pace, and implications scholars continue to debate mightily.[26] Yet, whatever its odyssey, by the fourth century BCE Aristotle could take alphabetic literacy for granted, noting that language, both oral and written, "consists of the combinations of the letters," and that letters were but "written marks, symbols of spoken sounds."[27] Letters made the words, and the words captured and caged as objects for domestication the flitting ephemera of our sensations. With the technology of writing converting sensations into information, Aristotle was free to concentrate his prodigious intellectual efforts on converting information to knowledge, on understanding the world about him. Those efforts revolved around the "understanding" itself.

Aristotle's "understanding" began and ended with definitions in the fullest scope of the term, and he explicated their significance, in a manner comparable to other scientific accounts, with a description of their "causes."[28] We shall defer a detailed treatment of Aristotle's ideas of causation until Chapter 2, where they fit more properly in an account of his natural philosophy. For now we need note only that, for him, "cause" meant "the 'why' of" something, which could be viewed from four different vantage points. He identified these as material, efficient, formal, and final causes, and we can discern them in the way Aristotle explained definitions. Depicted as a "phrase signifying a thing's essence," a definition incorporated words (its material cause), dialectical reasoning (its efficient cause or agency), logical demonstrations (its formal cause or logical structure), and purposive explanations (its final cause or *telos*). These were the four perspectives through which we view and understand "the 'why' of" anything, applied reflexively to the understanding itself.[29]

The process of defining began with words, the material " 'why' of" or underlying substance "out of which a thing [understanding] comes to be and which persists." Words were the "matter" that reason shaped into "form" and thereby made intelligible. As with many earlier expressions, this sounds somewhat strange to the modern ear, but on reflection does make sense, for, standing alone, a word is not really a word in and of itself but only a sequence of letters. Letter-symbols initially have no correlations with anything beyond individual sounds; only when they are given their semantic referents and syntactic structures do letters become a word in relation to other words and things. In this regard and citing Saint Augustine, Aquinas commented that a word is "formable before being formed."[30] Once formed as a proper sequence of letters, the word then lives on with others as a substratum all the way through to the end of understanding.[31]

From words arise definitions, and, accordingly, ensuring their correct letters became critical from the outset of any investigation. As we shall discuss in Chapter 2, Aristotle identified the matter occupying the heavens as "ether." The word in question, *aithēr,* he wrote, had been "handed down right to our own day from our distant ancestors." They in turn had derived it from the terms for "always" (*ai*) and "to run" (*thēr*), with an allusion here as well to the divine (*theos,* the origin of our prefix 'theo-'). Thus, the literal meaning of *aithēr* was "running always," which Aristotle rendered as "eternally in motion." By contrast, his predecessor Anaxagoras (ca. 500–428) had gotten the derivation wrong, misreading or mishearing the letters with the claim that *aithēr* had descended from *aithō,* which meant "ignite," "burn," or "shine," and thus misconstruing the heavens as composed of fiery matter.[32] Getting the letters and words right was the necessary first step toward understanding.

The "primary source" or agency in the process of definition (its efficient cause) lay with the mind's critical reasoning, which entailed both "empirical" and "dialectical" inquiry.[33] Empirical inquiry shaped and filled words with their sensory, informational content, supplying their semantic referents, as, for instance, Aristotle's observing the circular and repetitive motion of celestial matter (or relying on the accounts of others' observations). Dialectical inquiry meant raising questions and drawing inferences, establishing the proper links with other words. Reasoning about the observable, circular, and repetitive movement of ether, he identified these features with the heavens and therefore as unchanging, eternally in motion, and "divine in nature." Because such matter was "different from any of the terrestrial elements," these inferences also explained why "they [the ancestors] determined to call it 'ether.' " The misguided Anaxagoras had not only made a spelling mistake; he had also reasoned incorrectly about heavenly matter when inferring its fiery composition from his orthographic error. Thus could critical inquiry be brought to bear on the objective character of the derivations and of the words themselves, as Aristotle further noted that "the same ideas . . . recur in men's minds not once or twice but again and again." Aristotle referred to the mental extract or idea captured by words, and teased into recognition by proper inquiry, as *horos* or "term" (later, *terminus* in Latin).[34] Like common sense, the *horoi* were there objectively, as mental objects intuitively distilled from sensations of physical ones. Anyone could see them if they but looked and thought a bit.

Although he had no specific expression for logic as a whole in the modern sense, Aristotle exploited the word *logos* to explain the formal cause or " 'why' of" definitions and understanding, the patterns of thinking in their formal, logical structure. First used philosophically by Heraclitus (ca. 535–ca. 475 BCE), *logos* came by Aristotle's lifetime to mean not only "word" but also and more expansively, "account," "description," "formula," or "reasoned discourse."[35] In the *Metaphysics,* he even defined "definition" as "a single formula [*logos,* sometimes translated here as "account"] and a formula [*logos*] of a substance."[36] From Aristotle's era forward, philosophers have struggled with the task of untangling the bramble of logic and metaphysics surrounding *logos.* For medieval scholars, in particular, sorting out the proper account of its constituents—"form," "being," "actuality," and "essence," among others—was of special concern, literally the essence (*logos*) of their spiritually driven, intellectual quest. But regardless of the religious and metaphysical beliefs they (or Aristotle) attached to the term, in its epistemic dimensions *logos* centered on "demonstrations," especially scientific demonstrations (more on this later). In effect, what was said about *logos* referred to the web of mental extracts or terms, and constituted logical theory. Logic therefore tendered understanding's "formal" cause or "'why' of." In modern parlance Aristotle's logic and system of classification articulated a logic of terms (vis-à-vis a logic of relations) that corresponded to and organized the things of the world.

The final perspective from which to view the cause or "the 'why' of" definitions lay with their purpose, their *telos* or "final" cause. This was found in the intrinsic goal of an innate human need and passion to understand, as Aristotle announced in his famous opening line to the *Metaphysics:* "All men by nature desire to know."[37] He accepted as facts of human existence that we have a capacity for "puzzlement and awe" and that our puzzlement arises from discontent at the absence of understanding. We experience the lack of understanding as wonder or curiosity, and we replace it with contentment only when we have found the understanding we seek.[38] Elsewhere, in his *Nicomachean Ethics* (named after or dedicated to his son, Nicomachus), Aristotle painted a comprehensive portrait of human happiness or flourishing, *eudaemonia,* the highest good or end "for the sake of" which humans passed their time on this mortal coil. His scholastic heirs would later transform *eudaemonia* into the *telos* of attaining "true beatitude"—the true happiness of the believer's blessed, salvific, and eternal fulfillment in the divine. Yet with either *telos* definitions were fashioned "for the sake of" human well-being and contentment, the fulfillment of understanding in the deepest meaning of the term.[39]

Scientifically speaking, therefore, definitions proceeded conterminously with understanding. Through the activity of defining, the information created and made manifest by writing was transformed into knowledge, a movement in Aristotle's terms from matter to form, from "potentiality" to "fulfillment." On the practical plane of their discovery and exposition, Aristotle developed what is often called the theory of "natural kinds," the theory that the divisions of any category were made in accordance with the actual divisions of objects in the world.[40] As John of Salisbury expressed it, "Aristotle's primary concern is with substantial definitions, which should so comprise genus and substantial differences that they are equivalent to their subject."[41] The genus 'animal,' for example, might be subdivided into three parts: 'terrestrial animal,' 'aquatic animal,' and 'flying animal.' Each difference ('terrestrial,' 'aquatic,' or 'flying') defines its respective subpart of 'animal.' Otherwise put, "an Aristotelian definition consists of genus plus differentia."[42] The differentia ('terrestrial,' 'aquatic,' 'flying') in turn correspond to the natural divisions among animals, which we can establish through empirical and dialectical inquiry. All animals must fall into one of these three classes. To know or understand a thing, thus, meant to define it and thereby locate it in its proper place in the objective hierarchy of all things. Aristotle did note that classifying animals, as well as other objects, admitted of "much cross-division." For example, "vivipara" included men, horses, and cattle, among others giving birth to their young alive, whereas "ovipara" designated the class of egg-laying animals. Birds lay eggs, while humans do not; yet both are also classified as "bipedal."[43] But although cross-divisions might be difficult, or perceived wrongly, or even impossible to discern at times, behind any specific categories of objects flourished a mind-set, the classifying temper, forever in search of the correct nouns and adjectives that defined and gave us to understand natural things or objects.

In its fullness, then, the process of defining words and relating them to things yielded knowledge of the natural world, yielded *scientia*.[44] This knowledge was objective in the parallel senses that through the abstractions of the alphabet words were independent of their context and, similarly, definitions were context independent, filled with the proper sensory material (referent) and connected logically to other terms. The objects of knowing existed independently of and antecedent to the knower, whose task was then to discover and get them right through the use of conventional language. Aristotle adopted the term "univocal" to describe this primary, objective meaning of words and their definitions in reference to the fact that the same word meant the same thing in all contexts. Inside the house, running wild, or digging in flower beds, a dog is a dog is a dog, a furry and barking animal, distinguished from other furry, barking animals—say, seals—by additional objective characteristics, such as paws. Objectively, dogs were animals, animals were sentient beings, and all beings were housed in the general class of 'being' itself. Other, figurative uses of language, such as allegories, metaphors, epithets, and analogies, were predicated on their descriptive and objective, univocal meanings.[45]

* * *

Throughout his *Metalogicon* John of Salisbury constantly extolled the virtues of logic, "the science of verbal expression and [argumentative] reasoning." An "interpreter of both words and meanings," logic, which "rejoices in necessity," assumed its pride of place as "first . . . among all the liberal arts." Though chronologically last in the order of discovery, behind the "other branches of philosophy," it preceded all subjects in the order of importance, binding together words and definitions and demonstrations in a coherent fashion to make sense of things. Without the "vital organizing principle of logic," John warned, philosophy itself (and essentially all thought) would lie "lifeless and helpless." Like philosophy, but even more so, logic took the abstractions made possible by the technology of literacy to their highest reaches, supplying form to matter, structure to content. For scholastic thinkers its principal use lay in framing "demonstrations," the spinal column of deductive analysis, which supported and linked together the classifying proclivities of medieval learning.[46]

Not long after John had completed the *Metalogicon,* new translations of Aristotle's logical works began circulating throughout Europe, supplementing the earlier manuscripts of Boethius and Porphyry on which John had relied. Among others, John's teacher, the dazzling and controversial Peter Abelard, helped stimulate this expansion of interest, especially with his influential *Sic et Non* (*Yes and No,* compiled 1221–1232).[47] In that work Abelard presented a collection of 158 theological topics, with arguments both for and against each of them, all taken from writings of the early Church fathers. He did so in the belief that ultimately reason would solve the problems reason had created, that apparent contradictions could be reconciled. Although some (including Bernard of Clairvaux [1090–1153], founder of the Cistercians) thought Abelard had pushed reason too far, supplanting the humility of faith with the arrogance of argument, others followed in Abelard's footsteps, relying on the presumptive harmony of faith and reason to guide their logic in sorting out matters.[48] The study of logic continued to thrive all the more during the following century in the newly created universities. It took the passage of time to make all of Aristotle's works safe for Christian thinkers (not all of his ideas, of course), but augmented by new translations, his logical works gave them the tools to make the accommodations, and in due course numerous scholars energetically applied Aristotle's logic to an examination of the Stagirite's own writings themselves, as well as other topics in their curriculum.

Much like Molière's fabled Monsieur Jourdain, who was speaking prose long before he charmingly discovered that fact, people in antiquity had been thinking logically for generations upon generations without knowing it, most likely ever since our species exhibited the first hints of human thought. But before Aristotle no one had actually codified logical thinking systematically into a formalized structure, complete with technical terminology and explicit rules of inference. It was a stunning and far-reaching accomplishment. Aristotle probably took his inspiration from geometers and the way they justified arguments in their young and evolving discipline by using a proof process of logical deduction from commonly accepted definitions and first principles or axioms. In fact, Aristotle defined an axiom as "that from which" explanations follow—a principle or starting point we often call self-evident.[49] But even as geometry supplied the pattern of correct thinking, it tendered only one example of it. The "axiomatic" method, as it is frequently termed, was even more fundamental and more general, more applicable to reason writ large and with words.

In the *Organon* Aristotle methodically crafted his multitiered theory by bringing together two essential modes of reasoning—"dialectic" (*dialectica*) and "demonstration" (*demonstratio*)—into a formal system.[50] As noted earlier, dialectic referred to "a process of criticism," the question and answer used for discovering the premises that lead to a conclusion, or those a disputant would likely accept. By medieval times 'dialectic' had come to signify the means of establishing probable truths or facts, based on sensations and common sense, and expressing them in sentences or statements. These supplied an argument's premises. 'Demonstration' in turn denoted the method of establishing correct deductions from the premises. It functioned exclusively within the realm of thought, articulated likewise through words and sentences. Combined, dialectic and demonstration supplied the canons of rational thinking created from conventional language, the most widely taught and accepted means of intellectual discussion and debate for centuries throughout the Middle Ages and beyond.[51]

These canons found their clearest and most useful manifestation in a form of reasoning known as the logical syllogism, a deductive argument that included a major premise, minor premise, and conclusion. (For a few of the syllogism's technical features, see Appendix A.1.)[52] Valid inferences, expressed in syllogistic form, combined with true (or acceptable) premises, established through dialectic, to constitute the proof of any claim, regardless of the subject matter.[53] Of particular interest to Aristotle and his scholastic devotees was the "scientific" syllogism, which he defined as "one in virtue of which . . . we understand something." Aristotle illustrated his meaning with a distinction between understanding the "facts" (through dialectic) and their causes (demonstrating the "reason why"), and he showed how this distinction could be expressed with a pair of syllogisms:

	First Syllogism	**Second Syllogism**
Major Premise:	What does not twinkle is near	What is near does not twinkle
Minor Premise:	The planets do not twinkle	The planets are near
Conclusion:	∴ (therefore) The planets are near	∴ The planets do not twinkle

Both syllogisms represent valid arguments in their formal structure. Their conclusions follow logically from their premises. But whereas the first simply yields the "fact" that planets are near, the second provides the "reason why" they do not twinkle. Here the more basic facts in the premises help explain the conclusion. In the first case, that the planets do not twinkle hardly explains their proximity; but in the second case, according to Aristotle, their proximity at least partially accounts for their lack of twinkling. In scientific syllogisms, Aristotle expounded with a mouthful to challenge the most conscientious of future parsers, the premises are "true and primitive and immediate and more familiar than and prior to and explanatory of the conclusion."[54] One has to slow down for this one, but it basically meant that proper scientific demonstrations followed from the well-established elements of reasoning and the facts of common sense.[55]

Demonstrations thus organized and shaped directly the information supplied by the senses, lending an empirical dimension to many arguments. But the empiricism involved deviated considerably from the modern variety. As we shall see in Chapter 2, the basic thrust of Aristotle's approach to "science" as the organized knowledge of natural things lay with explaining the facts supplied by the senses and commonly agreed on, not in discovering new ones. One began with commonsense facts, those bits of sensory information available to anyone. The question then became why? What explained those commonsense facts? And the answer found its way into the syllogistic form in which one of the premises conveyed the sensory information. In the aforementioned second syllogism of Aristotle's illustration, the minor premise states the empirical, observable fact: the planets are near. The major premise states the general definition of the "near thing"—that is, "it does not twinkle." The conclusion that the planets do not twinkle, then, is an inference drawn from the definition (the quiddity or "thingness" of near things) and the placement of an empirical fact (near planets) within it.[56]

Thinking and demonstrating with syllogisms provided scholastics their main mode of analysis. It was, as termed here, a deductive analysis, in contrast to the reverse-engineering analysis of modern science (to be discussed later), and it had already possessed a lengthy history by the time the schoolmen were turning to Aristotle.[57] Usually paired with synthesis, deductive analysis functioned exclusively within thought, generally indicating the direction thought would follow. One began with an idea of some sort, cast as a philosophical statement, logical proposition, or geometrical construction. In order to prove the validity of such ideas, one subjected them to a series of deductive inferences from simpler statements or propositions. To that end the idea was analyzed, broken down into its "component parts," as John called them, and likewise each of the components was broken down, until one came to foundational assumptions, premises, or terms known to be true. These could not be demonstrated, Aristotle had noted, only accepted as given (*simpliciter* was the Latin term for it).[58] From such starting points one then worked deductively back to the original proposition, hence proving it. Similarly, in the case of disproving a proposition by a "reduction to absurdity" (*reductio ad absurdum*), one arrived at a foundational element known to be false. The twofold movement of thought in analysis and synthesis was described as inferring backward to the premises and then forward to the original propositions.[59]

Extremely subtle all this, to be sure, logical demonstrations and deductive analysis attracted some of the finest minds for centuries. Still, if explanation is where the mind comes to rest, the foregoing manner of explaining facts leaves us pacing restlessly. It seems enormously unsatisfying, convoluted, and quite irrelevant to understanding natural phenomena under investigation. Such awareness reveals our own science-based grasp of common sense and causality. We infer causalities from numerous empirical pieces of information (more on this in Chapter 2) or from crucial experiments (Isaac Newton's *experimentum crucis*). By contrast, the medieval sciences of demonstration observed them directly and then expressed them by means of conventional language and its classifying propensities. Such accounts provided a philosophical and logical method of both identifying and organizing the knowledge of natural things. Embedded within the information technology of literacy, premodern, deductive analysis served as the model for scientific, philosophical, and theological thinking. And with the link between logic and common sense firmly established, Aristotle's logocentric vision of knowledge prevailed in the works of the schoolmen of medieval universities, who extended his logical webs with numerous and often ingenious gossamers of their own.

* * *

For John and subsequent scholastics in the Middle Ages, the logical architecture upholding our condensed history was straightforward: letter symbols made words; words tendered definitions; definitions formed statements; statements organized demonstrations; and demonstrations yielded knowledge of *scientia* and *religio* alike. Laced through the entire process was *logos,* the spiritual glue that held together all the technicalities of science and reasoning. This led scholastic followers of Aristotle to attach an even greater significance to proper definitions than for simply perceiving the realities of nature, for the *logos* interwove science and religion. Not only was it the form of understanding, and therefore intelligibility in the natural, divinely created world, it was also the means by which God revealed himself directly to the faithful, as declared in the opening of the Gospel of Saint John: "In the beginning was the Word; and the Word was with God; and the

Word was God. He was in the beginning with God." This was the *logos* incarnate, the "Word made flesh." No one pondered these warps and wefts more probingly than Thomas Aquinas, whose prodigious memory and penetrating intellect belied his classmates' appellation of "dumb ox," given him in student days for his size and slowness to speak. In his commentary on John's Gospel, Aquinas began his first lecture, as did the Gospel, "in the beginning" with an investigation of the "name Word." Citing Aristotle, he used words to define words as "vocal sounds," the "signs of the affections that exist in our soul."[60]

From this base definition, Aquinas then proceeded to explore a critical distinction between the "exterior, vocal word" and the "interior word," each of which afforded a perspective on "the 'why' of" words and understanding. Both the exterior and interior words, he explained, stem from the "two operations" of the intellect, the aforementioned "empirical" and "dialectical" inquiries of Aristotle. In the first, the mind (or "intellect") "forms a definition." As should be evident by now, Aquinas followed the Philosopher in asserting that definitions apprehended the "properties of things," properties that were "disclosed in sensibility."[61] Words and their definitions thus connected one to the world of nature, to the world outside the mind, to the things "exteriorly expressed in words." The second of the mind's operations he termed "an enunciation or something of that sort," creating "an interior word which . . . understanding forms when understanding." He elaborated this vague phrase by noting that "enunciation" denotes an "operation by which it [the intellect] unites and separates."[62] In one sense this observation seems quite trivial, for all thought that uses words in conventional language entails making connections and drawing distinctions, uniting or separating. Yet in another and deeper sense for Aquinas, "enunciation" opened the door to the vast world of the interior word, the *logos* of God's creative activity in the souls of humankind.

Interior words tendered the means by which and in which human creative activity expressed itself in understanding. Words "conceived in the mind" represent "everything . . . actually understood," Aquinas remarked, and, accordingly, as humans we have "different words for the different things we understand." But unlike us, God expresses "all creatures." Pure Form or Actuality or Being, "by one act [He] understands Himself and all things" at an instant, a moment of no time lapse, at once present and eternal.[63] In his perfection, he needs but one word, *Logos,* rather than the many employed by frail humans. As contingent beings, humans shared in the divine Being; as purveyors of words in understanding, humans shared in the *Logos* that surpasses all understanding. Through such abstract comments, Aquinas highlighted the intrinsic connection between the activity of human thought in creating words and the activity of God's thought in "His one only Word." Just as the perspectives or fashions of "the 'why' of" allowed one to understand natural phenomena teleologically as divinely created processes, so too could "the 'why' of" enable one to understand the human processes of understanding through words as permeated with divinity. The words we humans actively sought and employed found their *telos,* their completion and fulfillment, in the divine *Logos.*

In ensuing sections of his *Commentary on the Gospel of John,* Saint Thomas proceeded to explain why the Word was first identified with the "Son" and subsequently with the "Father," who was "never alone without the Son or Word." He then extended this exposition to embrace the Holy Spirit, which possessed the "same glory and substance and dignity as the Father and the Son" and was likewise identified with the Word.[64] In this manner Aquinas formulated an extensive interpretation of and commentary on the Holy Trinity that had commenced with an account of the word "Word" itself. Further pursuit of the topic here would take us too far into the orbit of theology and away from the cusp of science and religion. For Aquinas, the ultimate goal in biblical exegesis and more generally in theology lay with apprehending what sinful humans could of God's wisdom, of *sapientia.* In these endeavors he employed the same tools and strategy that he exhibited in understanding the natural world. From words to inquiry to definitions to Word, our spiritual pilgrim's progress displayed the human yearning for the *vita beata,* the true happiness found only in "the love that moves the sun and the other stars," the same love that animates human souls.

* * *

Over the centuries it has become commonplace to characterize definitions and classifying schemes, including those of the Middle Ages, as being purely arbitrary, despite what their creators might have thought about them. The common sense of one society or culture, it appears, may well be the nonsense of another. Skepticism about definitions, classifying systems, taxonomies, and taxonomical thinking runs deeply in the European tradition. From the nominalism of the fourteenth century to the critical claims and taxonomies of the great eighteenth-century naturalists, the Comte de Buffon (Georges-Luis Leclerc, 1707–1788) and Carl Linnaeus (1707–1778), and into our own times, defining and classifying have been thought by many to be useful practices but not intrinsically connected to the outside, real world.[65] Umberto Eco, among others, has given us a plethora of amusing and trenchant historical illustrations of this arbitrariness, while the nineteenth-century logician Hermann Lotze (1817–1881) branded the random nature of classifying in perhaps its most simplified and drastic fashion. Grouping cherries and meat under the categories 'red,' 'juicy,' and 'edible,' he pronounced, gives us not a valid logical concept but rather a meaningless combination of words.[66]

Medieval scholars and intellectuals too contended regularly and frequently over how to classify specific content, where to put the details of *scientia* and *religio.* But while disputing which precise meanings went into which boxes or cubbyholes, they all shared the impulse of the classifying temper to organize the knowledge about them, even after nominalism challenged the underlying assumption that classifying schemes mirrored the real world. This was, after all, the way conventional language worked. Words defined the things brought to the mind by common sense. They were the means through which the information of common sense was arranged into common categories. C. S. Lewis summarized the medieval compulsion quite inimitably: "At his most characteristic, medieval man was . . . an organizer, a codifier, a builder of systems. He wanted a 'place for everything and everything in the right place.' Distinction, definition, tabulation were his delight."[67] Underlying the delight at classifying was a profound commitment to the exploration and use of conventional language as the proper way of expressing scientific knowledge and religious belief. The information technology of writing, and alphabetic literacy in particular, had given rise to the classifying temper and its universe of words and things.

CHAPTER 2

Demonstrable Common Sense

Premodern Science

Nothing is in the intellect that was not first in the senses.

—Thomas Aquinas, *Disputed Questions on Truth* (1256–1259)

EARLY IN HUMAN HISTORY, the impulse to order and make manageable the disparate and constantly changing phenomena that surround us led to various ways of discerning and devising patterns in the flux of experience. For our progenitors the invention of writing served this impulse well, providing the first information technology and paving the way for the first "information age," that of literacy. By the time we reach Aristotle and other Greeks of the Hellenistic age (beginning in the fourth century BCE), the idea of demonstrable common sense had emerged from the world of "words and things" to capture the empirical content of natural phenomena. For Aristotle, science would organize and explain data taken in through the senses. Data would be manifest as information through the robust technology of the phonetic alphabet. The alphabet provided the symbols for the framing of nouns and adjectives, the means of creating definitions, which correlated to objects and processes of the world and which enabled the fullest development of conventional language's classifying potential, topics discussed in Chapter 1. These developments in turn gave Aristotle the intellectual tools for forging the concepts, distinctions, and definitions that accounted for it all. His was a method created not to discover new facts or laws about nature but rather to make sense of existing common sense—everything in the world as we already perceive it.[1]

For centuries now, ever since the rise of modern science (and even some time before), it has been intellectually fashionable to pillory Aristotelian science and philosophy, along with that of his scholastic followers. We still laugh at Molière's delightful parody of causal "explanation" tied to definitions: opium causes sleep because of its dormitive virtues or powers ("Opium facit dormire . . . quia est in eo Virtus dormitiva"). Or we snigger at intellectuals debating the vacuous and puffed-up issue of how many angels can dance on the point of a needle. Or we chortle and shake our heads at the attempts to "save the appearances" of the circular movements of planets and stars orbiting the earth in a geocentric cosmology.[2] And, to be sure, there are good reasons why such images have become stock-in-trade burlesques. Yet it behooves us to set aside our ready dismissal of early scientific and philosophical opinions, at least for a while, by acknowledging their accomplishments. For the simple fact stands before us that Aristotelian thinking about human experience and the natural world endured for some two millennia. Aristotle got something right, or at least something useful, and we need to give him his due. Many in the ancient and medieval worlds certainly did.

Notwithstanding their eventual, thirteenth-century exaltation of the Philosopher, Christian thinkers only slowly adopted Aristotelian philosophy after long years of struggle among philosophers and theologians, mostly professors staffing various faculties of arts, theology, and medicine. In retrospection and from a distant point of view, this was not surprising. From the earliest days of their emerging theology, the Christian fathers (patristics) had exhibited a love-hate relation with "Egyptian gold," as pagan philosophy was sometimes called. Some, in the manner of Tertullian, genuinely hated it and would have no truck with the "doctrines of men and daemons," fearing the consequence of a tainted soul. "What has Athens to do with Jerusalem?" he asked famously, summarizing his beliefs that philosophy was the mother of heresy, that philosophers wasted their time with "profitless questions" and words that "spread like a cancer," and that Christian principles descended from the "porch of Solomon," the Gospel, and not the "porch of Zeno," the Stoic.[3] Others, such as Clement of Alexandria (ca. 150–ca. 215), loved philosophy, but at arm's length, believing that reliance on classical elegance and pagan thought would aid in proselytizing Christian verities among the educated. After all, he insisted, "philosophy is knowledge given by God."[4] Still others suffered it as a guilt-racked love—like a moth drawn to the fire but fearing immolation. Clement's successor in Alexandria, the "steely ascetic" Origen (184–254), not only castrated himself to remove the temptation of women but also sold all his non-Christian books to remove the temptation of philosophy. His prodigious memory and love of learning, however, would not let him forget passages learned in his younger years, and he employed a large measure of philosophical thinking in defending his version of Christian orthodoxy.[5]

Amid such fluctuating, love-hate attitudes prevalent during the transition years of the late Roman Empire into the early Middle Ages, a meeting of the minds gradually occurred between Athens and Jerusalem. Christianity was made philosophical; philosophy Christian. For medieval thinkers before the thirteenth century, this mainly meant Saint Augustine of Hippo. The onetime Manichean had confronted varieties of paganism in his own lifetime, rejecting and assimilating ideas as he saw appropriate for a synthesis of philosophy and theology that intellectually embodied his own conversion to Christianity. Moreover, he had also faced down rival movements within his newly adopted religion to become a bulwark of orthodox opinion for subsequent generations. Through Augustine's works a sophisticated body of philosophical doctrine streamed into medieval Christendom, derived somewhat from Stoicism but more notably from "those books of the Platonists," as he recorded in his *Confessions*, books that encouraged the "search for incorporeal truth" and an understanding of God's "invisible things" by means of "those things that are made."[6] Thus, even though the works of Plato went missing during these centuries (excepting the *Timaeus*), Neoplatonist themes upholding an ideal of reflective contemplation animated philosophically minded theologians—Anselm, Thierry of Chartres, and others, as well as, later on, Franciscans such as Saint Bonaventure.[7]

Just as Aristotle had challenged Plato in the sunrise of written philosophy, so too did their noonday heirs replay some of the same struggles in the Middle Ages. A Christianized Neoplatonism had lent itself to doctrines of an immortal soul, of cosmic creation, of a personal god, of divine illumination of the intellect, of "spiritualized" ideas emanating from God (the One, or Unity), and of an ideal reality beyond the sensory world of our immediate habitation, all doctrines that comported comfortably with the truths of divine revelation as recorded in scripture. Aristotelian thought—much of it funneled through the Arabic philosophers Avicenna and Averroes—challenged many of these notions with its own ideas of a naturalistic and sense-bound approach to learning, of an eternal universe (uncreated and unending), of an impersonal God who was little more than a first principle, of a soul perishing with the body at death, and of restrictions on God's creative activity in the world, all of which were seen initially as inimical to Christian beliefs. In fact, these were among the 219 propositions denounced by the Parisian bishop Stephen Tempier with his Condemnation of 1277.

But by then the more egregious of Aristotelian threats to Christian thought had already been tamed. Following his own teacher, Albert the Great (ca. 1200–1280), who had brought Aristotle's works into mainstream scrutiny, Thomas Aquinas accomplished much of the domestication. Some accommodations between science and religion were made in accordance with the truths of "natural knowledge" that reason itself could secure, thereby framing an early, medieval version of what would later be known as natural theology. Natural theology could prove that God exists, for instance, but "the will of God [could] not be investigated by reason"; it could only be "manifested by revelation, on which faith rests." As regards the eternity question, Aquinas argued that reason could neither prove nor disprove it. Therefore, "by faith alone" one could rely unquestioningly on the Church's belief that God created the universe from nothing.[8] In this way, as the various challenges to Christian doctrine were parried or absorbed, the attractions of Aristotle's philosophy became all the more pronounced.

Beyond his broad influence in matters of formal reasoning per se, Aristotle's allure and utility for the schoolmen spread steadily from logic to natural philosophy and metaphysics, to ethics and aesthetics, and to a host of other subjects. Eventually it reached into theology as well. The sacrament of the Lord's Supper, for instance, had ever embraced a mystery, but Aristotle's account of substance and accident tendered a rich, philosophical means of rationalizing the mechanism of the Eucharist with the doctrine of transubstantiation. According to this doctrine, when the priest uttered the words "Hoc est corpus meum" ("This is my body," from which the modern 'hocus pocus'), the inner, essential substance of the bread was mysteriously transformed into the body of Christ while its outer, accidental appearance remained the same. So too with other sacraments and mysteries of the faith, such as the Trinity and the Incarnation.[9] Matters of faith and ultimately not answerable to reason, nonetheless they could be rationally explicated, at least up to a point, as Aquinas's mammoth *Summa Theologica* and numerous other writings amply demonstrated. The newly arrived Aristotelian thinking thus grew to be doubly attractive, both for its approach to reason and natural philosophy and for its aid in elucidating the tenets of faith. Little wonder it became so thoroughly entrenched as the instrument for uniting science and religion among medieval schoolmen.[10]

While virtually all of Aristotle's works generated interest among his scholastic readers, three dimensions of his extensive and organized learning commanded the greatest measure of attention. Two of these will guide our brief introduction to his intellectual corpus and allow us to appreciate its allure for medieval intellectuals. First, based on common sense, conventional language, and logic, there was Aristotle's extensive and methodical scientific depiction of the natural world, which gave rise to the claims of natural theology. Second, supplementing his descriptive science were more general, theoretical reflections regarding explanation and causality, many of which led theologians to reinterpret their own doctrines in a rigorous and exacting fashion. Some of these latter reflections still generate lively, albeit secular debates among philosophers. (A third area of attraction, although lying beyond the scope of our current discussion, lay with Aristotle's ideas of the "good life," ideas that have continued to provoke a great deal of attention since their formulation some 2,400 years ago.) All these themes of Aristotelian thought were subject to extensive scrutiny by the schoolmen of the Middle Ages as they interwove their Christian beliefs into the most systematic science and philosophy available to them in order to create their own idealized magisterium of science and religion.

* * *

In the premodern world, "common knowledge" began with observations that were shared by everyone "in common," as it were. Everyone could see that some bodies were heavy, that they fell down. Solid materials—animal, mineral, or vegetable—all displayed their gravity (*gravitas*). We release an object—say, a rock—from our hand, and it falls to a solid piece of matter below it. Similarly, liquids flow down or downhill because they possess *gravitas*. Such *gravitas* constituted weight, with 'weight' understood generally as an adjectival, qualitative term, as, for modern instance, when we give 'weight' to sentences, opinions, or matters of importance, or when we speak of 'grave' matters, matters not only of the cemetery but ones that are heavy, ponderous, serious, as opposed to light-hearted comments or jokes and the like. Other bodies were light, possessing *levitas*. They tended to rise, rather than fall. Key examples included fire (the flames go upward) and air, which was thought generally to rise, air bubbles rising up through liquids being one illustration. The *levitas* of lighter bodies was likewise understood to be qualitative, as, we might say, with the unbearable *levitas* of being in love.[11]

The visual means of perceiving the qualitative properties of grave and light bodies were complemented by another of the sensations—namely that of touch. The sensation of touching something yielded information about the "haptic" (tangible) qualities of matter that otherwise could not be readily seen. A body might be hot or cold or any variation thereof, for example, or it might be wet or dry, which often we cannot ascertain visually but which we can sense by touching. Brought to us through sensations of sight and touch, the information about matter led Aristotle to infer four "contraries" of hot, cold, wet, and dry as "fundamental" elements behind all of the variegated phenomena surrounding us. Combinations of these elements explained "simple" or "primary bodies": the sensations of hot and dry made up fire, those of hot and wet, air; cold and wet composed water, while the sensations of cold and dry combined to form earth.[12] In turn, associations of simple bodies—those divine brothers and sisters with whom Francis communed daily—underlay all the compound bodies, all the material reality surrounding us.[13]

Bodies moved, of course, and when with everybody else Aristotle observed heavy bodies falling down and light ones rising, he also noted that they did so, naturally, in a straight line, unless "constrained," just as naturally, by another body. Upon release, a rock falls straight down from our hand until it strikes a hillside, whereupon it tumbles and swerves until coming to a place where it can fall no farther, a place of rest. "Natural movement," he concluded, must be vertical and linear, and simple bodies rise or fall according to their nature until achieving their "natural rest."[14] Were movement stilled, simple bodies would settle into their respective resting places, with earth at the center, followed by layers of water, air, and then fire. Compound bodies, too, seek their rest. Comprising nearly all the ordinary matter we perceive about us, they behave according to the preponderance of simple bodies in their composition. By itself air is lighter than water and will rise through water as bubbles; earth is heavier and will sink like . . . well, like a stone. But a pumice stone mixes both air and earth, sinking through the air, rising through the water, until it reaches its natural resting place, floating on water (at least for a while).

All other motion that departs from the vertical or natural Aristotle termed "forced" or "violent." Such motion required an external cause or force. The hand violently throws the rock; the bow violently impels the arrow. Left to follow their own paths without the external force, both "heavy" objects, rock and arrow, will fall directly down to their natural resting places. As projectiles, however, the path to their resting place is exceedingly indirect. Explaining the violent motion of projectiles would give Aristotle and his successors headaches, as they backed and filled this account with all sorts of ad hoc reasons to explain why the rock or the arrow would not immediately fall straight down after leaving hand or bow.[15] Here we only need emphasize again the qualitative perceptions about moving, natural phenomena. A body projected or falling through a medium, such as air or water, will pass through that medium according to how much resistance the medium provides. But as with 'weight,' 'resistance' did not mean a quantitative measure. Rather it meant something like "more body," as when we speak of a wine—say, a zinfandel or port—as having more body than another—say, a chardonnay. Now, informed by our modern history of quantitative analysis and a much different version of common sense, our own understanding of physical phenomena balks at such qualitative perceptions. But for centuries they lay, nonetheless, at the heart of traditional commonsense descriptions.[16]

Sensations, then, provided the instruments for acquiring information about the natural world about us. Consonant with them were the ideas that all sensations were in motion and accordingly conveyed phenomena in motion. Haptic qualities "differentiated" bodies, made them distinguishable from one another, because they themselves were in motion. Hot "moved" or tended toward cold and vice versa; similarly with wet and dry. Each of these categories provided a limit or endpoint of a certain kind of change, and all the other physical contraries (such as "heavy-light, hard-soft, viscous-brittle, rough-smooth, coarse-fine") were derived from them.[17] Otherwise stated, sensations provided the idiom of relations and motion, relating the sensor to the moving thing sensed. As a parlor trick, close your eyes and hold your thumb and forefinger together, barely touching. Hold them still, with no movement. After a short time it becomes virtually impossible to detect that the two digits are actually in contact with one another. Now slightly rotate your finger around your thumb; the sensation returns. Lovers holding hands will often caress with subtle motions; the motion enlivens the sensation, literally animating it and making it possible. The constant motion of sensations and phenomena led poets—Geoffrey Chaucer and John Milton, for example—and academic scholars alike to speak of the four contraries as possessing "sympathetic and antipathetic" properties, another indication of medieval animism.[18]

From commonsense observations about matter and motion, a host of immediate questions arose for Aristotle. If natural motion is vertical, what prevents the light matter of fire and air from flying off into the heavens? Conversely, what keeps the matter of the heavens from collapsing into the earth, and, closely related, what does "down" mean when heavy bodies fall to the earth? Answers to these and related questions led the Philosopher directly to one of his most enduring, bedrock conclusions in science: the separation of celestial and terrestrial physics. Here too, observations helped guide him toward this critical claim. The moon orbited the earth, as did the sun and other planets, as did the stars—all phenomena observed. Moreover, the motion of the heavenly bodies could be seen as regular, repetitive, and circular. Of course, Aristotle recognized that the sun's rising and setting depended on perspective, and that it and other sidereal movements might be explained as resulting from the earth's diurnal rotation on its own axis, perhaps accompanied by the earth's annual orbiting of the sun. This had been suggested by his contemporary Heraclides Ponticus (ca. 390–ca. 310 BCE), following in the footsteps of the Pythagoreans.[19] But the theory of a heliocentric cosmos raised too many questions that flew in the face of common sense, especially as regards the physics of bodies and motion directly surrounding us, leading Aristotle to opt for a geocentric theory.

"Down," then, Aristotle argued, meant toward the center of the earth and, by extension, the center of the universe. "Up" referred to vertical, linear movement that continued until it reached the orbit of the moon.[20] Material bodies composed of earth, air, fire, and water displayed their natural and linear motion, as well as violent motion when forced, only within the sublunary realm, below the orbit of the moon. From the moon outward, away from the center of the universe, a different type of "simple motion" prevailed, as did a different type of matter. Recognizing that "our senses enable us to perceive very few of the attributes of the heavenly bodies," Aristotle reasoned his way to this conclusion.[21] Finite, linear motion in the sublunary sphere existed as a function of the "contraries" "up" and "down," he explained, but those contraries could not extend, limitless, into the heavens. Were rectilinear motion to continue "up" without an end point, it would be infinite, and thus not part of a contrary. Moreover, in such a case moving bodies would keep traveling along a rectilinear path without end, and there would be effectively no coherence in the universe. Therefore, there had to be a stopping point, and because it occasionally eclipsed the sun and other planets, the moon showed itself clearly as the heavenly body orbiting closest to the earth. It followed as well that celestial motion had to be other than rectilinear and contrary; it had to be circular, "simple motion . . . about the center." Continuous and finite, the circular motion of the heavens not only was the sole logical option to rectilinear motion but also was corroborated by "inherited records" of observations that charted the unchanging nature of the "whole scheme" of the heavens and their "proper parts."[22]

With perfectly circular motion came, perforce, a corresponding sort of matter, one incorruptible, not subject to the changes and vicissitudes of the motion we experience in our contrary-laden, sublunary sphere. Because earth, air, fire, and water could not extend to the heavens, Aristotle named celestial matter a "fifth" type of matter or "essence" (the origin of our 'quintessence'). This simple, elemental matter was ethereal, made up of ether, and completely unlike any of the bodies surrounding us. Such extreme differences notwithstanding (and undetectable through sensations), he still thought of it generally as a type of matter, albeit one "pure and divine."[23] With this conception Aristotle labored to rework in physics the mathematical theories of his contemporaries Eudoxus of Cnidus (ca. 408–ca. 355 BCE) and Callippus of Cyzius (ca. 370–ca. 300 BCE). Following Plato's lead, both had sought a geometrical account of the complex movements of celestial bodies. These included the regular, diurnal motions of all heavenly bodies from east to west and the longer, irregular orbits of the sun, moon, and planets through the ecliptic, west to east, against the backdrop of the zodiac's fixed stars (more on this later). Eudoxus and Callippus had shown how apparent celestial irregularities could be produced by simple, circular movements of a planet within several concentric spheres, each rotating on its own axis. They thus provided an early instance of the later-termed practice of "saving the phenomena."[24] Wanting to explain the appearances rather than merely save them, Aristotle converted these geometrical systems into a physical, celestial space that was filled with concentric, ethereal spheres (sometimes called "crystalline"), invisible and transparent. This matter occupied the entire superlunary universe, which accordingly had no vacuum, and extended to the outer sphere of the fixed stars, the outer limit of a finite universe. Centuries later, in his widely used textbook *Treatise on the Sphere,* John of Sacrobosco (ca. 1195–ca. 1256) provided a popular, two-dimensional depiction of the standard, Aristotelian cosmology, which was frequently reproduced, as in the later and enhanced rendition by Peter Apian. (See Figure 2.1.) Armillary spheres, whose early invention was attributed to the Greek mathematician Eratosthenes (ca. 276–ca. 194 BCE), made three-dimensional versions available to Europeans of the premodern era.

As Aristotle formulated his view of the heavens, accounting for the origin and transmission of motion between the spheres became a paramount issue. For logical, physical, and metaphysical reasons, which we can only list, he concluded that there must be an "unmoved mover" at the source of it all. Logically, to avoid falling into an infinite causal regress whereby a cause caused a cause, caused a cause . . . , all the way to unintelligibility, Aristotle held that there must be an uncaused cause, which was one and same as the unmoved mover, the origin of motion and change. Physically, motion in both the celestial and terrestrial regions had to have begun somewhere. That place he identified with the "empyrean," the blank realm beyond the fixed stars. The motion of the entire cosmos, originating with the "prime" or unmoved mover in the empyrean, then generated the circular movements of both the fixed stars and the planets. From there it entered the sublunary region of terrestrial change and animated all the natural and forced movements we see about us.[25] In his *Metaphysics* Aristotle considered all "the heavens and the world of nature," all "being," as "of necessity" eternally depending on the "mover which moves without being moved." This was "God . . . a living being, eternal, most good . . . [to whom] life and duration continuous and eternal belong."[26]

It comes as no surprise that as many medieval scholars read these conclusions in the Philosopher's writings, they found their own beliefs reinforced and enriched by his elaborate description of the universe. The comprehensive order it revealed gave substance to a natural theology, the doctrine that

FIGURE 2.1. The Aristotelian cosmos, from Peter Apian's *Cosmographia* (1524).

God's universal presence was displayed in the things and events of the natural world, which our senses and the "natural light of the intellect" allowed us to understand.[27] This same premise underlay the natural theology later promulgated in the seventeenth and eighteenth centuries. But the difference in content between the two versions was stark. For medieval thinkers, natural theology grew out of a thoroughly verbalized nature, so to speak, captured by the words, definitions, categories, logic, and common sense of Aristotle's comprehensive philosophy. For natural philosophers in the early years of the modern era, nature was already methodically becoming mathematized, a profound development epitomized by Newton's laws of motion and modern, "classical" physics, as laid out in his famous *Principia*. It was the difference between an account of nature based on the information technology of literacy and one established with the foundation of modern numeracy.

* * *

Heavy and light matter, natural and violent motion, celestial and terrestrial spheres—descriptions tendered the first step toward science, but to gain understanding of these and other natural phenomena, one needed, in Saint Thomas's words, to proceed from the senses to the intellect, to devise explanations of the phenomena described. For the reflective Aristotle, the connection between sensation and motion (or, more generically, "change") revealed an underlying reality: "Nature is a principle of motion and change," he declared. In short, common sense reveals a world in which everything we perceive changes constantly. To explain natural processes, therefore, we must determine the "nature of motion," after which "our task will be to attack in the same way the terms which come next in order."[28] Thus did Aristotle begin spinning out of sensations a spidery web of definitions and distinctions, a web of nouns and adjectives spun with the technology of the alphabet to ensnare the flux of experience. From the standpoint of literary style, Aristotle's prose is, to put it mildly, crummy, like "eating dried hay," wrote poet Thomas Gray (1716–1771). This stems in large part from the versions we have of his writings, which are either his or his students' notes, not polished sentences ready for publication, as were the works of Plato.[29] Even surmounting his parser's nightmare of clunky prose, however, as well as facing the normal problems associated with translating, reading Aristotle can be a daunting challenge, especially for the philosophical novice. In its breadth, stunning and compendious, his mammoth work comprised an entire grid of interlaced concepts thrust between the mind and the moving world of nature.[30]

One way for us to picture how he approached this far-reaching enterprise is to imagine a menu-driven decision or logic tree, such as utilized in many contemporary computer programs. The tree usually begins with a question, often the most general pertaining to a topic, and then offers alternative responses, frequently "yes" or "no" in computerland. Based on the alternative one selects, another set of options arises, answers are chosen, and the process continues, narrowing the topic until a decision has been reached. For us such trees are arbitrary and depend on the purposes for which they have been designed—medical diagnosis or ordering take-out foods, for example. But for Aristotle such a tree, when formulated with the proper questions, accounted for the phenomena of the natural world and explained them scientifically.

Based on Aristotle's approach, scholars during the High and Late Middle Ages developed an entire "questions literature," intellectual procedures bearing some resemblance to our logic trees, but more closely related to the practices of "oral disputation" prominently found earlier in university education. One began with a question ("Let us inquire whether . . ." or, more simply, "Whether . . ."), then followed it with "principal arguments" (possible solutions supporting either the affirmative or negative positions), proceeded next to the author's own opinion, and, finally, concluded with his counterarguments against the opposing views. Many permutations can be found on this pattern, but the overall scheme has come down to us as the medieval "scholastic method."[31] For present purposes we need not venture far into this method or the details of Aristotle's system; following him but a short distance with some scholastic questions and answers of our own allows us to appreciate his approach to scientific knowledge, to see the resultant and coherent worldview, and to engage the classifying temper in operation.

Aristotle initiated his "attack" on the "nature of motion" in the broadest imaginable terms. He first asked whether there could be "such [a] thing as motion over and above the things." His answer: no. Motion could not be conceived without some thing actually moving. But because "things" fell under the "categories of being," and because the term 'being' per se perched atop the abstract tree, designating voluminously "everything that is," the next question followed—namely, how can one describe in general any change involving any kind of being? Aristotle responded: change must be somehow tied to "coming-to-be" and "passing-away [out of being]."[32] But this formulation prompted other, more troubling questions, for on the surface of it how could change (dare we say it?) . . . unchangingly be? Parmenides of Elea (fl. ca. 500 BCE) had laid down the law. Whatever exists cannot come into being from nothing, and nothing cannot be: "It is impossible to think or to say that not-being is."[33] This meant for Parmenides that real change was but an illusion. On one plane Aristotle could not refute the logic, for changing into and out of nothing (like the emperor's new clothes) made no sense. But on another plane, finding an equivocation in Parmenides's use of 'is,' he could and did divide the question, separating the substantive and logical meanings of the term. From which Aristotle concluded that, correctly understood, we can say being exists in many manifestations or guises ('form' is the technical term), and 'change' or 'motion,' then, simply refers to being's assuming a different form. All being may well endure, but it does so in different ways, and in natural philosophy one tries to explain the different processes whereby it gets from one form to another.[34]

In this account, the critical word becomes (again, how else to say it?) 'become'—that is, literally "coming-to-be." The next limb down in Aristotle's logical tree, then, supported the key divisions of "becoming." These he found in the categories of "potentiality" and "fulfillment," what something might turn into and what it already is.[35] In characterizing his concepts as they pertained to natural philosophy, Aristotle often illustrated them with examples drawn from the arts or crafts, analogies made on the premise that human and natural processes shared essential similarities. Once demonstrating the common components, he then frequently noted differences between the two. Thus "bronze is potentially a statue"; it is the "starting point" of change. The artist then shapes and molds it into a figure, whereupon it can be seen in fulfillment, not "as bronze" per se but as a statue.[36] With this example, change comes about externally by the artist acting on the bronze. In turning to things "constituted by nature," Aristotle noted, each thing "has within itself a principle of change and rest," and change here is understood as occurring internally.[37] He illustrated the point with examples of organic processes: the acorn becomes an oak tree; "man is born from man."[38] Rounding out this theory required one final notion. Some sort of substratum had to endure throughout change as any subject altered its form from potentiality to fulfillment. The bronze remained bronze all the way to the statue; the organic matter of the acorn all the way to the oak. If not, the absence of a substratum would lead to a violation of Parmenides's dictum about being in general, with something either passing into or coming out of nothing.

Heady stuff all this, highly abstract, revealing how quickly Aristotle soared from common sense, through the strata of reflection, into rarified concepts and distinctions. The schoolmen loved it. Centuries later, however, Francis Bacon (1561–1626), among others, would chide them for flying "at once from the sense and particulars up to the most general propositions."[39] But at this juncture, having accounted for change in its most abstract, philosophical sense, Aristotle dropped down to another, slightly lower branch in his own logic tree, easing up a bit in order to explain how causes functioned in the actual processes of nature. He thereby returned more closely to the world of direct, sensory information, where one could simply take "for granted" that things are in motion, as is "made plain by induction."[40] To know a thing on this branch of our understanding, he insisted, means to "have grasped the 'why' of it," to have explained the information our commonsense perceptions have provided.

Somewhat cumbersome to our ears, Aristotle's expression "the 'why' of" has traditionally been translated as the "cause" of something. In Chapter 1 we followed his account of definitions, which explained the "understanding" through his well-known portrayal of the four causes: material, efficient, formal, and final (all of which allowed one "to grasp . . . [a thing's] primary cause," Aristotle had added).[41] The traditional translation misleads us, however, both for being anachronistic in projecting a modern notion of 'cause' into Aristotle's thought and, more significantly for our narrative, for obscuring the conceptual affinity of the Philosopher's scientific causality and divine creation for the schoolmen of the Middle Ages. In the scholastic mind, such as that of Aquinas, religious thought was not merely grafted onto Aristotelian science; rather, it grew out of that science intrinsically, for the language was the same. Explaining "the 'why' of" things, whether divine or natural, lay at the heart of inquiry and understanding.

To realize this, let us set aside for the moment our contemporary sense of cause, which comes to us from the practices of scientists and from the philosophical pen of David Hume (1711–1776). In a nub, we moderns infer causality; we do not observe it. We see two separate events, A and B, in conjunction with one another on repeated occasions. Eventually experience leads us to expect that in the presence of A, B will follow, at which point our feeling or belief ("natural instinct," in Hume's terms) establishes the causal connection between A and B, a causal connection that at best can be highly probable but not demonstrably certain, as with a geometrical proof.[42] In contrast, Aristotle considered causes as observed in phenomena, not inferred. We observe the sculptor molding his bronze, the seed growing into the plant, the hand throwing the rock. Substances, not events, have causes, but substances move and change continually, and we observe those changes directly—all the flux of experience. Explaining "the 'why' of" phenomena depends on the perspectives or "tropes" (*tropoi,* also translated as "fashions") through which we observe these changing objects. Aristotle's four causes, therefore, represent four different fashions or vantage points from which we can and do understand "the 'why' of" something.[43]

When unpacked from its original Greek, "the 'why' of" can be seen to perform a "curious double duty," so notes Aristotle scholar Jonathan Lear. It reveals an interrogative, as well as an indicative function, suggesting both question and answer. We probe into the things of the world, and in its answers the world displays to us its ultimate and objective intelligibility, which we can grasp (an early version of the so-called weak "anthropic" principle). Aristotle assumed that the world indeed is objectively intelligible and that "the 'why' of" something, through its various fashions, allows us to penetrate to the world's most basic reality.

Chief among the fashions are "form" and "matter," the formal and material causes. The many logical and metaphysical nuances of these terms are topics for another occasion, but for natural philosophy, "form" designates the basic organization or pattern assumed by matter as it changes, matter's potentiality becoming fulfillment. Aristotle called "form" "the archetype" or "the definition [*logos*] of the essence," by which he meant a thing's inner and dynamic principle of change. We see the bronze assume the form of a statue, the acorn the form of an oak tree, the arrow the formal arc of its trajectory. Here and in other cases, form displays an object's intelligibility, the natural (or humanly created) pattern our equally natural intellect can grasp. When we have our definitions right, nature's intelligible patterns correspond with them; then we know something, both the natural process and the definition. Notice, "form" does not simply designate a realized state of completion wherein motion or change has ceased. Rather, it indicates also a striving state (a conceptual pilgrim's progress for the schoolmen) whereby we recognize and understand ongoing processes, such as the regular, functioning activities of organic matter—to wit, the grinding action of a gizzard.

For its role in natural, as well as the corresponding cognitive, processes, "matter" designates the underlying substance, "out of which a thing comes to be and which persists." Again with our examples, the processes remain the same; only the vantage point differs. Matter becomes form when the bronze takes its shape as a statue, or when the seed becomes a sprout and grows into a tree before our eyes, or when in local, violent motion the arrow travels its trajectory to its natural place of rest. Matter endures throughout these processes; the bronze, organic matter of the seed, and wood and feathers of the arrow all remain the same stuff while undergoing their various changes. By itself matter is not intelligible and only becomes so as it acquires form. Nonetheless, matter remains a "cause" or part of "the 'why' of" because without it form would have nothing to shape or to . . . form.[44] Without organic matter and material grit, a gizzard could not grind anything.

Aristotle invoked similar lines of reasoning to explain the efficient and final fashions of "the 'why' of." Usually translated as the "efficient" cause, "the primary source of change" was his way of describing what we often nowadays term "agency." But to capture the full process of "becoming," Aristotle paired "agency" with "patiency," the latter indicating the recipient of the action initiated and propelled by the source of change, much like the distinction between "active" and "passive." Once more, the explanation turns on the perspective one assumes in examining a process. In the pedagogical craft, he illustrated, a teacher is the agent (efficient cause) of the student's learning, the student the "patient" (recipient) of the teacher's instruction. But so too does the student's own learning provide the agency of the teacher's patient pedagogy, for the student "causes" the teacher to fulfill (or actualize) his potentiality as a teacher. Both describe the same process, but from different angles. And though it initially sounds odd to say a teacher's instruction must occur in the activity of the student, where else could it happen? Without the student, the teacher would be talking to a vacant room, or gesticulating in an empty stoa, or perhaps seeking to impart knowledge to a duck. (Indeed, this is the insight embodied in the old adage that "teachers never teach; only students learn.") Still, "teaching is not the same thing as learning, or agency as patiency, . . . though they belong to the same subject, the motion." As in the crafts, so too in natural phenomena, agency and patiency must be understood as complementary vantage points for examining "the 'why' of" a process or, "more scientifically, the fulfillment of what can act and what can be acted on." The internal, organic development of the seed, plus the nutrients of the soil, the water, the sun—all are part of the efficient cause of a plant's growth to maturation. The nutrients provide the agents of the seed's growth, while the seed serves as the agent of the nutrients' fulfillment as nutrients.[45] The gizzard is the efficient agent ("primary source") of digestion.

As Aristotle discussed the "agency" and "patiency" of efficient causality, he was led inexorably into the realm of purposeful activity, the fourth fashion of "the 'why' of," reaffirming both the interconnectedness of the modes of explanation and their perspectival nature. For things may also be seen from the viewpoint of the "end [*telos*] or that for the sake of which a thing is done." Now uniformly discredited among scientists, the teleological mode of explanation did not represent a separate type of causality for Aristotle but merely a different way of referring to nature. It is "by nature and for an end" that the swallow makes its nest, the spider its web; plants grow leaves for the sake of their fruit.[46] The purpose of the nest is to incubate the swallow's eggs and to protect the young into maturity. Likewise with the spider's web: it traps food for the sake of the spider's survival. Purpose here requires that the end somehow govern the process resulting in the end's own fulfillment. Again the comparison between human and natural motion becomes critical. Humans act "for the sake of" all the time; so does nature.[47] This did not mean a specific agent behind every natural process or end, ideas we tend to lump together, and certainly not a universal agent. No mind of a divine craftsman could be discerned in nature; Aristotle's unmoved mover was not the Christian's God. Nonetheless, nature is not mindless. And as part of nature's intelligibility, the teleological perspective provides another way of seeing that natural forms typically develop from their potentiality to their actuality and in so doing exhibit purposefulness in the world.[48] An organ in most birds, gizzards (material cause) provide the source of digestion (efficient cause) through their regular grinding patterns (formal cause), but they could also be said to exist for the sake of digesting food (final cause).

It's hard for us not to see Aristotle struggling here to capture and fix the ephemeral processes of nature in the definitions of conventional language. Tied to modern, scientific experiment, our notion of efficient causality has become so firmly entrenched that his focus on the linguistic pairings (for example, agency and patiency, potentiality and fulfillment) appears excessively abstract and tortuous.[49] But scholastics had little difficulty in understanding him because of the creationist underpinnings in his conception of causality. To them it made perfect sense, for nearly everyone was a creationist in the Middle Ages.[50] Not only did creationism comport with biblical accounts, wherein God created the universe and all in it, but the very processes of nature were understood as analogous to the same creative activity that Aristotle explained with his examples from the arts and crafts. As mentioned, Aristotle did not believe that any conscious design expressing divine purpose lay behind nature. Yet his insistence that the world is intelligible, that processes everywhere are directed toward realizing forms, that purpose and agency pervade these processes in human activities and nature alike all pointed to an ongoing creative activity in the universe, which for schoolmen meant the presence of a divine Creator throughout his creation. In effect, scholastics simply elicited another layer of perspective to account more fully for "the 'why' of" phenomena, the perspective of God *sub specie aeternitatis.*[51] In this picture nature was not to be understood as a projection of the mindful or purposeful activity of humans. Rather, the reverse held, for Aristotle and scholastics alike: "Action for an end [*telos*] is present in things which come to be and are by nature."[52] The purposeful agency of humans—as, say, in crafting a statue or teaching a student or doing a good deed—merely reflects the objective, purposeful patterns that structure the natural world. The "fashions" of explanation mirrored the true fashioning of things themselves.

We cannot overstate that all four causes reflected different vantage points from which to grasp the same process. In technical language, Aristotle's view of causality has often been described as "transitive" causality, a concept that desiccates its fulsomeness and richness for medieval thinkers. The skeletal idea here (remembering Parmenides) is that there must be some substratum of being that exists in both the cause and effect. "Every effect in some degree represents its cause," Aquinas wrote, with "such a representation . . . called a *trace.*" Sometimes termed "essential likeness," the substratum travels or transits, as it were, from cause to effect, while it undergoes changes in form.[53] This notion breaks down from our modern point of view, but the picture is quite coherent in keeping with the ideas that causation is observed and that "the 'why' of" can be explained from different perspectives. As the substratum of all existence, Being, or God, which Aquinas identified as Pure Form (Actuality), could be seen in all of the processes of nature, as matter acquired form, exhibited agency, sought ends. The divine trace was everywhere.[54]

Aristotle's thought, then, found a welcoming, receptive audience in Aquinas and other schoolmen, especially with the notion that God's creative power permeated all of nature. In Aquinas's expression, "We designate by the name of *creation* . . . the emanation of all being from the universal cause, which is God."[55] Moreover, God was not merely the starting point, the "prime mover" of natural philosophy pushing motion ahead; he stood also "for the sake of which," the end point drawing all motion to him. Again, Aristotle had shown the way, intellectually speaking, and the signposts were written in "the 'why' of." It had seemed initially that Aristotle faced a problem. Explaining "the 'why' of" required a material fashion or cause, among others, for matter was required to move matter. But if the prime mover residing in the empyrean were nonmaterial, then, Aristotle wondered, how could he have set in motion the material, ethereal heavens and the sublunary nature about us? As he elaborated his description of God, Aristotle answered his own query with what can only be deemed as poetry . . . in his own fashion:

> The first mover . . . of necessity exists; . . . it is good, and in this sense a first principle. . . . On such a principle . . . depend the heavens and the world of nature. And its life [the first mover's] is such as the best which we enjoy, and enjoy for but a short time. For it is ever in this state . . . , since its actuality is also pleasure. . . . And thought in itself deals with that which is best in itself. . . . And thought thinks itself because it shares the nature of the object of thought. . . . If, then, God is always in that good state in which we sometimes are, this compels our wonder. . . . And God *is* in a better state. And life belongs to God; for the actuality of thought is life, and God is that actuality; and God's essential actuality is life most good and eternal.[56]

Poetry surely resides in the ear of the listener, but regardless, this was an earful, well beyond the soul of wit. In its distilled version, the processes of nature demonstrate their proper ends of goodness and pleasure, and so likewise does human cognition in discovering them. The proper *telos* of the natural world is reflected in the *telos* of our own thought, its wonder and curiosity and pleasure and goodness, as we strive for better understanding of nature and of ourselves. There could be but one word for such a great *telos,* such a great allure: love. Aristotle left no doubt. God, the prime mover, he concluded, "produces motion by being loved, and it moves the other moving things."[57]

How must the Angelic Doctor's eyes have brightened and brimmed upon first seeing this passage. Here was the perfect non-Christian, later to be stuck forever in Dante's Limbo, depicting God as love, a theme animating the heart and soul of Christian belief. Nor was Aquinas to stand alone reveling in the realization that love makes the world go round. Poets, playwrights, songwriters—indeed, any one of us who has fallen in love and walked on air—has since shared in the sentiment. One thinks of Dante's closing lines of the *Commedia* ("l'amor che move il sole e l'altre stelle"), or of the anonymous French song ("L'amour, l'amour fait tourner le monde"), or of Gilbert and Sullivan ("It's love that makes the world go round").[58] But scholastic theologians inferred more, much more, for the notion carried the full, authoritative weight of pronouncement from the world's greatest natural philosopher. Theology and philosophy and poetry and science merged into one, as Aquinas devoted an entire question in his *Summa Theologica* to the topic of God's love.[59] The most profound biblical expression of this love lay in the Beatitudes, those blessings Jesus conveyed to followers and the curious alike in his Sermon on the Mount: "Blessed are the poor . . ."

We have now come full circle to Francis.

* * *

In his papal encyclical of 1998 ("Fides et Ratio"), John Paul II reaffirmed the central medieval vision of *scientia* and *religio*, the "profound and indissoluble unity between the knowledge of reason and the knowledge of faith." Citing Aquinas, who "had the great merit of giving pride of place to the harmony which exists between faith and reason," the pope explained how theology and philosophy could bolster and reinforce one another when properly understood. Theology, the "noblest part and the true summit of philosophical discourse," should explicate as much as humanly possible the "mystery of the Incarnate Word," an "understanding of Revelation and the content of faith," tasks it cannot accomplish without the "supernatural assistance of grace." For its part, philosophy, which "depends upon sense perception and experience," had especially the challenge of exploring the "use of language to speak about God," thereby continuing as theology's handmaiden. Much like Huston Smith, the pope also took to task "scientism" and a "scientistic outlook" for its narrow "impoverishment of human thought," although he did acknowledge and praise the accomplishments and activities of scientists as long as their rationality remained directed toward the "contemplation of truth" and the "search for the ultimate goal and meaning of life." With these latter comments and throughout the encyclical, John Paul II reiterated in a modern context the medieval hierarchy of learning: from the pinnacle of theology to its handmaiden of philosophy to the subordinate natural philosophy, what we now term science. Both faith and reason, he noted, fell within the Church's "magisterium," and historically it had been necessary for the Church to intervene in philosophical matters from time to time in order to "reiterate the value of the Angelic Doctor's insights."[60]

Up to now we have explored how the tools of reasoning underlying this vision were forged from the raw materials of letters and words, from conventional language and the creation of writing. Sensations carried the givens to the mind, which then went to work correlating data symbols with sounds and fixing the letters into words. Abstracting words from sounds by means of letters, the mind continued refashioning its own products from a critical distance, defining words by their semantic referents and arranging them according to the syntactical rules of grammar. The latter themselves were the products of linguistic data processing and management. Thus was speech converted into the first information technology, whose acme was the literacy captured and conveyed by the phonetic alphabet. The alphabet's simplified and systematic use of letter-symbols elicited the classifying potential of speech and gave expression to the human urge for organizing phenomena. Socrates talked; Plato wrote; Aristotle structured writing into systematic science or knowledge of the external world, using common sense, definitions, taxonomical arrangements, logical and syntactical rules. His work still stands as the philosophical culmination of centuries of developments in writing, a spectacular accomplishment in its own right. Scholastic intellectuals in the Middle Ages took these powerful instruments of literacy and created their own magisterium of science and religion. John Paul II was right in this respect: the pinnacle of these achievements was the "enduring originality of the thought of Saint Thomas Aquinas." More than any single individual he bore witness to the summons of Proverbs 4:5: "Acquire wisdom, acquire understanding." As much as humanly possible (perhaps with divine aid?), *scientia* and *religio* were united in a single *logos*, a single Word.

Laudable though it may be for many, the pontiff's recitation of our intellectual history in the service of reaffirming the concord of reason and faith overlooked a matter of profound significance, one also born of extensive labors and serious reflection over the centuries. Earlier we saw (in a note) the contrasting philosophical perspectives of Hume and Kant regarding causality. Hume based it in the "natural instinct" we display in inferring causality from a constant conjunction of events; Kant saw it as an a priori category the mind brings to the investigation of phenomena. Both perspectives shared a common assumption—namely, that neither of their philosophical accounts affected in the slightest how scientists went about their business. The point may be made even more starkly. In the modern world, philosophy and science are both distinct from and largely irrelevant to one another. Science does not treat itself on the whole as a science, although individual scientists may don their philosophical caps and do so. Similarly, philosophies are "invariant to all possible scientific theories," although some philosophers link their remarks on epistemology to key methods and practices of the sciences. In short, summarizes philosopher Arthur Danto, "the world as we live it and know it is consistent with all possible philosophies of knowledge."[61] And as with philosophy, so too with religion; both have been uncoupled from science. This is a separate universe from the intellectual world of Aristotle and the scholastics, where religion and philosophy, including natural philosophy, drew on one and the same vehicle of understanding, employed the same linguistic tools in reasoning about matters.

As we observed at the outset of this extended essay, the medieval harmony has been abandoned, replaced by a rift between science and religion. Of course, we still use conventional language for all the conveyances of life, everything from our most profound words of wisdom, creativity, and advice to the silliest of jokes. We also use words to describe scientific practices, procedures, and results and to integrate science's "external" dimensions into the rest of our lives, often through various and convenient classifying systems. Scientific reasoning, however, takes place elsewhere, in a symbolized universe apart from the conventions of written and spoken language, as Galileo incisively grasped centuries ago: without the "language of mathematics . . . one wanders about in a dark labyrinth."[62] We must now turn to the second phase of our story, the new and slowly emerging alternative information technology developed during the Renaissance. This technology, at whose foundation lodged a new relational numeracy, would provide the underpinnings of a new relation-mathematics, and with it a completely novel form of analysis, which we are here identifying as that of reverse engineering. This new information technology would give rise to the analytical potentiality of mathematics, just as alphabetic literacy had made possible the extensive classifying of speech. As thing-mathematics gave way to relation-mathematics, a new analytical temper was born, along with its twin: the first rumblings of a tectonic shift separating science and religion.

PART II

From the "Imagination Mathematical" to the Threshold of Analysis

Teeming Things and Empty Relations

Numbering is a more cunning thing than I took it to be.

—Robert Recorde, *The Declaration of the Profit of Arithmeticke* (1540)

LONG BEFORE SCHOLASTIC philosophers and theologians had completed their "Church intellectual," their magisterium of science and religion, tiny cracks in its foundations had already begun to make their appearance. These went largely unnoticed at the time, and they were introduced from sources that had nothing to do with challenging the presumptive harmony in the imagery and intellectual architecture of the era, nothing to do directly with science or religion: a business contract, a polyphonic motet, a perspective drawing, a clock. Collectively these small innovations betokened a much larger shift from the information technology of literacy to that of a new, relational numeracy, from a universe of teeming things to one of empty relations. No one explicitly set out to invent a new information technology. But urged by the guile of history, at least those portions of it we can recognize through hindsight, small novelties and happenings became fraught with irony: general, unintended consequences issuing from parochial concerns.

Given the central role that mathematics eventually assumed in the surfacing of the analytical temper, modern science, and the Great Rift, it would seem reasonable to look to the "science of patterns" (as mathematics is now often labeled) for evidence of the small cracks that would widen over time. Reasonable, perhaps, but not terribly fruitful. The internal developments in mathematical reasoning per se yield precious little in this regard, for few medieval or Renaissance intellectuals actually did mathematics.[1] In part, there wasn't much to work with. Most ancient texts on mathematics either had vanished or lay hidden and moldering in a yet-to-be unearthed cache of manuscripts. This was certainly true in western Europe and likewise in

Constantinople, which housed hundreds of Greek manuscripts but did little with them. It was not true in the Arabic world from the ninth through at least the thirteenth centuries. There, Islamic scholars not only translated and preserved works of Hellenistic science but at the time used them as foundations for producing arguably the greatest mathematics and science of the age.[2] As earlier noted, some of these materials began filtering into the Latin West only during the twelfth- and thirteenth-century wave of translations, highlighted by the works of Aristotle but including some among the mathematical ancients (Euclid, Ptolemy). Writings of major Greek mathematicians (Archimedes, Apollonius, Diophantus, Pappus, among others) followed much later, many appearing only in the Renaissance.

Ancient texts were lost, yes, but disappearing along with them was the critical idea, never very widespread in any event, that one should actually do mathematics and use it.[3] In "ancient civilizations," summarizes the doyen of early mathematical systems and reasoning, Otto Neugebauer, "mathematics and astronomy had practically no effect on the realities of life." His words are equally apposite for medieval and Renaissance Europe, with allowances made for a few later innovations such as calendars and clocks.[4] Today, even if we know little or nothing of mathematics ourselves, we are fully aware that it drives the science, technology, and industry of the world around us. "Do the math," "crunch the numbers," "figure the odds," and a host of similar clichés punctuate our popular culture. From social security cards at birth to the certificates issued at our death, numbers traverse our institutional lives as well; for some they define us. We take numeracy for granted, with all its sophisticated techniques of information management and highly advanced modes of mathematical analysis. Such a perception was lacking in the premodern era—nay, unimaginable.

Before the advent of modern science, to put it bluntly, mathematics was not important in any practical sense of the term. Or, to phrase the point a bit more delicately, it was framed in cultures whose practicality and expectations regarding mathematics differed markedly from our own. Aristotle was simply reiterating a prominent assumption of the ancient world with his observation that the "mathematical arts" were a leisurely, and hence aristocratic, form of "recreation" that did "not aim at utility."[5] Even the problems subsequently posed and solved by Archimedes (ca. 287–ca. 212 BCE), the greatest of Greek mathematicians and by consensus among the greatest ever, were done largely in the service of an "intellectual game," much as the Olympics were part of a Hellenic, agonistic culture of competitive play.[6] The Greeks loved to argue, and within their "polemical culture," making compelling arguments in the form of mathematical proofs stood out, like oratory, as an "activity of great prestige," albeit with a much tinier and specialized audience. Virtually none of these proofs had any direct economic or technological application, our version of practicality. Even Greek mechanics, with but few exceptions (such as war machines), was predominantly a "contemplative mechanics," a form of cultural entertainment, not utility.[7] Thus, although unveiling a novel, exemplary, and formidable model of abstract reasoning and deductive proofs, Greek mathematics, and especially its shining jewel, theoretical geometry, kept its distance from the world of practice, including real-world, "hands-on" geometry.[8]

This separation persisted even as mathematics traveled a different cultural odyssey throughout Christendom until the advent of modernity. As in antiquity, basic numeracy itself scarcely existed, and only a handful of individuals knew anything of mathematics at all, scattered scholars with idiosyncratic interests. Most of those interests allied with the tradition of Neoplatonism, in which mathematics was often esteemed for its religious and spiritual values, and for expressing many divine harmonies: of the heavens, of music, of personality, of life. Not infrequently it was thought to border on astrology or superstition or even heresy. According to Hugh of Saint Victor (ca. 1096–1141), for instance, *mathesis* (mathematics) was at best an "instructional science," but only when "the 't' is pronounced with the 'h.'" Otherwise, "when the 't' is pronounced without the 'h,'" *mathesis* meant "vanity" and referred to the "superstition of those [astrologers] who place the fates of men in the stars and who are therefore called 'mathematicians.'"[9] One had to be careful not to lapse into category sin with a fricative slip of the tongue. Moreover, on the practical plane, neither royalty nor aristocrat nor their minions could keep close tabs on their wealth. The bourgeois stalwarts in new towns fared somewhat better, but even there accounting remained primitive.[10]

Behind the foregoing, general observations lodged a yet more profound reason for the continued marginalization of mathematics, especially in its higher manifestations, before the modern era: mathematics lay deeply embedded in the information technology of literacy. Out of their own, ongoing oral traditions, the Greeks had devised a literate culture and its practices.[11] Grounded in the alphabet, literacy came to dominate the intellectual culture of Greco-Roman antiquity, not only with Muse-inspired literary creations but also with a science manifest in proper definitions of words and things. Even soaring abstractions, great though they were in logic and metaphysics, remained rooted in classifying the information created and captured by writing. Mathematics was a subordinate feature of this culture, a "literary genre," writes Reviel Netz, "a form of literature in [its] own right."[12] Like other literary genres, it too had its own specialized lexicon and logical ways of proceeding, both of which in this case were keyed to diagrams and their expression of spatial objects in and through a "formulaic language," which gave it identity and coherence.[13] And as did other genres, it too found its niche in the classifying mentality.

Educated schoolmen in the Middle Ages inherited this literacy-based, classifying ethos. By then the practices of reading and writing had undergone a material transformation from papyrus to parchment, from the scroll to the codex, and with their handwritten books scholastics delighted in pigeonholing their thoughts. They generally followed Aristotle in defining mathematics as one of the three theoretical subject matters, along with theology and physics, which they classified with all the other defined objects of learning, rather than treating it as a commonly practiced discipline. In a curriculum of the seven liberal arts, mathematical topics encompassed four of them, the quadrivium of arithmetic, geometry, music, and astronomy. The first two made up the purely theoretical branches of the subject, and although the latter two seemingly gave nods toward practical applications ("subalternate" sciences they were sometimes called), both music and astronomy were equally timeless, which kept them also in the realm of mathematical theory. Nor, critically and recalling Aristotle, did the major branches overlap; one could not extend arguments or claims from geometry to arithmetic or vice versa. In technical terms, the incommensurability of geometry and arithmetic and the even more universal "incommunicability of the genera" held sway. Aristotle and his scholastic followers did speak of a *mathematica media,* which pertained to a sort of middle ground—applied mathematics, we might term it—whose importance would grow during the Renaissance.[14] But for the most part, physics and mathematics, mutually exclusive categories, eyed one another as a pair of junkyard dogs, suspiciously and from afar.

The former accounted for change in nature; the latter constrained itself to the universe of eternal, unchanging verities. We thus should not be misled by the number of mathematical subjects or their role in education. Tucked away in their cubbyholes, more talked about than done, the mathematical disciplines remained for Europeans thoroughly subservient to the classifying impulses of literacy.

Literacy's impulses of naming, identifying, and grouping gave voice to another pronounced feature of mathematics in scholastic thought—namely, its "enchantment." Scholars frequently use the term 'enchantment' in referring historically to a medieval world of "spirits, demons, moral forces which our predecessors acknowledged," writes Charles Taylor, a world in which the "line between personal agency and impersonal force was not clearly drawn."[15] Many, notably Thomas Aquinas, wrote pages upon pages about angels and demons, defining the various powers of incorporeal beings and their places in the created order of things. Charms, superstitions, and spells likewise had their proper entries in the glossary of understanding.[16] As we discussed in Chapter 2, portions of this world could be explained by appealing to Aristotle's account of causality. Different perspectives of "the 'why' of" revealed the intertwining of personal agency and natural forces, of purpose (*telos*) and attractions (active and passive features of movement), and of qualities throughout nature. In addition to the "univocity" of descriptive, scientific terms, words had their metaphorical, analogical, and allegorical uses and meanings. And from there it was but a short step to envisioning a corresponding, animated world of teeming things, any of which could be interpreted as "signs" revealing "meanings" that lay hidden in natural phenomena. Francis saw God in the sparrow. Dante began his allegorical journey in sin's "dark wood." Pilgrims traveled great distances to bathe in the healing aura of relics—spirited pieces of wood, bone, cloth. The world was enchanted and, for most, well organized.

Medieval mathematics was enmeshed in this enchanted world, conveying multiple "meanings" far surpassing any of its technical operations. Throughout the period an enchanted mathematics continued to provide "the 'why' of" teaching the quadrivium in the arts faculties of medieval universities. Especially popular was the work of Neoplatonist Boethius, which circulated in numerous manuscripts and encouraged study of the "fourfold way" (quadrivium) to spiritual insight. Alongside the manuscripts of Boethius and others, several works of practical mathematics eventually made their way into the curriculum. More mathematical than most, the mystical Hugh himself authored an often-copied text on practical geometry, probably designed for classroom use.[17] A bit later, in the thirteenth century, John of Sacrobosco penned an even more popular work, the *Algorismus,* which became widely used as an elementary text on numerals and arithmetic, and which introduced students to many practicalities of basic computation and which complemented his equally popular *Treatise on the Sphere.* Yet even these and similar texts were used primarily in teaching the subjects of the quadrivium to arts students for their spiritual and philosophical enlightenment, not for practical projects.[18]

A disputatious lot, scholastics certainly debated about mathematics, its nature as well as its purpose and place in the hierarchy of learning, and most, like Hugh, agreed as to its pedagogical utility. But their attention lay elsewhere. Completely self-contained, the quadrivium was said to have found divine favor and rationale as it reflected the fourfold progression of the soul, a pilgrim's journey taught by "number itself." We begin with unity and trinity, Hugh wrote, one and three. Three times one are three, times three are nine, times three again for twenty-seven, and times three a fourth go around for "eighty-*one*" (italics in the original). In the fourth multiplication "unity recurs," and one would see the "same thing" happen repeatedly were the multiplications carried out "towards infinity." In the full blossom of allegory, the first progression begins with the soul's incorporeal, "simple essence," a unity "symbolized by the monad," and descends into a "virtual threeness" of concupiscence, wrath, and reason, which describe the soul's powers in its earthly sojourn. A second progression descends further from this threefoldness to "controlling the music" of the human body, a harmony "constituted in the number nine" because there are nine openings in the human body through which, according to "natural adjustment, everything by which the body is nourished and kept in balance flows in and out." The third progression sees the soul pouring itself out through the senses on all visible things, "symbolized by 'twenty-seven,' a cube number . . . dissipated in countless actions," a Boschian riot of earthly delights: sensuality, temptation, and sin. Finally, freed from the body, the soul "returns to the pureness of its simplicity" in the fourth multiplication, in which the number one reappears in the arithmetical product. This makes it "glowingly evident that the soul, after this life's end, designated by 'eighty,' returns to the unity of its simple state." (Rather than the usual three score and ten, Hugh conveniently took eighty years as the measure of a human life, citing the prophet in Psalms 90:10.)[19]

Yes . . . well, maybe one has to have been there in order to appreciate and experience fully the enchanted world from the inside. Hugh's voice might have been distinctive, but it was not at all unusual among the schoolmen, who continued teaching the quadrivium's brand of philosophical and allegorical mathematics until well into the early modern period. At Chartres, for example, Hugh's contemporary Thierry the Breton (ca. 1100–c. 1155) not only taught the quadrivium but used it to create an elaborate "mathematical theology" by delving into the mysterious workings of the Trinity and exploring the "four kinds of reasons which lead humankind to the knowledge of the Creator: namely, the proofs of arithmetic, music, geometry, and astronomy." For Thierry "number" (*arithmos*) rivaled and in many ways surpassed "word" (*logos*) as a pathway to the divine.[20] And while others didn't travel as far as Thierry into such a full-blown world of theological mathematics, they certainly invested arithmetic and geometry with enchantments of their own.

Occasionally enchantments themselves could lead to mathematics. In the fourteenth century, a group of scholastics known to historians as the *calculatores,* centered largely at the universities of Oxford and Paris, invoked mathematical arguments and demonstrations to represent the "quantity of a quality." For instance, a discussion of angelic motion—to wit, how many angels could dance on the head of a needle—led some to consider the composition of continuous quantities and thus raise the troubling issue of actual versus potential infinity.[21] Others used mathematics to compare degrees of qualities like hot and cold, wet and dry, white and black, health and sickness—what was then termed the "latitude and longitude" or "intension and remission of forms." These excursions into quantification produced some early, "immature" examples of visual graphing with horizontal and vertical axes. Analyses of local motion too fell under the purview of investigating the degrees of qualitative change. At the University of Paris, Jean Buridan and Nicole Oresme, among others, produced the clearest formulations of the impetus theory and of the mean speed theorem, respectively, the latter being a forerunner to Galileo's law of free fall. Yet despite the appearance of some quite sophisticated and seemingly "modern" calculating techniques, these scholars stayed firmly within the framework of literacy, subordinating quantity to a classifying, logical analysis and exploring the meanings of words and their qualitative definitions rather than actually investigating bodies in motion or other natural phenomena.[22]

We moderns tend to look on many of these medieval expressions with puzzlement. To our eye, not only mathematics but scholastic thought writ large appears fraught with paradox everywhere we look. For on the one hand it harbored its mysteries and enchantments, its "things that go bump in the night"; yet on the other hand, especially with the arrival of Aristotle's philosophy and its utilization by Aquinas and other schoolmen, it became, as one scholar has it, an age of "reason."[23] Recognizing the overwhelming preponderance of an information technology of literacy in the age's intellectual life goes a long way toward dispelling our perplexity. It helps us appreciate both the age's reason, tied as it was to literacy and the classifying temper, and its own brand of magic, its meaning-laden enchantments articulated with the same tools of linguistic expression. Moreover, the domineering presence of literacy allows us to appreciate, as well, just how thoroughly sidelined mathematics remained before the modern era. The few forays into mathematics that scholastics did exhibit were made, so summarizes historian Michael S. Mahoney, "with their own intellectual concerns in mind: the union of faith and reason."[24]

With the unhurried arrival of the Renaissance centuries, two broad, intertwined series of developments—both associated with technologies—would radically challenge and in many ways transform the classifying temper of the High Middle Ages, teasing apart the enchanted and material worlds commingled in popular imagination and in refined minds alike. The first of these was an information explosion, powered by the printing press, that relentlessly overwhelmed the ordinary, commonsense categories used by scholastics and eventually, for those like essay inventor Michel de Montaigne, even vanquished the idea of classifying itself, giving warrant to skepticism in the process. The onset of the information surfeit accompanied the quests of humanists. Beginning in the fourteenth century, Renaissance scholars mounted a vanguard search for new manuscripts and materials to feed their burgeoning interests in the classics of antiquity. The classical revival began with Latin works before spreading to Greek ones, and it focused mainly on literary texts, such as Virgil in poetry, Livy in history, and Cicero in rhetoric and moral philosophy. In fact, the *studia humanitatis* (roughly, "studies worthy of human beings") entailed a purely literary curriculum: grammar, rhetoric, history, poetry, and moral philosophy.

Perusing their newly discovered texts, humanists frequently took notes on them, which they wrote down in "commonplace books," a practice falling within the long tradition of rhetoric that itself dated to antiquity. Aristotle had separated philosophy from rhetoric, the former dealing with truth, the latter with probability, those gray areas of life, such as politics, public policy, law, moral behavior, and the like, whose subjects were amenable to public debate and persuasion. The Roman Cicero (Marcus Tullius Cicero, 106–43 BCE) had extended rhetorical practices in part by advocating that public speakers create a storehouse of common topics, categories one could retrieve in the service of argument. These topics were akin to our notion of commonplace maxims or proverbs, except that for us they've often become clichéd and trite expressions ("Neither a borrower nor a lender be," for instance, or "A stitch in time saves nine"), whereas Cicero used them as basic categories of thought, often contrary to one another, which a public speaker could weave artfully into his orations as needed. In subsequent centuries, rhetoricians furthered the practice of storing common topics in memory, frequently by keying them to visual images and arranging them in mental categories or "places," such as a temple with many columns, objects in a room, or rooms in a house—"memory palaces" they were called.

By Renaissance times the "art of memory" had cultivated its counterpart, that of writing down the commonplaces in notebooks, from places in memory to "places" (*loci*) on paper.[25] These were often organized into headings under which contrasting topics would be entered as a reader encountered them in his studies—for example, headings such as 'vice' and 'virtue,' 'love' and 'hate,' 'war' and 'peace,' and similar themes. Commonplace notebooks thus served humanists essentially as personalized encyclopedias of classical culture, storehouses for what one could know, those moral norms organized by topic in the form and substance of concepts society took for granted. Cumulatively, the newfound literature conveyed a plethora of different human behaviors, customs, and moral examples from pre-Christian antiquity—countless illustrations of honor and pride, cleverness and wit, courage and cowardice, humor and pathos, and other human qualities. With all the new exemplars, the consequences of these humanist practices of learning and note taking showered themselves with irony. For the more students and authors gleaned such diverse and voluminous information from classical texts, the more they swamped their own moral categories, bursting their classifying mind-set and setting their world adrift.

New world discoveries and the turn to nature added to the accumulation of fact and the rupture of classification. Information about diverse and unusual (from a European perspective) peoples and cultures, about new lands and their inhabitants, about a whole new range of flora and fauna, about a new and emerging picture of "aqua-terra" (the size of the earth and its division into land and water)—all exploded "old world" consciousness. From the "shock of discovery" to "inventing discovery" itself, Europeans embarked on new examinations of nature, new natural histories that far exceeded the provincial limits of, say, Pliny (23–79) and other writers from the ancient world.[26] Nature became increasingly scrutinized. University of Padua professor Andreas Vesalius (1514–1564), the founder of modern anatomy, epitomized the process when dissecting a human heart for his students according to Galen, the great medical authority of antiquity. Galen had said that in the heart's septum, the central wall separating the right and left ventricles, there are holes, which allow for the passage of blood from one side to the other. Vesalius pointed from book to heart to demonstrate the claim to his anatomy class but then discovered: there are no holes in the septum. The authority of the book yielded to the authority of careful observation. Using artists from Titian's studio in Venice, Vesalius then created new engravings covering the new anatomy and published them to broad acclaim and sales. His new "information" soon found its way to all the corners of Europe.[27]

As typified by Vesalius's experience, the printing press was the most important technological agent of the information explosion. Johannes Gutenberg's first Bible appeared in 1455, although he had been printing texts for at least the previous five years. The publishing offspring of Gutenberg soon inundated Europe with a conservatively estimated eight million books in the period of the incunabula alone (before 1500), a quantity surpassing scribal manuscript production of the previous millennium. (Some scholars, among whom Lucien Febvre and Henri-Jean Martin, put the number at an even greater fifteen to twenty million copies.) Of incunabula editions, some 45 percent were religious in nature (including Bibles, the Church fathers, and scholastic theology), and about 30 percent were literary (classical, medieval, modern), with another 10 percent or so devoted each to legal subjects and natural philosophy. By the end of the following century, a thriving publishing industry had issued some 150,000 to 200,000 separate editions of written works, totaling the astronomical sum of 150 to 200 million books! From pamphlets and flysheets to massive tomes and early encyclopedias, from short runs to repeated editions of best sellers, printing overwhelmed Europeans with "too much to know."[28] Categories fractured under the onslaught of all this new information; classification systems broke apart into a hodgepodge of disparate and jumbled lists of facts. At the time, the cumulative effects of the information explosion were captured and displayed by the inimitable insight of poet John Donne: "'Tis all in pieces," he lamented, "all coherence gone."[29]

Coevally with the category-swamping volume of facts, there appeared a second series of developments, the piecemeal arrival of a new information technology, which would eventually begin to bring some "coherence" at least to portions of the information pertaining to nature. In effect, the rupture of classifying created an intellectual vacuum into which this new technology trickled bit by bit.[30] These developments took place largely outside the hallowed halls of universities during the Renaissance centuries, roughly from 1250 to 1600. New practices transformed mathematics in each corner of the quadrivium. In arithmetic, commercial expansion brought wealth and luxury not seen since antiquity and with them a new counting system based on Hindu-Arabic numerals to manage the affairs of businessmen. In music, new, contrapuntal harmonies filled the air, but more critically an abstract notation captured the time-bound information of sound, allowing musicians and composers untold expansion of their musical imagination and sensibilities. In geometry, the sheer beauty of drawing, painting, sculpture, and architecture competed favorably with nature (some might say improved on it) as the invention of linear perspective brought forth new, functional rules of geometry in its ways of seeing. And in astronomy, practical innovations in calendars and clocks transformed timekeeping into time measurement and symbolic abstraction, unifying the movements of the heavens with those on earth.

In all these areas—commerce and arithmetic, polyphonic sound and music, art and geometry, timekeeping and astronomy—a common theme crisscrossed topics as a new and emerging means of creating and managing information made its appearance. The new *techné* ("the way things are done around here") was based on symbols, empty and simplified yet full of instructions for those who knew how to use them.[31] These were not the animated and enchanted symbols of allegory associated with words and meanings. They were the functional instruments that embodied new ways of encoding, storing, and manipulating bits of information. Mapping onto various dimensions of our experience (business transactions, musical sound, artistic perspective, linear time), these encoding systems framed new understandings of phenomena. They provided the tools for a new layering of abstraction upon abstraction in many walks of life, as empty relations correlated with increasingly complex events. And they laid the groundwork for the extension, expansion, and transformation of ancient mathematical thought from thing-mathematics to relation-mathematics, which followed the recovery of its texts.

By the mid- to late fifteenth century, humanist interest in the classics had expanded to include dialectic (the "logic" of rhetoric), especially with Lorenzo Valla (1407–1457) and Rodolphus Agricola (1443–1485), but the humanist program of education—the progressive educational system of its day—had generally excluded mathematics. Alongside the *studia humanitatis*, however, there flourished in Italy another series of schools, now known as the abacus tradition, which would complement and eventually rival humanistic studies. (The earliest intimations of the modern rift between "two cultures" find their origins in these separate pedagogical traditions.) In abacus schools students learned the new art of numbering with Hindu-Arabic numerals, which they often applied not only to the expansion of commercial practices (accounting, investing, and the like) but also to music, art, and timekeeping, as new ways of managing information with symbols seeped into various practices. Following in the wake of the broad humanist urge to get classical, literary, and philosophical texts right, abacus teachers and scholars also sought ancient mathematical authors, texts, and problems, a quest that gathered momentum in the sixteenth century. The newly recovered, newly translated, and newly printed texts found a receptive audience among those whose collective learning was steeped in the new technology of information, in the emerging, relational numeracy.[32]

The new technology fomented two great consequences of historical moment, which we shall explore in the following chapters. First, it served to "disenchant" medieval mathematics, emptying it of moral, allegorical, and philosophical content—the various enchantments prized by medieval mathematicians.[33] Second, and even more significantly, it created the instruments for devising a new analytical temper of reverse engineering, which would become the mainspring of the Scientific Revolution. For this part of our story we shall be veering away from orthodox treatments of religion somewhat, even though the period under consideration—the thirteenth through the sixteenth centuries—witnessed tumultuous developments in the religious history of western Europe. From the Avignon papacy to the Great Schism to the worldly Renaissance popes and to the sixteenth-century Protestant Reformation itself, Western Christendom was rocked to its foundations, an upheaval that dramatically transformed the religious landscape. These large events have been chronicled and interpreted for centuries.

The lesser known and less observed, quieter events alluded to earlier will guide our narrative. They emerged against an ongoing backstory of the presumptive harmony between science and religion of medieval times, and also of a centuries-enduring, premodern thing numeracy that had remained embedded in the classifying temper and its medieval enchantments. We shall need to revisit briefly the place and nature of this information technology in the ancient and medieval worlds (Chapters 3 and 4). In so doing, we shall see how the practices of early numeracy diverged from those of literacy but also remained subsidiary to them, and we shall note some of the historical restrictions that early numeracy had consequently imposed on higher mathematics, restrictions that persisted with the continued embedding of thing-mathematics in medieval thought. From there we shall turn our attention in Chapters 5 through 8 to the emergence during the Renaissance of the new information technology in arithmetic, music, geometry, and astronomy.

In one of the many perceptive remarks that bespoke his genius, the German philosopher-historian Ernst Cassirer stated that before the Renaissance, mathematics had been considered an element of thought, but that afterward, from Leonardo and Galileo on, it gradually assumed the position of a central "force" in Western intellectual life.[34] As long as mathematics remained subordinate to classification, it could not become an independent "force" in intellectual life. The harmony of science and religion could be maintained by use of conventional language, which underlay both natural (causal) descriptions of the phenomena in the universe and metaphysical descriptions of the divine, in the fullness of its presence and its purposes for human life. All this was predicated on the alphabet. Only with the preconditions of a new information technology, relational numeracy, in place could relation-mathematics, and with it the analytical temper, surface and embark on its destiny as an independent "force" in Western intellectual life. Only then would the stage be set for the analytical breakthrough that first appeared with Galileo, and with it the first major episode in the modern world in which science and religion followed their divergent paths. Like the "Birth of Venus," appearing on her shell, slightly coy but radiant and alluring, the birth of the analytical temper heralded a new era.

CHAPTER 3

Early Numeracy and the Classifying of Mathematics

> Mathematics is an abstractive science. . . . Abstraction is the apprehension of the form of any kind of a thing whatsoever.
>
> —Domingo Gundisalvo, *On the Division of Philosophy* (ca. 1150)

FROM ITS ADVENT as an information technology, numeracy—the abilities to count and to measure—tracked a historical path somewhat similar to that of literacy. As noted earlier, at both ends of the Fertile Crescent writing entailed the invention of symbols, marks of one sort that stand for things of another, in order to create and store information, a process that emerged under the pressure of early business practices and record-keeping needs. Likewise, numeracy developed as symbols came to stand not only for the names of things but also for collections of items under the same name.[1] Over time the growth of symbols permitted numeracy's expansion beyond a general, descriptive sense of quantity as the counting ability broadened to mastery of larger collections of things and to measuring, calculating, and manipulating spaces, as well as objects. For the latter operation Herodotus adopted the term "geometry" (from the Greek, *geo,* "earth," and *metris,* "measure") and attributed its discovery to the Egyptians.[2] Notwithstanding his authority as the "father of history" (Cicero's appellation), there were likely several independent birthings of geometry as a means of surveying settled lands and of measuring celestial movements, the latter in order to predict planting and harvesting seasons. As with writing, from very early on the knowledge of "all dark (mysterious) things"—arithmetic and geometry—intertwined intricately with religion, generally under the purview of a priestly class whose occupations ranged from practicing astrology and divination to constructing pyramids and ziggurats.[3]

Issuing from practical needs, over the course of centuries both information technologies enabled humankind to soar into the stratosphere of abstract thought as the written word stretched from mundane lists to otherworldly philosophy and as numbers grew into what would eventually be known as higher mathematics.[4] Throughout these developments, just as writing systems evolved with ever more sophisticated syntax and semantics, culminating in alphabetic literacy, so too did numeracy develop increasingly complex systems of counting, measuring, and calculating. But there comparisons end, for the abstractions of each diverged markedly. Before the Common Era, moreover, numeracy never achieved its technological breakthrough comparable to the invention of the alphabet. Not until the advent of the Indian or Hindu numerals (ca. first through fourth centuries CE) did a simplified and functional notation emerge that would remove the ambiguities in expression, whose continued presence placed a great interpretive burden on readers. And not until the thirteenth century did that notation, now known as the Hindu-Arabic counting system, enter significantly into the consciousness of Europeans, the topic of a later chapter.

Throughout their long, premodern odyssey, the abstractions of early numeracy at times generated enormous, even breathtaking achievements in higher mathematics, especially in geometry under the aegis of the Greeks. Yet when seen with the hindsight of modernity, these triumphs occurred in a historical context equally constrained by internal as well as external limitations. Internally, the lack of a fully developed, positional system of symbol notation and operations played no small part in the continuing separation of arithmetic and geometry, in the ongoing difficulty of performing complicated calculations, in the inability of mathematicians to resolve problems surrounding infinity, and in the limiting of geometry to methods of synthetic construction with ruler and compass, notwithstanding the depths of Greek insights and proofs.[5] Externally, the growth of higher mathematics was constrained by both the separation of "theoretical" and "applied" or "practical" mathematics and the continued association of the former with religious mysteries and cosmic harmonies, obstacles deeply embedded in the culture of Greco-Roman antiquity.[6]

Some of these impediments were seemingly overcome for a while during the "golden age" of Hellenistic geometry (third to second centuries BCE). Centered largely in the cultural and scientific center of Alexandria during the reign of the Ptolemies, the flourishing of mathematics saw the merging, consolidation, and furthering of previous theoretical achievements, achieved largely apart from any religious concerns. The works of Euclid, Archimedes, and Apollonius (ca. 262–ca. 190 BCE) all bespoke a level of abstraction and mathematical superiority previously unseen in the annals of early numeracy. Some evidence, though quite scanty, also suggests that these abstract accomplishments occasionally extended directly into the world of practical creations and applied mathematics as mathematicians sought to solve problems under the impress of real, practical issues—matters of mechanical invention and experiment.[7]

Yet, despite their apparent autonomy and hints of even greater achievements, the geometrical accomplishments of Euclid and others could never fully break free from literacy and classifying as the dominant mode of abstract thought. The reasons for this are quite complex and still debated, but an emerging consensus stresses the interlacing of Hellenistic mathematics and key features of Greek culture.[8] The oral practices and traditions attending early Attic history had expressed and furthered an "aristocratic," "ludic," and "agonistic" warrior society, a Homeric ethos of competition (the *agon*), display of skills (*arete*), and reward (*timé*). Many of these features carried over into the content and context of later literary expressions, including those of drama, poetry, political oratory, and philosophy, as literacy steadily crept deeper into Greek culture. They extended, as well, into the Hellenistic period and into "geometrical proof—the genre where the art of persuasion [was] brought to perfection."[9] Proofs in geometry were thus an outgrowth of this literary and ludic culture, a demonstration of reasoning that was competitive and entertaining for a "small, scattered group of genteel amateurs."[10] After its relatively short-lived burst of "literary" autonomy, mathematics became predominantly reabsorbed into other intellectual mainstreams of the day, a subordinate category in Aristotelian philosophy and a religious, contemplative ideal associated especially with Platonic (and, later, Neoplatonic) philosophy. The astronomer Claudius Ptolemy embodied both these intellectual currents throughout his massive mathematical compilation, the *Almagest,* a work destined for well over a thousand years to anchor Islamic and Christian cosmologies in the classificatory learning of antiquity.

The story of early numeracy, then, emerges as a tale of how "thing-mathematics" remained a subsidiary of the classifying temper, the backdrop against which modern "relation-mathematics" would eventually make its appearance. In sketching this story, briefly as a hurried prelude to later developments, we need to bear in mind that although our focus will fall rather negatively on what the ancients did not do in numeracy and higher mathematics, we should not thereby dismiss their actual and quite remarkable accomplishments. The ability of humankind to master massive amounts of the world's information by means of the abstract symbols of early counting and the reasonings of higher mathematics, especially geometry, must be considered among the most distinctive and impressive of antiquity's feats. Moreover, when these achievements resurfaced in the context of a new information technology, that of modern, relational numeracy, the mathematical world itself was transformed.

* * *

In all early counting systems, from Egypt to China to Mesoamerica, two fundamental features permeated the abstractions of numeracy wherever it appeared, separating them in turn from those of literacy: correspondence and recurrence. These characteristics eventually became known as the cardinal and ordinal principles of abstract number, principles derived from the late nineteenth- and early twentieth-century investigations into the logical foundations of mathematics.[11] But long before they became principles, the cardinal and ordinal features of number were utilized unreflectively in number languages and counting systems.[12]

Numeracy began with the recognition that by correlating strokes, marks, or ciphers with material objects, one could keep track of collections of things. This basic correspondence comprises the central idea of what were later called cardinal numbers, or the cardinal concept of number, most often distinguished from the ordinals, or the ordinal concept of number. The cardinal feature of number refers to the entire sum of units in any collection of units or objects. Taken strictly by itself, it does not involve counting, only making one-to-one matchups between the discrete members of collections. One collection is larger than another if at the end of making correlations between them some members of it remain. A stadium, for instance, contains some number of seats; people enter and claim them, one seat per person. If there are neither standing people nor empty seats, then the cardinal numbers denoting the collections of people and seats are equal. If there are empty seats, the seat collection is larger than the people collection; if standing people, vice versa. Note, we have not counted or organized the members of these collections serially in any way, only correlated seats and persons. Thus the cardinal aspect of number refers simply to the total number of a collection of member objects, correlated with another collection and identified with a number word or symbol.[13]

Although very little understood until well into the early modern period of the sixteenth and seventeenth centuries, the cardinal correlations of numeracy differ fundamentally from the semantic referents of conventional language, even while they both probably stem from the same practice of making visual comparisons.[14] Through the medium of sight, the mind derives from comparisons of several collections of things a numerical mental extract—say, 'seven'—which corresponds to the same number of objects in a collection (whether sheep, sins, or wonders of the world is irrelevant). In a like fashion, by visual means the mind elicits a classifying concept—say, 'vertebrate'—whose defining properties are then understood to correspond to any animal with a backbone. Both the numerical term and the classifying term are therefore created by an *act* of visual comparison. But the *results* of the act differ. The term 'vertebrate' itself is defined by its visual content; the content prods us to imagine a backbone, and thus armed with this visual image, we correlate a seen object and a category. The term 'seven,' however carries no such visual content. In imagining seven we really have no image whatsoever. We may imagine seven marks, or strokes, or pebbles, or sheep, but these are all groups of specific objects, not 'seven' per se; the objects serve as visual placeholders or reminders for completely abstract, nonvisual units, which can be correlated with any things whatsoever.

In classifying systems, by some accounts, it is equally hard to imagine adjectives like 'red' in the abstract. Empiricist philosophers maintain that our images of abstract red must originate in an object of that color; we derive 'red' from red objects. Likewise, the argument runs, with 'seven.' Both 'red' and 'seven' are abstractions taken from images of concrete objects or collections, an affinity that helps us understand the early association of adjectives and the general sense of number. But as numeracy moved beyond early adjectival description to more complex systems of correlations and numbers, so too did it become more abstract. It may be that philosophically, as well as historically, we cannot at first imagine 'seven' without seeing seven somethings of one sort or another. But seventy times seven has no sense for our imagination, even in the derivative way that abstract red has. A red rose, a red pony, a red sky in the morning—in the classifying scheme, we derive from them 'redness' or the general idea of red, which we still imagine visually. But we cannot derive abstractions for larger number collections (usually those beyond three, by most accounts) from the images of smaller ones; the larger abstractions remain without visual image; to reach them they must be counted or calculated.[15]

The abstraction inherent in correlating things numerically thus led unsurprisingly to enumerating or counting larger collections, and to the second critical feature of numeracy, ordinal numbers, or the ordinal concept. In fact, a moment's reflection will tell us that cardinal correlations, though necessary, do not suffice for a counting system. To count, we need not only a means of correlating marks or strokes but also the capability of ordering them in some kind of sequence or series. Ordinal numbers do this. Whereas a cardinal number refers to the total number of members belonging to a collection, the ordinal number designates the ordering of those members within the collection. Ordinal number, in short, refers to a number's "rank." When '12' indicates donuts in a box, it is a cardinal number, referring to a collection. When it designates a seat on an airplane, it is an ordinal number, identifying one particular seat because of its ranking in the order of a collection of seats. Ordinal ranking appears in early counting systems as soon as number words extend beyond their general, descriptive use as adjectives.

Such ranking was achieved by devising an ordered series of numbers, one that "progresses in the sense of growing magnitude, the natural sequence: one, two, three. . . ."[16] The natural sequence means that one can always follow a number with its unit successor, adding one stroke or marker, and that one can repeat this process ad infinitum. This is the principle of recurrence. Reaching and managing larger collections of things were made possible by devising a "collective unit" or "base" of a counting system, whose repetition enabled the rank ordering and calculating of indefinitely large numbers. Historically, all counting systems have coalesced around bases 2, 5, 10, 20, or 60, the most widespread being our familiar denary or decimal, base-10 system. Mathematicians refer to early counting systems as "additive," meaning that the numerical value of a collection derived from totaling all the values of the individual symbols or ciphers, the bases plus the increments.[17] Most additive systems employed different number words and symbols for the integers up to the base, for the base itself, for the base and its additives, for multiples of the base and their additives, and for powers of the base. Again, the total numerical value of any collection depended on adding the units, groups, and ciphers irrespective of their positions.[18]

As counting evolved, numeracy grew increasingly intricate and complex. Sophisticated and varied techniques were devised for the basic operations of addition, subtraction, multiplication, and division, as well as for manipulating fractions, computing rapidly with an abacus or counting board or fingers, and figuring powers and roots (squares, cubes, and the like). The latter were most frequently arrayed in tables because of the difficulty and tedium of calculating them individually. The number realm expanded from whole numbers to fractions to rational numbers, and even to the irrational numbers, such as the square root of two (horrifying the Pythagoreans). To describe these developments in any detail would take us too far afield from our focus on the identifying, cardinal and ordinal, features of number, correspondence and recurrence. Here we need note only that both these features were implicit in the calculating techniques of early numeracy and permeated all additive counting of the ancient world. Numbers denoted "a multitude composed of units," wrote Euclid, abstract collections that stood for things or objects.[19]

Generally speaking, the cardinal correlations of numbers and things allowed information about the things to be recorded or stored—the size of an army, for instance—while the ordinal features of counting systems permitted manipulations of the information by means of calculating methods devised for various purposes, such as figuring amounts on tax rolls. The counterpart to the semantics and syntax of written language, these features made early numeracy an effective system of information management for myriad practical matters. Yet despite the enormous accomplishments of early number systems, there persisted assorted, built-in limitations to the abstractions and competency of early counting. Some systems, Greek and Hebrew for example, used letters of the alphabet to represent numbers. With their emphasis on the ordinal or ranking feature of number, these systems of lettered notation provided distinctive ciphers for easy enumeration on the first pass through the alphabet.[20] Once the letter cycle ended, however, letters had to be repeated or supplemented with other symbols and quickly became ambiguous and cumbersome in dealing with larger collections and more complicated calculations, as well as fractions. Other systems, such as the familiar Roman numerals and the less familiar Egyptian hieratic, or priestly, characters, were devised as combinations of strokes, groupings, and ciphers; these systems emphasized cardinal collections over ordinal ranking. They could be effectively employed to designate large collections, but calculating with them proved extremely awkward in the absence of both symbols for operations and place-holding value for ciphers. Such difficulties led to the invention of the counting board or abacus as a physical means for accomplishing the fundamental operations of arithmetic. As with writing before the phonetic alphabet, early counting relied heavily on the script's context and the reader's interpretive skills to ferret out the correct values of ciphers and computations, opacities intrinsic to premodern numeracy.

* * *

As principal indicators of numbers and early numeracy, the cardinal and ordinal features of correspondence and recurrence are usually restricted to discussions about counting collections of unitary, discrete things and objects. Yet the same features extended as well to geometry. The abstractions of geometry yielded spatial information about objects, such as the size and area of a plot of land or the relative locations of the wandering stars. In so doing, they utilized visual abstractions, essentially outlines, to represent figures, forms, shapes, and positions. These lines were not symbols in quite the same manner as number-ciphers, which stood for abstract, countable units, but they served a similar and abstract, place-holding purpose, signifying the size, shape, position, and configuration of objects. Further, these abstractions often correlated with the countable features of numeracy in the course of measurement, particularly as shapes were compared in their various dimensions (lengths, widths, angles, ratios, proportions, and the like).[21] Although certainly not alone, the Greeks furthered the earlier techniques of their Egyptian and Babylonian predecessors and stood at the forefront of developing geometrical methods for handling mathematical problems, and particularly for devising ways of proving theorems or propositions. Early numeracy reached its most abstract expression with Greek, especially Hellenistic mathematicians.[22]

The birth of Greek mathematics lies in shadows almost as dark as those enveloping the early stages of alphabetic literacy. Beginning with the semi-legendary figure of Pythagoras (ca. 570–ca. 480 BCE), who appears only slightly less shadowy than Homer, the Greeks interwove their mathematics with religion, much like their Egyptian and Babylonian forbears.[23] In this they were furthering a tradition that would extend well into the Christian Middle Ages and beyond. Classical sources describe Pythagoras as a young wanderer who left his birthplace on the island of Samos and journeyed throughout the (modern) Middle East, especially Egypt, where he fell under the spell of Egyptian priests and assimilated many of their religious customs. Upon settling in Croton in southern Italy, a region of Greek colonization, he established a cult of his own, a kind of utopian community with a way of life bearing some similarities to the mystery cults of Orpheus.[24] Pythagoras himself was depicted, adoringly or satirically, as a religious teacher (sometimes a paragon of virtue, sometimes a charlatan), a "mystagogue" who promoted doctrines of the transmigration of souls, reincarnation, and a mystical, cosmic unity organized by number, especially the *tetractys*, a holy "fourness" made up of the numerals 1 through 4.[25] A mysterious, secretive sect, the Pythagoreans wafted in and out of contemporary accounts over the next several centuries. Behind closed doors, their interest in number mysticism also gave rise to what became known as a Pythagorean, mathematical tradition that treated number theory, proportions, geometry, musical harmony, and associated topics, occasionally breaking into a wider, more public context.

Within the century after his death, Pythagorean doctrines were on their way to becoming thoroughly intermingled with those of Plato (429–347 BCE), and early mathematics henceforth seldom escaped the latter's influence. Although current with the state of mathematics in his own time, Plato was not himself a mathematician, but more of a mathematics enthusiast.[26] As is well known, he stressed its value for training the mind: "Let no one ignorant of geometry come under my roofs," read the inscription over the entry to his Academy. Such training is suited for a philosopher-king not because of its managerial, practical possibilities, he held, but because it "draws the soul toward truth and produces philosophic thought."[27] According to ancient tradition, Pythagoras had coined the word 'philosophy,' the "love of wisdom," which Plato exalted even further into a way of life. He himself had visited Pythagoreans in Sicily and had likely modeled much of his famous *Republic* after their communities. In some accounts, his Pythagorean friend Archytas of Tarentum (428–347 BCE) personified the philosopher-king in all but title. Both as a way of life and as a discipline of mind, Plato absorbed the legacy of Pythagoras, which in turn he bequeathed to Western thought and after which, for Alfred North Whitehead among others, there have remained only footnotes.[28]

Within Pythagorean communities themselves, current scholarly consensus identifies a tension of long standing between two rival groups, both claiming guardianship of the cult's secrets: the *mathematikoi* and the *akousmatikoi.* Sometimes referred to as the "inner circle," the former, which has given us 'mathematician' (from the Greek *máthema* or "science"), comprised those followers to whom Pythagoras had entrusted the truths and mysteries of numbers. The latter carries no ready translation or cognate (the ungainly 'acousmatics' is sometimes used), but it denoted those who preserved the master's "oral sayings" (*ákousma* in Greek), especially as regards matters of religious sacrifice, worship, diet, ritual, and the like that governed the members' quotidian and spiritual life.[29] Some scholars have surmised that high-level mathematics developed among the inner circle of *mathematikoi,* while remaining predominantly hidden from public view. Because there is virtually no direct evidence from the likes of those traditionally called Pythagoreans, such as Eudoxus of Cnidus, Plato's student Theaetetus (ca. 417–ca. 369 BCE), or his pal Archytas, the argument for this point hinges on later reports and on the presumption of antecedent developments that "must" have occurred before Euclid and the third-century flourishing of mathematics. In brief, the accomplishments of Euclid and others would not have been possible without a precursor, a Pythagorean-Platonic tradition of mathematics.[30]

By Euclid's day, the busy streets of the recently founded Alexandria were peopled with, among others, a small cluster of mathematicians, drawn from throughout the Greek world to this research haven of learning and science. Noteworthy for its famed library and museum (a building dedicated to the nine Muses), the city benefited even more from the patronage and encouragement of the early Ptolemies.[31] In Alexandria, mathematicians practiced their craft in a more public venue and created a network of corresponding colleagues throughout the Eastern Mediterranean. They posed and solved both theoretical and, occasionally, practical problems with geometrical methods and, most critically, devised methods of proof in a fashion mathematicians still employ—all with virtually no mention of the religious dimensions of the earlier, and later, Pythagorean tradition. This blossoming of Greek mathematics proved to be of relatively short duration, however. Over subsequent centuries the arrival of the Romans and consolidation of their empire, the transformation of Hellenistic culture, and the slide into what historians call "late antiquity" was accompanied by the waning of Hellenistic mathematics from its sophisticated, third-century heights, with the noteworthy exceptions of the later Alexandrians, Diophantus (ca. 201–ca. 285), Pappus, and Ptolemy.[32]

Throughout all this mathematical activity, Greek geometrical practices appeared to reflect explicitly both the cardinal and ordinal principles. Numbers themselves were often depicted by geometrically arrayed points, actually pebbles in the Pythagorean tradition, as with triangle (1, 3, 6, 10 . . .), square (1, 4, 9, 16 . . .), and oblong (2, 6, 12, 20 . . .) numbers. (See Figure 3.1.) Thus classified, these "figurate" numbers were combined in various ways as Pythagoreans developed *arithmetica*, now called number theory. Particular attention was paid the properties of odd, even, prime, and composite classes of individual numbers, and especially the "perfect numbers," those numbers that equaled the sum of their divisors—for example, $6=3+2+1$, and 3, 2, and 1 are all divisors of 6. Conversely, the figurate numbers also came into play when constructing the geometrical figures themselves. A point was typically associated with the number 1 or "unity," while a straight line, being determined by two points, was tied to the number 2. Continuing in this fashion, it took three numbers to make the smallest plane figure (a triangle), and thus 3 meant a plane, whereas the "square" 4 was the critical number identified with area or space. In this way the holy *tetractys* of 1, 2, 3, and 4 supplied the foundational numbers of geometry.[33] Further, cardinal correlations often took the form of a one-to-one correspondence between numbers and line segments or points on a line, while numerical methods of calculating were occasionally used in treating geometrical figures (for example, the familiar 3-4-5 right triangle associated with Pythagoras), in devising ratios (for example, approximating the constant π, the ratio of a circle's circumference to its diameter), and in computing areas, volumes, angles, congruencies, and a whole host of geometrical problems. The cardinal correlations between line

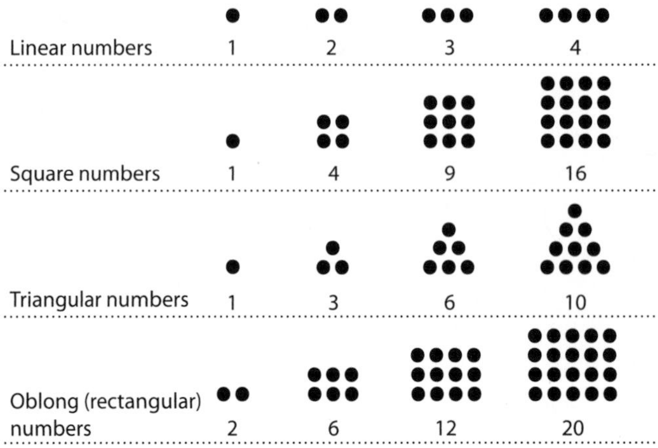

Linear numbers	1	2	3	4
Square numbers	1	4	9	16
Triangular numbers	1	3	6	10
Oblong (rectangular) numbers	2	6	12	20

FIGURE 3.1. Greek figurate numbers.

segments and number units connected the geometrical and arithmetical worlds. Or at least it would seem so at first regard, but incorrectly, as we shall soon discover.

We can also discern the ordinal feature of recurrence in the ideas of continuity and infinity embedded in Greek geometry, beyond the ordinality already noted in Greek letter-numerals. Recall, ordinal recurrence not only refers to a number's place or ranking in the natural sequence of magnitude but implies as well the indefinite continuation of counting, adding successive units without end, or calculating, such as carrying pi (π) to the nth degree. To imagine the unending repetition of an operation such as adding units requires an enormous mental leap into the unknown—in fact, into the infinite. This was particularly wrenching since early numeracy was created mainly to keep count of finite collections of things, or to measure and calculate figures of finite size and shape, no matter how large the collection or figure.[34] The diagonal of a square, for example, is a finite line, perhaps a quite short one, but can only be represented by an irrational number, the $\sqrt{2}$ in our notation, whose calculation is never ending. Still now, wherever confronted, infinity boggles the sensible mind, even one that counts.

Seemingly then, numeracy's hallmarks of correspondence and recurrence suffused Greek geometry, as well as the Greek counting system and arithmetic. Yet curiously enough, we find that geometry and arithmetic came to be organized as completely separate disciplines of knowledge. Why was this so, given the common characteristics that apparently united them? Shrouded in myth, a part of the story has to do with the very discovery of irrational numbers, and thus incommensurable ratios, among the Pythagoreans. Some classical sources speak of the cult's horror at encountering the irrational, which shattered their belief in a rational universe whose reality was structured by the whole number ratios of the holy *tetractys*. Horrified, they withheld the soul-rending discovery for over fifty years. Other sources attribute the discovery to one Hippasus of Metapontum, a Pythagorean of the fifth century BCE, who then made it known to the world. For revealing the secret, he was drowned by the gods, or perhaps other Pythagoreans.[35] Fables notwithstanding, the Pythagoreans' recognition that the relation between the sides and diagonal of a square could not be expressed in simple integers was critical in establishing the incommensurability of multitude and magnitude, of arithmetic and geometry as distinct categories of knowledge.

Another part of the story can be explained by recalling a bit of Greek philosophy. From the Milesians and Eleatics on (beginning in the sixth century BCE), philosophers dealt with a problem known then and still as the one and the many. Among the earliest of philosophical issues, the question was whether there existed a single reality or stable substratum (the one) behind all the diversity and change we experience, or whether each moment of our diverse and changing experiences (the many) was unique and different from every other moment.[36] Zeno (ca. 490–ca. 430 BCE), the Eleatic philosopher and student of Parmenides, whom we met earlier, presented the problem mathematically. In so doing, he perceived an incongruity between the one, which he expressed as the continuity of a line (a whole), and the many, which he identified as discrete points (the parts) that composed it. He exploited this incongruity to create his famous paradoxes. One known historically as the "dichotomy" demonstrates the issue. To cross any finite distance, one must first traverse half of it before crossing all of it, but first one must traverse half of the half, half of that half, and so on. This repeats an infinite number of times; so, it follows, one can never cross a finite distance, or indeed move at all. As with the dichotomy, Zeno's other paradoxes revolved around the apparent irreconcilability of the infinite repetition of spatial points or temporal instants with finite continuities of length and duration.[37]

How could finite, continuous spaces be reconciled or correlated with an infinity of successive, discrete repetitions or points? The Greeks had no answer, nor did they possess any language outside conventional literacy with which to formulate one.[38] And from this absence emerged the notion that separate perceptions of continuity and discreteness produced distinct and incommensurable categories or genera of knowledge. Put another way, mathematical procedures themselves were subjected to the classifying abstractions of language and literacy. One began with a perception of continuity or magnitude, a feature common to things such as lines, spaces, shapes, objects, figures, time, and place. For Aristotle, continuity possessed, further, the identifying characteristic of a "common boundary" that joined a continuous whole (say, a line) and its parts (the segments that composed it). This feature made all continuities "infinitely divisible," since lines forever divided in half would still be lines . . . short ones, but lines nonetheless. Phenomena so identified as having common boundaries among their parts and as being infinitely divisible were then subsumed under 'geometry' as that category of mathematical knowledge treating continuous spaces.

By contrast, discreteness was understood as a commonly perceived property of distinct numbers, individual objects, collections of units, counting, and the like. This perception permitted placement of all such objects in the class of arithmetical phenomena, of multitude. Aristotle actually claimed that both number and spoken language were subclasses of 'discreteness' because neither numbers nor word syllables could be joined "together at a common boundary." Each number and each syllable remained self-contained, separate, determinable. Four anythings were always demarcated as 'four,' never 'five,' whereas line segments and points could never be so demarcated. Ortega y Gasset has termed these perceivable qualities the "extension-thing" and the "number-thing" in reference to antiquity's penchant for classifying real objects or things.[39]

Continuous lines or spaces, therefore, remained incommensurable with the ratios constructed from finite, discrete, self-contained numbers.[40] Once more, the master classifier, Aristotle: "Arithmetic is about units and geometry is about points and lines." And one cannot "prove anything by crossing from another genus—e.g. something geometrical by arithmetic."[41] Even though the cardinal and ordinal features of numeracy permeated both these mathematical subdisciplines, the Greek classifying mind kept them separate from one another, with momentous consequences. Distinct genera of the yet more general mathematics, they were later codified by Boethius and sent into the Middle Ages as the quadrivium, a grouping of mathematical topics that dated from the time of Pythagoras.

* * *

A final corroboration of mathematics' subordination to the classifying temper comes to us from a rather unlikely source, on the surface at least, the Alexandrian astronomer Claudius Ptolemy. The reigning astronomer in Islam and Christendom until the Copernican revolution, Ptolemy's major work was *The Mathematical Compilation* (*Hē Mathēmatikē Syntaxis*), better known historically as the *Almagest,* "the greatest," as it was called by Arab scholars.[42] Like Euclid, Ptolemy was not so much an original thinker as a systematic synthesizer, a highly proficient geometer who set out to "recount what [had] been adequately established by the ancients."[43] For him that meant the findings of Apollonius of Perga and Geminus of Rhodes (fl. first century BCE), among others, but primarily the work of Hipparchus of Nicaea (ca. 190–ca. 120 BCE), generally considered the "greatest astronomer of antiquity."[44] Heir to the Hellenistic "golden age" in geometry and astronomy, Ptolemy folded its mathematical accomplishments and vigorous debates about the heavens and planetary motion into the classificatory science of Aristotle, albeit with some noteworthy variations in emphasis as regards both the cosmology and the categorical divisions of knowledge. In his eclecticism, he also bore witness to a strongly tinctured Pythagorean and Platonic heritage with an insistence on studying "divine and heavenly things" through the medium of mathematics.[45] For Ptolemy, the mathematical mapping of the heavens and the study of celestial order entailed at once the pursuit of "aesthetic, moral, and physical characteristics" of the divine cosmos, all of which elevated mathematics as the primus inter pares among theoretical sciences.[46]

Ptolemy began his great work by separating the "theoretical part of philosophy from the practical," reaffirming a tradition of some eight centuries' standing in the Greek world, the exception being perhaps the aforementioned mathematics of the third and second centuries BCE. Moreover, the division carried moral and religious overtones reminiscent of both Platonic and Aristotelian manners of philosophizing. One should value the theoretical subjects, Ptolemy declared—echoing Plato's advice to the philosopher-king—for the "moral virtues" associated with a "progress" in striving for a "noble and disciplined disposition," a "love . . . of the eternal and unchanging" that elevates the soul. In these passages he spoke somewhat disparagingly of the practical side of philosophy, insinuating that "continuous practice in actual affairs" impedes the "instruction" required for a theoretical understanding of the universe.

The religious tenor of these remarks was augmented by Ptolemy's affirmation of Aristotle's three-tiered cosmology. In the empyrean, "somewhere up in the highest reaches of the universe," resides "an invisible and motionless deity," he wrote, "eternal and unchanging," the "first cause" of all motion, heavenly and earthly alike. The deity was the proper study of theology, the first of the theoretical divisions of knowledge according to the master, Aristotle. At the center or "middle" of the universe was situated our familiar earth, "sensibly spherical" and stationary. Here, below the "lunar sphere," dwelled "material and ever-moving nature," the order of "corruptible" bodies, the world of "up and down" (linear movement), the proper study of physics, a second branch of theory.[47] Departing from a strict adherence to Aristotle, Ptolemy insisted that while both these divisions of our learning were indeed theoretical, they should "rather be called guesswork than knowledge," theology because of its invisible and "ungraspable nature," physics because of the "unstable and unclear nature of matter." Of both subjects, he declared, there could be "no hope" of agreement among philosophers.[48]

Not so with mathematics, which stood as the largest, centrally positioned panel of Ptolemy's cosmological triptych. Physically between the empyrean and earth lay the heavens, and Ptolemy's eyes sparkled like the stars themselves as he gazed on their movements, which he could actually see, unlike the invisible deity. But like the deity, the heavens too were eternal and worthy of our most supreme attention. Aristotle had explained why. Circular motion was perfect, heavenly matter ethereal, celestial phenomena unchanging. Light and quintessential, heavenly bodies were the ideal objects to be examined by mathematics, the division of theoretical knowledge corresponding to the cosmic space between the divine and the earthly. Three cosmological subjects, then (empyrean, heavens, earth), corresponded to three theoretical classes of knowledge (theology, mathematics, physics). For Ptolemy, the mathematical study of the heavens was the most special, all the more so for its "unshakeable knowledge," so long as one approached it "rigorously."

Because it investigated the "eternally unchanging," mathematics was itself "eternal and unchanging . . . in its own domain." The technical manipulations used in astronomy thus uplifted us to "contemplation" of the divine, a beautiful and noble purpose served ideally by mathematics.[49]

Pondering the heavens meant for Ptolemy abstracting from our sight the geometrical patterns we find in the celestial sphere. Euclid had constructed an entire science with the information abstracted from space—visual drawings of shapes, figures, and forms. Ptolemy applied these abstract tools to the visual information captured from the movements of heavenly bodies. Their patterns of motion were perfect, in the sense Aristotle intended when speaking of celestial motion. To make them evident, Ptolemy formulated exact accounts of the positions of heavenly bodies, both of the fixed stars that circumnavigated the earth daily and of the planets that additionally moved along the ecliptic.[50] For us, these are all motions of bodies in time; for Ptolemy, they were motions of bodies out of time, at least in our earth-bound sense. They were in time with celestial harmonies and depicted as positions in space. Time and eternity joined together in the heavens, so to speak, and being eternal, heavenly time was therefore timeless, the "moving image of eternity."[51] The images and conceptions generated by Ptolemy's cosmology permeated the religious worldviews of Christianity and Islam for centuries to come.

The fly in Ptolemy's ointment, as in that of every other ancient astronomer who sought explanation of celestial circular movements, resided with actual observation. Babylonian stargazers had created numerous catalogs of star movements, which had been updated by Hipparchus and subsequently by Ptolemy himself. Ptolemy even devised an early type of armillary astrolabe, which he used to measure sidereal and planetary movements and which allowed him more accurate sightings of the discrepancies other astronomers had previously observed. Theory notwithstanding, the planets did not move in perfect circles but exhibited slight deviations from the round. They also seemed to speed up and slow down as they pass through the perigee and apogee of their orbits. Some even appeared to jump backward in the course of their periodic journeys, exhibiting the phenomenon known as retrograde. Even to the naked eye the heavens certainly did not look perfect. The task at hand was to figure out how to make observations fit the Aristotelian, geocentric theory and its commitment to perfect, circular motion in the heavens.

To this end Ptolemy applied the constructions of Euclidean geometry and with them a commanding assumption: mathematics need not—in fact, could not—discover anything new, but only "save the appearances" or "save the phenomena." Although a theoretical science with its sights on divinity, mathematics thus came equally to serve as an instrumental means of rationalizing what was already known through definitions based on observations—namely, that the stars and planets moved in circular and eternal orbits. While the first use of the actual phrase "saving the phenomena" actually comes from Plutarch (ca. 46–120), not until Ptolemy's work did the practice gain general currency.[52] Throughout his treatise Ptolemy used mathematical (geometrical) models or "hypotheses" to account for observable planetary phenomena. His models utilized a number of geometrical devices—in particular, the eccentric, the equant, and the epicycle—with a high degree of precision and predictability in their plotting of the circular motion belonging in theory to the heavenly bodies. In Ptolemy's hands, "saving the phenomena" fortified both the cosmological separation of celestial and terrestrial physics and the categorical "incommunicability" of mathematics and physics, keystones of the Aristotelian, premodern, and classifying world.

This use of mathematics was a distant cry from the "Eureka" moment that had led Archimedes to investigate the mathematical properties of buoyancy or the static forces of the lever and balance. And it was an even farther cry from Viète's "analytic art" that would reveal the launching of the Scientific Revolution centuries later. In furthering the categorical separation between mathematics and physics, it was the mathematical reinforcement and explication of Aristotle's dictum that "all teaching and all intellectual learning come about from already existing knowledge."[53] A flying buttress to the religious, soul-elevating purpose of mathematics in the Middle Ages, "saving the appearances" would remain the reigning objective of scholastic mathematicians and astronomers throughout the Renaissance until Copernicus's challenge.

* * *

Masters of space, slaves to time, the Greeks in their mathematical world used ruler and compass to construct and manipulate a world of fixed spaces, constructions rationalized through a series of demonstrative proofs, which were based on definitions formulated in conventional language.[54] Throughout these operations, the idea of space was abstracted through visual diagrams from the sensory experience of space. In their quest for eternal, timeless truth, the independent reality of this "symbolic space" became literally an ideal arena for the practice of Greek rationality. Not so with time, which remained cyclical and "crushingly static" for not only Greeks but also humans throughout the premodern world. Time, of course, was experienced and kept track of in various ways, but not abstracted, not considered an independent reality, and certainly not measured or analyzed.[55] Time accompanied motion, and motion came through the senses, whose abstractions yielded the fixed and timeless classifications of knowledge. Aristotle elicited the logical conclusion: "In mathematics . . . motion . . . is a fiction."[56]

As previously mentioned, the abstraction of information from the momentary ephemera of the flux surrounding us involves the twofold movement of (1) drawing away from experience, observing it from some critical distance or perspective; and (2) pulling or dragging out of experience some indwelling form or pattern, a "mental extract." Complementing literacy, early numeracy provided basic abstractions, fixing information in the symbols of numbers and shapes (figures or diagrams). Further abstractions ensued as mathematical minds reworked these products from a critical perspective. For Euclid and others, in particular, this meant drawing out "theoretical" or "deductive geometry" from "mensural" or "empirical geometry."[57]

The Greeks were marvelously and uniquely adept in abstract logic and deductive reasoning, in ways other ancient cultures simply were not.[58] Still, as long as the higher-level abstractions of mathematics were keyed to reworking the existing lower-level constructions of counting and measuring, there were limits to how far their mathematical and logical prowess could take them. A number system based on the symbolic notation of the alphabet and a geometry based on the linear constructions of ruler and compass remained incapable of contriving functional relations between symbols, and thus imposed restrictions on the manipulation of information. Our hindsight allows us to recognize that absent from Greek conceptions were symbols for operations, a notational system of place-holding ciphers, instructions from the void (the empty cipher '0'), the technical means for connecting arithmetic and geometry, the abstraction of time as a symbolic and measurable reality, and the related notion that mathematics could be an instrument of discovery, not simply one of post hoc rationalization, when applied to physical phenomena.

All these limitations were occasioned by the information technology of early numeracy. But the most stringent and profound of early numeracy's confinements lay with its subordination to the core assumption of literacy and writing, that abstractions elicited 'things' from the flux of sensation. "Number," said Aristotle, could be "used in two ways: . . . what is counted . . . and that with which we count." But "whatever a number is," he continued elsewhere, it is "always a number of certain things."[59] Similarly with lines and spaces: "In the line which encloses the circle, being without breadth, two contraries somehow appear, namely, the concave and the convex." In this and similar passages, lines were "pulled" or "dragged" out of experience as quasi-objects of their own, not merely as the boundary or edge of a physical disc or surface. Mathematics historian John J. Roche summarizes here for Aristotle, Euclid, and other Greeks concerned with the abstractions of geometry: "Points, lines and surfaces were also understood as geometrical objects in their own right and not simply as structures in idealized physical bodies or processes."[60] In short, numbers and spaces, multitudes and magnitudes, all were intrinsically tied to things, whether "ideal" or "material." Different kinds of things, often defined in Aristotle's wake as discrete and continuous, comprised the two main divisions of mathematics. Because of these associations, mathematical objects always carried linguistically based "meanings" in premodern worlds, and on this deep premise—that its theoretical objects were intrinsically connected to words and things and meanings—mathematics continued its odyssey into the Middle Ages.

CHAPTER 4

Thing-Mathematics

The Medieval Quadrivium

> Mathematics rides proudly along on the four-wheel chariot of its Quadrivium.
>
> —John of Salisbury, *Metalogicon* (1159)

ON THE EVE of the first millennium, and just before an army of the teenaged emperor Otto III (980–1002) promoted him from the snake pit of ecclesiastical, imperial, and Roman politics to the papacy as Sylvester II (999–1003), Gerbert of Aurillac (ca. 946–1003) wrote a letter to a friend and former student, one Adalbold of Liége (later appointed bishop of Utrecht). In it the monk, abbot, archbishop, and soon-to-be pope took some pains to address Adalbold's earlier request for more "geometrical figures" and, presumably, explanations of how to understand some of their features. Gerbert laid out a problem involving the area of an equilateral triangle. Manuscript sources had supplied two ways of figuring it, he noted, one arithmetical and one geometrical, but these ways had netted widely divergent areas for the same triangle, "which is impossible."[1] Until recently he had followed the "arithmetical rule," a procedure cadged from Boethius's account of "triangular numbers" whereby one multiplied two of the sides, added the third, and then divided the results by two. When applied to a triangle with sides of 30 feet, an example drawn from an earlier letter, the arithmetical rule yielded an area of 465 square feet $\left(\frac{(30\times 30)+30}{2}=465\right)$. But evidently Gerbert had seen how bizarre and bogus this was, for he had come to prefer the "geometrical rule," which he thought more closely tied to the famous theorem of Pythagoras and certainly believed more accurate. In another letter (now lost) he had apparently invoked the same geometrical "theory of the height" to calculate the area of the same triangle as roughly 390 square feet, closer to our modern computation of the product of the base and height divided by two.[2]

Not content merely to compute the answer, Gerbert proceeded "to exemplify . . . in smaller numbers" the method for finding the height and thus area of this "mother of all figures."[3] He used a triangle with sides of seven feet each (one foot for each deadly sin and each salvific sacrament?), and he explained with pastoral and professorial patience the "little diagram" accompanying his exposition. "I believe that you know . . . we . . . use only square feet" for measuring areas, he began. Accordingly, one could see the total number of one-foot squares when they were superimposed on the triangle. Counting the squares produced the area after the arithmetic procedure of Boethius. There are twenty-eight squares, and $\frac{(7 \times 7)+7}{2} = 28$. (See Figures 4.1a and 4.1b.) But here lay the mistake, for the diagram also plainly showed the twenty-eight squares of feet "extending beyond the sides" of the triangle, demonstrating that the arithmetical rule "takes the half with the whole [of a square foot]," not to mention the extra square foot on top. By contrast, the "geometrical discipline" allowed one to compute "what is shut in the lines [of the triangle]." Further, with the same diagram one could also easily see and count the "perpendicular" (height), which is six squares tall. "Lend your eyes to it," he intoned, and you will recognize the "universal rule," which is to consider the ratio of the height to the base as six is to seven. With this ratio in hand and any given units of measurement for a side—say, feet—one simply applied the Pythagorean theorem and calculated the area.[4]

That Gerbert was the era's premier mathematician speaks volumes about the low level of mathematical learning from the early Middle Ages through at least the twelfth century.[5] Others were at an even greater loss when it came to numbers and spaces. Around 1025, for instance, Radolph and Ragimbold, two learned monks of the cathedral cities of Liége and Cologne, respectively, exchanged a series of letters on mathematical topics they had come across in reading Boethius's commentaries on Aristotle. They were not mathematicians, but as former students of Fulbert of Chartres (960–1028), who himself had studied with Gerbert, both were inquisitive and keen to advance their learning. They puzzled over Boethius's words. In particular, they struggled with the theorem that the interior angles of a triangle were equal to two right angles. "Interior angles" of a triangle? Neither had the slightest notion of

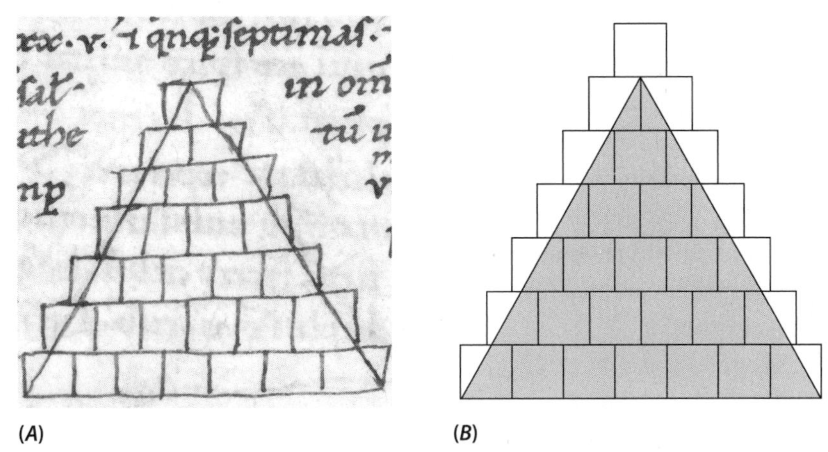

FIGURES 4.1A AND 4.1B. Gerbert's attempt to solve the area of an equilateral triangle. Figure 4.1a is from a twelfth-century manuscript containing Gerbert's geometry. Figure 4.1b is a modern reconstruction based on his letter to Adalbold (999). Note the discrepancy. In Figure 4.1a the copyist has extended the apex of the triangle to the top of the squares, essentially creating an isosceles triangle (had the squares been drawn accurately), whereas in Figure 4.1b, following Gerbert's description, the top square extends above the triangle. Gerbert explained this as the means of obtaining the six-foot height of an equilateral triangle with a seven-foot side.

what that meant, something even the mathematically averse among us recognize. Moreover, they had likely come across an early astrolabe in their education and even though the geometry of the instrument lay far beyond their combined ken, Radolph wanted one and tried to have it made. It was a sign of the age.[6]

Gerbert stood out in this context. Rising from humble origins, probably a free-holding peasant family in the Auvergne, he established himself as one of the period's brightest intellects and most sought-after teachers.[7] His early education at the Benedictine monastery at Aurillac revealed his precocity, and was followed by three formative years in Catalonia, which exposed him to Arabic science and culture and to the quadrivium.[8] His own skills and talents catapulted him from there into a teaching career at the cathedral school of Reims, where he remained off and on for over two decades, and into the highest echelons of ecclesiastical and imperial service and controversy. At Reims he taught the entire quadrivium, along with the trivium, shining in the former with his own abacus (first displaying in Europe the Hindu-Arabic numerals from 1 to 9), an armillary sphere (or early version of an astrolabe), a single-stringed monochord (for music instruction), and other practical teaching aids for introducing young students to mathematics. Although the last years of his life were given over almost entirely to providing advice to a succession of Ottos in what would later be known as the Holy Roman Empire, to helping establish the Capetian lineage of kings in France, and to serving as the titular head of Latin Christendom, he retained an abiding interest in the science of his era, earning him a contemporary sobriquet as the "Scientist Pope."

Despite its apparent practicality, Gerbert's recipe for mathematical pedagogy reaffirmed the fundamental features of antiquity's "thing-mathematics" for the medieval, classifying temper, a mind-set that continued to separate theory from practice and arithmetic from geometry, all the while interweaving religious significance into the world of number. The Greeks had created the categories. Practical computations were sharply distinguished from theoretical mathematics. The former went by the names of "logistic," a "science which dealt with numbered things," and "geodesy," the practical measurement of surfaces and volumes. The latter embraced arithmetic proper (basically the theory of numbers) and geometry (the theory of spaces).[9] Moreover, as we saw in Chapter 3, the Greeks had planted theoretical mathematics firmly in the garden of philosophers and their concerns with words and things, within the technological framework of alphabetic literacy. Plato held that mathematical ideas, like all others, were generated from within the soul or mind, as a sort of intermediary between the "pure forms" of intelligibility and the perceptibles of sensation; Aristotle argued they were elicited from our sensations as we interacted with the world without. Yet, regardless of their source, ideas were associated with definitions and things, material or intelligible, and along with the things of words (*logos*), the things of mathematics (*arithmos*) were the play, the discrete things of arithmetic multitude and the continuous things of geometric magnitude.

As in Plato's Academy, where philosophers used mathematics to pursue higher-level learning, so too in Gerbert's classes were the practical topics directed to lofty goals. Problems were not taught for the principal purpose of applying mathematics to everyday, mundane matters, such as measuring the height of mountains and towers or the depths of wells and rivers might suggest to us.[10] Rather, such topics offered an introduction to the quadrivium, that "place where the four roads meet" (from the Latin *quattuor,* 'four,' and *via,* 'road'), the "fourfold way" to understanding the "wisdom of things," a mathematical roadway to heaven, as it were. Boethius had coined the Latin term, but the four subjects themselves—arithmetic, music, geometry, and astronomy—reached far back to the ancestral days of Pythagoras and to the religious and philosophical implications he had conferred on mathematics in the subsequent Platonic and Neoplatonic traditions. Gerbert followed these conventions, leaving little doubt that the main objective of mathematics instruction was to illuminate the "power of numbers[, which] contained both the origins of all things in itself and explained all from itself."[11] A divine power indeed.

Nor did Gerbert shy away in the slightest from what we recognize as the enchantments of numerology. In a legal case, for example, he argued that the Nicene Council was such a "privilege of the Divinity" that any deviations from its "resolutions" were to be considered as "absolutely invalid." To justify the claim, he cited the opinion of Saint Ambrose (340–397), who had written that "not by human effort" or "arrangement" did 318 bishops convene at the council. The numbers themselves proved it a divinely sanctioned synod, a "proof" based on the alphanumerical system of Greek counting, which used letters for numerals. In Greek, the number 318 was written as *τ ι η*. The first letter (tau) equaled three hundred and suggested the ancient cross; the second letter (iota) was ten and stood for the initial letter in Jesus's name (Iesu); and because the third letter, eta or eight, was considered the first "full" number, it symbolized that the bishops were in a full, ecumenical council—the Council of Nicaea in 325 CE.[12] That such numerological arguments as this one could nestle comfortably with other strategies in a legal brief attests the thorough intertwining of divine enchantments and numbered things.

Practical or theoretical, arithmetical or geometrical, natural or supernatural, the world of number thus continued its historical embedding in literacy's world of words and things well into the Middle Ages and beyond. The first-century neo-Pythagorean Nicomachus of Gerasa had encouraged the study of mathematics as the way to wisdom, the "science of the truth in things." Shamelessly pirating from Nicomachus's widely popular *Introduction to Arithmetic,* Boethius sent the idea of "number . . . [as] the principal exemplar in the mind of the creator" into the classrooms of medieval instructors by means of his own heavily copied and used textbook (*De Institutione Arithmetica*).[13] Even more fundamental, Boethius reinforced the ties between mathematical study and things or objects by referring to arithmetic and geometry alike as expressions of the "denominatives of numbers." Used frequently in medieval discussions of mathematics, the term 'denominative' (*denominativa*) and its cognates meant giving a "meaningful name to an object." The big "thing" or "object" for Boethius was "number" itself, which in turn brought its "logical force" to the rest of all the named objects of mathematics.[14]

Associating "things" naturally and inherently with mathematical thinking serves up a medieval notion difficult for us moderns to digest, at least initially. For if we think about mathematics at all, it's generally to imagine an abstract, logical universe built of odd and mysterious symbols (at least odd for nonmathematicians), which have only remote and extrinsic connections to our immediate experience and surroundings. Not so for the few medieval thinkers who contemplated number. Boethius had stressed the "substance" of numbers; his scholastic successors perceived the connections between substances and numbers as apparent, immediate, and intrinsic.[15] A deep and nearly instinctive impulse among humans to attach thought to the world outside the mind probably helps explain these associations, if we may conjecture a bit, for accurate correlations of thought and surroundings have long provided a critical advantage in our species' survival and evolution.[16] Yet whatever our causal speculation, scholastics certainly believed that "being" of some sort was woven into the fabric of abstractions, a highly powerful and robust notion, which captured what philosophers today describe as the "aboutness" or "intentionality" of thought.[17] "Being" and "beings" bound thought to the world. This remained true even for nominalists who denied the existence of "universals" but correlated singular statements with singular, external objects. In this milieu, mathematicians too absorbed "being" into their abstractions. And the assumption of an intrinsic connection between mathematical thought and the outside world continued well into the Renaissance and early modern period with the insistence of many intellectuals that numbers and shapes attest extramental, frequently divine realities. These they found in many places: in the harmonies of music and the heavens, in the mysteries of the holy *tetractys,* in the golden number, in the meanings of people's names, in the numerical significance of the liturgical year, in deciphering the numbers in scripture, or even more plainly, as the Bard put it, in Falstaff's search for good luck, that "divinity in odd numbers."[18]

External things or realities, then, permeated premodern thing-mathematics. The quadrivium complemented the study of the trivium's linguistic arts as a means for understanding the world of things; for a few scholars and theologians, *arithmos* (number) shouldered an intellectual and spiritual weight comparable to *logos* (word). In their compulsion to organize and classify, to name and define, the schoolmen approached mathematics and its things in two distinct ways, the themes of this chapter. First, several philosophers and theologians, most notably Thierry of Chartres and Nicholas of Cusa, focused on the religious significance of the quadrivium, offering in the process nothing new at all about mathematical thought per se but developing "mathematical theologies" or religious implications from the world of number. Their elaborate definitions of sacred, mathematical things often overlapped with more common, metaphorical and allegorical interpretations of number that gave voice to a widely ranging host of enchantments, a medieval magical realism, as it were. If mathematics now can be understood as the very antithesis of magical thinking, in the Middle Ages it was often the handmaiden of magic, astrology, divination, and even pursuit of that holy grail of thought and reality known as the philosopher's stone, a substance sometimes held to be a "prima materia" generated by number itself.

Second, in the thirteenth and fourteenth centuries another coterie of academic intellectuals developed what we may call a qualitative mathematics as they sought to refine definitions of things by introducing quantitative means for determining the degrees or intensities of various substances or qualities. Although deeply ensconced in the classificatory world of words and things, scholars such as the Oxford Calculators and their Continental counterparts actually did formulate out of their "home-grown interest" innovative techniques of quantitative analysis.[19] And some of these techniques suggested new directions for investigating nature, especially as regards questions pertaining to motion. But these new directions would be pursued only when the context had shifted decidedly, only when a new numeracy had been devised in the arena of practical mathematics and subsequently assimilated by natural philosophers.

Whether the objects of mathematics were expressly divine things, enchantments, or the refinements of earthly substances and qualities, the underlying assumption of the "quiddity" or "thingness" of mathematics was strengthened at every turn in the Middle Ages. This assumption imposed a formidable and conservative impediment, we might even say barrier, to the creation of modern, abstract and relational, mathematics. As we shall explore in later chapters, breaching the rampart of "thingness" in mathematics required an enormous wrenching of consciousness as regards the creation and use of abstract, empty symbols and their connections to the world. Thus, all the while an information explosion and the rupture of classification were steadily chipping away old ways of thought from the inside, a quite noisy occurrence within the kingdom of letters especially after the appearance of the printing press, a stealthier, external assault on the bastion of quiddity was being mounted by the incremental creation of a new information technology. Until the moment of breach during Galileo's lifetime (and even for quite some time thereafter), thing-mathematics and its associations with divinity and enchantments would continue to supply an ongoing chapter in the chronicle of science's harmony with religion, the backstory against which the rupture between them would occur.

* * *

As with so many arenas of Christian thought, the writings of Augustine set the stamp on later endeavors to understand God through the auspices of number.[20] In the early, religiously syncretistic years of Christianity's history (second through fifth centuries CE), some writers had sought to bridge Athens and Jerusalem by relying on themes drawn from Pythagorean number mysticism and Neoplatonic perspectives in order to forge their version of an orthodox theology. These and comparable, later undertakings found inspiration in the Bible's book of Wisdom. There, the writer (not Solomon, as was long held, but probably a well-educated, Hellenized Jew living in the first century BCE) had spoken at some length of God's providential presence in history, proclaiming that among other actions God had "ordered all things by measure, number, and weight" (Wisdom 11:21). The passage lent scriptural support to decoding the structure of the universe through number, of understanding *arithmos* as the mediating reality between humans and the divine. Augustine took note.

As a young man of thirty-two, having recently fled from heresy and lust into the arms of Christianity, Augustine fell under the sway of number in his formative writings. Declaring that "the liberal arts lead the mind to God," he explained that one can come to knowledge of the "hidden divine order" by understanding the "simple and rational numbers." By means of these numbers, one could also gain a grasp of music, geometry, and astronomy, and achieve the same end.[21] Number itself was a "mental construct," he asserted, in effect implanted by God in the souls of humans, "ever present in the mind and understood as immortal."[22] At its origin was an irreducible unity, "one and unchangeable," the "monad" in Pythagorean tradition, Yahweh or God in the Jewish or Christian.[23] The one or "unity" was not actually a number itself but rather the source or generative power behind all numbers. Nor was two, the "dyad," a real number; rather, it was the basis of diversity and multiplicity in the world, the introduction of different things. (Recall that two indicated a line geometrically because a line connected at least two points or objects.) The first "real" number was three, the triune principle of God and therefore the Trinity, but also for Augustine one cornerstone of a spiritual-temporal duality. The twofold nature of Christ as divine and human expressed the dyad within the triad, while writ large the archetypal numbers three and four (the latter being the quadruple principle of mankind, associated with the first square number and space) combined to produce "seven, which is commonly used as a symbol of universality" or everything that exists. (Sometimes the Latin *universitas* translates as "the whole thing" or "totality.")[24] The numbers one through seven thus brought a soul from unity to universality. And with the numbers seven and twelve in hand as the sum and product of three and four, other symbolic correlations fell quickly into place: seven sacraments, seven deadly sins, seven days of the week, seven planets of the universe, twelve (4 × 3) signs of the zodiac, twelve hours of the day, twelve tribes of Israel, twelve disciples, and still reverberating in our own day, the twelve days of Christmas. Pythagoreans had long held the number ten as the image of cosmic unity, but Augustine gave it a Christian twist by combining the threefold Trinity of the Creator with the sevenfold hebdomad of the created.[25] In this version, as Augustine glossed the previously cited passage from Wisdom, God was supreme: "Measure without measure . . . Weight without weight . . . a Number without number, by which all things are formed."[26]

In his *Confessions* and other later writings, Augustine turned his back on the potential for idolatry he came to see in *arithmos* (including astrology) in favor of explicating a more Christocentric theology, one closer to *logos*, the word incarnate. At times he equated number and wisdom, *arithmos* and *logos,* and saw them as "identical," for both were "immutably true."[27] But number begat pride, encouraging belief in a determinism that subverted both God's will and human freedom, not to mention an exalted view of one's own intellect. Pride stood as the greatest of sins, the reason for mankind's fall from grace in Eden, the unvarying result of an enfeebled free will stained with original sin. One had ever mindfully to guard against it, including the pride of a purely philosophical *logos.* Thus did Augustine come to emphasize that Christ was the flesh-and-blood, human mediator between God and the world and not the emanation from a Neoplatonic monad. The humility of Jesus's appearance and sacrifice put all philosophical thinking into its proper, illuminative but not soul-saving place, subordinate to a "theology of grace."[28]

Nevertheless, numerological associations and explanations continued to pepper his exegetical writings, as well as his accounts of human and divine history, and he incorporated them frequently throughout his highly influential tome *The City of God.* By the time of his death in 430, Augustine had consolidated for future generations a twofold bequeathal connected with number. First, there were hints at the possibility of a full-blown, deeply mathematized theology drawn from the Pythagorean tradition and affiliated with the quadrivium, although these suggestions retained for Augustine a substantial degree of "ambivalence," which he had never fully sorted out and had nimbly sidestepped.[29] Second, in many of his works he passed on a host of earlier associations, as well as ones of his own invention, that metaphorically tied the world of number to a divinely animated universe, a spiritualized mathematics whose number symbolism resonated throughout Christendom.

Pushed to the side after Augustine's synthesis of Christian beliefs and classical, logocentric philosophy, mathematical theology appeared only sporadically during the medieval centuries.[30] It could be found in the somewhat isolated thinking of John Scotus Eriugena (literally, "John, the Irish-born Gael"), an erudite Irish monk who served at the French court of Charles the Bald, part of the Carolingian Renaissance. Eriugena was a singular exception to intellectuals in the early medieval period in that he read Greek and actually translated several Greek works. The most notable of these was the corpus known now as Pseudo-Dionysius, believed at the time to be Dionysius the Areopagite, the first convert of Saint Paul at Athens, but actually an unknown Christian Neoplatonist of the late fifth or early sixth century, likely a Christian follower of Proclus. Drawing on Greek as well as Latin sources, especially Augustine and Boethius among the latter, Eriugena crafted a cosmological vision in which all of reality "emanated" from an infinite, transcendent, and ultimately unknown God. Numbers flowed eternally out of this uncreated monad and spread further into its "self-created" reality through a hierarchical series of ratios and harmonies. These divided into two channels, one streaming directly into nature, the other into a dependent and "created" monad, identified as the human mind and its reason, which in turn could grasp the "arithmology" behind creation. As part of Eriugena's own reading of Wisdom, he even invented a "mathematical angelology," parsing celestial hierarchies with Pythagorean, harmonic ratios.[31]

Eriugena's construal of faith seeking understanding through a mathematized theology was but one of many versions of Neoplatonism's influence on medieval religious thought, and a minor one at that.[32] Few followed his lead into the mystical world of number until a few centuries later when several masters at the "school of Chartres" assumed the task. At the pinnacle of this group stood Thierry the Breton, or Thierry of Chartres, as he is more commonly known. Little is known of Thierry's life other than he probably studied at Chartres, taught in Paris during the 1120s and 1130s, and returned to Chartres as chancellor in 1141. He was apparently both brilliant and impenetrable, at once a devoted mentor to favored students and an irascible lecturer who did not easily suffer fools. A rumor circulated that Abelard had to leave Thierry's lectures because they were too difficult to follow, and he was known to expel the less capable students from his classes—a medieval version of the modern "teach the best, dismiss the rest" practice of some instructors. His best, among whom John of Salisbury for instance, praised him in glowing terms.

Unlike Eriugena, who did not venture very far into the mystical realm of number, or Gerbert, whose allusions likewise skirted a deeper exploration, Thierry plunged into a dense, metaphysical thicket with the categories of the quadrivium at the center of his theology. As had Augustine and Boethius, he too began with arithmetic, but went further and used it to explain the Trinity's three persons (technically, "hypostases") of Father, Son, and Holy Spirit as "unity," "equality," and "harmony" or "concord" (*concordia* in

Latin). Another word for 'divinity,' 'unity' (*unitas*) defined God the Father as the Creator and preserver of all being, the source of change, which always preceded created things just as the unity of "one" preceded and generated all numbers. In this way the arithmetical principles governing unity also explained God's creativity in the world. Numbers were derivative unities or entities participating in the One, and so too did creatures depend on their Creator to sustain them. Moreover, "unity" generated "equality," the second term of the Trinity, or God the Son, which introduced the universe of created things. The multiplicity of things occurred because unity was either multiplied by itself ($1 \times 1 = 1$) or multiplied by its "other" ($1 \times 2 = 2$, $1 \times 3 = 3$, and so on), operations that produced equalities. And knowing the equality of a created thing meant grasping its "definition" or "determination," understanding that the thing in question neither increases or decreases but rather remains in "proper measure" identical to itself, a thing of multitude or of magnitude.[33] Here was the link between the wisdom found in defining words (*logos*) and that pervading number (*arithmos*). Finally, Thierry depicted God the Holy Spirit as the harmony of unity and equality, which united the entire number realm and all of creation, harmonizing science and religion through number. Thus interpreted, for Thierry these Trinitarian principles far exceeded metaphor; they were the "essence of creation" itself.[34]

Starting from this account of the Trinity, Thierry then elaborated an even broader, more complex metaphysical rendering of the entire quadrivium. He built on Boethius's claim that arithmetic was first in the mathematical disciplines and that music, geometry, and astronomy all issued from it.[35] Combining this premise with his own spiritualized arithmetic of the Trinity, Thierry continued to explain the universe as a densely and divinely structured series of mathematical arrangements, reaching from the operations of arithmetic to the harmonies of music to the figures of geometry and to the heavens themselves. And he was only getting started. In his later works he propounded an even more opaque theory of "folding," which described how the "mutability of . . . matter unfolds through the diversity of plurality, [and] the divine form likewise enfolds into one and recalls to the simplicity of a single form in an inexplicable way." "Inexplicable"? Certainly. Unfathomable? For most, most assuredly. Behind this and comparably baffling formulations stood the reciprocal notions that God was the "source and ground" of the quadrivium and that the quadrivium in turn served as the prime mediation between God and humans, a "univocal hermeneutic identifying the divine One and the numeric One."[36]

It takes a special temper to get inside these dense thoughts and images. Very few in Thierry's day could manage it (not so many in our own times, either). Not surprising then, Thierry had little influence and but a tiny number of followers among philosophical and theological minds of succeeding generations. One of the more prominent few was Nicholas of Cusa, who drew heavily on Thierry and others in the Pythagorean and Neoplatonist traditions. Like his intellectual forbears, Cusanus, as he is generally known, created a thick, metaphysical rendering of reality tied to number, which integrated scientific and religious perspectives. One of his central tenets was the notion of "learned ignorance," or "how knowing is not knowing," which was a bit like the Zen koan of one hand clapping, especially given the paradoxical fact that, without irony, Cusanus wrote many learned volumes to propound his own ignorance. His central message with the phrase was that God could not be limited in any way and could be contemplated only through the "maximum learning of ignorance." With this opening, Cusanus spoke of God as "unity," the "simply and absolutely maximum, than which there cannot be anything greater."[37] He used the term 'absolute' in its Latin-based, etymological sense of "free from" (*absolutus*, from which our 'absolve') to depict God's infinity as freed from all restrictions whatsoever. It followed in his apophatic or negative theology that "there is no proportion between finite and infinite." Our finite ken has no cognitive concord with the infinite God. (Galileo, as we shall see, would turn this premise on its ear.)

Influenced by Thierry, Cusanus expanded the same "folding" imagery employed by the master of Chartres. Two processes governed reality, "unfolding" and "enfolding." Unfolding began with the unknowable God or unity and expressed the divine presence in uncountable manifestations as he unfolded his being mathematically in the world about us, the Creator in the creature. (For this reason, some thought Cusanus's views essentially a form of pantheism and thus heretical.) But this was not a one-directional process, for from the moment of their creation, the "unfolding" of God in time and space, all beings were simultaneously being "enfolded" back into the undifferentiated oneness of their divine source. God therefore was the infinite and "absolute maximum," both immanent in the world and transcendent beyond it, its alpha and omega. Even with the aid of mathematics, frail human thoughts could only measure the limited, unfolded realm of more and less, and thus could provide no knowledge of the true, enfolded essence of any created thing.[38] Number might well define reality as we can know it, but knowledge of its divine essence lay beyond our own learned ignorance.

As suggested, a full-blown mathematical theology, hinted at by Augustine, did not represent a mainstream of religious thought during the Middle Ages, requiring as it did a special sort of mind to penetrate its idiosyncratic, mystical, and spiritual opacity. But Augustine had also encouraged the more prosaic practice of seeing symbolic meanings in numbered things. In the richly animistic and enchanted world of the Middle Ages, number symbolisms and numerology abounded in parlance and practice. Even Thomas Aquinas, that fervent follower of the Philosopher, could rather casually toss off a numerological description of the 144,000 marked for ultimate redemption found in the book of Revelation (7:3–8, 14:1–5). As a multiple of 10, the 1,000 signified perfection, while the 144 resulted from 12×12, with one 12 denoting faith in the Trinity, distributed through the four parts of the earth (3×4), and the other 12 understood as the twelve apostles or twelve tribes of Israel.[39] To be sure, the prophetic book of Revelation easily lent itself to a plethora of number symbolisms, especially with its abundance of fours and sevens—the four horsemen of the Apocalypse, the seven-horned and seven-eyed lamb of God, the seven seals, the seven trumpets and bowls, the beast with seven heads, and the like. But in truth, one could select virtually any natural number and tease various religious meanings out of it, meanings attached to things or events.

Take the uncommon numbers 17 and 153 as quirky illustrations. In classical antiquity 17 was sometimes connected with warfare, heroism, and conquest, while in the Hebrew Bible it was paired with the Flood, which began on the seventeenth day of the second month and ended on the seventeenth day of the seventh month. Augustine drew out its theological implications for Christians, describing it as a combination of the Ten Commandments and the seven gifts of the Holy Spirit, which medieval theologians referred to as the unification of God's law and his grace. Things grew yet more intriguing when 17 was scrutinized in tandem with 153. The disciples of Jesus caught exactly 153 fish, as recorded in John 21:11. Why this number? Well, arithmetically, 153 is the triangular figure of 17, which can be expressed as $1^3+3^3+5^3=1+27+125$. It is also the sum of the integers 1 through 17, and, further, it can be interpreted as $3\times3\times17$. The schoolmen had choices. For 153 could be read as a symbol relating divine law and grace or as one relating the Trinity, the law, and the seven gifts of the Spirit or similar combinations of deific realities.[40] Many minds pondered such meanings.

Number symbolisms permeated philosophy, theology, and the literary arts, but they also assumed a prominent place in quotidian practices, in rituals, in magic, and to a lesser extent in astrology. Prayers written in the *Nun's Rule* (*Ancren riwle,* thirteenth century) mentioned the "four gospels," the "five wounds of Christ," the "seven gifts of the Holy Spirit," and the "ten commandments." Repetitions counted. A fourteenth-century popular writer and ascetic, Henry Suso (1295–1366), swallowed five times every time he drank, venerating the wounds of Jesus, and he cut his fruit into four pieces, signifying the Trinity plus a "fourth part for Mary." Doing or saying things three times figured commonly in religious rituals: stepping three times over a grave, then dripping holy water on it thrice and saying three Paternosters; using the sacramental oil of extreme unction to trace three crosses on the forehead of the dying. Magical incantations often appeared in threes, and threefold repetitions could be found time and again in Anglo-Saxon magical rites. Sevens, tens, and twelves likewise invested comparable meanings in ordinary practices.[41] Some in our own age continue to single out thirteen, still thinking it an "unlucky" number, but in enchantment Europe, virtually all numbers bore marks of meaning for the credulous.

Astrology too frequently carried such marks, although its presence in the Latin West often appeared controversial. Augustine and other Church fathers had denigrated the practice for its associations with mystery cults, its idolatry of astral divinities, and its determinism, which eclipsed human free will. Horoscopes were nonetheless frequently cast, increasingly so under the influence of sciences introduced from the Islamic world, where astrological traditions were more common. Most thought of astrology as an extension of astronomy, based as it was on the juxtaposition of stars and planets, whose positions and alignments were discovered through the geometry of the heavens. Its subject was influences, mainly of the stars over the course of human events, life, character, and, most commonly, health. As such, portions of it were believed amenable to the causal accounts provided by natural philosophy, particularly as applied to medical diagnoses and treatments, while other of its dimensions stressed the more highly contested issue of astral influences over the predictability of fate. The periodic resurgence of interest in Neoplatonism, with its belief in the divine powers of arithmetic and geometry, invigorated the practice of astrology well into the Renaissance and beyond. Yet behind its claims lay the dominant assumption of an embedded thing-mathematics: its objects and their meanings could only be captured and expressed by the words of conventional language.[42]

Further examples of the metaphysical and religious, allegorical and metaphorical, meanings of numbered things could be multiplied, producing a world (and book) virtually without end. Yet this brief sketch suffices to lead us to three unholy but inescapable conclusions. First, however they might be understood theologically by the likes of Thierry and Cusanus, or even Gerbert and Hugh of Saint Victor, the religious and enchanted uses of number contributed nothing to mathematics itself. Intellectuals and practitioners alike relied solely on those "simple and rational numbers" that Augustine had claimed as adequate for understanding the "hidden divine order." They needed go no further. Second, and to the contrary, the conceptual center of their explorations lay with definitions and correlating them with divine things. Thoroughly and deeply rooted in the classifying temper, the religious treatment of number became a way of cataloging ideal, divine entities, of creating, as Amos Funkenstein has winsomely put it, "an inventory of mathematical objects and their absolute properties."[43] Conventional language conveyed the meanings of numbers and their corresponding things, whether ideal or material. Thus 'three' referred to a collection of objects, but in religious minds that meant the Trinity, a collection of the most divine threeness of things: Father, Son, and Holy Spirit. The words interpreting the numbers mattered, not the numbers themselves. Thierry and Cusanus, among others, surely believed these entities had their counterparts in the real world outside the mind, a God-created world structured by number. But viewed from our own perspective, their definitions served primarily to buttress the quiddity or thingness of medieval thought. Finally, as a result mathematics remained, therefore, thoroughly subordinated to the information technology of literacy and to the classifying temper of medieval minds. From the substrate, letter symbols of data to the symbolic meanings of its terms and definitions, literacy governed the things of mathematics just as it did the things of words.

* * *

On reflection, it makes sense that medieval scholars, with their impulse to classify, would seek to put the objects of mathematics into their proper cubbyholes. Theirs was a world not of making new discoveries but of finding the correct categories for what was already known. And while new discoveries themselves may not have stimulated intellectual activity, refining categories surely did. Central to the scholastic method, "dividing the question" launched many investigations into terms and their scope of application. As they looked at their own words and definitions from a critical distance, the schoolmen, with their classifying temper, constantly recast and modified them, adding divisions, distinctions, and clarifications to the taxonomical arrangements they had invented. Critical adjustments were made constantly as they pursued the various logical, theological, and philosophical subjects that preoccupied them in their enchanted world of words and things. In these circumstances, some of the dons came to invoke mathematical arguments and demonstrations, and even to invent new mathematical techniques, the second theme of this chapter.

Many of these innovations occurred in the fourteenth century, when several lines of questioning led to the *via moderna*—as the emerging "nominalist" movement associated with William of Ockham was then termed—a new way of thinking about things. Although Ockham showed little interest in mathematical reasoning per se, his writings and those of other nominalists contributed greatly to the context within which the "quantification of quality" flourished. One of the central issues addressed by nominalists sprang directly from Aristotle's account of change, namely, explaining more precisely the alterations undergone by any given body. Ockham himself wondered about understanding the "fiction of abstract nouns" and how they correlated with the changing realities outside the mind.[44] How does one explain what happens when a body becomes hotter or colder, wetter or dryer, heavier or lighter, blacker or whiter, healthier or sicker . . . or when it moves from one place to another?

Comparable questions had been raised in earlier generations when Peter Lombard (of *Sentences* fame) asked whether "charity or grace, which is caused by the Holy Spirit, could not vary in humans because this would imply a change in the Holy Spirit." In response, most theologians held that individuals possessed greater or lesser degrees of such qualities because of their participation in them. In this "doctrine of participation," the quality remained the same, but the degree of one's involvement in it varied. John Duns Scotus developed an alternative approach, holding that a given quality itself, such as grace or charity, could be increased (or decreased) formally by adding (or subtracting) new and distinct "forms" or elements of the same quality. Here, the quality itself, not the degree of participation, became the variable. This was the "intension and remission of forms," which provided the logical language and framework for expressing the questions about change raised by Ockham and others.[45] Additionally, the terms "longitude and latitude" of qualities often encouraged a visual representation of variations. Such phrases, Nicole Oresme wrote, indicated "more or less in some way," as in "more white" or "more rapid."[46] In modern philosophical terms, matters of kind gave way to matters of degree.

With their inquiries, some natural philosophers turned their attention to a welter of often-intertwined problems that we associate with the physics and mathematics of local motion, for motion too was amenable to the quantifying of a quality.[47] Two centers of this activity predominated: Oxford and Paris. At Oxford's Merton College, the group of "Anglici" or "Britannici" (as they were called at the time; now scholars know them as the "Oxford Calculators" or "Mertonians") included Thomas Bradwardine, Richard Swineshead, William Heytesbury (ca. 1313–ca. 1372), and a host of lesser-known figures.[48] During the second quarter of the fourteenth century they produced a series of manuscripts in mathematics, natural philosophy, and logic, all of which rose out of the "standard practice of logical disputations" for undergraduates, a part of the arts curriculum at Oxford.[49] These manuscripts circulated quickly and widely, influencing another group of scholars at the University of Paris, a group that counted Jean Buridan, Oresme, and for a while, Albert of Saxony (ca. 1316–1390) among its members. Buridan and Oresme produced the clearest formulations of the impetus theory and of the mean speed theorem, respectively, while Albert of Saxony developed some of the most sophisticated premodern mathematical descriptions pertaining to the motion of free fall. Collectively, both schools exercised considerable influence over natural philosophy throughout Europe for the next two centuries.

To appreciate both the innovative mathematical techniques that emerged from these forays into qualitative mathematics and, more importantly, their limitations, we can look briefly at Oresme's account of the mean speed theorem and its depiction in early graphic form. The theorem itself states that a body accelerating its velocity evenly or uniformly over a period of time will travel the same distance as if it were moving at a mean, constant rate of speed during the same period.[50] As with his slightly older Oxford contemporaries, Oresme began with definitions in order to arrive at this conclusion. And, adjusting for terminology, the definitions seem quite modern. Swineshead, for example, designated "uniform local motion" (velocity) as motion whereby in "every equal part of the time an equal distance is described," which Oresme echoed as motion "of equal intension in all its parts," a "continuous quantity."[51] Thus (for a modern example) an automobile moving at a constant velocity of 60 miles per hour for three hours will have gone 180 miles, with an equal distance of 60 miles being traveled in each hour. From the uniform motion of continuous velocity, Mertonians proceeded to define the simplest sort of variable speed, that of uniform acceleration. For this, in Heytesbury's words, "motion . . . is uniformly accelerated if, in each of any equal parts of the time whatsoever, it acquires an equal increment of velocity." Oresme rendered the definition yet more succinctly with the pithy phrase "uniformly difform quality."[52] The same automobile, starting from rest and accelerating uniformly at a rate of increasing its velocity 1 mile per hour each second it travels, will reach a speed of 60 miles per hour after one minute.

As a way to depict these changes visually, Oresme devised an early graphing system to aid in his investigations and to offer a means of proving their results. A sort of halfway step between the ancient and modern, his pictures differed greatly from the lettered diagrams that accompanied Greek geometrical problems and proofs. Likewise, surface similarities notwithstanding, they varied from the later coordinate analyses of numbers, points, and curves introduced with Cartesian, analytical geometry.[53] (See Figures 4.2a and 4.2b.) To depict constant velocity Oresme used a rectangle. (See the rectangle EGBD in Figure 4.3.) In the Figure 4.3 reconstruction, the velocity ("intension" or "longitude") was represented by a vertical line, AB, and

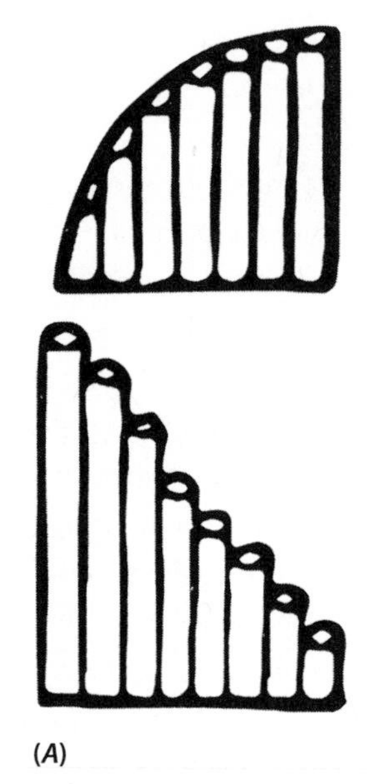

(*A*)

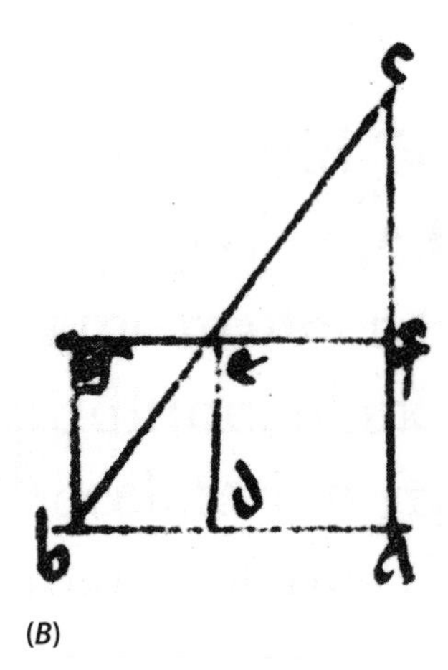

(*B*)

FIGURES 4.2A AND 4.2B. Diagrams from Oresme's *Treatise on the Configuration of Qualities.* Figure 4.2a depicts equal increments of a latitude of qualities. Figure 4.2b is taken from a fifteenth-century manuscript showing his diagram for the mean speed theorem.

introduced an actual measurement into the diagram, a quantity that combined distance and time (how fast a body travels equals the distance it covers in a given amount of time). The horizontal line, BD, denoted an amount of time, a "latitude" divided into equal parts, through which a body moved. Further, the area of the rectangle EGBD (its height times its width, or BE × BD) could be seen as representing distance traveled, for velocity multiplied by time equals distance. The lines BD and EG therefore could also be interpreted as the distance covered by a body moving at a specified velocity over a specified length of time.

Oresme continued with his diagrams to illustrate uniformly difform motion, or acceleration. This he captured with a right triangle. In the reconstruction (ΔBCD in Figure 4.3), the line extending from B to C represents uniform acceleration, while each of the lines reaching vertically from BD to BC denotes the addition of an equal increment of time. As with the rectangle, the line BD could also represent the total distance traveled in uniform acceleration. In an ingenious move, Oresme then joined the two diagrams and

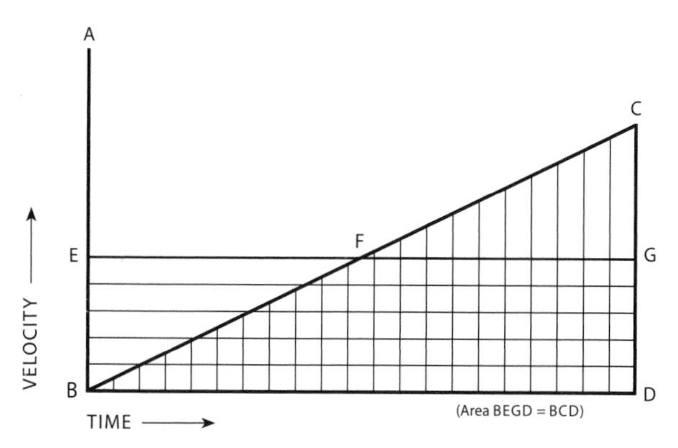

FIGURE 4.3. A modern reconstruction of Oresme's geometric proof of the mean speed theorem.

proved the mean speed theorem geometrically by showing that the area of the rectangle EGBD equaled the area of the triangle BCD. This he did by using arguments and proofs from Euclid.[54] Voilà: the distance a body travels in uniform acceleration is the same it covers at a constant, mean speed velocity in an identical amount of time.

Because the quality or intensity of velocity in acceleration varied from instant to instant, Oxford and Parisian Calculators alike were led to grapple with the difficult concept of "instantaneous velocity" as they sought to explicate the nature of the theorem and its proof. Here they foundered, and it is instructive to see why. For their limitations had nothing to do with their intelligence or with any external restraints imposed on them from religious sources. They were, rather, the built-in limitations of thing-mathematics and the definitional bases on which it was erected, the limitations of literacy. Notice, the "mean" of the uniform acceleration (BC) was a singular point (F), as were all the other increments of acceleration represented by the vertical velocity lines erected on the base (BD). But the acceleration itself was a continuous quantity, as Oresme noted (so too velocity). The Greek dilemma of incommensurability had resurfaced: How could one correlate discrete, individual points with continuous lines, instants of time with temporal time flow, multitude with magnitude?

Definitions were wanting, even when complicated. Heytesbury, for example, cast "instantaneous velocity" as "the distance which would be traversed by . . . a point [or moving body], if it . . . moved uniformly over such or such a period of time at that degree of velocity with which it moved in that assigned instant."[55] As Edward Grant has noted, this was "hopelessly circular" because it defined "instantaneous velocity" by means of a uniform speed equal to the very instantaneous velocity that was being defined.[56] Note as well, the only other definitional option would have been a self-contradiction residing in the words themselves. An instant is a point of no time lapse, while velocity is a measure of distance over lapsed time. A narrow tautology or a self-contradiction: try as one will, logically these terms simply don't mesh . . . so long as they are included in a definition of the same thing. For Aristotle such reasoning had kept arithmetic and geometry on the opposite sides of a conceptual divide. And here was the rub. The Calculators clearly struggled with expressing velocities and accelerations in infinitesimally small time intervals, but they had no mathematical notion of limit. Calculus lay far in the future. Nor did they have available an effective means for handling the proportionalities involved in their calculations. Perhaps above all, even as they sought redefinitions of velocity and explanations of acceleration, projectiles, and freely falling bodies, the dons lacked any conception of a real and abstract, measurable time. In sum, they had no grasp of what we recognize as mathematical definitions, especially those pertaining to motion, which since the time of Galileo have been increasingly understood as the calculating procedures used to manipulate abstract symbols, matters we shall take up in the forthcoming pages.

Moreover, none of their quantification of qualities had anything to do with investigating natural phenomena per se. Oresme was not alone in insisting that his diagrams and proofs of the mean speed theorem dwelt "only in imagination," not in reality.[57] These were mental fictions, logical exercises, the conundrums of *sophismata*. They were the products of a classifying temper, which abstracted things from experience, capturing and conveying them in definitions. Space remained the continuous or divisible thing abstracted by geometers, number the discrete or indivisible thing abstracted by practitioners of arithmetic. These two abstractions had remained incommensurable for centuries. The same conceptual issues surrounding infinity and continuity cut deeply into the ponderings of fourteenth-century natural philosophers, and were the same issues that nearly two centuries later would confront and challenge Galileo.[58] The Calculators sought to bridge the incommensurability of number and space, arithmetic and geometry through refinements in definitions. But in the end, their efforts amounted to "little more than intellectual exercises reflecting the subtle imagination and logical acumen of scholastic thinkers."[59] Deep within the mathematical thoughts of the Mertonian and Parisian dons lay the dominant, limiting assumption that mathematical definitions, like all others, referred to things.

* * *

With their own brand of religious meanings, enchantments, and logical exercises, the intellectuals of medieval universities furthered and reinforced the subordination of theoretical thing-mathematics to the classifying temper, a pervasive outlook they had inherited from antiquity. Meanwhile, back at the farm and market, one might say, folks measured. From the earliest days of settled agriculture and commercial exchange, people counted and weighed, sifted and sorted, bargained and computed. They used available, agreed-on standards as measuring "sticks"—a digit, a foot, an arm (*braccio*), a stride, a stone, a carob (carat), a grain, and so on. Folks didn't care so much about precision or exactitude in such negotiations; convenience was their aim. Roman *agrimensores* measured the lands of Roman senators; the question of an incommensurable difference between, say, the diagonal of one's square plot and its sides mattered not a whit.[60] Such practical measurements continued into the Middle Ages as well: try to imagine cutting and fitting stones for massive cathedrals without them. Still, despite occasional references to "middle" or "mixed" mathematics, the pervasiveness of classifying largely kept theory and practice separate in classroom mathematics.

The confines of thing-mathematics would be superseded during the Renaissance with the advent of modern, relational numeracy, a central topic of our ensuing chapters. It emerged in practical arenas outside the university's quadrivium, arenas where mathematical theory and quotidian measurement would steadily converge. The hands-on creation of a new, symbolic world of information mastery in arithmetic, music, geometry, and astronomy would provide a new and different array of mental extracts for mathematical and scientific minds to rework into layers of hitherto untold abstraction. These developments would take place virtually in the shadows of the period whose intellectual history has been typically characterized by other, more prominent and concurrent themes: the Renaissance-inspired turns to the past and to nature, the latter of which increasingly stimulated investigation of natural processes; the Reformation-induced revival of skepticism, which challenged the traditional categories of both religious and secular learning and ways of thought.

Succinctly put, the cultural and intellectual dynamic behind these concurrent themes, both their cause and consequence, lay with an information explosion and the rupture of classification, which we've discussed earlier. Not only did commonplace categories and ways of organizing knowledge break down under the tidal wave of new information. But for some, among whom many nominalists and Protestants, conventional language itself grew steadily disenchanted. Themes of fideism, mysticism, apophatic or negative theology, or pietism tended to push God out of words, leading eventually to the notion of a "Hidden God."[61] This didn't make their proponents any less religious; they just went about it in different ways.[62] But especially as regards knowledge of nature, the information technology of alphabetic literacy had reached its limit. It would be replaced eventually by a new technology of relational numeracy and by the relation-mathematics the new numeracy midwifed into existence. These developments would strip number of its divine and metaphorical qualities, steadily disenchanting it and awakening minds to the prospect of thinking with abstract, empty symbols.

Nicomachus had called arithmetic the "mother and nurse" of the other branches of the mathematical sciences.[63] In a sense he was right, although from his perspective in a completely unintended and unimaginable fashion. For the multifaceted activity that we are identifying as the creation of a new information technology can be detected initially in arithmetic. Accordingly, we shall first turn to the recasting of the cardinal and ordinal principles of abstract number in a simplified and operational notation, which has come down to us as the Hindu-Arabic number system, and to the lifetime labors of one Leonardo of Pisa, more commonly known these days as Fibonacci.

CHAPTER 5

Arithmetic

Hindu-Arabic Numbers and the Rise of Commerce

All things which have existed since the beginning of time have owed their origin to number . . . [and] are subject to its laws.

—Anonymous, *The Treviso Arithmetic* (1478)

SOMETIME AROUND the moment workers were laying the cornerstone of its most famous, though flawed landmark (August 9, 1173), the Tuscan town of Pisa witnessed the birth of the first of its two historically illustrious sons, Leonardo (the other being Galileo, born nearly four centuries later). Scholars know far more about the life of the leaning bell tower than they do that of Leonardo Pisano (Leonard of Pisa), as he was traditionally called. The bookend dates of his life carry the uncertain circas (ca. 1170, perhaps as late as 1175–ca. 1250), and few confirmed details have been found beyond the brief autobiographical sketch in the "Prologue" to his enduring work, *Liber Abaci* (*The Book of Calculation*).[1] He referred to himself in Latin as *filius Bonacci,* "son of Bonacci," although Bonacci was likely his grandfather's or extended family's name, giving later rise to his current nickname of Fibonacci, a moniker coined by the noted Italian historian of science Guillaume Libri in 1838. Shortly thereafter, in the 1870s, a French mathematician, Édouard Lucas, applied the name to one of Leonardo's series of numbers—the Fibonacci series—which, from its whimsical origins in calculating rabbit reproduction, has long intrigued mathematicians and scientists.[2]

Unlike his sainted contemporary Francis, who had rebelled against his cloth-merchant father, Leonardo initially followed his own father, Guilielmo (William), into the business world. A wealthy tradesman, Guilielmo had taken up duties as a "public official" at a customs house in the North African commercial port of Bugia (the modern Béjaïa, in Algeria), near the western end of the famed Silk Road. There Guilielmo's position involved safeguarding and promoting the rights of the *fondaco,* an inn and business establishment where merchants met to discuss prices, merchandise, politics, government levies, and related concerns. There too, in addition to the many silks, spices, and luxury items from the Far East, the trading community had received over the centuries a product of far more lasting value: a positional counting system, developed by Indian (or Hindu) mathematicians, complete with the zero, its place-holding symbol for nothing.[3]

Joining his father in Bugia, Leonardo very likely spent considerable time in the dimly lit rooms at the rear of the *fondaco,* where bookkeepers kept accounts, where apprentices learned their craft, and where he was given a "marvelous instruction" in the new counting system, the "art of the nine Indian figures." He then furthered his education in subsequent travels throughout the Mediterranean, mastering Greek mathematics and methods of proof, the algebra of Al-Khwârîzmî (ca. 780–ca. 850), and Hindu-Arabic techniques for arithmetic operations. By the time of his return to Pisa, he could "cast out nines" with the best mathematicians of the day.[4] Once home, Leonardo set about writing his primer on arithmetic, which he finished in 1202 and which summarized his learning for the benefit of Italian businessmen.[5] All his works, but especially the *Liber Abaci,* helped launch the "abacus tradition" of practical instruction in the art of calculation for generations of Italian merchants who were taught how to do their maths in abacus schools and keep their books with the Hindu-Arabic numerals.[6]

The sheer volume of abacus schools, students, and texts reveals an emerging and remarkable culture of numeracy during the Renaissance. The schools themselves began appearing in the thirteenth century, with the earliest institutional record found in statutes of Verona, dated 1277. As urban and commercial life expanded, demand rose steadily for those trained in basic skills of mathematics and bookkeeping. In Florence alone some twenty abacus schools flourished between 1340 and 1510, and in a single year, 1338, the city could boast as many as two hundred abacus students (although chronicler and Europe's first statistician, Giovanni Villani, reported over a thousand, clearly an exaggeration). Nearly a century and a half later (1480), of the roughly 680 students in "formal schools" (about 30 percent of boys ages ten to fifteen), one-third (229) attended abacus schools—this in a population estimated at 41,590 souls. As in Florence, likewise in Venice, Genoa,

Pisa, and other mercantile centers; students learned their craft from arithmetic teachers (*maistri d'abbaco*) who also often taught adults to use the new counting system in the commercial world.[7] Leonardo's text itself spawned literally hundreds of similar manuscripts, and the volume of books on arithmetic grew precipitously after the invention of printing. In 1478, a scant two and a half decades after Gutenberg's first Bible, an untitled, anonymous work, today known as the *Treviso Arithmetic* (*Aritmetica di Treviso*), was published at a small press in a small town of the same name, near Venice, opening the floodgates for a torrent of *liber abbaci*.[8] First in Italy, subsequently to the rest of Europe, the new counting system spread slowly but steadily, surmounting relentlessly the entrenched interests of Church and guild, and the conservatism of custom. By Shakespeare's day the merchant of Venice was using the new numerals to calculate and record his pound of flesh.[9]

Timing, of course, is everything in history, and though not the first to expose Europeans to the Indian counting system, Leonardo was certainly the most significant instigator of the "European arithmetic revolution."[10] Feeding the youthful appetite for mastery and expansion of commercial activities in Italy and beyond, his opportune work was fated to transform the arithmetic of the quadrivium, a remarkable accomplishment for essentially a primer in basic counting and computation. Contemporary observers and historians ever since have chronicled in detail the burgeoning of Renaissance business practices—new banking techniques, partnerships and corporations, investments and securities, borrowing and collateral, and the double-entry bookkeeping that balanced credits and debits and demonstrated a business's viability. The importance of mercantile or commercial capitalism for the modern world can scarcely be overstated, nor along with it the rise of a new number system to manage its affairs.

But another implication of this activity, perhaps more subtle though equally profound, may be unearthed from Leonardo and the abacus tradition. In using the new, Hindu-Arabic numerals and positional counting, many Europeans were acclimating themselves to analyzing problems and manipulating symbols in ways that deviated markedly from the technology of alphabetic literacy and its embedded thing-mathematics. Along with the rise of commerce, a new information technology, modern relational numeracy, was entering its infancy, born of this symbolic turn. Within the new technology, symbols for numbers became simpler and more abstract, more so even than the letters of the alphabet, which always represented vocal sounds, the building blocks of words, speech, and writing. First standing for things or objects, the new numerals would also come to display a functional or operational feature, particularly important in proportions, as relations between ratios were then generally called. These were the tools for analyzing profits, sales, supply and demand, and all the other forerunners to hedge funds and derivatives.[11] Put another way, the new symbols harbored codes of instruction, designating algorithmic procedures of data acquisition and manipulation.

Ultimately, when fully mature, the new numeracy would elicit the analytical possibilities of relation-mathematics (vis-à-vis thing-mathematics) and with them the mainspring of the modern, mathematical sciences, initially appearing in physics and astronomy. The scientific knowledge midwifed by the new analytical methods of this information technology would steadily displace itself from the narrations of religion. But when the bell first sounded in Pisa's campanile, these eventualities lay far in the future. Beginning with arithmetic and the newly arrived Hindu-Arabic numerals, the piecemeal adoption of new symbols occurred in a European context densely laden with religious meaning (the "default" setting of religious belief, in Charles Taylor's synoptic gloss). Even in the practical arenas of mixed mathematics, where the informational preconditions of modern mathematics and science were forged in arithmetic, music, geometry, and astronomy, many protagonists of our story continued to invest new mathematical symbols with old, predominantly religious meanings.

Inexorably, it seems, those meanings were expunged from the new information symbols as their functionality became increasingly apparent, refined, influential, . . . and useful. But traditional beliefs always die hard, and in this story they would die all the more slowly given the wrenching to one's consciousness posed by the new information technology. For with relational numeracy, a strange underpinning of the modern world began to surface: empty symbols would tell people what to think and what to do.

* * *

As though explaining a foreign language to students, Leonardo began his work from the ground up by introducing the new letters or symbols, the

Brahmi		—	=	≡	+	𑁖	𑁗	𑁘	𑁙	𑁚
Hindu	०	१	२	३	४	५	६	७	८	९
Arabic	٠	١	٢	٣	٤	٥	٦	٧	٨	٩
Medieval	O	I	2	3	ꝗ	ϛ	6	ʌ	8	9
Modern	0	1	2	3	4	5	6	7	8	9

FIGURE 5.1. The evolution of Hindu-Arabic numerals, from the most ancient, Brahmi ciphers to the most recent, European. Leonardo's symbols were derived from "Arabic-Indic" notation, depicted here on the third line.

"nine Indian figures" plus the "sign 0," which, he wrote, the "Arabs call *zephir*" (or *ṣifr,* from the Sanskrit *śūnya,* meaning "empty"). The symbols he used were actually western Arabic numerals, distinguished from those of the eastern reaches of the Islamic world. (See Figure 5.1.) In the early pages, he included tables of comparison with Roman numerals and reiterated the long-standing definition of a number, calling it a "sum" or "collection of units." Quickly, however, he asserted that through the addition of units, "numbers increase by steps without end," thus moving to the more modern, functional idea of serial succession involved in counting and generating the natural numbers.[12] He then noted that the writing of numbers begins "at the right" and proceeds "ever to the left" as "place follows place." Each new place designated a new value assigned to the digit, and in a lengthy description, familiar even to the arithmephobes among us, he explained how columnar residence (in the tens column, or hundreds, or thousands, . . . and so forth) determined the value of each "figure" or digit. As he proceeded to describe basic arithmetic, he repeatedly stressed the need to keep columns straight, insisting that the place of one number be clearly "written . . . under the place of another."[13] Columnar position was critical to all computations.

The importance of a digit's position in a number was not new; it could be found in many ancient counting systems. Recall, for instance, the relative positions of the I and the X in Roman practice. With the I to the left of the X, the number is 9 (IX), whereas on the right it equals 11 (XI). Other systems utilized columnar position more extensively, including the Babylonian, base-60 arrangement, first conceived sometime around 2000 BCE.[14] Indeed, digit position facilitated calculating large numbers more so than any other means available in the ancient world. (To appreciate this, try multiplying two large Roman numerals—for example, MCMXCIV and DCCLXXXVII—without benefit of columns or place holding.) It would also seem to have provided a great impetus for creating a zero or something comparable to serve as a placeholder for an empty column. The Babylonians certainly had an understanding of an empty column: they left it blank. But this produced myriad practical ambiguities because numerical values depended on the gaps between stylus impressions. Gap sizes varied from scribe to scribe, and often fractions could not be readily distinguished from whole numbers on the basis of the strokes alone.[15] As with early scripts, the task of correctly reading numbers placed a hefty burden on the reader's ability to interpret the symbol clusters in context. Eventually, around 300–200 BCE, a place-holding symbol for zero was introduced, but it never carried the meaning of 'nothing' denoted by our modern zero, never indicated a result in subtraction, such as $8-8=0$. When the Greeks took over much of Babylonian astronomy, they introduced a small circle to denote the cipher, and from it we derive the actual symbol of our modern zero.

Yet neither the Greeks nor the Romans after them built on Babylonian, positional counting, which remained a cultural cul-de-sac. We must look elsewhere, to India, for the direct genealogy of modern numeracy.[16] Originating in the Indus valley civilization of the third millennium BCE, and considered by some "the most successful intellectual innovation ever made on our planet," early Indian numeracy had evolved by the second century of the Common Era into a truly positional counting system, one that combined several advantages only suggested by other systems and that marked its superiority over them.[17] It was denary, or decimal throughout, much simpler than the Babylonian base-60 arrangement, and possessed a limited number of unique, genuinely abstract symbols (unlike Mayan pictographs, by contrast) for the digits 1–9. It utilized place-value notation in which the value of a digit was completely determined by its position in an array of digits. Furthermore, it incorporated the concept of zero systematically in number word and symbol. The cipher for zero stood as a placeholder in empty columns, eliminating the ambiguities harbored by Babylonian notation, and, more importantly, it carried the idea of nothing (originally called the "void") in calculations, including subtraction.

Leonardo understood thoroughly the advantages of fusing the "three elements—notation, place-value, and zero—into a harmonious whole," as he explained how to compute with the system throughout the first seven chapters of his book.[18] With scores of pedagogical examples, he demonstrated the ways Hindu-Arabic numerals simplified, clarified, and extended the abstractions of earlier positional counting. The new symbols simplified the values in each column by reducing clusters of marks to a single symbol, unlike the Babylonian system, in which often several wedge marks filled a column. In turn, this clarified the value of an entire number, assigning columnar significance to each of its digits and separating unmistakably whole numbers from fractions (Leonardo was among the earliest to introduce a bar between the numerator and denominator). And by deriving their value from their columnar relation to other numerals, the number symbols extended abstractions by reflexively focusing on the system itself. As an information technology, the new numeracy was more self-contained and abstract than earlier counting systems; in linguistic terms, syntax prevailed over semantics; in mathematical terms, ordinality over cardinality. One's attention fixed more directly on the "algorithms" involved in manipulating numbers, techniques so called after the Latinized name of the Arab mathematician Al-Khwârîzmî. In this process, and most striking, the new place-holding numerals ascribed a functionality to the number symbols themselves, in effect delivering instructions for when one "puts" (writes down) a digit and when one "keeps" it (in "head and hand") for the next column, the practice of borrowing and carrying remainders from column to column as it is now described.[19]

To facilitate and expand the functional instructions provided by the symbols, Leonardo provided lists of simple computations of whole numbers and fractions, and urged that they "most zealously . . . be learned by heart," the sort of exercise elementary school children do now in memorizing their sums and times tables. He included as well a detailed description of "computing with hand figures, a most wise invention of antiquity," and accompanied it with a page of diagrams. (See the illustrations of finger counting in Figure 5.2.) Temporarily storing numbers "in hand" or using a "chalk table" (before the days of paper) helped one through the intermediate steps in the

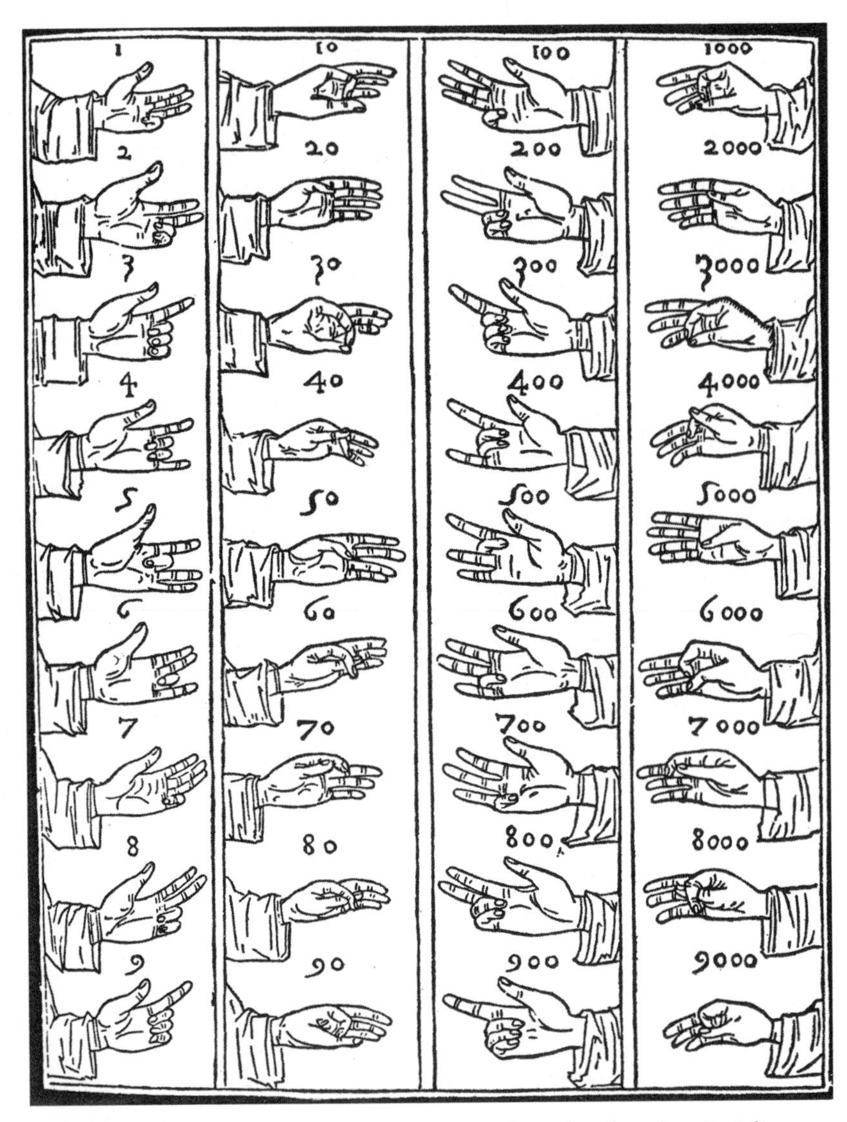

FIGURE 5.2. Finger counting, or dactylonomy. Woodcut taken from Luca Pacioli, *Summa de arithmetica, geometrica, proportioni et proportionalita* (1494). Identical to the system advocated and used by Leonardo Pisano, this method of finger counting enjoyed widespread popularity throughout the Middle Ages.

longer calculations and problems.[20] With these rudiments in place, Leonardo moved on to more elaborate computations involving larger numbers and fractions. He then devoted a series of chapters to demonstrating how this system actually worked in business situations.

The premium in this section of his manuscript was placed on what we now call, rather quaintly, "story problems," situations in which existing numerical information had to be manipulated in order to discover further, unknown bits of information deemed important. These included "finding the value of merchandise" (for example, the price of a "hundred canes [lengths] of cloth"); the "barter of merchandise and similar things" ("pepper for cinnamon," "pepper for saffron"); "companies and their members" (companies of "two men," "three men," and "IIII men"); and the "alloying of monies" (such as "when some given quantity of silver is put in a pound of money").[21] Leonardo's book basically laid out the plan of attack for mathematics education ever since—a "bait and switch" strategy whereby beginning students divvy up pieces of pie or so many cookies, as well as many other folksy, intuitive, and accessible examples of arithmetic functions (the bait). At a certain point the teacher takes away the story (the switch) and one faces the abstractions directly, creating either a budding mathematician or an arithmephobe, depending on whether one "gets it"—that is, the process of abstraction.

Throughout the applications of counting and computing in business situations, Leonardo stressed the operations of "proportional numbers," teaching variations on a "universal rule" in order that an "unknown number [might be] found from the known." Typically in these problems there were four proportional numbers, three in hand, a fourth sought. The first three might, for example, correspond to (a) merchandise, a fixed number or measure of items; (b) the price of the merchandise, calculated in one of the existing currencies; and (c) a different amount of the same merchandise. The unknown quantity was the price for the latter, the object of one's calculations. To find it, Leonardo counseled arranging the numbers in proper order, "as the first is to the second, so is the third to the fourth." He then proceeded to explain how with knowledge of any three of the numbers one can find the unknown fourth by combining multiplication and division (an algebraic technique now recognized as "cross multiplying" and solving for *x*). Long familiar to Arab businessmen, this method was known as the "rule of three" or the "merchant's rule" and dated from Chinese practitioners of the first century CE, Indian texts of the fourth.[22] Leonardo described several other procedures involving simple proportionalities, such as the "rule of five" and the "rule of false position," which proved equally useful for business calculations.

The focus on proportionalities directed Leonardo's attention to the serial or relational dimensions of number. True, in their cardinal correlations, numbers still corresponded primarily to things with their informational content of prices, merchandise, shares in investment, interest on loans, and other objects of commerce. But increasingly he concentrated on the number relations themselves as he ventured into topics in arithmetic series, square roots and irrational numbers, proofs in Euclidean geometry, and number theory, the last addressing countless, often strange properties of numbers as related to other numbers. At play in all these topics was a feature in the process of abstraction alluded to earlier—namely, the separation of the mind's products from experience, which allows the mind to look at its own creations from a new and critical vantage point. Once the cardinal correlation between object and number symbol has been made and a numerical "mental extract" established (say, between a pair of rabbits and the unit symbol 1), then the ordinality of counting and computing begins. After each computation we are freed to observe the new, abstract relations from a critical perspective and to rework or reconfigure them in various ways (producing, for example, the Fibonacci series). Beginning with the basic operations of arithmetic, we can step by step create or discover an infinitude of more complicated relations. Proportions revealed a major step in this progression, for they entailed relations between relations.

In effect, a proportion behaved as a small equation in Leonardo's treatment. Operations on one side of his proportions were dependent on those of the other side, a functional dependency that introduced both a kind of movement and a contingency or conditionality to the abstractions of numeracy. The unknown variable sought was contingent on the ratios of the known numbers, and as those numbers changed in different numerical examples, so too did the unknown. We must bear in mind that Leonardo had no symbols for operations, even for the basic arithmetic functions (+, −, ÷, =, and many others). Those conventions still lay centuries in the future, and it remains quite cumbersome to describe these mathematical innovations without them. Nonetheless, conceptually and practically he was forming a mind-set capable of operating with functional abstractions, of viewing its own abstract products with a critical eye toward further abstractions. A symbiosis had begun between abstract symbols and the algorithms that drove them. And each—symbols and algorithms—would lead to advances in the other. As we shall see, Galileo relied almost exclusively on proportions in his physics, in no small consequence of their having been in practical use for over three centuries before his birth.

Leonardo's discussion of proportions was a far cry from Aristotle's linguistic *analogia* ("proportion" or "analogy") or even, surface similarities notwithstanding, from Euclid's definitions. It appeared on the face of it that Leonardo had adopted Euclid's definition of proportionality—the relations read similarly, after all. In book 5 of the *Elements,* definitions 5 and 6, Euclid had spoken of "magnitudes in the same ratio, the first to the second and the third to the fourth . . . , which are called proportional." In book 7, definition 20, numbers are "proportional when the first is the same multiple, or the same part, or the same parts, of the second that the third is of the fourth."[23] Now recall Leonardo: "As the first is to the second, so is the third to the fourth." Moreover, in his introduction to algebra (chapter 15 of the *Liber Abaci*), he did employ some letters, generally preceded by a dot (for example, ".ab," ".bc," ".cd," and the like) to capture both numbers and line segments as he applied some of Euclid's geometrical proofs to his own treatment of proportions.[24] But different intentions lodged behind the words; the intellectual context here had changed quite discernibly because of the new number technology. Even as he referred to line segments, his computations dealt with numerical proportions. And in introducing them, Leonardo spoke not of definitions or things of any sort but rather of the "method of proportion," which he followed by promising "to show . . . how this method proceeds."[25] This provided his linguistic opening into the algorithmic procedures of calculating with the new, Hindu-Arabic numerals and thence to the algebraic abstractions derived from his numerical equations.

The abstraction begun in his treatment of proportion extended even further with Leonardo's steps into algebra. From our vantage point, we see the basic move in algebra as a generalization or abstraction from numerical proportions or equations, with letters substituting for numbers. As in the foregoing example of the rule of three, Leonardo used manifold numerical illustrations and equations, enough of which would convey to his readers the general operation. Merchants could intuitively take the rule and apply it to many situations other than those specific examples by which it was learned. Algebra simply made this chore easier by abstracting further from the numerical equations and making them general. Instead of treating each numerical computation or proportion independently, it became possible to create an abstract and functional relation, an equation, that could apply literally to an infinitude of specific number relations.[26] Again, Leonardo did not as yet have algebraic symbols, did not use letters for numbers in his equations; like the symbols of arithmetic operations, those lay on future horizons, awaiting François Viète and others. But Leonardo was training and preparing his followers to manipulate functional relations, relations involving dependent and independent variables.

Succeeding generations saw businessmen and merchants apply increasingly the abstractions and computational techniques of Leonardo and other *abaci* authors. Among the advisers to the men of commerce, a small number of specialists known as "cossists," living primarily in Italy and Germany, paid direct attention to advancing the calculating procedures themselves. Thoroughly conversant with the arithmetic and algebraic work of Leonardo, cossists also drew heavily on the writings of the great Arab mathematician Al-Khwârîzmî, whose famous book, *Kitāb al-Jabr wa-al-muqābala,* had been translated in the twelfth century by Robert of Chester into the Latin *Liber algebrae et amuncabala,* from which we derive our English word 'algebra.'[27] Al-Khwârîzmî's technical term for an unknown quantity was *shai'* ("thing"), in Latin rendered as *res* and in Italian as *cosa.* Because most of the early algebraists were Italian, the name "cossists" stuck. In one of history's smaller ironies, these so-named "thingists" contributed a great deal to the emerging world of abstract relations, which would supplant many of the defined things of nature.

Within the ever-fluctuating world of investments and returns, profit and loss, cossists like Luca Pacioli (ca. 1445–1517), the "father of accounting," or Niccolò Tartaglia, solver of cubic equations, were often hired as number practitioners, a sort of mercenary, financial condottieri.[28] Their arithmetic and algebraic skills gave a burgeoning commercial class the technical means of carrying out and keeping tabs on progressively complex trades and exchanges, and their analyses, not yet algebra but beyond arithmetic, supplied an abundance of "operational recipes," a "problem-solving approach to mathematics."[29] Beyond furnishing practical advice, cossists often spent much of their free time tackling purely theoretical questions, competing among themselves to solve new and different classes of equations, such as cubic and quartic, as well as problems in number theory. (Success in such competitions enhanced one's reputation and made employment easier and more lucrative.) One upshot of their work was to nudge early algebra from its purely "rhetorical" phase, in which all problems were written out in conventional language, toward a more "syncopated" phase in which some symbols were occasionally substituted for words. These were small steps toward Viète's eventual breakthrough into a more recognizable, symbolic algebra.

At Oxford, physician-cossist Robert Recorde (ca. 1510–1558) both summarized and epitomized many of these developments in a series of books on arithmetic and the "Arte of Cosike nombers" (the latter contained in *The Whetstone of Witte,* 1557, the earliest book on algebra in English). Having first detailed the arithmetic workings of "abstract number" and "denominative number" (numbered things), he subsequently proceeded to distinguish two types of the latter: the one called "nombers denominate vulgarely"; the other known as "nombers denominate Cosikely." Both types referred to numbered "things," but the "vulgar" denoted objects straightforwardly ("10 shillings, 10 men, 20 shippes, 100 shepe, and the like"), while the "cosike" things referred to were the "signes" of other numbers and procedures—"absolute" numbers, roots, squares, cubes, the "square of squares," and continuing into greater powers. In this way Recorde recorded a seamless transition from algorithmic operations on numbered things to operations on the relations between them, operations and relations designated by abstract signs or symbols. (He also invented the equals symbol, =, still used today.) Even as he continued to extol the virtues of Nicomachus, Plato, and other writers who insisted on the "mathematical artes" as the divine font of "unfailible knowledge," and even as he continued to extol the virtues of various numerological enchantments (of the sort discussed in Chapter 4), he was emptying his mathematical abstractions of beings. Some of these abstractions were even "surdes"—irrational roots, such as $\sqrt{2}$, that could have no expression in rational, whole numbers. They were thingless, absurd. But one could calculate with them and correlate results with external phenomena. Such was the "cunning" of number.[30]

Collectively then, the abacists and cossists spread the new numeracy throughout the Renaissance, first in Italy, then in Germany and the rest of northern Europe. In addition to the business world's recognition of numeracy's commercial appeal, many others—artists, musicians, engineers, artisans, and craftsmen of all sorts—came gradually to perceive its benefits and to extend even further the practical reach of the new arithmetic substrate of basic symbols and their information-harvesting and information-managing capabilities. Employing the merchant's rule and manipulating proportions in general steadily grew indispensable to composers, to perspective artists, and to calendar and time reckoners as abstract relations infiltrated their disciplines and activities. As we shall explore, many of these activities involved moving phenomena, apprehended by the perceptions of sight, sound, and time, which were themselves in constant motion. The new technology tendered novel ways of identifying, collecting, and processing bits of information in this incessantly moving world. The functional flexibility of the new counting system and its computing capabilities helped promote hitherto untold mastery over such information in flux.

Euclid's definitions came out of a universe of words and things. His big "thing" was space (the "extension-thing" Ortega called it) and the fixed, spatial relations that could be constructed with a ruler and compass, which were then justified logically through a series of deductive inferences. These inferences rested on a foundation of definitions, those axiomatic building blocks of the entire system. In his proofs he brought his demonstrations back to these definitions. Much as Aristotle had used the information tools of alphabetic literacy to abstract and construct a logical system of words and classes, so too had Euclid constructed a logical system for the abstractions of space. To be sure, Aristotelian logic and Euclidean geometry were not the same.[31] Still, they shared some features. Both were essentially deductive systems beginning with definitions, foundational assertions, or axioms. Both systems were static, their definitions fixed, their logical relations (complicated and sophisticated as they were) unchanging. The procedures of the abacists and cossists revealed not a fixed world of words and things but one of changing relations, of variables, of functional dependencies, of an informational world in motion, elements that steadily became commonplace in mathematical definitions. In contrast to the substantive things referred to by definitions in conventional language, mathematical definitions increasingly grew to describe procedures one could follow in order to solve problems or to explore further the movements and relations of serial magnitude.[32]

An immediate consequence of these new practices came to be appreciated at the time. Before the use of Hindu-Arabic numerals and positional counting, calculations were made with an abacus, a counting board, as we noted earlier. Once a computation with the counters had been completed, the results were then recorded elsewhere, generally in Roman numerals. The acts of computing and recording were thus distinct, with the counters carrying out the computations mechanically and number symbols noting down the information about things or objects. The new, positional counting system, by contrast, merged calculating and recording into a single set of operations, designated by a single set of symbols, written on the same page. In short, Leonardo simplified the information symbols of counting and made them functional, capable of myriad permutations. Symbols and functions, new expressions of cardinal and ordinal number, had midwifed a new numeracy.

* * *

Despite the innovations and practical focus of the new system, the traditional philosophical and religious dimensions of arithmetic never strayed far from the minds of most number masters, and were even enhanced during the Renaissance. In the wake of humanism's interest in antiquity, there followed a renewal of curiosity about and attention given to Plato and Neoplatonic strains of thought, to the Hermetic tradition and magic, to astrology and numerology, and to the cabbala and gematria—all topics encouraging belief in the mysticism of number. Among other stock Renaissance icons, the Platonic and eclectic Hermetist (one thinks of Marsilio Ficino [1433–1499] and Giovanni Pico della Mirandola [1463–1494]), the alchemical wizard (Paracelsus [1493–1541], for example), the mathematician-magus (John Dee [1527–ca. 1609]), and the prophesying astrologer (Nostradamus [1503–1566]) cast their spells through an interweaving of science and magic, frequently through the mystery and enchantment of divine number. Even while it grew instrumentally abstract and practical, number still reigned over a qualitatively dense and allegorically animated universe.[33]

Take the practice of "casting out nines," for instance. As taught by Leonardo and many others in abacus texts, this was a long-standing method of verifying basic computations. He patiently explained the operation with an example, one of many: 506,789 + 4,321 = 511,110. To check this addition, one begins with the "residue" of a number (the "digital root" it is now termed). This is found by adding the digits and eliminating the multiples of nine along the way. Beginning on the left of the number, $5+0+6=11$, and "casting out" 9 yields $11-9$, or 2. Continuing, $2+7$ (the next digit) $=9$, which is purged, or cast out. That leaves the remaining 8 and 9, which total 17, and $17-9=8$, the residue. (Nowadays, the same results are achieved more economically by iteratively summing all the digits in a number until only a single digit remains. The digits in 506,789 add up to 35, whose digits in turn sum to 8, the digital root.) To complete the verification, one sums the residues of the addends; if the addition is correct, they will equal that of the total. As we just saw, the residue of 506,789 is 8, while that of 4,321 is 1. Together they total 9, whose residue after casting is 0, the same as that of the total, 511,110 ($5+1+1+1+1+0=9$, and $9-9=0$). Similar checks may be performed on subtraction, multiplication, and division.[34]

But this was not merely a nifty technique for checking a computation. Perhaps as far back as Pythagoras, casting out nines had been used to determine the numerical value, and thus worth, of a person. In the third century, Hippolytus of Rome (170–236), an important theologian among the sainted patristics of the Western Church, called to task arithmeticians (essentially astrologers) who "suppose that they prophesy by means of calculations and numbers." He then described a version of casting out nines as applied to various historical and mythical figures.[35] This was possible, especially in Greek and Hebrew, because with both languages letters also served as numerals and thus each name had a corresponding numerical value. That corresponding value, its digital root, was termed *pythmēn* in Greek and meant "fundament" or "foundation" or "bottom" (perhaps originally applied to the keel of a ship). A nineteenth-century translation has it as "monad," the fulsomely metaphysical notion we encountered earlier. The digital root, then, was the essence of a character, and in the agonistic struggle between combatants, the character with the higher number would prevail. Thus did Achilles, with a corresponding number of 1,276, vanquish Hector, whose number was an inferior 1,225. (Achilles's digital root is 7, while Hector's is only 1; a ratio of 7 to 1 meant no contest.) And is it only coincidence that in rescuing his kinsman (perhaps his nephew Lot), Abraham and his household chief, Eliezer, led forth a retinue of 318 servants, the same numerical total as Eliezer's name? The number 318, recall, represented the number of attendees at the Council of Nicaea in 325 CE and, properly interpreted, supplied a divine legitimacy to the council and its promulgations.

That Hippolytus criticized the prophesying pretensions of casting out nines as illegitimate should not screen us from recognizing its widespread and continuing tradition. To the contrary, it remained in use for centuries as a pillar of the common practices of onomancy, divination based on the numerical value of person's name, and gematria, a broader form of numerology centered on the art of interpreting letters and numbers. The *Eadwine Psalter* from Canterbury, circa 1160, for example, contained onomantic tables of the "victorious and the vanquished numbers" and the "letters of the alphabet and their numerical equivalents." Both were used not only to determine the outcome of battles by casting out nines from the names of commanding generals but also to divine which partner of a married couple would outlive the other (although in this case one cast out sevens, not nines). The same practice shaded into astrology, with additional tables correlating numerical equivalents and planetary positions on days of the week; such tables were used in divining whether a sick person would live or die, depending on his or her digital root and on the day of the week he or she fell sick.[36]

Somewhat later, religious writers during the Reformation utilized the techniques of gematria to identify the beast of the Revelation, the devil incarnate, whose number was generally given as 666—either Martin Luther or the pope, depending on one's proclivities. The Catholic Peter Bungus, for instance, engaged in some adept maneuvering to convert "Martin Luther" into the number 666, while Protestant Michael Stifel (1487–1567), with even greater contrivance, explained how the number referred to Pope Leo X.[37] (For the details of their numerical maneuvering, see Appendix A.2.) We may scratch our heads at these and comparable practices, such as scapulimancy (divination according to the geometry of sheep shoulder blades), but Renaissance and early modern minds had little trouble with reading religious meaning in the objects of mathematics, of intertwining the allegorical and functional dimensions of number.[38] Medieval enchantments had not yet been discarded.

From the commercial world and its new arithmetic, then, a new information technology steadily made its way toward forging the analytical temper. But this was not all. Another implication of the new numeracy was waiting to be teased out of its functionality. For the reverse engineering of modern analysis to develop, its functions required movement, and movement required time. Before the Renaissance time was only kept track of, not measured. Measuring time would require that it be abstracted and perceived as an independent reality, much as space had been for the Hellenistic Greeks and Romans. But Greco-Roman antiquity was timeless; time is modern. Its emergence takes us to the second branch of the quadrivium, music, whose developments shadowed those of commerce, providing another arena for uncovering the use of abstract and functional symbols, information symbols that would measure and map—and analyze—time.

CHAPTER 6

Music

Taming Time, Tempering Tone

A number, having no body, cannot be sonorous.

—Vincenzo Galilei, "Il primo libro della prattica del contrapunto intorno all'uso delle consonanze" (1589)

IN HIS COMPULSION TO CLASSIFY, Hugh of *Didascalicon* fame had labeled music as the part of mathematics dealing with "multitude" or "discrete quantity" that "stands in relation to another," vis à vis arithmetic, which stands "in itself." Arithmetic and music in turn were paired opposite "magnitude" or "continuous quantity," which embraced geometry and astronomy, while three subcategories further identified the "varieties" of music: "music of the universe," "music of man," and "instrumental music." When we scrutinize his categories in some detail (and there were more), they initially appear quite foreign to the modern reader. Of the three music divisions, for example, only the last actually concerned itself with sound. Music without sound? Strange, indeed, until we reflect further on the harmony between science and religion, nature and God, that underlay Hugh's classifications. It was all of a cloth. The "theoretical arts" (including music) were intended as "remedies" against the "evil" of ignorance, just as the "practical arts" of virtue fought vice and the "mechanical arts" supported human "needs against life's weaknesses." (Hugh did like his trinities, although with some exceptions: "Food," he noted, "is of two kinds: bread and side dishes.") As one of the three theoretical arts, along with theology and physics, mathematics may not have been the royal road to learning, but it had surely positioned itself as a pathway to "Wisdom," participation in the "living Mind," the "divine Idea or Pattern, coming and going in us but standing changeless in God." Against this backdrop, harmony revealed the divine order in the nature of things.[1]

The associations were straightforward: harmony was order; order was divine; and through "proportion" one understood its manifestation in music. Filtered through Hugh's sources, especially the Neoplatonic Boethius, Cassiodorus (ca. 485–ca. 585), and Isidore of Seville (ca. 560–633), these ideas and associations traced their way back to Plato and through him to the quasi-mythical Pythagoras.[2] Not only did Boethius bequeath to Hugh the definitions and classifications of music, he had also arranged them neatly in an ascending hierarchy, starting with instrumental music, rising to the consonance found in the human soul, and finally soaring in a soundless crescendo to the music of the universe, the "harmony of the spheres" it was often termed.[3] This was the silent and beautiful music of the philosophers, measured against which, for Plato, the sounds of instruments offered little more than the cacophonous gossip one gathered from "trying to overhear what the neighbors are saying." At best, instrumental sounds might provide a stairway to the greater, philosophical comprehension of cosmic harmony; at worst (generally the case), musicians "put ears before understanding."[4]

The origin and nature of cosmic harmony had been depicted in Plato's *Timaeus,* which offered Hugh and company a metaphorical supplement to the Christian story of creation.[5] There Plato had narrated a tale of how the "craftsman" (the Greek *dēmiourgos,* from which 'demiurge') fashioned the "body of the universe," having first created the "world's soul" to bring about order from disorder, "to be the body's mistress and to rule over it as her subject." A metaphysical "mixture of the Same, the Different, and of Being," the world soul was crafted according to musical intervals found in the pure, harmonic ratios of the Pythagorean *tetractys* (1:1, 2:1, 3:2, and 4:3) and in their extensions.[6] These musical intervals in turn accounted for the positioning of the various parts of the world body: the sphere of fixed stars, the seven wandering stars (each in its own orb), the earth, and a spurious "counter-earth" for a total of ten, the convenient sum of one through four.[7] So constituted, Plato's image of a divinely created and finely tuned universe resonated deeply within many medieval minds and remained an integral part of Western culture well into the seventeenth century, even beyond.[8] In this context, a musician, wrote Boethius, "adopts the science of music not in the servitude of work [say, singing or playing an instrument] but in the rule of contemplation."[9]

Such was the "theoretical music" taught in medieval universities, an intellectual and allegorical journey into the presence of the divine through philosophical and religious meditation on harmony.[10] But another tradition, equally religious and thoroughly practical, echoed the actual sounds of music from the walls of monasteries, cathedrals, and churches throughout Christendom. This was the tradition of plainsong or plainchant, whose sketchy origins date perhaps from the chanting and psalmody of ancient Jewish worship but are more generally believed to be situated primarily in early Christian liturgical rites, especially those of the Western Church.[11] By the ninth century the mainstream form of plainsong had come to be identified with Pope Gregory I (ca. 540–604), under whose auspices its regular use during the liturgy had become widespread.[12] For centuries Gregorian chant had been handed down from singer to singer in a predominantly oral tradition. (Contemporary sources complained that students were "barely able in ten years" to master plainchant from their teachers.)[13] Then, in the post-Carolingian age of the ninth and tenth centuries, monks, clergy, and scribes addressed matters and began writing down the chants with accompanying, abstract notation, musical signs called "neumes."[14]

Essentially diacritical symbols, neumes served as mnemonic devices directly associated with the words of a chant. (See Figure 6.1.) Originally such marks had specified emphasis placed on particular syllables, much the way accents still function in Romance languages—French, for instance. Neumes far transcended pronunciation aids, however. They indicated the rising and falling changes in tonal pitch, thus reminding their reader of the actual melodic flow of the chant. At the same time, the signs carried no information about the timing or rhythm of tones and left the melodies free-flowing rhythmically. From neumes yoked to words, eventually words and symbols separated, and purely abstract musical notation was launched. Once begun, there followed over the next half dozen or more centuries (ca. 1000–1600) a veritable flood of diverse musical symbols, systems, and variations. Driven by the needs of the "most significant event in the history of Western music," the rise of polyphony, these innovations brought musical notation to the cusp of modern conventions, creating in the process an information technology of sound.[15]

The centuries-long evolution of abstract notation furnishes one of two foremost themes bearing witness to the advances in practical music as an

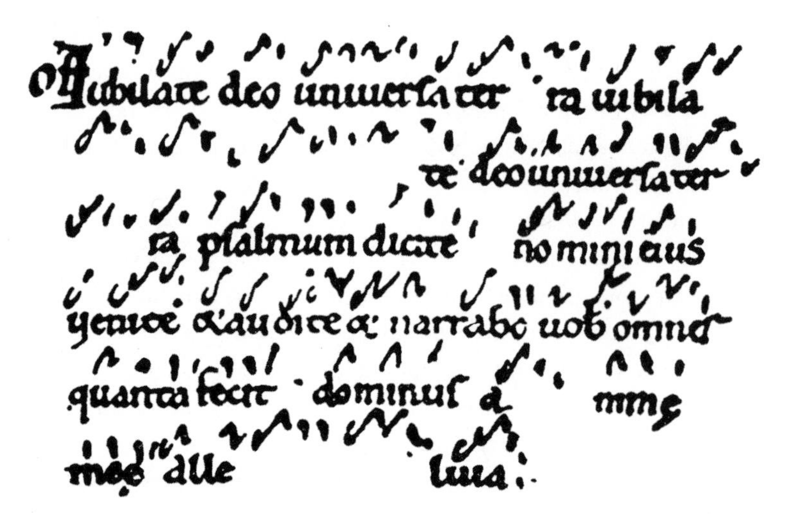

FIGURE 6.1. Neumes written in *campo aperto* (without staff lines) over words from a psalm, “Iubilate deo universa terra,” sung in monophonic Gregorian chant. Circa the tenth century.

information technology and alternative to the theoretical music of the schools. The second, known as the “problem of tuning,” surfaced in the sixteenth century with the growing popularity of nonvocal music, especially compositions involving keyboards and fretted instruments. The tuning question forced musicians to rethink the nature of harmony using complex and variable proportionalities, generally involving the irrational and real numbers, which described physical sounds actually heard, rather than relying solely on the prescriptive ratios of philosophical, Pythagorean number. The young Galileo grew up amid this controversy, which featured his father, Vincenzo Galilei (ca. 1520–1591), and Gioseffo Zarlino (1517–1590), Vincenzo’s one-time mentor and the greatest theorist of polyphonic counterpoint. In their dispute over tuning and sound, Zarlino held fast to the contemplative ideals of music—the natural, divine harmony of the spheres and the Pythagorean-Platonic tradition of consonant numbers—while the far less philosophically minded Galilei maintained that musical intervals were “altogether artificial” and infinite in number. Tunings, he believed, arose from the practical necessities confronting musicians and from the physical characteristics of sound. Numbers could measure intervals, but the human ear, not “sonorous” number, determined consonance and dissonance.[16]

Throughout their controversy, Zarlino and Galilei joined with other musicians in a long tradition of collectively forging a new way of thinking practically and instrumentally about musical intervals, notation, and harmony. Even Zarlino, clinging heroically to the Pythagorean tradition (with Ptolemy's alterations), promoted his own practical tuning compromises. Along with Hindu-Arabic numerals, abstract musical notation and the disputes over harmony helped habituate generations of scholars and humanists alike, as well as practicing musicians, composers, and theorists, to using and manipulating empty symbols, to interpreting such symbols as carriers of instructions, to calculating variable proportions, and to perceiving sound as an independent, time-bound series of fluid relations. As the technology of musical literacy became both more abstract and more functional, a trajectory that would ultimately wind up in the full "functional harmony" of eighteenth- and nineteenth-century music, the symbolic sands of practical time and tone steadily and inexorably eroded the fading medieval image of philosophical music.[17]

* * *

"Unless sounds are held by the memory of man, they perish, because they cannot be written down." Thus did Isidore of Seville in the early seventh century pose the central challenge facing creators and performers, as well as hearers of music, then and now.[18] Like the will-o'-the-wisp hovering over a marshy pond, sound floats elusively, often hard to recall and difficult to mimic, especially for the untrained or tone challenged. In musical notation lay the answer to Isidore's and our problem. Notation captures sound's effervescence and channels it into symbols that in turn can guide one back into the world of fleeting, acoustic sensation and its attendant emotional range and intensity. As with their ancestral letters and numbers, musical symbols offer marks of one sort that stand for objects of another—bits of sound—in order to create and store musical information.[19] And in loose parallel with letters and numbers, musical notes look in two directions, outwardly and inwardly. Facing out, as it were, they correspond with sounds, thus representing pitch or tone, while within their own symbolic universe notes dwell in intervals, relations with other notes. These relations express melodic changes in tone (plus chordal harmonies since the seventeenth century), and of even greater, further-reaching significance, they measure rhythm or time.[20]

With the alphabet, vowels and consonants (meaning "with sound") had been created as the constituents of phonemes, the spoken syllables forming words. Because musical "consonance" (also meaning "with sound") was contingent on intervals between sounds, it lent itself early to treatment with the abstractions of number as a means of capturing the intervals. Fixed, Pythagorean ratios from the *tetractys,* for example, were used to express tonal unison (1:1), the octave (2:1), the perfect fifth (3:2), and the perfect fourth (4:3).[21] The interweaving of early numeracy with other abstractions, including the Greek practice of using letters to designate musical sounds, meant that musical literacy was crafted from its inception in the ancient world as a hybrid information technology.[22]

In the period of our concern, the hybrid assumed new forms as practical advances in notation marched point for counterpoint in their own historical rhythm with the emerging, modern numeracy. Much like the evolution of writing from its earliest appearance through the creation of the phonetic alphabet, medieval musical symbols passed through several phases of development. From plainchant, neumatic notation to the rhythmic, modal notation of the twelfth and thirteenth centuries, and then to the "mensural notation" prevalent during the Renaissance, musicians progressively reworked their initial signs and marks in the now familiar, twofold pattern of abstraction. In each phase they observed their symbols from a new, critical distance or perspective and elicited new mental extracts that corresponded to fleeting, auditory experience. All through this ongoing process of abstraction, numeracy supplied the symbols of proportion and measurement, while for its part, musical notation applied these tools to determining intervals of tone and time, the latter understood as independent and separable from other natural processes. Thus did musicians in the practice of their discipline pare away ambiguities from symbols, transforming and simplifying them into a pure distillate of pitch and duration, the two dimensions of musical sound.[23]

Symbolic mastery of pitch preceded that of duration, but the rise of polyphonic music in the West supplied the catalyst for domesticating both. The term 'polyphony' refers to the simultaneous sounding of more than one melodic line, a practice itself begun as an elaboration of plainsong. Throughout its early history, Gregorian chant had remained predominantly monophonic, with vocalists singing one melody, all following the same pitch, or occasionally chanting in parallel intervals, such as octaves, fourths, or fifths, but without introducing any new melodies.[24] Gradually new melodic variations were added, leading to a more sophisticated and adventuresome form of chant called "free organum," when voices began moving in separate, sometimes even opposite directions. One melody might rise in pitch while another descended. By the twelfth century further advances had produced a "melismatic" or "florid" organum, an even freer and more complicated manner of intertwining voices. For instance, the original chant might sustain a single note, usually called a tenor (from the Latin *tenere,* "to hold"), while other melodic lines in either a single voice or several voices wove around it variations, or "melismas," of up to twenty or more notes, frequently above the tenor line.

As with many sorts of invention, necessity mothered improvements in the techniques of chant notation once the diacritical, mnemonic neumes had become commonplace and as the music made further demands on symbolization. To clarify the rise and fall of pitch, scribes began etching straight lines on parchment, with the placement of notes above and below. Then ink was added to the lines, marking the beginnings of the musical staff. Notes became simplified and standardized as diamonds or squares, some with stems or plicas (folds), while a host of ancillary marks rounded out the neumatic notation of Gregorian chant still relied on today. (See Figure 6.2.)

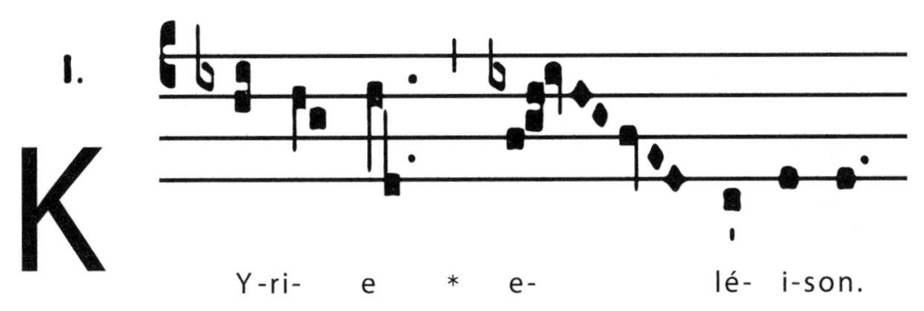

FIGURE 6.2. Gregorian chant written with neumatic, square-note notation. The chant is the beginning of the "Kyrie." On the top line, left, is a "clef" or "key" sign, which indicates the note C in modern music and which orients the remaining staff lines. Depending on the chant, the clef could be placed on any of the four staff lines. Clef notes could also be designated as F or, later on, G, the anchor notes of our modern bass and treble clefs.

Many of these early innovations were compiled by a Benedictine monk about whom little is known beyond his name, Guido d'Arezzo (ca. 990 / 992–ca. 1050). Guido is generally credited with standardizing the four-line staff of plainchant notation, which was in principle the modern musical staff (and the earliest of graphs in Europe). His explanation stands today as it did nearly a thousand years ago, circa 1025: "Each sound . . . in a melody is found always in its own row. . . . Some rows of sounds occur on the lines themselves, others in the intervening spaces. . . . The sounds on one line or in one space all sound alike."[25] To remember changes in pitch on or between lines, this musical monk also invented what later became known as "solmization." Long before the Rogers and Hammerstein show tune "Do-Re-Mi" set the modern standard as a homophonous mnemonic for the musical scale, Guido derived a comparable device from a popular hymn to Saint John, which gave us the original:[26]

Ut (Do)	***Ut***queant laxis	**Re**	***Re***sonare fibris
Mi	***Mi***ra gestorum	**Fa**	***Fa***muli tuorum
Sol	***Sol***ve pollute	**La**	***La***bi[i] reatum,
Si	***S***ancte Iohannes.		

Guido claimed that his method of mnemonic instruction could produce a "perfect singer" in about a year or two, and that having committed to memory the "properties of single sounds, . . . of all descents and ascents," one would even possess a "well-tried method" for learning "unheard songs." He further remarked that letters could stand for musical notes, of which there were but seven.[27]

In the wake of Guido's innovations and more precise pitch notation, plainchant became increasingly polyphonic and more complex throughout the eleventh and twelfth centuries. Accordingly, it forced composers to tackle the question of timing or rhythm, for coordinating voices singing more than one melody required some way of ensuring that singers would not jump ahead or fall behind throughout the course of several, simultaneously proceeding melodies. Composing polyphonic music, in short, obliged "careful thinking about temporal units," in order to construct a "temporal structure whose components have to fit together."[28] Such thinking came to prominence during the latter years of the twelfth century in Paris, where there emerged the most significant medieval school of polyphonic music. Associated with the University of Paris and the Cathedral of Notre Dame, whose construction began in 1163 (a mere decade before that of the Pisa bell tower), the school attracted avant-garde musicians from throughout Europe who pushed back the frontiers of polyphony.[29] A century later, in a remarkable manuscript, which historians have simply named "Anonymous IV," a young, probably English music student at the University of Paris narrated details about the formative development of the Notre Dame school. With it he also included an extensive, theoretical treatise on early polyphony and its "appropriate notational symbols," which summarized the groundbreaking work of the era's composers and theorists.

In his manuscript, Anonymous IV described the "measurement of durational values" that governed the "length and brevity of music," creating a system of what are now termed "rhythmic modes."[30] These modes marked historically the first appearance of metric time, time measured by a symbolic process that compares elapsing time intervals with abstract units or standards. In all likelihood the units, and hence the modes, derived from poetic meter, which had been discussed in some detail by Saint Augustine in an unfinished text (*De Musica*), a manuscript available to the musical scholars at Notre Dame.[31] The basis of musical rhythm lay with correlating various lengths of sounds and the silent spaces between them. Poetic meter had done this for the spoken word, separating stressed and unstressed, long and short syllables. Notre Dame musicians translated such divisions into organized sounds—Anonymous IV christened them "durational units"—which created a series of temporal patterns with long and short tones called *longa* and *brevis*. In so doing and despite ambiguities, the breve ("one durational unit") effectively served as a measurement standard of equal intervals of time, with the *longa imperfecta* its double and the *longa perfecta* its triple.[32] Using these measures, composers were able to construct a half dozen or so modes of rhythmic clusters. (See Appendix A.3.) The most common, called trochaic or first mode, for instance, paired a double-length note, *longa imperfecta*, with a short one, the breve, producing a phrased rhythm of long-short, long-short, and continuing. Additional symbols were used to indicate repetitions, changes in mode, and phrase endings.

Critically, in forming these rhythmic patterns of notes, composers also introduced symbols for the musical rest, described by Anonymous IV as "a pause or omission of sound for a definite length of time or durational units."[33] With this seemingly simple but quite revolutionary observation, Anonymous IV identified explicitly the detachment of time intervals from their surroundings and their measurement by means of a humanly invented, symbolic system of notes and rests. To be sure, all singing, from Greek melodies to plainchant, was performed in some sort of rhythm, although early cadences remain largely unknown to us. But with notated music, longer and shorter notes combined precisely with longer and shorter rests to create and encode a "metrical wave, whose uniform pulsation is perceptible through all the changes of the tonal surface." Music thus became explicitly a "temporal art," both in the "banal" sense that tones are presented in symbolically notated, temporal succession and in a far more profound sense that time "reveals itself" as the "content" of human experience. Full recognition and appreciation of this deeper insight would be unearthed only centuries later, but it lay embedded in the groundbreaking notations of polyphony: the creation of an exact standard for measuring time and with it a means for depicting musical rhythm symbolically.[34]

Measuring time diverged sharply from the earlier practice of "keeping time" or "keeping track of time," explains physicist Géza Szamosi. Early sundials and clocks, for instance, kept track of time by shadowing the course of the sun or by modeling celestial movements; they followed time, so to speak, through various demarcations devised for human benefit, but they did not measure it. By contrast, through the efforts of practicing musicians who created the notation of polyphonic music, metric or metered time came to exist independently of natural phenomena, including the solar system. Musicians didn't follow time; they denoted it with their symbols, correlating them with regular intervals of duration in a systematic fashion, what we call musical beats. Even today we don't set our clocks by the sunrise; rather, we say that "sunrise occurred at such and such time." In short, "clocks could measure time only *after* the idea of a measurable time became established."[35] The musical practice of measuring time indicated a change in perception; it was not yet a scientific concept. But from our perspective, although it required another three centuries, it was a short step from measuring time to reworking its measurements functionally into the reverse engineering of analysis, a conceptual step made possible by the new numeracy in symbiosis with musical notation.

Symbolic, measured time, then, structured rhythm. Put another way, musical phrasing followed from the presence of "strict temporal units which discipline the free flow of rhythms."[36] Once under way, measuring musical time grew steadily in simplicity and clarity. Around 1260, a German theorist, Franco of Cologne (fl. mid-thirteenth century), supplanted the rhythmic modes with what became known as "mensural notation," the reigning system of musical symbols until the seventeenth century. In his widely circulated and influential manuscript, *The Art of Measured Song* (*Ars cantus mensurabilis*), Franco systematized the relative values of notes and rests, reaffirming time as "the measure of actual sound." To the long and the breve he added the semibreve (the equivalent of our modern whole tone) and the duplex (a double-long), all of which composed classes of time-bound notes broken down into further time units. And to these note values (seven in all), each with its own symbol, Franco added six symbols for rests of different lengths, also designated as longs, breves, and so on. His additions transcended the limitations of the six modes, converting mnemonic reminders into fully developed information symbols. By assigning a relative duration to every single note and rest, the Franconian system made it possible in principle to capture any imaginable melody, just as the letters of the phonetic alphabet had made it possible to write any word regardless the spoken language.[37]

Franco's scheme of notes and rests produced an exacting musical chronometry from which there was no retreat. Spurred on by the demands of practical music, not philosophy, subsequent composers and theorists devised further subdivisions and their corresponding terminology as measuring units, terms such as *minim* and *tempus*.[38] Points or dots were introduced on some scores to indicate the separation of measures; bar lines first made their appearance for the same purpose in the fifteenth and sixteenth centuries (although not associated with time signatures until the seventeenth). In this setting the symbols for rests functioned analogously to the newly arrived zero in the Hindu-Arabic counting system, in effect giving instructions from the void for operations with adjacent quantities or notes. Franco put it this way: "The rests have a marvelous power for through their agency the modes are transformed from one to another."[39] Moreover, the strait jacket of measured, symbolic time liberated composers as new musical ideas of meter and rhythm flowed from the *Ars nova* and the *Ars subtilior* of the fourteenth century through the artistic height of vocal polyphony in the fifteenth and sixteenth. Long before then practical music had also spilled over the confines of the church and into the courts and streets. As music danced in air, from sacred motet to secular madrigal, symbols for time intervals joined with those of pitch to anchor the basic notational system that in its essentials remains the one we use today.

With their steadily spreading popularity during the Renaissance, Hindu-Arabic numerals seeped inexorably into the times and proportionalities of mensural notation, further broadening exposure to the symbols of modern, relational numeracy. The first actual time or *tempus* signs (the forerunners of our time signatures) appeared on motet scores written in the early fourteenth century. Shortly thereafter an anonymous theorist wrote the earliest known account of them, explaining that "the round circle" (O) indicated "perfect time" (dividing a note into three beats), while the "semicircle" (C) or (Ɔ) stood for the "imperfect" (dividing it into two).[40] Later, French and Italian musicians extended the pattern, creating a "system of [temporal] note-values in which each [note] is either a multiple of the next smaller one or a divisor of the next larger one."[41] Based on the divisions of notes, time signs thus designated fractions or ratios (for example, C = 1 / 2, or 1:2, one note divided into two beats). Composers initially expressed these fractional time measures with the common, widely used symbols of Roman fractions, drawn from ancient weight measures and monetary units.[42]

Using a symbol to assign a time value to notes at the beginning of a score worked fine as far as it went, but matters quickly compounded when musicians sought a change in tempo from one part or section to another in a composition, a frequent occurrence. This entailed both introducing a different time sign or some other means of indicating the shift, and comparing the new one with the original. Such comparisons involved relations between fractions or ratios, and from Euclid forward were understood as proportions. Hence calculating these changing time proportions, the "diminution and augmentation of metrical values in certain arithmetic ratios," became an integral feature of Renaissance musical compositions. Given such concerns, replacing mensuration signs with Hindu-Arabic numerals and the fractions of the new counting system greatly clarified and facilitated the process, expanding the range of rhythmic possibilities. A simplified example allows us to illustrate. Consider the following sequence of notes:

The O indicates that the initial time is perfect, with a breve note divided into the three semibreves shown here. The fraction means that the four notes (or beats) following it occupy the same time as the three preceding it. In early scores, the denominator generally referred to the notes before the fraction, called the *integer valor* (without proportion), while the numerator indicated the ones succeeding it, the *proportio* (proportion). In this example the numerical fraction supplanted a fixed or "noncumulative" mensuration time sign.[43]

Time changes grew even trickier when proportions became cumulative, building on each preceding section. In these new calculations theorists often relied on the "rule of three," which they borrowed from commercial arithmetic. We can best demonstrate this rule's application by extending the previous example. To do so we add a second row of smaller notes called minims (modern half notes):

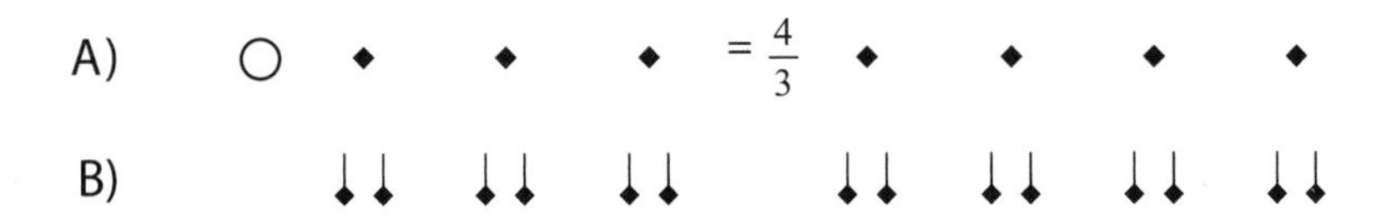

Row A, recall, correlates the three semibreves (modern whole notes) of the perfect long with four semibreves sung in the same amount of time. On row B, in duple arrangement, each of the semibreves is shown subdivided into two minims.[44] We can now see the ratios of the proportion. The first three semibreves are to the four after the fraction as the six minims are to eight (3:4 :: 6:8, or $\frac{3}{4}=\frac{6}{8}$). Leonardo of Pisa had pioneered the model for Europeans, showing how knowledge of any three numerical values in a proportion enabled one to find the fourth using the merchant's rule of three. Among countless others, composer Johannes Tinctoris (ca. 1435–1511) followed this procedure on one of his scores, shifting from a proportion of $\frac{2}{1}$ perfect breves to one of $\frac{3}{2}$. He then calculated the number of minims (understood as overall beats, or notes plus rests) contained in the three breves of the new section. In modern, algebraic form, that proportionality was $\frac{2}{3}=\frac{18}{x}$, with $x=27$. Thus two perfect breves with eighteen minims were proportional in their amount of musical time to three perfect breves with twenty-seven. (See Appendix A.4 for a more detailed description of Tinctoris's notation.)[45]

Many theorists had a field day with the exciting and new, arithmetic means of calculating proportionalities of time and rhythm, often creating fractions that far exceeded the relatively small number of proportions actually implemented in musical performances.[46] Yet, despite the lack of direct performance benefits issuing from these exuberances, there ensued other consequences, slower in manifestation but further reaching. The practice of using ratios and proportionalities to calculate and manipulate variable, durational units of time both enhanced and revealed a growing facility with the new, relational numeracy and with the territorial expanse of its purview. Time intervals were captured and brought under control, tamed by a young and energetic information technology.

By the sixteenth century, mensural notation had interwoven its own "signs or characters" with those of the new numerals to extract from musical sound a symbolic distillate of tone and time, pitch and duration. It had become the lingua franca of late Renaissance discussions of consonance and dissonance, whose determinants were the intervals of tone and their dynamic qualities. Those discussions centered on an issue known somewhat picturesquely to contemporaries and also to us as the "problem of tuning."[47] A perennial concern, and the second of our themes chronicling developments in practical music, the tuning question became the subject of more urgent debate after the 1480s, and reached a "point of crisis" in the following century with the explosion of purely instrumental music, unprecedented anywhere before then.[48]

Coordinating different voices in polyphonic music, clerical and secular alike, was one matter. Voices could bend intonation to different instruments in the course of performance, or when they sang unaccompanied, to ensure their harmonic interplay and complementarity. (Even so, they often went astray under the exigencies of new and intricate, chromatic melodies.) It was entirely another matter, however, to harmonize with instruments whose tonal intervals were fixed, such as with frets (lute or viola da gamba, for example) or keyboards (organ, harpsichord), and harder yet for members of instrumental ensembles to harmonize with one another, especially through key changes in a score. Adjustments were needed, compromises to be negotiated. Musicians were forced to rethink seriously the harmonic intervals of pitch and scales, and to do so using irrational and real numbers in calculating intervallic proportions. New or revived tunings were produced as the Pythagorean scale gave ground to "just intonation," to "equal temperament," and to countless idiosyncratic arrangements of intervals. No longer unassailable as the fixed, prescriptive, and mystically divine ratios, harmonic proportionalities gradually became a function of the tuning system in which they were devised and the keys in which scores were written.

Behind the courtly civility generally displayed in their public dispute over tuning and harmony, composer-theorist Gioseffo Zarlino and his one-time student, lutenist-musicologist Vincenzo Galilei, squared off over an issue of great depth and fierce rivalry: fundamentally, how to distinguish music from noise.[49] The theoretical context of the disagreement derived from the Pythagorean scale.[50] (For construction of the Pythagorean scale, see Appendix A.5.) Pythagoras and his followers had correlated the pleasing (harmonic) intervals of pitch with fixed ratios, based on the sonorous numbers of the *tetractys,* one through four. The octave, that "mother of all consonances," according to Anonymous IV (others dubbed it "queen"), denoted a harmonious pair of tones eight steps apart, its ratio of 2:1 produced by the vibrations of any string and another one of double (or half) its length.[51] A pair of tones five steps apart produced the perfect fifth, a different but pleasing sound that blended well with the original. Its ratio, 3:2, stood for three times the original string length divided by twice the original. Like the fifth, the perfect fourth (ratio 4:3) blended too, albeit with some tension. Nonetheless, it resonated compatibly with the original pitch. These were the "pure," Pythagorean tones, which in modern times are presented as the first four notes of the "harmonic series." (See Figure 6.3.) Further, smaller intervals followed for the entire octave and actually connected the larger ones in a melody, binding them together, as it were. But the smaller intervals—adjacent notes separated by a whole tone or semitone, for instance—produced pairs that did

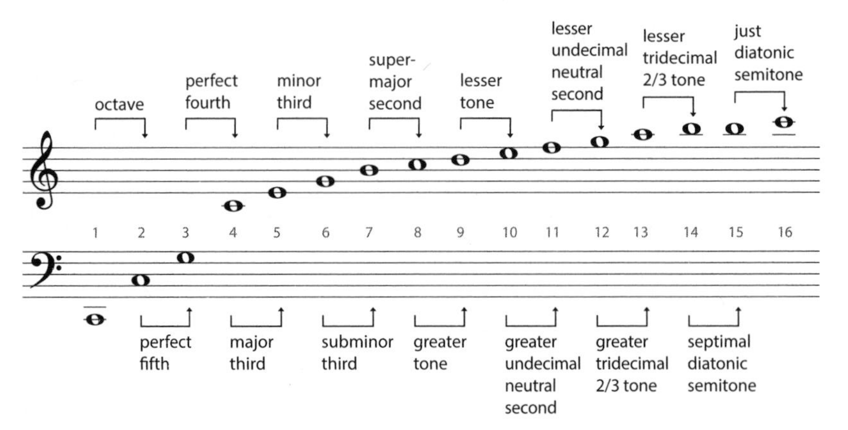

FIGURE 6.3. The harmonic series represented with modern notation and labeling, beginning with the note C, two octaves below middle C on the piano. The harmonic ratios are termed "superparticular," meaning that the excess of the greater term over the lesser is a unit—2:1, 3:2, 4:3, 5:4, 6:5, and continuing.

not blend well together. For the Greeks, these intervals were irrational. They sounded dissonant, even harsh, especially tones of the third and the sixth notes of the scale.

Zarlino was certainly not the first in the sixteenth century to address the problem of dissonance or to appreciate the numerical irrationalities that caused it, but his four-volume study, *Le Institutioni harmoniche* (1558), established him as the preeminent authority of the day, the most "lucid and perspicacious" proponent of the "golden age" of counterpoint.[52] In large part his treatise targeted the chaotic state of music theory and practice, which he found described in music books filled with "so many dots, rests, colors, ciphers, signs, ratios, and other strange things that they appear to be the books kept by an intricate business house." These volumes had little to do with "harmony and melody"; their authors had forgotten that "music . . . is concerned only with the sonority . . . that springs from pitches and tones, and with nothing else."[53] Zarlino seemingly brought order to this chaos by reviving and building on Ptolemy a tuning system that has become known as "just intonation," then called "syntonic (or tense) diatonic." In fact, Zarlino accepted most of the age-old assumptions of theoretical music, even while explicitly seeking to combine the theoretical and practical sides of the discipline in his own writings.[54] At the very opening of *The Art of Counterpoint,* part 3 of his opus, he maintained that the "harmonious intervals" of melody were to be "commensurable," that true consonance required "rational intervallic proportions and temporal measurements."[55] He had no intention of overturning the Pythagorean tradition.

Rather, Zarlino thought to correct its shortcomings by extending the harmonic series to include the numbers five and six. In particular, the number six (*numero senario,* "senary number") provided the linchpin to consonance. It was the first perfect number, Zarlino believed, because it possessed the virtue of being the sum of all the numbers of which it is a multiple ($1+2+3=1\times2\times3=6$). It was thus the perfect candidate for governing perfect harmony, manifest not only in art but in "nature" itself.[56] This adjustment added the ratios of 5:4 (major third) and 6:5 (minor third) to the ratios of the Pythagorean *tetractys,* as well as folding in the sixth more sonorously. Previously designated "imperfect consonances," these tones had grown increasingly popular in polyphonic singing, and were now brought into greater acoustical accord with tradition. With the pure fifths and thirds leading the way, sonorous number still determined harmony.

But on the practical plane this remained a strategy of robbing rational Peter to pay irrational Paul, as Galilei sought to demonstrate in a multipronged attack. Start with the basics, he insisted: "Nature has neither hands nor mouth. It cannot play music or sing, and our playing and singing is [*sic*] totally artificial."[57] We must use our ears and listen in order to ascertain what is harmonious. Listening to singers, for example, reveals that they do not sing "any intervals in their true ratios apart from the octave" but vary their pitch as the melodies require.[58] All tunings, including Zarlino's, were at best only an approximation of harmonic intervals. Even more to the point in Zarlino's case, the intervals in the syntonic diatonic themselves led directly to many dissonances, which Galilei demonstrated in painstaking detail, proceeding from note to note in careful analyses of the ratios involved. The minor third from D to F, for example, yields a ratio of 32:27, a distortion of the minor third from E to G, which is 6:5. So too with the major third of A to C# (81:63) when measured against that of C to E (5:4), and so too even with the perfect fifth from D to A when compared with C to G.[59]

Galilei punctuated his argument with an illustration based on the so-called spiral of fifths, the irrational and dissonant results of a Pythagorean or just tuning system constructed from only the perfect fifths (ratio 3:2) and octaves (2:1).[60] As one proceeds through several octaves, intervals of the fifth expand ever so slightly, so that in returning to the starting note, the tones don't match but eventually are quite far apart and exceedingly dissonant. (For the mathematical calculation of "irrational" dissonance in tuning, see Appendix A.6.) To counter this, the fifths had to be flattened, or "blunted" in Galilei's terms, and equalized along with all the other intervals. Tied to the piano and fixed-interval instruments, modern convention has followed Galilei by converting the spiral of fifths into a circle, creating equal intervallic distances within an octave and making semitones equivalent. For example, G# is the same note as A*b* in equal temperament; in Zarlino's system they were different.[61]

Even more dramatically, Galilei proved experimentally that the ratios for the octave, the fifth, and other pure tones, long considered irrefutable, were not eternally established harmonies. They worked only in limited cases, for strings of equal material and tension. Changes in pitch that resulted from tightening or loosening strings, which Galilei determined by hanging weights from lute strings, produced the traditional tones only when the squares of the ratios governed string relations: 4:1 gave the octave, 9:4 the fifth, 16:9 the fourth, and so on. Comparable variations occurred as well when considering the columns of air in organ pipes and in wind instruments. Galilei elicited a general observation from these investigations: numbers were significant only when they represented measurements of specific physical characteristics—length, tension, cubic volume, and the like—that is, when they described pitch rather than prescribing it.[62]

In fairness to Zarlino, for some music the system of just intonation continued to offer advantages overlooked by that of equalized temperament, which generated dissonances of its own, and it continues to have advocates among a small group of modern composers and musicologists.[63] In fact, it has largely gone out of favor due primarily to the widespread popularity of the piano and other instruments of fixed intervals used in both classical and popular music today. In short, equal temperament endures essentially as a matter of convenience. But in reality, nature trumps both Zarlino and Galilei, for there are no definitive, mathematically rational answers to the problem of tuning, no regular and fixed intervals between octaves that guarantee harmonious tones regardless of the tuning system used. For this reason, the final word goes to Galilei: the ear, not sonorous number, determines consonance.

* * *

By the young Galileo's day, the information symbols of musical notation lay in front of critical minds, available for reworking and reconfiguring, available for analysis. Broadly speaking, the practicality of notation was twofold: it systematically captured the relations of musical sound in marks used by vocalists and instrumentalists, and it fed into the expanding compass of thinking instrumentally with the newly emerging numeracy. Both these processes steadily chipped away at the hallowed beliefs in philosophical number, allegorical mathematics, and theoretical music as taught in medieval schools. Yet, even more significantly for developments in science, such practicality challenged the classifying temper and its underlying alphabetic literacy by reawakening in the realm of music the hoary and slumbering Greek bugbear of incommensurability, the categorical separation of discreteness and continuity.

Anonymous IV had relied on Hugh of Saint Victor's classification of music as "discrete quantity" in defining the "relative proportions" of music as "equal" ("unisons") and "unequal," with the latter further subdivided into five different classes of fixed proportions ("multiplex," "superparticular," and so on).[64] But as performing musicians and composers came to know well, there is nothing static in the dynamic flow of musical sound. Even the absence of sound, captured by rest symbols, drives expectations forward. While one can surely single out and pay attention to discrete intervals of time and tone, and advance one's own musical practice and playing in so doing—say, with a metronome—music itself streams forever in ephemeral motion. Vincenzo Galilei himself made the point pellucidly: "The subject of music, which is vocal and instrumental sound," he highlighted, "is a continuous and not discrete quantity."[65]

Of course, a single observation did not shatter the classifying temper, did not unify the inviolable categories of arithmetic and geometry, discreteness and continuity. That awaited the more fully developed technology of relational numeracy, whose story we are unearthing. Galilei's comment, however, does reveal a growing, perceptual recognition of the quadrivium's irrelevance among a smattering of forward-looking thinkers who evoked the promises of sunrise, a dawning of new ways to approach problems. In Germany, for instance, the same Michael Stifel who had unmasked Leo X as the devil incarnate tackled head on the commensurability of irrational and rational numbers in the context of music theory and practice. (He actually became one of the foremost mathematicians in sixteenth-century Germany.) Fully aware of the spiral of fifths and the irrationalities harbored in tuning, Stifel showed how a full tone could be "precisely taken as halved" by using square roots and adding a "comma," which he called a *schisma* (whose value he calculated as the miniscule ratio of 531,441:524,288): "Thus: 8:√72, [is a] minor semitone with a schisma; √72:9, [is a] minor semitone with a schisma. Moreover, you will prove this halving is precise by duplicating [squaring] the parts, because it is necessary that either part when duplicated gives precisely a tone. Thus, the proportion 8:√72 duplicated gives 64:72 or 8:9. Also the proportion √72:9 duplicated gives 72:81 or 8:9." The ratio 8:9, we recall, is the interval of a full tone between whole notes—for example, between C and D on the piano (see Appendix A.5). Despite his willingness to consider irrational and rational numbers together in the practical computations of musical harmony, however, Stifel continued to cling doggedly to the theoretical separation of "rational" arithmetic from "irrational" geometry, claiming of the latter that its "irrational numbers" were not "true" numbers at all but only "fictions," which remained "hidden under a cloud of infinity."[66]

Others, including the Italian cossist Girolamo Cardano, would go further toward "rationalizing the irrational." Widely known for his disputed "discovery" of the solutions of cubic equations and other mathematical accomplishments, he was also a musician and composer of some note, and he wrote several unpublished manuscripts on the tuning question. Cardano paid particular attention to modes of ornamentation, such as trills and vibratos, those tiny tonal shifts used by singers and instrumentalists alike to enhance their melodic lines. Among these, he latched onto the interval of a "diesis" (half a semitone, or quarter-tone), which produced "such a movement [that it] titillates the ear and increases its pleasure." Dividing a tone into two equal semitones or a semitone into two quarter-tones "correctly and arithmetically," he noted, required a "true calculation" involving an irrational root.

But in practical situations one actually adopts a "rational approximation" of a tone, which is "closer in perception" and pleasing to hear. To find this tone, one computed the integers, "greater and less, which most nearly satisfy the equation" (such as one might do in figuring a square root), generating in effect a converging series of approximations, ever closer but not exact. Through this procedure, one could "undoubtedly arrive at an insensible difference," so fine the ear could not detect any variation between the approximation and the precise value. "This is universal reasoning," he concluded, "and needs no other rule."[67]

Again, these forays into the mathematics of tuning were not science. Or at least not yet.[68] Nonetheless, hindsight allows us to see that in employing an information technology of abstract, functional symbols to capture musical relations in motion, musicians were building practical bridges across the chasm between theoretical categories (arithmetic and geometry), even while ironically and unknowingly helping to create other fault lines (science and religion). Moreover, as we shall next encounter, comparable practices and developments were occurring in other branches of the quadrivium, including advances in perspective and geometry furthered by painters, sculptors, architects, and others in the arts who sought visual, as well as auditory harmonies.

CHAPTER 7

Geometry

The Illusions of Perspective and Proportion

> You arrive at [a likeness] when you have certain information about all the angles, and all the lines and their according measurements, and where they run together.
>
> —Leon Battista Alberti, *On Sculpture* (ca. 1460s)

WHEN HUGH OF SAINT VICTOR and his fellow scholastics pigeonholed the divisions of mathematics into four subcategories, they were revealing the centuries-old predilection for grouping words and things, the classifying temper. In their definitions of arithmetic and geometry they were also revealing a deep, conceptual divide dating to Aristotle and before him to Zeno: the incommensurability of continuity and discreteness, the ordinal and cardinal dimensions of early thing numeracy. In our own age we too confront countless issues centered on the gap between continuity and discreteness, perhaps most recognizably in processing analog and digital versions of audio-visual information.[1] Since the days of Galileo and Descartes, however, the technology of modern, relational numeracy and its offspring, higher-level relation-mathematics, have enabled us to bridge this divide in practical, extremely useful ways, with contemporary computers eventually transforming *techné* into hitherto unimaginable dimensions. The medieval and Renaissance eras possessed no such technology, and in its absence intellectuals had no means of resolving the division between arithmetic and geometry. Bridging the conceptual gulf between the major branches of mathematics was simply too much for even the greatest of minds to overcome. Then, as now, such a feat lay beyond the logic of conventional language and its underpinning of alphabetic literacy, beyond the technology of the written word.

Only bit by bit, as it were, did Europeans invent a new information technology of relational numeracy out of the cultural materials and historical demands of the day, presaging a major, tectonic shift in our intellectual landscape. While the new technology would ultimately close the fault between discreteness and continuity, it would also contribute to opening one of equal or greater depth between science and religion. Our story follows these shifting fault lines. On the discreteness side of the premodern incommensurability divide, we have already seen how smaller fissures began surfacing with the new, abstract, and relational symbols of Hindu-Arabic counting and with the new, abstract musical notation that came to measure time. In both patterns of abstraction a discernible separation appeared between the categorically defined, philosophical arithmetic and music of the schools and the actual practices of merchants and musicians.

On the opposite, continuity side of the incommensurability fault—which for Hugh and company harbored geometry and astronomy—comparable developments marched apace. Although Euclid's *Elements* had for the most part disappeared with the eclipse of classical learning, a practical sort of geometry with its origins in the traditions of Roman land surveyors (*agrimensores*) had found its way into the early quadrivium of monastic and cathedral schools.[2] Hugh himself had written an often-copied manuscript on the subject, probably for teaching purposes, where he distinguished between "theoretical (speculative)" and "practical (active)" geometry, and where he defined for the latter yet another trinity of subsets: planimetry (measurement of planes and surfaces), altimetry (measurement of heights and depths), and cosmimetry (measurement of spheres, including celestial bodies).[3] With the new translations of Euclid in the twelfth century, theoretical geometry entered the university curriculum, but the revival of Aristotle in the thirteenth relegated it to a minor part of the educational program, somewhat on the order of the "math requirement" in many modern liberal arts colleges. Nevertheless, both within and beyond university walls, the religious and speculative dimensions of geometry continued in force throughout these centuries with a prevalence similar to that displayed in arithmetic and music. Geometrical proportion and symmetry, for instance, allowed one to glimpse "transcendental" beauty, considered along with truth and goodness an intelligible attribute of God. Visible beauty was born of contrasts in proportion, just as the harmonic mixture of opposing sounds midwifed auditory consonance. Both soared beyond the sensible realm into a subliminal and cosmic, mathematical noumena that enveloped the universe and at times bordered on the mystical.[4]

The leitmotifs of medieval geometric proportion and symmetry were voiced largely through vocabulary inherited from the Roman architect and author Vitruvius (ca. 80–70–ca. 15 BCE), whose work was cited by authors steadily from the ninth century onward, most notably by the great Dominican compiler Vincent of Beauvais (ca. 1190–ca. 1264).[5] In his mammoth encyclopedia, *The Great Mirror* (*Speculum Maius*, 1244–1255), Vincent explicated the Vitruvian principle of proportion as it pertained to the human figure, a theory known at the time as *homo quadratus* (literally, "squared man"). Vitruvius had taught that four was the number of an ideal man because the distance between the ends of his outstretched arms matched that of his height, thus yielding the base and height of a square. At the center of symmetry, such geometrical archetypes often came to convey religious symbolisms as well as an aesthetic essence. In this instance the cruciform shape and proportions of *homo quadratus* appeared in the design of numerous churches and cathedrals, also reminding communicants of Jesus's suffering and sacrifice on the cross. Other mystical numbers and shapes (including the pentad, the five-pointed star, or five-petaled rose) afforded symbols uniting the microcosm and macrocosm in the minds of many during this summertime of allegory.[6]

Centuries later, but not too many, Leonardo da Vinci depicted many of these same associations in his most famous drawing, the *Vitruvian Man*, where he combined the even-sided square with the perfect circle to frame the exemplary human figure. (See Figure 7.1.) Vitruvius's and Leonardo's proportions stemmed from a fascinating, often metaphysically and prescriptively charged number that has come down to us through history as the "golden ratio," also known as the "divine proportion" or "golden section" and frequently designated by the Greek letter phi (φ).[7] In relying on the golden ratio Leonardo displayed as much attachment to a medieval theme, the centrality of speculative geometry, as to one of Renaissance innovation, the centrality of man. Speculative geometry became even more pronounced in the Renaissance years of Leonardo and others with the resurgence of Neoplatonism, a flourishing and eclectic Hermetic tradition, astrological gleanings, and a host of comparable topics that occupied the likes of Prospero

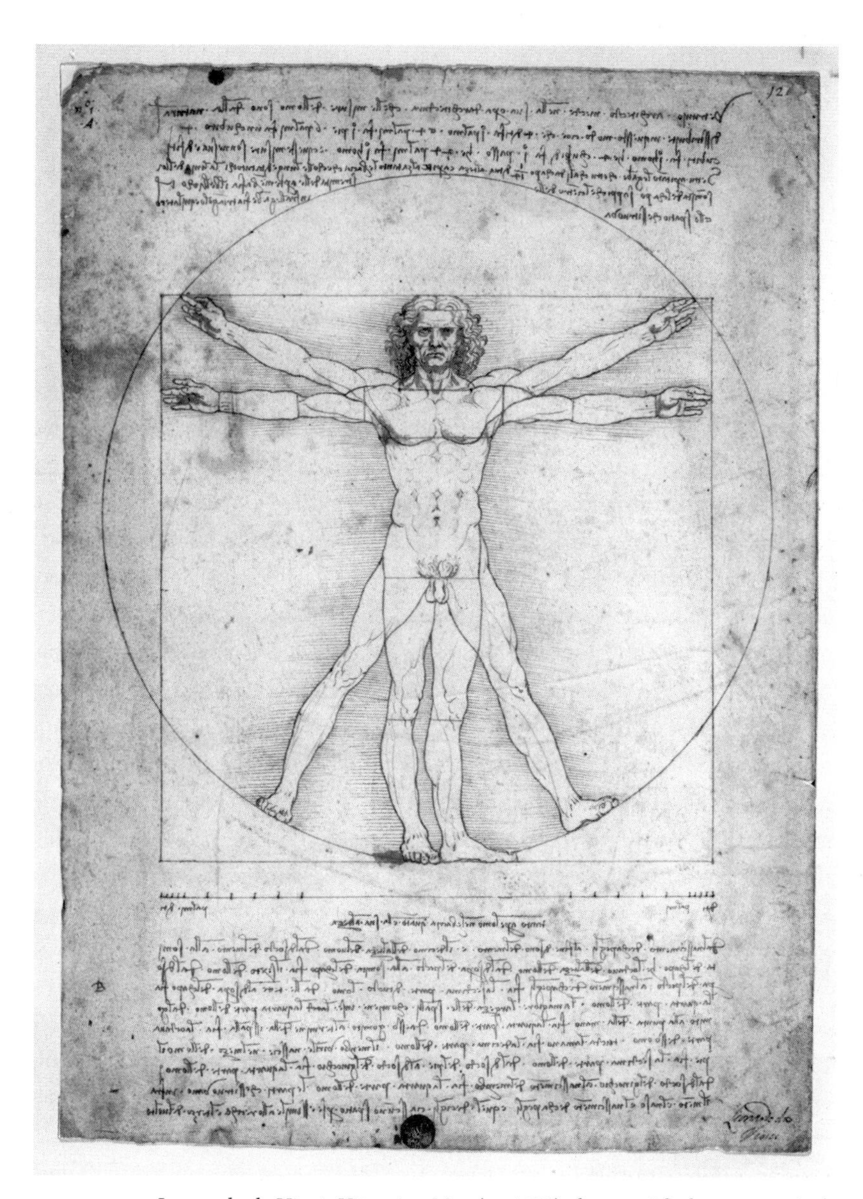

FIGURE 7.1. Leonardo da Vinci, *Vitruvian Man* (ca. 1490), drawn with the symmetrical proportions keyed to the "golden ratio as described by the Roman author-architect Vitruvius.

and his real-life avatar, the English mathematician-magus John Dee.[8] Still, notwithstanding the attention paid its metaphysical flights, geometry's re-emergence and presence in European learning produced virtually no major theoretical advances beyond Euclid before the seventeenth-century analytical geometries of Descartes and Fermat and the projective geometry of Girard Desargues (1591–1661).[9]

At the same time, alongside and outside the philosophical treatment of geometry, its practical utility for various trades and occupations blossomed profusely as agents of commerce and construction, exploration and battle, and worldly sights and sounds expanded the medieval *scientiae mediae* (or *mathematica media*). In scholastic taxonomies the "middle" (later called "mixed") sciences and mathematics occupied an equivocal place between the branches of mathematics and physics, and included astronomy (occasionally), optics (called *perspectiva* in Latin), and mechanics (also known as statics, or sometimes the "science of weights").[10] They neither apprehended causalities in the "real" world of physics enough to be called a bona fide part of natural philosophy nor did they address the eternal verities of pure mathematics. Rather, they treated matter from the standpoint of quantity, without any concern for causes. But the applied sciences and mathematics, especially when supplemented by the "mechanical arts," were simply too useful to be ignored. Thus, even though conceptual fault lines persisted between arithmetic and geometry alike, as well as between physics and mathematics, the tradition of *mathematica media* encouraged an active, everyday traffic among physics, geometry, and arithmetic in tangible, often engineering, applications.[11]

Embedded in these nearly three centuries of developments in practical geometry lay the piecemeal creation of a novel means of capturing and processing visual information. As we have seen, the ancients, Aristotle chief among them, had used the technology of writing to harness the constantly fluttering ephemera of our sensations, especially sight, into scientific knowledge through a process of "understanding" that began and ended with definitions in their fullest extent: a universe of words and things. In a departure from classifying, practical men in the Renaissance corralled the fleeting phenomena of vision differently, using the points, lines, shapes, angles, figures, and other data of geometry to elicit new and distinctive "mental extracts." Their attention shifted away from definitions and the abstractions of a symbolic thing-space to investigating just how vision itself functioned, just what was entailed in a point of view, and just how such information might be brought to bear pragmatically on projects at hand. Whether constructing a fort, aiming a cannon, surveying a plot of land, navigating the seas, devising a map, designing a building, sculpting a statue, or painting a picture, visual information became encoded into numerous types of geometrically organized grids or matrices. In this way practitioners could situate abstract points, visual figures and shapes, or real objects in the fluctuating field of sight. Such grids served as a conceptual framework for determining the relations of perspective and proportionalities between objects and, more critically, for following positional changes as viewpoints shifted. Theirs was a world of the "winged eye," visual processes and relation space, not of words and things.[12]

Processing visual information entered its most encompassing and conspicuous arena in the visual arts themselves, where the discovery of linear perspective transformed the world of painting, drawing, printmaking, sculpture, and architecture. Like its musical counterpart, polyphony, linear perspective was the singular product of Western, European creation, one whose significance for some can scarcely be overdrawn.[13] One lineage of its ancestry can be traced directly to the optics of antiquity, but the distinctive innovations of situating bodies and movements by means of coordinates, graphs, scales, and grid lines must be documented as best one can with scattered and only occasional references before the fifteenth century. Then, during the century and a half prior to the birth of Galileo and concentrated in the environs of Florence, several remarkable generations of artists, humanists, engineers, and craftsmen cultivated what some modern art historians identify as a unique and distinctive "gaze," a new "act of seeing."[14]

Whatever the cultural meaning of "gazing" may suggest (art historians offer a number of competing, sometimes far-flung and mercurial interpretations), the new practices of linear perspective displayed a down-to-earth, technological significance in the means of depicting, understanding, and manipulating the objects of vision.[15] Initially discovered by Filippo Brunelleschi (1377–1446), what artists and writers came to call "legitimate construction" (*costruzione legittima*), a series of techniques used in painting and sculpture, was introduced and explained by Leon Battista Alberti (1404–1472), the "first theorist of perspective."[16] Later in the century, the works of Piero della Francesca (ca. 1412–1492), paintings and mathematical writings alike, perched at the acme of "seeing through" (the literal meaning of *perspectiva*). In both artistic depiction and investigative understanding, these men, along with many among their contemporaries, opened new vistas onto the most ancient and still abiding of mysteries, that of light.

The new way of thinking about vision and light involved more than merely applying Euclidean rules to fixed objects and spaces. In artistic practices, generic class was being converted into general dimension and specific differences into singular points. Seeing objects as a function of the visual process meant propounding a new "logic of likeness," to enlist Alberti's trenchant phrase, halting footfalls toward the geometrization of matter.[17] Artists and commentators alike were opening the door to a prototype of reverse engineering, of modern analysis. Like the magician's legerdemain, there was something mesmerizing in these visual creations . . . and not just a little paradoxical. For they appeared to represent natural objects far more realistically, far more scientifically said some, than had previous depictions, and yet they did so by creating and managing the illusions of reality.

* * *

In the year 1413, representatives from the Florentine Guild of Linen Makers and Rag Dealers (Arte dei Linaioli e Rigattieri) met to inspect a statue the association had commissioned from a young sculptor, Donato di Niccolò di Betto Bardi (ca. 1386–1466), better known by the diminutive Donatello. Competing to enhance their presence in Florence's politically powerful guild structure, the guild masters of this "middle guild" (*arte mediane*) had agreed to supply a statue of Saint Mark for the Orsanmichele, a former granary converted to church, which stood in the center of the city.[18] But chagrin and disappointment showered the guild masters as they gazed at the sculpture. It was all wrong. Before them stood a malproportioned man shape, a caricature with an elongated torso and head, not the majestic, triumphal figure of the apostle they had been promised. (See Figure 7.2a.) Donatello would have to start over. The shrewd, young sculptor demurred, however, convincing his commissioners to let him work on it another fortnight, and to place it in its proper niche, whereupon he then covered the statue during the "work" period. Of course, Donatello did nothing in the interim. He merely positioned

(*A*)

(*B*)

FIGURES 7.2A AND 7.2B. Donatello, statue of Saint Mark (1413). Figure 7.2a is of the original, now standing in the Orsanmichele museum, while Figure 7.2b is a copy, elevated and placed in its intended niche on the outside of the building. The elevation allows the perspective of the viewer to foreshorten the torso, making it look normal.

the statue in its elevated alcove, providing the unaware guild masters with a gorgeous piece of art, which left them elated.[19] (See Figure 7.2b.)

Donatello had given the linen and rag folks more than a statue; he had given them a new perspective, a new point of view. Granted, for centuries reaching back at least to Phidias of Athens (ca. 490–ca. 430 BCE), sculptors had recognized that foreshortenings and elongations were required in statuary (and friezes as well) to enable naturalistic viewing of a work from a given position. Donatello, too, likely invoked visual intuition rather than precise geometry to make his adjustments. But as sometime protégée and slightly younger companion of Brunelleschi, he had also been exposed to a more elaborate appreciation of viewpoints and construction lines.[20] With Brunelleschi and Donatello, and soon painters such as Masaccio (Tommaso di Ser Giovanni di Simone, 1401–1428) and Paolo Uccello (1397–1475), as well as the "Renaissance man" himself, Leon Battista Alberti, perspective was making its move from artistic intuition to geometric construction, a new technique for processing visual information.

It began with space, this new perspective: not the finite, physical space occupied by and between objects, the thing-space of Euclid's abstraction into geometric symbols and diagrams, but the visual relation-space, continuous and infinite, that was a function of what one saw. Already known by 1413 as "the perspective expert," Brunelleschi launched investigations into visual space with an experiment shown to friends and associates sometime in the early fifteenth century, a demonstration perhaps devised as an outdoor parlor trick, an optical illusion.[21] His biographer Antonio Manetti described the project, which was based on a painting of Florence's famous, octagonal Baptistery of Saint John. Brunelleschi began with a small, wooden panel, about a foot square, on which he painted the façade of the baptistery, including its inlaid marble, with such "great precision" that "no miniaturist could have done it better." For the sky he used "burnished silver," which served as a mirror, reflecting the actual clouds as they drifted by. In the center of the panel, Brunelleschi then drilled a small hole, the size of a "lentil bean," widening "conically like a woman's straw hat" to about the "circumference of a ducat" (small coin) on the back side. (See Figure 7.3.) From the same, reverse side, one could look directly through the hole at the baptistery from a position just inside the entryway of the cathedral facing it. The pièce de résistance of the experiment ensued when, still looking through the hole, Brunelleschi held a mirror in front of the painting at arm's length.[22] The mirror reflected the image (in reverse), complete with its moving clouds, so accurately that it was virtually impossible to distinguish the painting from the reality: "The spectator felt he saw the actual scene when he looked at the painting." Thus did the experiment demonstrate, Manetti concluded, how one could depict artificially "figures and objects in correct proportion to the distance in which they are shown."[23]

Brunelleschi never recorded how he "originated" his "system of perspective," leaving generations of commentators, beginning with Alberti, ample scope to fill in the blanks with infrequent details and their own imagination. Part of the story stems directly from the early science of optics, Dante's "handmaiden of geometry" and a topic of much currency among intellectuals, artists, and artisans surrounding Brunelleschi and his associates.[24]

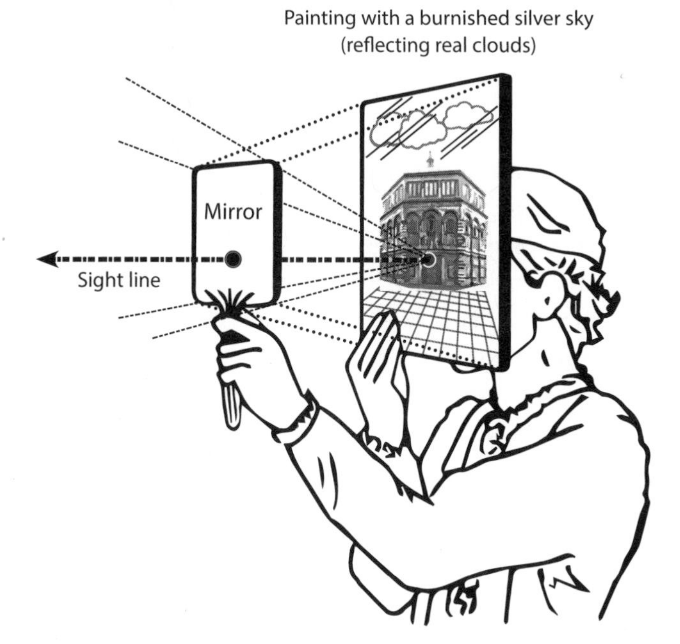

FIGURE 7.3. Depiction of Brunelleschi's experiment, which was designed to show identical views of Florence's famous baptistery, either by seeing the perspective painting reflected in the mirror or by removing the mirror and looking directly at the baptistery.

Using scant empirical evidence, ancient writers had formulated the study of optics in search of the right definitions of visual motion, the essence of seeing.[25] Their theories had been reworked by Muslim natural philosophers beginning in the ninth century, which culminated in the work of Ibn al-Haitham (Alhazen or Alhacen in the Latin world, ca. 965–ca. 1040). Translated into Latin as *De aspectibus* or *Perspectiva,* Alhazen's treatise on optics amalgamated physical, physiological, and mathematical elements from antiquity into a "single comprehensive theory" and became the key source for the optics theories of a few scholastics, above all John Pecham (ca. 1235–1292). Pecham's own trim volume, *Perspectiva communis,* in turn became the most widely read and influential work on optics during the Renaissance.[26] Throughout all these discussions the scientific objective remained the precise definitions of vision and light.

But the definitions of natural philosophers could only go so far, and Alberti dismissed their arguments as "without value" for the "purposes" at hand—namely, accounting for the "power of vision" and the construction of linear perspective.[27] Of more "value," though of less certain parentage, was the practice of exploiting grid lines for organizing visual space. Like the identity of the first person to eat a raw oyster, the earliest use of such lines lies deep in the well of the past.[28] The aforementioned and often-cited Vitruvius had supplied a few remarks of his own on the role of sight lines in designing and painting scenes in theaters, but like ancient sculptors he possessed more of an intuitive reckoning than any geometrical grasp of scenographic depth.[29]

Euclid offered more, supplying Renaissance minds with geometrical diagrams of rectilinear, "visual rays" emanating from the eye. In his *Optics* he began with the notion of a visual "cone," a central viewpoint, such as Brunelleschi's peephole, from the "vertex" or eye of which the "rays of vision" radiate. He then introduced rules and constructions from his own *Elements* to devise claims about the cone and about various measurements and angles associated with sight. For instance, he demonstrated that objects of the same size appeared larger when closer to the eye because of a wider angle of vision.[30] But as Vitruvius would do later, Euclid stopped short of where matters might have become interesting for an understanding of perspective. With his final proof he drew a square, then proceeded to demonstrate that if the eye rests on the "perpendicular" of the "meeting point of its diameters," the "sides of the squares will appear equal," and so too the diameters. In other words, when looking down on a square from a point above its center, one will see, yes, a square square! Sometimes singling out the obvious can be a useful prelude to greater insights, but not so in this case. Euclid handily proved the rather simple symmetry of the proposition, and then . . . nothing further. Renaissance artists started here, from where he had withdrawn, by looking at the square from a different point of view—say, from off to one side. How might it appear then? Clearly not square. With Brunelleschi leading the way, the purveyors of linear perspective sought to solve the problem of the square's construction, and hence of representing a building or landscape or simple square floor (*pavimento*) from the position of the viewer.[31]

Euclid's devoted follower Claudius Ptolemy supplemented his master. As with his astronomy, Ptolemy synthesized much of ancient learning concerning optics, combining Aristotelian descriptions of discrete objects in physical space with a geometrical account of "visual radiation" as a "continuous rather than discrete" property.[32] Visual angles continued to provide the basis for judging the size of objects, but they were supplemented by additional factors, such as the nature of visual rays.[33] Distance and orientation, too, played a role in judging the size and place of objects, as did, of special note, the innate sense of the viewer's own motion or rest. The "visual flux" is governed by the "sense of touch," Ptolemy remarked, which allows one to determine whether the viewer or the object seen is in motion. While Ptolemy's geometrical optics remained firmly planted in the Aristotelian world of things, his emphasis on the primacy of the viewer's position (stationary or moving) in situating visual objects added grist to the Renaissance perspective mill.[34]

Ptolemy's most enduring contribution to linear perspective, however, lay not with his optics but with his innovations in cartography and in particular two original elements from his *Geography*.[35] The first of these was the development of a world map (*mappamundi*) based on a coordinate system using the parallel lines of latitude and longitude, both of which Ptolemy derived from his investigations into astronomy.[36] Initially applied to the celestial sphere that circled the earth daily, Ptolemy extended the lines to the earth's sphere as well. Moreover, in using earthly latitudes and longitudes as coordinates, he was forced to address the tricky matter of representing a three-dimensional sphere on a two-dimensional surface, the second of his accomplishments. Ptolemy knew there would always be some distortion when depicting a spherical surface on a plane, and he devised three different ways of addressing the difficulty of mapping cartographic projections.[37] Most influential for artists in particular, the third of these utilized the horizon line principle in constructing the projection as though viewing the earth through a visual cone from a distant point in space, which rendered the latitudes above and below the sight line as ellipses. With latitudes and likewise meridians formed in proper perspective, "in an oval shape," all entire lands of the known world (what Ptolemy called the *oikoumenē*) could be mapped in correct proportions to one another. Regional maps could then be drawn up as subsets, based on the same coordinates as the world map.[38]

What, if any, of the foregoing actually entered Brunelleschi's mind before presenting his marvelous demonstration will remain forever in doubt.[39] As an architect studying ancient ruins in Rome, he certainly came across the physical remains of plots and layouts, which he measured on his own, and that may have sufficed.[40] The situation differed with the university-educated, classical scholar, humanist, and "master builder" Alberti, who had far more extensively assimilated much of the deeper background and conveyed it in his writing.[41] He recognized the fresh and inventive nature of Brunelleschi's experiment and other accomplishments, even dedicating the Italian version of *On Painting* to the "genius" whose "wonderful merit [would] obtain everlasting fame and renown."[42] This was a worthy observation from a worthy observer, for with Alberti the new visual space, linear perspective, found its pioneering theorist and expositor, and visual information one of its foremost early expressions.

From the outset of his treatise *On Painting,* Alberti made it clear that he would encode and process information from the field of vision through a series of geometric "signs," taking "from mathematicians" only "those things which seem relevant" to the world of painting and visual experience. In classical, rhetorical fashion, he introduced a series of definitions to explain himself. A "sign" was "anything . . . on a surface . . . visible to the eye." Thus a "point is a sign," one "not divisible into parts," and so too a line, "whose length can be divided into parts . . . but is so slender in width that it cannot be split." Further, lines woven "closely together like threads in a cloth . . . create a surface" itself, the most important of visible signs. A sign, it followed, could represent either a part of the surface or the surface as a whole. Surfaces in their stead displayed two "permanent properties." The first was the "outer limit" of a body, its "outline," which could take any combination of shapes—circular, triangular, quadrilateral—and which we recognize only "by width and length," not "depth." Depth enters visual experience from the second permanent property of surfaces, which Alberti introduced with the image of "a skin stretched over the whole extent of the surface." A surface could appear either "plane," "spherical," or "concave," or some "composite" of these, lending the solidity of three dimensions to visible objects and spatial depth to the relations between them.[43]

Understood geometrically, then, signs delineated the abstracted features of bodies as they appeared to observers, visible marks standing for visual properties. Alberti proceeded to explain how this visual information could be organized and reproduced (or constructed), beginning with the "visual rays" of Ptolemy and others, which gathered into a "pyramid" (his variation on the Euclidean "cone") containing the entire visual field and having its vertex at the eye. The field was framed by the picture plane, best understood as a window placed arbitrarily some distance between the eye of the viewer and whatever scene it observed. The visual rays in effect diverged from the

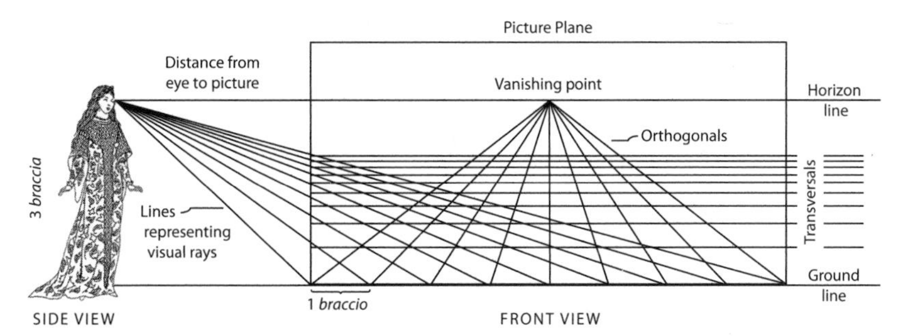

FIGURE 7.4. A depiction of Alberti's perspective construction, showing front and side views. The picture plane frames the observer's view and the vanishing point structures its contents. The orthogonal lines are perpendicular to the picture plane, while the transversal lines are parallel to it and to the horizon. Thus is created an organized grid of visual space within which objects are situated and proportions between them established.

eye, but once reaching the plane of the picture, they then focused on a "centric point" where all the lines behind and perpendicular to the plane (orthogonals) converged. The centric or vanishing point (so called nowadays) was opposite the viewer's eye and on the same horizon line with it. Lines running behind and parallel to the picture plane and the horizon (transversals) were spaced proportionately closer to each other as they receded toward the vanishing point. (See Figure 7.4.) Their actual number and spacing were determined by the arbitrary divisions one imposed on the picture plane. Alberti used three *braccia* as the height of the viewer and the viewer's distance from the plane, then used *braccio* units to segment the plane's width.[44]

The resulting geometric grid, which Alberti called the "pavement" (*pavimento,* the name given to tile floors popular at the time), allowed one to plot the proper size, shape, and positions of all the objects to be painted as blocks of empty, abstract space in perspectival and proportional relation to one another.[45] The vanishing points were vital in these operations; along with the eye point, they were the keys to ordering visual space. The counterparts to a zero in the Hindu-Arabic numeral system or to a rest in musical notation, the empty vanishing points delivered the spatial instructions that governed the placement and shape of all the grid lines within the picture plane. Accordingly, they established the possible relationships among visible surfaces in all their configurations and parts, as well as between the picture plane and the viewer, a hypothetical and abstract world of visual relation-space. Alter the vanishing point or viewpoint and one changed the geometric grid of the entire picture plane, and with it the shapes, positions, and relations of its contents.[46]

From the Florentine century forward, a trio of perceptions—motion, continuity, and infinity—accompanied the new, geometric techniques of contriving visual space. Alberti's "winged eye" saw objects in this space as continuously in motion, a function of ever-moving sight and ever-changing points of view. Paintings "move" us emotionally, he said, engaging in wordplay, because "feelings are known from the movements of the body." Horizons, picture planes, and vanishing points organized a "composition" by harnessing the flux of light, a stop-action picture framing a scene in a motion-filled universe.[47] Lines, it was assumed, would continue without end until they vanished in infinity, cut by a horizon, or a picture plane, or the plane of another object in the field of vision, an unending extension oftentimes referred to as "privative" infinity. Piero della Francesca, for instance, spoke of "a line without end" as one whose end points had yet to be determined.[48] More generally, visual perception often was seen as peppered with infinities, an idea Leonardo da Vinci reiterated in his *Notebooks:* "Every surface is full of infinite points," he observed. "Every point makes a ray, [which] is made up of infinite . . . lines . . . filling the air with . . . infinite images . . . and infinite [visual] pyramids," each of whose "center line . . . is full of an infinite number of points of other pyramids."[49] A continuous horizon line by itself hosted an infinite number of vanishing points.

As suggested earlier, creating a perspective grid with such lines and points and pyramids entailed arbitrarily framing and ordering a block of space in radically different fashion from the Greeks, for whom space was finite, the physical place occupied by material objects.[50] The symbolic abstractions of Euclid's geometry were predicated on this fixed, finite world of things. Lines, visible and determinate, had ends, and were expressed by a simple pair of letters—for example, AB denoted the line A______B. In modern parlance, by contrast, AB designates a "line segment," on the Renaissance assumption that a "line" per se continues ad infinitum, an idea foreign to the Greek world.[51] In short, linear perspective came to express measurable, limitless space just as abstract, musical notation expressed measurable, limitless time.[52]

Undergirded by an abstract, pictorial geometry, changes in perspective could even make objects appear and disappear, and spaces along with them;

witness *The Ambassadors* of Hans Holbein the Younger, where a blob of light and dark when observed from the front becomes a skull when viewed from the extreme side. (See Figures 7.5a and 7.5b.) Commonplace now with today's technology, such visual metamorphoses drew attention at the time to the illusionary quality of viewpoints and perspectives.[53] A prototype of modern analysis attended these constructions as artists broke apart and re-created a

FIGURES 7.5A AND 7.5B. Hans Holbein the Younger (1496–1543), *The Ambassadors* (1533). The white mass at the center bottom of the painting (Figure 7.5a) is actually an anamorphic skull, which can be seen as 'normal' (Figure 7.5b) from a viewpoint nearly at the edge of the painting.

(*A*)

(*B*)

point of view, reverse engineering visible objects according to the grid lines of the view and vanishing points. This superseded mere realism in depicting natural objects, evoking a deeper and emerging idea that objects are best understood in their positions vis-à-vis one another. Constructing objects visually meant manipulating illusion, deceiving the eye into believing the truth of what it saw, while for its part deception accustomed one to seeing the truth of things in relationships, not in the collective and classified qualities of their *quidditas,* or "thingness."

* * *

As far as we know, no one's zeal for proportion matched Paolo Uccello's fixation on perspective. "Oh, what a sweet thing this perspective is," he reportedly replied to his wife's entreaties "when she called him to bed." We don't know what he was eyeing in those moments, but the obsessive Paolo, "searching" for vanishing points "at his desk all night," put everything into perspective: chalices, caps, buildings, people, scenes . . . his marital bed. (See, for example, Figure 7.6.) More than anyone, he epitomized the exciting new world of visual geometry, although Giorgio Vasari held him in rather low artistic esteem, believing that he had turned a "fertile and effortless talent into one . . . sterile and overworked."[54] Still, it would not surprise us if Paolo's doppelganger in proportion were to emerge from a yet unread archive. Proportion was omnipresent in the Renaissance, permeating the truck of business, the tempo of music, and here the perspective of art. It was also subject to numerous interpretations, as old and new ways of looking at number and proportion often huddled in the same mind. For example, as he was devising a modern-appearing, technical description of the workings of linear perspective, Alberti also continued to enlist many traditional, prescriptive and Pythagorean, "simple ratios" of "harmonic proportion" in his discussions of architectural design and practice.[55] Yet, even here, Alberti's work on architecture found its practical and descriptive counterpart to, say, Vincenzo Galilei's attack on "sonorous number" with his insistence that the test of a building should be the very practical maxim "By your fruits shall ye know them."[56] Listen to it, live in it: the results of the new techniques were to be determined by experience, not by ideal, mathematical prescriptions of what one ought to hear or how one ought to live.

FIGURE 7.6. Paolo Uccello, perspective drawing of a chalice (1450).

While 'proportion' had many interpreters in Renaissance visual arts, its practical, mathematical descriptions stemmed mainly from the new perspective, not from the manifestations of idealized, divine number. Alberti had laced his treatises on painting, sculpture, and architecture with practical proportion, and so did the generations after him, artists and commentators alike. Among them, painter-mathematician Piero della Francesca gave mathematical footing to perspective's new means of calculating and manipulating proportionalities. (He also painted some hauntingly and dazzlingly beautiful pictures.) Both he and Alberti formulated techniques for creating and applying proportionalities based on a one-to-one correspondence, or "mapping" between line segments, between points on models and sculptures, and between numbers and points on an image. Evoking the cardinal principle of number, these mappings enabled correct and accurate scaling in both two and three dimensions and helped bring closer together the symbols of Hindu-Arabic arithmetic and the grids of geometric perspective.

As a painter, Piero's reputation has been buffeted by popularity's tidal ebbs and flows until achieving new, apparently lasting heights within the last century.[57] In his own lifetime he was known as much for his mathematics as for mastering the craft of painting, and while accurate biographical information remains in short supply, most scholars agree that it was in his later years when he wrote three works on mathematics, including *On Perspective in Painting* (*De prospectiva pingendi*).[58] Throughout his study, Piero sought to show not only how artists could "degrade" objects (the common idiom for converting objects from a "perfect" form into one seen in perspective) but just how the "force of lines" shaped Alberti's "power of vision" in giving formal structure to the "true science" of the visual world, the core of which was proportion.[59]

Through a series of careful how-to propositions, Piero expanded Alberti's one-point, "legitimate construction" with the earliest known description of the "distance point method," although it had likely been a "work-shop technique" handed down for a few generations.[60] Only implicit in Alberti's account of perspective, this method emphasized the viewer's eye as a unique reference point from which to construct the grid lines of the pavement. The pavement's square tiles, receding in perspective, could only be fully justified geometrically from one particular vantage point and from one specific eye distance in front of the picture plane. (In normal viewing, visual psychology often allows the mind to compensate for geometric distortions when looking at a picture from a different angle or distance.) Further, with the viewer's reference on the horizon, the introduction of additional distance points also led to constructions of more-complex grids, using two or more vanishing points at distances outside the picture plane. More-complex grids permitted the accurate depiction of objects at various angles to the picture plane and brought the whole matter of scaling and proportionalities to the forefront.

Piero drew extensively from Euclid for proofs in justifying his grids and explaining proportion, relying frequently on the use of similar triangles and our now-familiar rule of three. These made possible a correspondence between the lines of smaller, more distant objects within the picture plane

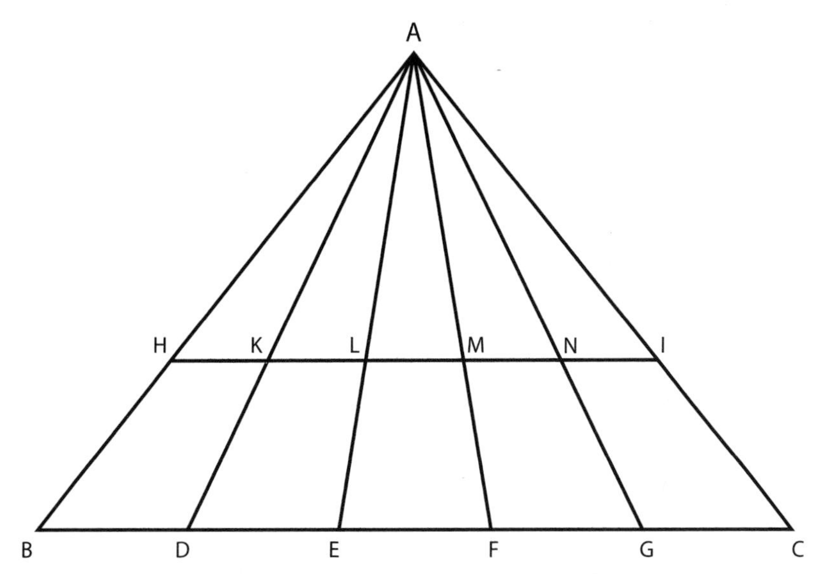

FIGURE 7.7. From Piero della Francesca, *De prospectiva pingendi*, book 1, proposition 8. The diagram accompanied his proof of correct scaling, based on the proportions of similar triangles.

and those closer to the viewer, thus allowing for precise scaling and perspective. Piero's own diagram in book 1, proposition 8, showed how it works. (See Figure 7.7.) With A as the vanishing point and lines BC and HI transversals parallel to the picture plane, the orthogonal lines (BA, DA, EA, FA, GA, and CA) create pairs of similar triangles. Thus ΔABD is similar to ΔAHK, and lines HK and BD are proportional to AH and AB, respectively (HK:AH :: BD:AB). Any objects constructed with these lines, or within the space they enclosed could be measured and reproduced exactly on a scale of the artist's choice. Proportionalities were used everywhere to establish the correct sizes of objects more distant in the plane relative to those nearer the "window." And from knowing the lengths of any three corresponding lines in a pair of similar triangles, one could calculate the fourth line, using the rule of three. Thus was created not only a grid allowing for the realistic representation of objects but one that enabled detailed and accurate measurements and locations of objects in relation to one another.

We mentioned earlier that Alberti labeled these proportional comparisons a "logic of likeness." This "logic," which Piero followed as well, had little or nothing to do with the "essential likeness" on which Aristotle, Thomas Aquinas, and many others had based their accounts of causality and classification; there were no metaphysical "traces" in the new visual likenesses.[61] Rather, this was a "techno-logic," so to speak, designed to explain the workings of proportionalities in a practical sense. Alberti even invented a tool for sculptors to use that could "perceive and record, and make definitive notation, of the features, positions, and postures of the parts of any body." He named this device a *diffinitore* (*finitorium* in the Latin version of his text), which in English we might render with the clumsy 'definer.' By fixing "dimensions" (linear measurements) and "terminal points" with great accuracy in a one-to-one correlation between points on a model and those on the sculpture material, the *diffinitore* was able to replicate any three-dimensional object employed as a model for sculpture.[62]

Piero applied comparable techniques to replicating complex, three-dimensional figures on a two-dimensional surface. As illustrated in his sketches of a head (see Figures 7.8a, 7.8b, and 7.8c), he devised a more intricate gridding, in a manner quite like the modern elevated drawings used by architects. Here, Piero drew front, side, top, and bottom views of a head, which he then crisscrossed with grid lines keyed to specific features and labeled with numerical sequences. For example, around the broadest circumference of the head (above the eyes), he marked off sixteen equal segments (top view). Then he showed the segments that could be seen from the front, side, and bottom views. After tilting his gridlines in the desired inclination, he transferred the position points from the elevated drawings to mark off the head in the "degraded," perspective image he sought. The one-to-one correlations transferred the correct proportions between the head's features from the elevated to the perspective versions of the sketch.

With their technical operations both Piero and Alberti avoided the "cult" of the golden ratio in figuring proportions. Instructing sculptors, Alberti explained how he "extracted . . . measurements and proportions" from "many bodies," which he used as the basis for comparisons, "laying aside the excesses of the extreme" and eliciting the "median" among them. His sample set was subjective and arbitrary, to be sure, selected from his own images and perceptions of "healthy and beautiful" bodies. But the uniqueness of this method lay with deriving a statistical, descriptive generality from the actual proportions found in his samples, not from the ideal of a philosophically

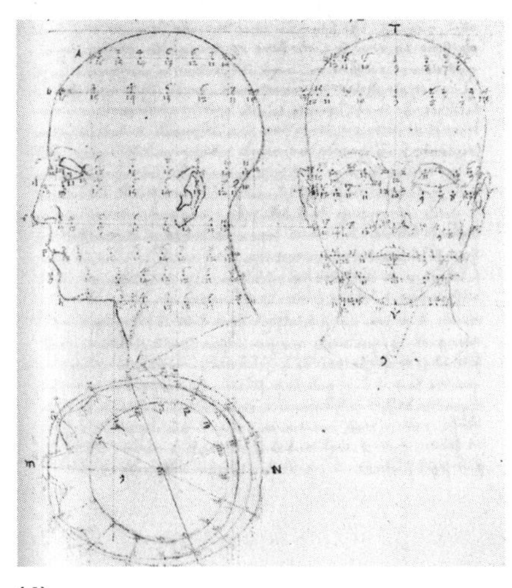

(*A*)

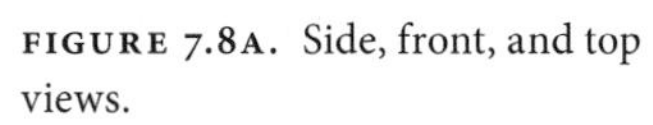

FIGURE 7.8A. Side, front, and top views.
FIGURE 7.8B. Side and front perspectives.
FIGURE 7.8C. Bottom and top and views.
Perspective sketches of a head. From Piero della Francesca, *De prospectiva pingendi.* In Figure 7.8c, Piero constructed and numbered sixteen equal, pie-shaped segments of the head. These numbers "mapped" onto the front, side, and perspective views in Figures 7.8a and 7.8b, illustrating the means of foreshortening the head proportionately when viewed through the picture plane.

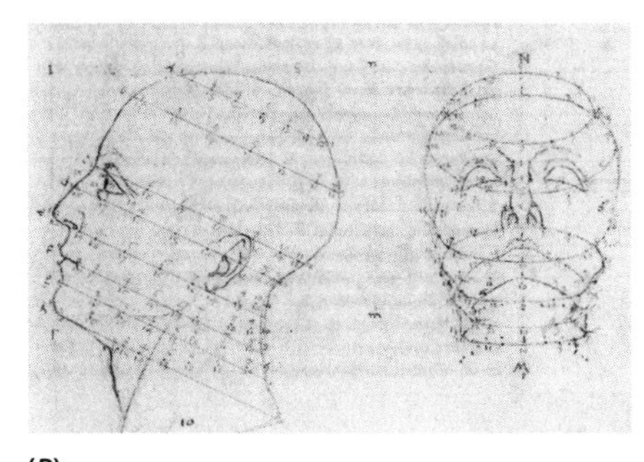

(*B*)

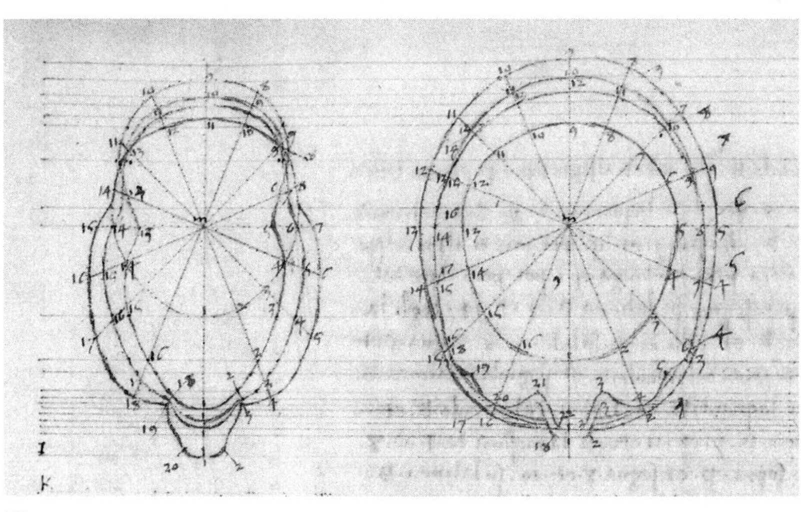

(*C*)

inspired, geometrical harmony imposed prescriptively onto the relations between body parts. Thus he took measurements from numerous ankles, noses, hands, feet, navels, waists, and so on to elicit his averages. From them he composed a typical "dimension" of bodies, and from the specific variations on the dimensions he detailed the "points" that permitted the sculpting and painting of particular figures and faces, as with the making of portraits.[63]

These technical means of manipulating proportions were striking and novel. As an idea proportion had been around for millennia but under the sway of the classifying temper. Often seen as divine, proportions and ratios were fixed, and their "thingness" pervaded their dimensions, delineated and determined by definitions. (Recall, the Greek term "analogy" also meant "proportion.") With the Renaissance, proportions became fluid and flexible. In business, percentage returns and profits fluctuated with markets and with prospects. In music, proportion in intervals functioned dynamically as tempi were changed from section to section in a composition, as tunings were altered, as singers and players bent their intonation to harmonize with one another. In art, perspective extended geometric proportion in the service of a point of view, which like music's time ratios could and did change even in a single painting. With their relentless dissociation from fixed structures and spaces, proportionalities and ratios were steadily becoming the instruments of measurement and calculation, the tools of analysis.

* * *

The more exaggerated claims of some art historians notwithstanding, linear perspective, in the words of Anthony Grafton, was neither a "new worldview nor a way of capturing the three-dimensional world as it really is."[64] Grafton's comment rightly cautions us against the intemperate, anachronistic misreading of evidence. A case in point comes from Alberti's *On Painting,* where he described proportion or "comparison" as "a power which enables us to recognize the presence of more or less or just the same." The last phrase reads in Italian "le cose qual sia più, qual meno o equale," and a more literal translation would be "the things that are greater, less, or equal." These terms express the three relations lying at the heart of the ordinal concept of number and its ability to extend the natural sequence 1, 2, 3, . . . ad infinitum. In fact, they are the identical words Descartes used nearly two centuries later in order to explain all abstract and philosophical thinking on the central model of mathematics, the "chief secret" of his method. But we are not yet to Descartes.[65] The explicit, Galilean, and Cartesian distinction between "primary" and "secondary qualities" ('quantitative' and 'qualitative,' in our idiom) lay in the future, as did the creation of coordinate geometry, which made possible the direct "mapping" of algebraic formulas onto space, so integral to modern mathematical analysis. Alberti, Piero, and their contemporaries were contributing to a new numeracy, not to a new mathematics and philosophical style of reasoning.

As we have insisted throughout this narrative, numeracy is not mathematics, any more than literacy is philosophy or literature. But mathematics grows out of numeracy, just as philosophical and literary works presuppose the alphabet (or other symbols for writing). Both literacy and numeracy designate information technologies, a means of informing our sensory experience through symbols, a "substrate" of "mental extracts" for identifying, storing, and transmitting the mind's sensory contents. Our lives as a "symbolic species" involve us in an ongoing process of abstraction, whereby the mind views the outside world from a critical distance, extracts certain features from it, and then looks again critically at the products of its own creation and previous abstraction. As a historical process, higher, more abstract levels of thinking assume these foundational operations, which condition (but do not determine) its subsequent contents. Early thing-mathematics grew out of early thing numeracy, just as modern relation-mathematics grew out of a modern, relational numeracy, two historical subsets of an information technology, each with a different array of symbols and operations. The Renaissance worlds of commerce, music, and art effected a radical transformation in numeracy, which in turn laid the foundation for modern mathematical analysis.

For classical and modern science alike vision has never been sui generis its own empirical touchstone, never the sole fount of scientific knowledge. Nor likewise in art, where "there is no neutral naturalism," writes E. H. Gombrich.[66] Rather, the sensory data of vision have always been funneled through an information technology situated between us and the world without. Medieval schoolmen spoke of "subtle species" emitted by objects, received through sight, distilled by the mind, and molded into words—their way of describing the technology of literacy and classifying. Viewed from the perch of "modern science," the Renaissance transformation of visual information into grids, perspective, and technical manipulations of proportion imposed a different screen between the world and the mind, one involving deception and illusion, yes, but one contributing to the technology of relational numeracy and analysis in the work of Galileo and beyond.[67] Before turning to these developments, the fourth dimension of the medieval quadrivium remains to be investigated, however, for woven into the fabric of astronomy was the mystery of time, the technological mastery of which paved the way for the merging of physics and mathematics, the terrestrial and celestial spheres of science. Such mastery became critical for Galileo's breakthrough, for it enabled the analytical uncoupling of time from motion; its expression as an independent, symbolical reality; and its reintegration into motion through mathematical formulas, such as the law of free fall.

CHAPTER 8

Astronomy

The Technologies of Time

'Tis but an hour since it was nine,
And after one hour more 'twill be eleven,
And so from hour to hour we ripe and ripe,
And from hour to hour we rot and rot,
And thereby hangs a tale.

—William Shakespeare, *As You Like It* (ca. 1599)

OF ALL THE DIMENSIONS OF HUMAN EXPERIENCE, time looms by far the most enigmatic, at once our most profound reality and a figment of our imagination. Past, present, and future dwell in the present, but the present doesn't exist, or does so only as an instant between future and past, a point of no time lapse, an "eternal now" for believers today as well as in the Middle Ages, a chimera for the space-time of modern science and its sensibilities. (One thinks here of Whitehead's well-known dictum "There is no nature at an instant.")[1] To speak of time means to mix metaphors (time's arrow, time's cycle) or to beg questions with tensed syntax, or to peruse volumes of early humanistic writings in search of the first stirrings of modern historical awareness, or to attempt, somehow, squaring the scales of an externally measured, "clock" time with our inner sense of passing time's variable pace.[2] Even our bedrock of day and night turns out to be the quicksand of cosmological accident: a gigantic planetesimal strikes the earth, splitting from it our moon and creating an earth-moon gravitational system that has slowed down the earth's diurnal rotation from a frenzied six-hour day to our more familiar, calm, and habitable spin of twenty-four hours, a day-night rotation itself decelerating gradually as the moon recedes farther from the earth (about two inches per year).[3] Zeno put the enigma objectively in the form of a fault line between mathematical discreteness and continuity, time's instants versus time's flow. In the matter of subjective perception, we still stand as helpless as Saint Augustine in one of his more folksy and mind-bending moments: we all know what time is until we have to explain it.[4]

And thereby hangs our tale. In our narrative, the history of time's mystery intertwines with astronomy, the fourth dimension of the medieval quadrivium. Most prominently, discussions of astronomy in our period center on the breakthrough work of Nicolaus Copernicus (1473–1543) and his famous "book that nobody read"—*On the Revolutions of the Heavenly Spheres* (*De revolutionibus orbium coelestium*).[5] Published in 1543, Copernicus's new astronomy entailed reversing the positions of the earth and sun in a (then) novel understanding of the heavens as a heliostatic system. Scholars have long chronicled the various contributions leading to this revolutionary publication and the plethora of ensuing consequences, including Galileo's trial and condemnation. The significance of the *Revolutions* is well established. But searching the medieval and Renaissance terrain for evidence of an emerging information technology as clues to the eventual science-religion rift leads us to look in less explored areas, to raise different questions about premodern astronomy. These center on the surfacing of time as an independent component of analysis.

In the practical developments of the quadrivium chronicled thus far, innovations in arithmetic, music, and geometry had all introduced new, abstract symbols: Hindu-Arabic numerals, musical notes and rests, converging perspective lines. The new symbols conveyed relations and functions; they correlated with phenomena; they issued instructions; they transmitted messages; they even corralled nothingness (the zero, the rest, the vanishing point) in useful and informative ways. Other, comparable symbols, often denoting functions, were added to these innovations, symbols such as the +, the −, and the = of arithmetic, all sixteenth-century novelties. In fact, the entire age was awash with symbols. Not all of them were of equal significance or abstraction, particularly from the standpoint of their subsequent contribution to the conventions of a new information technology. Culling, reshaping, and transforming symbols was a persistent and multifaceted task, one as demanding as their creation, especially since their inventors seldom knew exactly what they sought. Certainly neither Leonardo of Pisa nor Franco of Cologne nor Piero of Sansepulcro nor a host of their colleagues and contemporaries essayed to invent an information technology. But invent one willy-nilly they did. And every time a businessman calculated returns on his latest investment, every time a composer scripted a new score, every time a painter sketched out his new commission, and every time hundreds of craftsmen from various occupations thought about their next projects using the new symbols, a virtually imperceptible bit was added to it.

The new information symbols shared a commonality: they captured and encoded information delivered to the mind by moving sensations, predominantly the givens of sight and sound, in a fashion that dramatically departed from the centuries-old practice, based on alphabetic literacy, of devising and classifying definitions.[6] Aristotle had grasped sensation and motion as intrinsically bound to one another, and by logical extension both were also intrinsically, if perplexingly, joined to time, for "we perceive movement and time together." Though a sensible enough observation, his reasoning behind it seemed out of joint because, as he also defined it, time was the "number of change," "the number of motion in respect of 'before' and 'after.'"[7] But number, as we noted earlier, dwelt in the timeless kingdom of mathematics, where "motion . . . is a fiction." Equally puzzling, we all see the heavens move, but for Aristotle they were eternal and unchanging. How so? He tried to solve such puzzles with further definitions. Whereas "all motion is in time and all time is divisible," he wrote, "the now . . . defines . . . time, . . . which is intermediate between nows." But then this rendered the "now" (much like a geometrical point) "indivisible and inherent in all time." Well, perhaps. Despite heavy-duty parsing and qualification, the tangle worsened, and time appeared, at times, both divisible and indivisible, continuous and discrete.[8] While Zeno hovered in the background with his paradoxes and Plato stuck to conceptual poetry with his striking metaphor of time as the moving image of eternity, Aristotle labored tortuously to explain a conundrum whose resolution lay beyond the definitional confines of classificatory thought.[9]

True, as Aristotle observed, time bathes our sensations with the inexorability of before and after, but abstracting it from all sensation and motion required a further distancing from phenomena than could be achieved by definitions. The new, emerging information symbols did this. They abstracted functional and relational forms or patterns (new "mental extracts") from the constantly moving ephemera of sensations; then, drawing away and observing these patterns from a further, critical perspective or distance, they abstracted time. They stopped sensory motion and fixed it symbolically. In so doing they rendered possible the analytical uncoupling of time from motion. As discussed previously, this uncoupling first appeared in music, where composers and musicians separated time from the movement of sound. The advent of musical notation and time measurement marks if not the first, certainly among the initial manifestations of analysis in the modern sense of reverse engineering. Early polyphonists broke apart the freely flowing, evanescent, and directional flux of tuneful sound, resolving it into its constitutive elements of rhythm and pitch, of time and tone. They then captured and recombined these elements in an abstract and functional notation that allowed them to understand and manipulate the proportionalities of sound, its dynamic relationships. For them time was information—measurable, isolable, regular, and message carrying—an independent, symbolized component of their analysis, the measure of musical beat.

So too, would time become . . . in time . . . an independent variable of the modern scientific analysis of motion in general, a major feature of Galileo's breach of the analytical threshold. For this to happen, time itself journeyed from definitions to symbols, carried by the interwoven technologies of the calendar and the clock. Both technologies tendered means of keeping time (as distinct from its later measurement) in Christendom. Both keyed their means of timekeeping to celestial movements, particularly of the sun and moon but also involving the fixed stars and constellations of the zodiac. Both bore long histories, dating from antiquity's Julian calendar and various early timekeeping devices (such as sundials and clepsydrae). And both underwent time-altering transformations in the Middle Ages and Renaissance. The result of these transformations, begun in music, was "metronomic" (metered) or "clock time," an isolable construction of measured and organized information, expressed by the new symbols, which opened the door to the analytical cleaving of time from motion.

While the symbolic abstraction of time was developing under the impress of new technologies, it also steadily bridged heavenly and earthly times. A central feature of the premodern, classifying temper, at least from the standpoint of later scientific learning, lay with the categorical separation within physics of the celestial and terrestrial realms. As heavenly time and earthly time became funneled into the technologies of calendars and clocks, the gap between these divisions shrank, just as the ethereal and soundless music of the spheres was giving way to earthly rhythm and pitch. Further, with their oscillating mechanisms, which combined the familiar ticktock of time's instants with its flow, the technology of clocks crept toward a practical integration of discreteness and continuity. Such mastery over the "inaudible and noiseless foot of time" helped close the gap between mathematics and physics, even while surreptitiously widening the eventual rift between science and religion.[10] For believers, time remained and would continue forever among the mysteries of existence; for scientists, it became and would continue a topic subject to the analytical temper.[11]

⁎ ⁎ ⁎

All time reckoning derives from some sort of synchronism—correlations of events, people, or natural phenomena. In antiquity reckoners fashioned time correlations for the years not by numbers or dates (there is no Greek or Latin word for 'date') but by significant events or kingships or consuls and the like.[12] Premodern time, "time without B.C. / A.D.," was not simply "dates under another guise," emphasizes classicist Denis Feeney: "Roman years did not have numbers; they had names." Their time horizons were "dotted with significant clusters of people in significant relationships with each other through memorable events."[13] Time was classified, put into cubbyholes along with other features of the natural and social worlds. During the Republican era, for instance, Roman time-governed activities were correlated with lists known as *fasti* (from the Latin *fas*, or "divine law," and from which we derive our 'fast,' as in 'fasting,' 'breakfast,' and so on), which were associated with specific pursuits.[14] The time blocks of the *fasti* were arranged according to months and days, and keyed to three fixed points of reference: Kalends, Nones, and Ides. Kalends denoted the new moon, the first day of the month, and because most of the population was illiterate, the events of the month's *fasti* were announced at the forum on that day. Our own term 'calendar' descends from this practice, known as the *calendaria*.

In Rome the classifying temper continued its dominance over time even after the introduction of a new, solar calendar. Throughout the Republican centuries, correlating civic activities with astronomical regularities had posed an ongoing, knotty problem for the Romans, who relied on a lunisolar calendar likely adopted from the Greeks. After the civil wars of the first century BCE, and with time reckoning having become a mess during the "years of confusion," Julius Caesar (100–44 BCE) proclaimed a new, solar or tropical calendar, which began on January 1, 45 BCE.[15] The Julian calendar appeared to standardize the practice of correlating time's passing exclusively with the movement of the sun along the zodiac, a measure of functionality the Babylonians had introduced into time reckonings.[16] (For a description of the Roman calendar and its relation to the "celestial sphere," see Appendix A.7.) But the new calendar primarily served only Rome and other major cities; elsewhere throughout the empire, local calendars generally prevailed in keeping track of time.[17] As well, the passing years remained classified as "eponymous" or "regnal," tied to specific kings, emperors, consuls, and other officials.[18] Long before the sixteenth century and the onset of modern chronology, Caesar's calendar had offered the prospect of a universal grid or framework for dating based solely on astronomical observations existing independently of human events. But like space in the ancient world, time remained embedded in the classificatory and prescriptive demands of social and religious contexts. Before time could be considered independent enough to be uncoupled from motion, it had to become information, to be separated and abstracted from these contexts.

This was especially true in Christendom, where over the centuries its calendar evolved from a liturgically centered, ecclesiastical reckoning of time into a universal and essentially secular, nonreligious dating system, the Gregorian calendar of 1582. The evolution of the Christian calendar was a lengthy, ongoing, and complicated process, which we can radically simplify by isolating three major episodes of consolidation, challenge, and correction. It was also a process enveloped in irony. Monks, schoolmen, and believers of all shades labored for the express purpose of improving the organized temporality of the Christian's movable feasts and earthly sojourn, the celebration of God's sacrificial presence in historical time and the hope of worlds unseen. But in their search for greater accuracy time reformers produced other, unintended consequences—namely, a universal chronology that had nothing to do with Christian events per se, notwithstanding the accepted convention of using BC / AD to keep track of the years. In so doing, they also laid the groundwork for the "scientific" measuring of time ("natural *computus*") in the service of astronomy and astrology, and with it a cornerstone of analysis.[19]

During the onset of the Christian era in the late imperial and early medieval years, the Julian calendar amalgamated with two additional time systems centered on correlations pertaining to the life of Jesus: the lunar cycles of the Mosaic tradition, used to establish Passover and other feasts (hence the Lord's Supper), and the designation of Sunday as the beginning of the week and the day of Christ's resurrection. The task of merging these time systems fell within a long-standing tradition of combining practical mathematics, astronomical observations, and local concerns. But a problem arose, for the three systems—the Julian year, the Mosaic month, and the Christian week—were incommensurable from the outset. The numerical ratios correlating them led to irrational proportionalities, much like those discovered in musical intervals.[20] This meant that Christians had to adopt compromises in order to establish the conventions of their calendar. Reckoning the days, weeks, months, and years produced a practical, how-to discipline known as *computus,* which required both the exegetical work of interpreting scripture and the computations derived from the movements of the sun, moon, and stars.[21]

In the first of our three episodes early *computus* practices for figuring time were consolidated in the eighth century by the work of an Anglo-Saxon "candle of the Church" (so named by Saint Boniface), a lifelong Benedictine monk and prolific scholar, the "venerable" Bede of Jarrow (ca. 672–735).[22] Among his many works, Bede devised a "little book concerning the fleeting and wave-tossed course of time" as a guide to determining the dates of Easter Sundays and other movable feasts of the Christian year. Until then these had been contentious, if not outright chaotic, with several early traditions vying for what most agreed was the ideal of having all believers celebrate Easter on the same day.[23] But with his watershed account, *The Reckoning of Time* (*De temporum ratione,* 725), his Venerableness ended centuries of earlier Paschal disputes.

Scripture specified that Passover (*Pesach* in Hebrew, from which our Latinized "Paschal"), the weeklong commemoration of the liberation of the Jews from Egypt, should begin on the night of the fourteenth day of the Jewish month Nisan, the "month of corn," which marks the early days of spring. This day also was generally understood as the day of the Last Supper and the onset of Christ's "suffering" or "Passion."[24] But spring is a solar season, and with the Bible silent on its exact beginning, third-century Christians supplied the vernal equinox, designated as March 21, for its astronomical starting point.[25] Working from these assumptions, and proceeding with an admixture of credulity and critical acumen typical of many medieval authors, Bede developed his *computus* for figuring Easter dates almost exclusively on the basis of existing astronomical data, not by perusing the skies for new measurements, although he did claim to have consulted a sundial for determining equinoxes.[26] Throughout his book he worked with average periods of solar and lunar cycles, mean tropical years and lunations, and his treatment involved rather simple arithmetic calculations based on these averages.[27] As Leonardo of Pisa would do years later, Bede even began with instructions for finger counting, an effective "manual language" for mastering the arithmetic necessary for such computations, or simply "for the sake of exercising one's wits."[28]

The trick in calculating Easter lay in correlating the dates on which the equinox, the Paschal or full moon, and Sundays fell in the overall Paschal cycle, a much more daunting task than the Romans had faced with their somewhat lackadaisical pre-Julian, luni-solar calendar.[29] This required determining the age of the moon, known as its "epact," on any specific day of its lunation and matching it to calendar dates. (Thus the new moon was luna I, while the full moon was luna XIV, and the like.) It required, as well, determining how often the days of the week repeated themselves on specific dates. In practice one could simply consult a table and compare a list of calendar dates side by side with the epacts and days of the week to see their correlations. Bede actually provided such a table for the "lazy or slow-witted" and those "rather less skilled in calculation."[30] But like the old saw about teaching a man to fish rather than giving him one, Bede explained how to compute a Paschal cycle of the correspondences and to predict future Easters. Then at least the more skilled could do it themselves.

Bede crafted his Great Paschal Cycle from two lesser series of computations.[31] In the first of these, devised by the astronomer Meton of Athens, 235 lunations correlated very nearly with nineteen solar years (see endnote 20). The second, known as the solar cycle of twenty-eight years, linked days of the week with calendar dates. Adjustments had to be made to align both sets of computations with one another. Calculating the nineteen-year Metonic cycle, for instance, meant intercalating months to make the solar and lunar sequences match. In a common solar year of 365 days, there were twelve lunar months or 354 days, a difference of 11 days. After three years the difference grew to 33 days, at which point an extra month known as an "embolism" was inserted into the year, creating an "embolismic" year of 384 days. This intercalation in turn disrupted lunar reckoning in embolismic and subsequent years, adding complications to aligning epacts and solar dates. The needs of considering whether the month was full or hollow (thirty or twenty-nine days) and of folding in the days of the leap years every fourth year imposed further wrinkles in the reckoning. The upshot was that by the cycle's end there would have passed some 6,939.75 days when calculated by the nineteen Julian, solar years (19 × 365.25). But computing the lunations over the same period yielded a total that was 6,940.75 days, or one day longer.[32] To compensate, Bede then intercalated a one-day "leap moon," the *saltus lunae,* in the eighteenth year of the cycle. Even so, a slight discrepancy remained in the Metonic cycle, for as noted earlier, the periods were incommensurable.

The 28-year correlation of solar date to day of the week, by contrast, was basically more straightforward. A 7-day week and 52-week year yield 364 days; add one and you have the 365-day common solar year. (The one-day difference explains why a given date on a day of the week, known as a "feria" in calendar-speak, falls a day later from the previous year, and in leap years two days later.) After 28 years, the number of weeks is 1,456 (28 × 52), to which are added a further 5 weeks for a total of 1,461. The additional weeks derive from the 28 extra days, one for each year, plus one week's worth of days from the 7 leap years in the cycle. This meant that every 28 years the same day of the week and the same calendar date would coincide with one another. (If January 1 fell on a Sunday in year Y, it would do so again in year Y + 28.) Combining the lunar-solar cycle with the solar-weekly cycle (28 × 19) produced a full-blown lunar-solar-weekly, or Paschal cycle of 532 years. Each day in the cycle was unique, with the entire cycle repeating after 532 years. For Bede's purposes, it meant that Easter Sunday dates repeated themselves, and, more importantly, that they could be predicted, along with the other movable feasts dependent on them.[33]

Bede did more than compute the yearly cycles of time for determining Easter. He also addressed the commanding issue of chronology, time's arrow. Building on the work of his predecessors, especially Augustine, he interwove three types of chronology—divine, natural, and human—in order to situate humankind's salvation.[34] On the basis of the interplay between the sun and moon, he retrospectively computed the date of creation as March 18, 3952 BCE. From that moment, writes medieval scholar Arno Borst, "time was like an arrow shot from a divine source at a heavenly target, and it was still in flight."[35] Of course no living observer could know its precise end, but Bede provided the figural structure for imagining it: "the Six Ages of This World," populated by the human species from Adam to the present, followed by a seventh age of "perennial Sabbath" and an eighth of "the blessed Resurrection," the "repose of paradise." All the six worldly ages were determined with a mathematical accuracy derived from arithmetical, astronomical, and exegetical computations.[36] A few years later, in his *Ecclesiastical History of the English People* (731) Bede applied his "human method of dating based on the Incarnation of Christ" to the history of salvation, filling his work with accounts of Christian martyrs and divine judgments on secular events. His "ecclesiastical" history was thus *computus* applied to chronology, combining time reckoning, historiography, and liturgy: a calendar and a chronology suffused with the medieval image of divine animation in natural and human affairs.[37]

For his few fellow "candles," those monkish lights pursuing learned ignorance rather than unlearned wisdom, Bede's writings were quite popular during and after his own lifetime, and throughout the medieval centuries his manuscript on time circulated in scores of copies. His computations and tables became a standard reference work for time reckoners (*computatores*).[38] But neither time nor the heavens stand still. Contrary to Aristotle's belief, the relations between fixed celestial bodies actually do change; in modern parlance, they "drift" and they "slip." In the present context, "drift" refers to changes in the vernal equinox because of a phenomenon known as the precession of the equinoxes, first discovered by Hipparchus. (See Appendix A.8 for a description of equinoctial precession.) For its part, "slip" denotes the moon's falling behind the Metonic cycle by roughly one day in 308 years, notwithstanding the *saltus lunae*.[39] New observations involving both drifts and slips challenged the findings of Bede's *computus* and led to a second episode in the evolution of a modern, secular dating system, one punctuated by the challenges of new empirical evidence and renewed astronomical thinking.

We have already spoken of the European revival of learning that began in the eleventh century, a revival whose hunger was awakened by emerging urbanization and fed by the diffusion of Islamic science and philosophy. Astronomy figured heavily in that diffusion and found a receptive audience among many practitioners of *computus*. One early example in particular was a Lotharingian monk named Walcher (b. ?–1135), who appeared in England around 1091 and eventually settled at Malvern, a Benedictine monastery in Worcestershire, where he became prior. "A good philosopher, a worthy astrologer" (so reads his gravestone), Walcher had become "distressed" with the accuracy of existing astronomical tables and *computus* as a guide for human activities, especially medical practices.[40] He had gotten hold of an astrolabe, recently introduced to Europe along with Islamic astronomy, and with it carried out in 1092 the first recorded, quantitative observations in the Latin West designed to improve astronomical predictions. Before dawn on October 18, he measured the height of the full moon at its darkest moment during an eclipse, whose exact time he was able to determine. From this reading, Walcher then calculated the time of the preceding new moon, and on the basis of his more accurate measure of a lunation he constructed a new set of lunar tables. Thus did he combine celestial measurement with *computus*, which "inaugurated a hitherto unknown genre of astronomical literature in the West."[41] That his tables too were flawed mattered less than the new precision he had introduced by measuring with the astrolabe.

Walcher's tables had a "catalytic effect" on other time reckoners who followed him, and depositories of medieval works today count among their holdings an astounding accumulation of over nine thousand pre-seventeenth-century Latin manuscripts containing *computus* tracts and tables. Many of these were made anonymously and, like Walcher's, many contained new observations or computing techniques introduced from the Muslim world.[42] These manuscripts attest a steady stream of new information about sidereal and planetary positions, which was spreading throughout western Europe, concentrating especially in Paris during the thirteenth and fourteenth centuries, and which heavily influenced calendar computations. It bears mention also that measurements in Islamic astronomy were expressed largely with Hindu-Arabic numerals (which could express fractions far more efficiently than could be achieved in Roman counting), a practice adopted steadily by astronomers and computists in Latin Europe.[43]

Among the new tables and star charts the most notable and widely circulated were those originating in Spain, which, recall, was a gateway of Islamic learning into Europe from the eleventh through the thirteenth centuries. Critical especially were the Toledo and Alfonsine tables, the former created by a group of astronomers at the Toledo school (previously discussed), the latter formulated at the court of Alfonso X, king of Leon and Castile (dubbed "the Wise" or sometimes "El Astrólogo," the Astrologer, 1221–1284). Throughout the Middle Ages few individual scholars either practiced or promoted empirical studies beyond sporadic observations, which were generally made to confirm a theory, not to test it. Nonetheless, the sheer volume of new astronomical information collectively illustrated much that was problematic with Ptolemaic star charts and cosmology. Although firmly grounded in Ptolemaic, geocentric astronomy, the Toledan and Alfonsine tables presented measurements and theoretical challenges to prevailing views not only of the cosmos but also of the Bede-based, computistic reckoning of time associated with it.[44]

The Alfonsine tables, for example, became the focal point for debates over the length of a solar year and how to measure and explain equinoctial precession more precisely. The issue turned on whether the sun's position at the equinoxes was to be measured from the earth (which produced a "tropical" or solar year) or from a distant star or constellation (yielding a "sidereal" year).[45] Alfonsine measurements of the apogees of the sun and planets differed by as much as seventeen degrees over those of Ptolemy and when applied to calendar reckonings yielded small but significant and cumulative differences in determining the length of the year. These, in turn, added up to a Paschal cycle that fell further and further behind tropical years.[46] In one sense, of course, the debates mattered not a whit. Calendars are simply agreed-on conventions, and whether one adopted the sidereal or tropical year as the point of reference for determining the ecclesiastical year was of little import. Except . . . except for the profound belief that the history of salvation, the only history that mattered, had a specific starting point at the first Easter and one had to get it right. And by any measure its key reference points, the equinox and Paschal moon, were drifting and slipping.[47]

Throughout the Renaissance new instruments added greater precision to celestial measuring and further confirmed equinoctial drift and lunar slippage. The names of these tools read like a modern list of hardware store (or pharmaceutical) products: "sundials, nocturnals, theodolites, sectors, armillaries, astrolabes [mentioned previously, discussed later], equatoria, and astronomical compendia." And there were more, like the "rectangulus" and the "albion" of master clock designer Richard of Wallingford (1292–1336).[48] The cumulative effects of new instruments, new readings, new tables, and renewed cosmological thinking all convinced time reckoners of the need for calendar reform. The vernal equinox, taken as March 21 in Bede's day, had drifted some ten days by the sixteenth century, to March 11. The third episode of the calendar's evolution was nigh. Corrections were to be made, and with them a universal dating system would be established: a linear succession of days and years synchronized only with the sun's movements. The Christian origins of the Western calendar would be pushed into the background for the sake of convenience and consistency with astronomical evidence.

Many clerics, scholars, and natural philosophers had been calling for calendar reform for nearly four hundred years before the sixteenth century, but until then Church politics had always intervened, and other, more serious interests had commanded the attention of the curia and the councils. The pattern appeared to repeat with the Council of Trent (1545–1563), convened in response to the Protestant challenge and to the demands of Catholic reformers. In its last session the issue of correcting the calendar was once again broached amid other pressing matters. Only this time, in the ensuing years committees were formed, opinions solicited, proposals weighed, and actions taken. The result was the papal bull of Gregory XIII (1502–1585), *Inter gravissimas,* promulgated in February 1582, which announced the new calendar nearly two decades after the cardinals and officials at Trent had closed up shop and gone home.[49]

Far easier to manage technically than politically, the reforms basically tweaked the Julian calendar, shortening the length of the year by about .002 percent. To bring the year back into conformity with a March 21 equinox, Gregory simply intercalated the ten days the calendar had drifted behind. Thus October 4, 1582, was followed by October 15, 1582. To slow down future drift, the number of leap years in four centuries was reduced to ninety-seven (from one hundred), a feat accomplished by considering centennial years common unless they were divisible by four hundred. (Thus 1600 was a leap year, as was 2000, whereas 1700, 1800, and 1900 were not, remaining common years.) This replaced the Julian practice of simply taking every fourth year as bissextile, including the centennial years, and kept dating much closer in line with the equinox. For contemporary Christians, the more significant portion of the reform lay with its improved methods of calculating epacts and Paschal moons, which were grafted onto the more fundamental adjustments in the solar year and lie beyond present concerns. Initially, there was widespread acceptance of the Gregorian reform in Catholic countries, with only occasional grumbling—Montaigne, for instance, said he was "constrained to be a bit of a heretic" when it came to the new calendar, with his "imagination . . . always casting itself ten days forward or back." Many Protestant nations, however, balked at it, as did most of Orthodox Christianity. Only as the years, decades, and even centuries passed would other nations and areas of the globe adopt it, won over by the sheer practicality of the new reckoning. (England and their American colonies made the shift in 1752.)[50]

From our perspective Gregory's reform made manifest the growing appreciation of a generalized, secular dating system, which soon became even more apparent in the work of another contemporary—French scholar, chronologist, and Calvinist Joseph Scaliger (1540–1609). Unlike Montaigne, his fellow countryman, Scaliger thought Gregory had not gone far enough in creating a universal time grid. A tireless researcher, referred to by his contemporaries as a "bottomless pit of erudition," Scaliger scoured all the examples of earlier calendars and chronology he could find in order to synchronize systematically every prominent dating system with one another. In his own hefty tome (*De emendatione temporum*, 1583), he combined all this learning into an even larger time grid, whose objective was a chronology freed "from moral and religious functions, . . . a purely numerical discipline."[51]

The basis of Scaliger's grid, which he named the "Julian period" after Caesar, was formed by joining the Metonic and solar cycles with yet another, fifteen-year civil cycle used by many medieval computists. This latter was the "indiction," which the Romans had introduced during Diocletian's rule for the purpose of assessing taxes. Years were tracked sequentially in this cycle (the first year of the indiction, the second, and so on), but the indiction cycles themselves were not so identified. Combining all three cycles ($19 \times 28 \times 15$) produced a single time sequence of 7,980 years, after which the period would repeat itself. Every year in the Julian period thus had a unique set of three identifying numbers, one drawn from each of the constituents, and together they gave each year what Scaliger termed its distinct notation and "character."[52] Counting backward and adding his own reckoning to that of Dionysius Exiguus (and Bede), Scaliger determined that the year 1, the accepted birth year of Jesus, fell on the confluence of the first year of a lunar cycle, the ninth of a solar cycle, and the third of an indiction. These computed to the year 4713 of the Julian period and meant that the period itself, starting with the Julian year 1 (when all three constituent cycles began concurrently), occurred 4713 years before Christ, or BC. Such a date was clearly earlier than the world's creation as determined by the accepted chronology based on the Hebrew Bible. But for Scaliger that was quite beside the point. His was an abstract, arbitrary, and artificial system, which he defended: "The intervals of time are like the intervals of space; just as spatial measurement can be combined, so can temporal measurement." And measured from within his system, the world was created in 3949 BC.[53]

Further disputes over the beginning of the world (and of time itself) followed Scaliger's work. But in 1627, the Jesuit Dionysius Petavius (1583–1652) cut the Gordian knot of time's beginning in his *Opus De Doctrina Temporum* by eliminating it altogether, separating chronology from theology, and producing a "chronological orientation which transcends the limits of custom and authority." Establishing the date of Jesus's birth had proved to be a matter of considerable and undecidable controversy, as sixteenth-century time reckoners realized. Petavius did not consider the birth date to represent an actual event that occurred at 0 on the calendar; rather, 0 was simply a convenient, agreed-on point from which to measure events, including Jesus's actual birth. (It is now believed to have occurred anywhere from 6 to 4 BCE, possibly as late as 1 BCE, but further precision remains speculative.) Petavius thus extended the correlations launched by Dionysius, Bede, and other computists, synchronizing AD with solar years backward with the addition of the "BC" and establishing the BC / AD grid as a dominant, secular convention for measuring universal time.[54]

With the creation of a dating system and calendar that existed independently of references to humans and events, time itself became correlated with an ordinal sequence of numbers, divided only by the convenience of

BC / AD (often expressed nowadays, in our more secular, "Common Era" as BCE / CE). The appearance of time tables and bar charts made visible the new chronologies, forerunners of the modern, linear timeline.[55] Theoretically the grid could extend endlessly backward or forward, although such speculations at the time were rare. The numbers in turn corresponded to critical moments in the movements of the sun and stars, the equinoctial points. Thus did the ordinal and cardinal features of numbers ensnare and manage time. Days, weeks, months, and years; dating, chronology, and even history; past, present, and future; the inexorability of before and after—all were brought within the fold of the calendar's technology and its relational numeracy. Time's mysteries were converted to problems, its passage to information.

* * *

"Littel Lowis my sone, I have perceived wel by certeyne evidences thyn abilite to lerne sciencez touchinge noumbres and proporciouns; and as wel considere I thy bisy preyere in special to lerne the Tretis of the Astrolabie."[56] Thus in 1391 did Geoffrey Chaucer (1343–1400) begin his treatise on the astrolabe, the oldest known technical manual in English. In all likelihood, Chaucer's ten-year-old son, "Little Lewis," would have been sorely challenged to comprehend his father's explanation, for at the time the astrolabe ranked among the most sophisticated of instruments in Europe. Essentially an analog computer, it could do a lot of things with its moving parts and scales, including mark celestial positions, survey earthly elevations, and perform basic computations.[57] (Appendix A.9 depicts the Chaucer astrolabe and provides a brief description of its parts and operation.) It could also track the time of day. The astrolabe was not quite a clock, that "fallen angel from the world of astronomy" in Derek de Solla Price's striking and apt metaphor, for keeping tabs on time required some sort of automata, some means of observing the passing of time when one's back was turned.[58] But the astrolabe bridged the calendar and the clock, subdividing the smallest calendrical unit of time, a day, into components known since antiquity as hours (from the Latin *hora,* in turn from the Greek root *yer-*, meaning "year" or "season"). Through the astrolabe and similar devices, time writ large calendrically merged with time writ small horologically. The crux of this historical process lay with an anonymous and uniquely Western invention: the mechanical, oscillating clock.[59]

A fallen angel the mechanical clock might have been, but its early history suggests a more earth-bound nativity. Clocks were launched into prominence primarily by the monastic demands of "time discipline," the need to establish and follow the "canonical hours" of prayer and work (*ora et labora*) set down by Saint Benedict of Nursia (ca. 480–543 or 547) in his famous and influential "Rule." Benedict had spoken of the hours only in general as a means for establishing the divisions of an orderly day, and so the practice continued for centuries, loosely governed by the "rhythm of collective conduct" rather than any modern sense of punctuality. There were no precise time rules per se, only widely varying time-organizing practices manifest in different monasteries and regions. By the High Middle Ages, amid these variations, communal prayers or "offices" had become regularized into seven daytime services (lauds, prime, terce, sext, none, vespers, and compline) plus a nighttime gathering (vigil, later matins).[60] Church bells typically announced the services; monks, like "Frère Jacques" of nursery rhyme lore, were designated to ring them and wake their confreres at night. In these circumstances, a more exacting version of time discipline steadily sidled into custom, nurtured by invention's mother, the necessity for marking time with greater regularity.[61]

A qualitative leap in timekeeping followed with the anonymous fabrication of an oscillating escapement mechanism, probably sometime between 1270 and 1330. Ever since antiquity, dividing days into hours generally had required observing time segments in the flow of continuous motion and correlating them with a name, activity, or number.[62] The earliest clocks tracked the daily movement of the sun (sundials), or water streaming into or out of a container (clepsydrae, water clocks), or wax or other flammable fuel burning at a steady rate (candles and oil lamps). Examples of all these devices could be found later in medieval monasteries, where they logged very roughly the "temporal" or "unequal hours," so called because the twelve equal divisions of days and nights varied with the changing seasons. By the thirteenth century, timing devices were sometimes connected with bells, converting them into a kind of alarm clock. Before escapements, however, timing still relied on flow.[63]

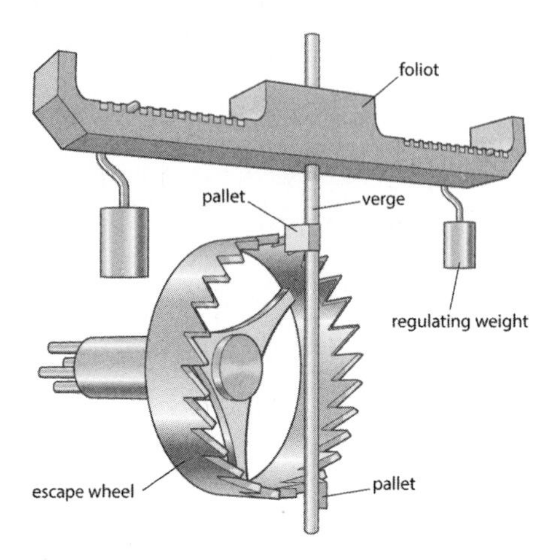

FIGURE 8.1. Verge and foliot escapement.

The escapement was a regulating mechanism, one of the three essential components of all early (and succeeding) mechanical clocks, which required additionally a power source and a means of registering the passage of time (such as a clock face or dial, first appearing in the fourteenth century). The earliest of escapement mechanisms was that of the "verge and foliot," which established the principles for those that followed. (See Figure 8.1.) In these devices, power was provided by weights hanging from a cord wound around a drive axle; as the weight descended, the axle turned and with it turned a saw-toothed drive or "escape" wheel (sometimes called a "crown" wheel). The wheel, in turn, interlocked with a vertical shaft, called the verge (from the Latin *virga,* "stick" or "rod"), on which were welded two "pallets" or metal tabs (initially at right angles with one another). Each of the pallets momentarily stopped the crown wheel, allowing it to advance or "escape" one tooth at a time, and simultaneously reversed the direction of the horizontal bar or foliot, causing it to oscillate back and forth. (In Italy balance wheels were employed in place of the foliot bar for the same purpose.) Adjustable weights on the foliot could be set to vary the speed of oscillations. The continuous motion of the drive wheel was thus converted into oscillations, the familiar ticktock of the device, breaking the continuous flow of time into equalized, discrete intervals. Gear trains, common in the Middle Ages, then connected the mechanism to a clock dial for indicating the hours.[64]

From the genius of these simple and anonymous beginnings, improvements and innovations tumbled over one another during the next three centuries: new drives, such as the mainspring and fusee; new escapements, such as the anchor; and new, ornate clock faces. Clocks became prestigious showpieces for cathedral, town, and court—intricate machines like the astronomical clocks of Giovanni Dondi (1330–1388) and Richard of Wallingford, which recorded heavenly movements and tolled the hours.[65] They grew smaller and movable, were placed on tables and shoved into pockets (watches date from the late fifteenth century). The monastic and canonical, unequal hours yielded to equalized units, twenty-four of them in a day, correlating with the fifteen-degree-per-hour advance of celestial bodies and independent of the seasons. From the monasteries, "time consciousness" infiltrated European, especially business and urban life.[66] Time became information, then money. Throughout all this, mechanical, oscillating clocks grew steadily more accurate and precise, a detailed history to which we can give but a passing nod.

Increasing mechanical precision provided greater accuracy in capturing and managing small and equal units of time: hours, minutes, and even seconds (as early as 1579).[67] But by Galileo's lifetime accuracy had accrued only so far. As we shall see in Part III, mechanical clocks were not yet precise enough for the time measurements he needed to make, and their introduction into scientific experimentation lay further in the future. In the meantime the clock had thoroughly entered European imagination, often serving as a metaphor for the heavens or the workings of nature in its entirety.[68] Beyond its mechanisms and imagery, the clock's contribution to a new information technology was of a different order. The creation of an abstract calendar and chronology had distilled time from references to humans and events and correlated it with an ordinal sequence of numbers that extended endlessly into the future and the past, extrapolating from the natural, numerical progression, 1, 2, 3, . . . ad infinitum. Clocks went the other direction, as it were, not forward or backward but in between, interpolating between numbers ever-finer intervals of time, likewise ad infinitum. From the calendrical sequencing of equal days in a year, to the equal hours in a day, to equal minutes in an hour, to equal seconds in a minute, time in whatever scale was mapped onto a single numerical string, technical mastery of which fell within the combined manipulations of calendars and clocks. The numbers marking time's intervals referred not to any things, even moving ones, but only to the abstracted relations of time itself. As information, clock time was uncoupled from definitions; in Galileo's hands it would be analytically separated from motion.

* * *

Whether or not familiarity breeds contempt, it surely sires abstractions. For over four centuries, largely outside the ivy-walled schools, Europeans became acclimated to practical innovations in each domain of the medieval quadrivium. Businessmen, musicians, artists, and time reckoners alike carved out an abstract, independent, and symbolic arena of information, which grew increasingly habitual. Theirs was a growing and collective technological sophistication, largely unrecognized at the time, coursing through a history perhaps best characterized as merely a few steps removed from a Hobbesian state of nature, wherein the lives of many were often "solitary, poor, nasty, brutish, and short." Were history logical and less interesting, we would be able to demonstrate rigorously and conclusively how each practical development in the abstractions we have chronicled led through this state of affairs directly to Galileo's discoveries, to the scientific method, and to the rift between science and religion. But storytelling, even with facts, doesn't work quite that way. True, it does bear noting that Galileo calculated with Hindu-Arabic numerals, that he learned music and musical notation (and perhaps a fair measure of acoustical physics) at his father's knee, that he had well absorbed linear perspective and proportionalities (his first job application was for the position of geometry instructor at an art school), and that from his youthful notice of a swinging chandelier in church (likely apocryphal, but true nonetheless) he thought deeply about the nature of motion and time throughout his long life. Each of the areas pointed in some fashion toward his accomplishments.

Yet logic leaves gaps, mysteries remain, and events often jumble together, a fact any historical account must acknowledge. By Galileo's lifetime, instrumental abstractions abounded in European life, cultural and intellectual products available for reworking. Driven by fate or irony, these empty symbols had purged themselves of religious, philosophical, or even metaphorical meanings and had harnessed the whirl of human experience in ways unimaginable in antiquity or the Middle Ages. Even so, in and of themselves, these developments did not directly spell the separation of science and religion, for many intellectuals continued to hold concurrently in mind the new informational uses of symbols and the more traditional ascriptions of meaning often associated with them.

The emerging appreciation of time illustrates the point. Bede saw dates as divinely created "for the sake of a certain symbolism," through which the "world's salvation [would] both be symbolized, and actually come to pass."[69] His was but one of many expressions of what Augustine had named "present time," which Hippo's good bishop had subdivided into the "present of things past, memory; [the] present of things present, sight; and [the] present of things future, expectation," and which Charles Taylor has interpreted as "gathered time." This present, gathered time was the eternal now, the eternally present fullness of time, not merely an instant of no time lapse between past and future but also a living manifestation of God's Being, union with which all humans profoundly desired. Within the gathered present, of course, we could measure our own past, present, and future according to days, weeks, months, years, or whatever yardsticks we might devise, all of which for Augustine marked the fragmented soul, its incompleteness and "distraction" from true ends. But throughout all our human measures time itself was God's present (both his gift and his now) and God's presence (both in the world and in souls).[70]

Centuries later Petavius considered dates and the time they measured simply as instrumental conveniences. The Benedictine Bede and the Jesuit Petavius were equally committed in their religious beliefs, as likewise was the Protestant Scaliger. But whereas Bede had intertwined his beliefs with calendar dates and chronology, with the allegorical world of number, both Petavius and Scaliger separated their beliefs completely from the "numerical discipline." Time for them had become what Walter Benjamin would later call "homogeneous, empty time," time as a container, not a constituent of reality.[71] This empty informational container had its counterpart in the empty mathematical symbols by which it was measured and to which its units corresponded in a cardinal, one-to-one manner. And as with Vincenzo Galilei's insistence that numbers carried no "sonorous qualities," so too had time's numbers emptied themselves entirely of qualities. The medieval image was well on its way to being discarded.

The new information technology might well have purged time of intrinsic divinity and enchantments, but it didn't follow that life occurring within the container of time was equally emptied or immediately desanctified. That would happen centuries later. Against the proximate backdrop of this consideration we see renewed concern for religious interpretations of events and natural phenomena occurring within time, which itself was becoming increasingly "historicized."[72] Prophecy, eschatology, astrological forecasting, and millenarianism were all widespread pursuits throughout the sixteenth and seventeenth centuries as intellectuals adjusted to a newer, "immanent frame" of temporally understanding the divine and its modes of operation. These practices, notes Charles Webster, were frequently associated with natural magic, producing "physico-theologies" with their "own style of interpretation of universal history," which many believed to be in its last, millennial stage.[73] Others looked optimistically ahead. In this regard Francis Bacon led the way in transforming many of these concerns into a "secularized eschatology," in which the progressive "advancement of learning" would not only improve the welfare of humankind but, of equal significance, would also repair "in some part" the losses of man's fall "from his state of innocency."[74] Still, however meaning might have been endowed in time, or however eschatology might have been imagined emptied at time's end, in the long run technology trumped trope. The container itself remained quite "indifferent to what fills it."[75]

In the last four chapters we have been exploring the "mixed-mathematics" preconditions that would eventually result in the shift from a premodern thing-mathematics to a modern and abstract relation-mathematics. We have gathered the threads of this tale under the rubric of a new information technology, one at whose core was a new, relational numeracy. Through its symbols and instrumentalities, sixteenth-century scholars and natural philosophers began reworking many ancient topics. With the sixteenth-century recovery and new translations of mathematical and scientific texts from antiquity, including the works of Euclid, Archimedes, Apollonius, Pappus, Diophantus, and others (many for the first time in over a millennium), older problems were revisited. The new technology opened the gates to solving many of these problems. As number metamorphosed from thing to relation through the new technology, its empty, instrumental symbols began eliciting the analytical potential of mathematics itself. Galileo would be the first to cross over this analytical threshold.

PART III

Galileo and the Analytical Temper

The Moment of Modern Science

Nature takes no delight in poetry.

—Galileo, *The Assayer* (*Il Saggiatore*) (1623)

IN THE YEAR 1606 the eldest son of Vincenzo Galilei, Galileo Galilei, then forty-two years of age, brought forth his first acknowledged publication: sixty copies of a small instructional manual titled *Operations of the Geometric and Military Compass.*[1] The manual was written to accompany a new computational device Galileo had developed during the previous decade, numerous copies and versions of which he had either sold or given to potential patrons. Most recently, in 1605, he had presented one to the young prince Cosimo de' Medici, son of Christina of Lorraine and Ferdinando I, Grand Duke of Tuscany. The duchess had gotten wind of the new instrument and had invited Galileo to instruct their fifteen-year-old heir apparent in its use. Thus began a series of summers that saw Galileo in Florence tutoring Cosimo in a variety of mathematical and practical subjects appropriate for an eventual duke. It was an alluring opportunity to court favor with the royal family and also to pursue the prospect of escaping Padua, where he had been employed for over a decade but where circumstances had soured somewhat, animating in him a desire to resume life in his boyhood environs. (In autumn of 1610, the move did occur, after the enormous publicity generated by *Sidereus Nuncius,* Galileo's first account of his discoveries with the telescope, which he dedicated to the Medicis.) Along with the compass, Galileo had promised Christina a set of written instructions, which he subsequently published, the limited number of copies being intended to aid in cornering the market for the new device and keeping a leg up on its competitors.[2]

The instrument itself was but one of numerous mechanical, computational devices whose popularity was on the rise in the sixteenth century. It was built on earlier models of "proportional" and "reduction" compasses, which Galileo had combined with other calculating tools. These included an artillery elevation gauge, the *squadra,* introduced by Niccolò Tartaglia, and a hinged implement for triangulation, the *archimetro,* probably invented by Ostilio Ricci (Tartaglia's pupil and Galileo's teacher of mathematics). Independently of Galileo, an English mathematician, Thomas Hood (1556–1620), published a book in 1598 describing the use and fabrication of a similar apparatus, a "sector" (as it's often called in English), which functioned comparably to Galileo's compass.[3] In yet larger scope, bystanders in the period witnessed the proliferation of artisanal and engineering accouterments of virtually every stamp—geometers' compasses, drawing and drafting instruments, surveying tools, sundials and countless clock refinements, orreries and astrolabes, construction and military equipment (the list grows ceaselessly)—as generations of practical-minded men learned "thinking with objects."[4] A profusion of instruction manuals, amounting to a publishing subgenre of its own, accompanied hardware innovations, with engineers, artisans, and mathematicians frequently joining talents to explain the workings of the new gadgets. Galileo was firmly ensconced and quite up to date in this world of instrument building, engineering problem solving, and manual writing.[5]

But as thoroughly grounded as he was in his own era, Galileo also ran ahead of it. For his gift to young Cosimo revealed as well that he had decidedly pushed matters forward in instrument design and practicality, transforming for the first time a dedicated and somewhat limited tool into a general, analog computing device. It was an augury of things to come, characterizing, emblematically at least, the threshold moment of modern science and Galileo's leading place within it, a theme we shall be advancing throughout the remainder of this book.[6] In brief, during the sixteenth century the preconditions for the "mathematization of nature" had reached a critical mass, which gave rise to the metamorphosis from thing-mathematics to relation-mathematics, and with it the Scientific Revolution. Those preconditions emerged from a variety of sources, many of which we've addressed and which we may here simply pass in review with an itemized list: the information overload and subsequent rupture of classification; generations of Renaissance commercial, engineering, and artisanal practices; the age's art and music; the religious impasses of the Reformation and the revival of skepticism; the turn to nature and the challenges to Aristotelian science; the recovery of ancient mathematics texts and problems; the burgeoning of mixed mathematics; and above all—our central story line—the gradual creation of a new information technology, that of relational numeracy, which both anchored and made possible new developments in abstract mathematics as a tool of analysis and discovery. This metamorphosis occurred during Galileo's generation, and while he was by no means its sole representative figure, he stood at the forefront of a group of "mathematical practitioners" working at the cusp of theory and practice, forging a path into the new intellectual terrain we now identify as modern science.[7] And as the new science surfaced from this activity, it would be intrinsically void not only of the earlier enchantments that had often permeated mathematical thinking but of religious concerns in general.

* * *

Galileo's compass and its accompanying instructions grew out of his own courses on military architecture and fortifications, subjects he began teaching shortly after his arrival at the University of Padua in 1592. Initially he had designed the device "only to speak to military men" in their calculations pertaining to fortifications and projectiles. But he continued to experiment with new scales added along the arms, and by 1599 he had effectively transformed the compass into a general-purpose, mechanical calculator capable of handling "infinitely many . . . problems of geometry and arithmetic." (See Figure III.1.) Hinged like an architect's drafting compass, the instrument served a similar function as the modern slide rule (without the yet-to-be invented logarithms), our contemporary engineer's shirt-pocket companion until the advent of electronic calculators.[8]

The scales of the compass based many of its computations on the proportionality between two similar triangles, creating a geometric means for solving a host of arithmetic problems, including those relying on the rule of three. Leonardo of Pisa had shown how the rule figured centrally in commercial transactions, and how it could be managed with Hindu-Arabic numerals. Musicians invoked it in changing tempos between measures, painters while situating objects in perspectives. With his compass, Galileo simply made "rule-of-three," as well as "inverse rule-of-three" computations purely mechanical, which he illustrated with explications of figuring "monetary

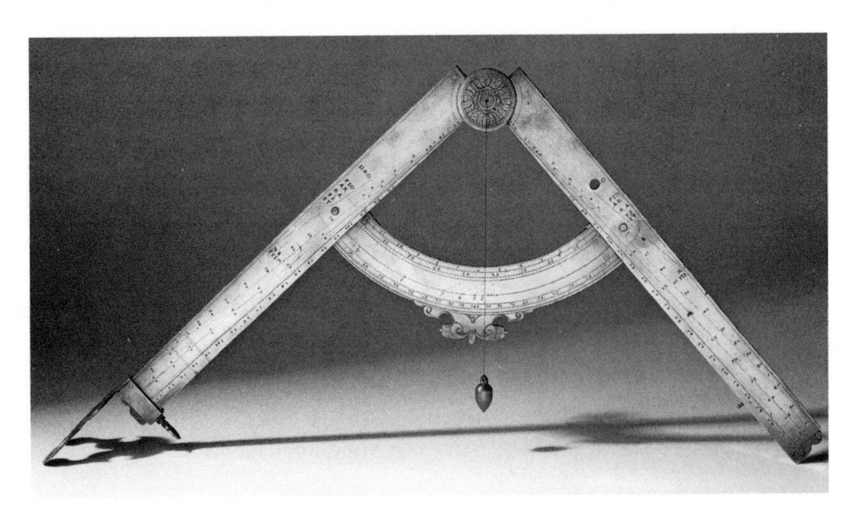

FIGURE III.1. One of Galileo's geometric and military compasses, built by Marc'Antonio Mazzoleni, Galileo's personal instrument maker, and possibly given to Cosimo II by Galileo, along with a copy of his 1606 operations manual. Shown here in locked, open position with the quadrant holding the legs at ninety degrees, thus permitting the plumb line to measure an angle of elevation. The instrument was used for ballistics, surveying, and star positions, among other applications.

exchange" and "New Year's Gain" ("compound interest").[9] After explaining the basic operations for using the "arithmetic lines," Galileo introduced other scales—the "geometric," "stereometric," "metallic," "tetragonic," and "polygraphic lines." These addressed a panoply of subjects in applied mathematics, besides those of commerce: mapmaking ("altering one map into another"); extracting roots; discovering the "mean proportional" (the geometric mean, which allowed one to calculate with irrational and real numbers); replicating and scaling geometric shapes and figures; determining ratios of composite metals; calibrating the trajectories of cannonballs "of any material and of every weight"; and even arranging "armies with unequal fronts and flanks." In the final, lengthy section of the *Operations* Galileo explained the triangulating features of the quadrant and its function as a surveyor's device, which yielded "various ways of measurement by sight."[10]

With such an extensive array of capabilities the compass was the perfect tool, both for and from Galileo's emerging, analytical cast of mind. Numeracy's hallmarks, the cardinal and ordinal principles, made it possible. The device utilized a cardinal, one-to-one correspondence between numbers and line segments (along the scales), which it could then manipulate in an analog fashion, using the ordinal—continuous and recurring—features of number to perform basic functions. The line segments correlated with spaces or objects, and manipulating the segments meant one could break the corresponding spaces or objects into their components, project them directly onto grids, and demonstrate their changing proportionalities, positions, shapes, locations, and the like—tracking what would become known as the "primary qualities" or quantitative features of matter. In its essence the compass expedited a new way of thinking about objects in space.

The compass thus extended the means of analysis that had accompanied the arrival of linear perspective. Earlier devices and practices had introduced novel techniques of seeing objects for artists. Leon Battista Alberti's *diffinitore,* for instance, allowed sculptors to follow a "logic of likeness" in transferring proportionalities, point by point, from a model to a full-size sculpture. So too did Piero della Francesca's projective grids and perspective drawings. Unsurprisingly, later versions of Galileo's apparatus, developed by others, carried scales designed expressly for artistic uses: in architecture, the *archiestro* by Ottavio Revesi Bruti (ca. 1570–ca. 1642), and in perspective drawing, the "optical" or "perspective compass" by French geometer Girard Desargues. Scholars often refer to Galileo's heightened emphasis on sight in his scientific practices (for Galileo, "seeing is believing," writes David Wootton).[11] But the innovation here was not simply a greater concern for vision per se. Artistic in inspiration, mathematical in rigor, his was a new kind of seeing, which would radically revise how one approached scientific investigation, an outgrowth of mapping objects onto spatial grids from a point of view. One didn't need definitions to capture moving objects; rather, one could measure and analyze them visually, reverse engineering them in all their various states.

In doing so, the compass did even more, far more. For it brought together in a single operation the functions of arithmetic and geometry, a practical and mechanical joining of those two, hitherto incommensurable disciplines. True, one can find occasional evidence of what some scholars still call a kind of "geometrical algebra" existing in antiquity, a manipulation of arithmetic and algebraic problems through the constructive and deductive operations of geometry, particularly during the golden age of Hellenistic mathematics.[12] But early forays into collapsing the division between number and space were overwhelmed by the classifying temper that kept them apart. Such ventures were apples to the oranges of Galileo's compass, which far superseded previous efforts. In place of the Euclidean, linguistic definition of fixed proportionality, Leonardo of Pisa had spoken of the "method of proportion" and "how this method proceeds," allowing him to interpret proportionality in a novel fashion, one that refashioned the definition of a mathematical thing into the procedures for manipulating symbols. The compass went further, embodying such procedures mechanically in brass and wood.

With such a tool in Galileo's hands "things" became "relations."

* * *

Galileo's compass materialized out of the most fertile and productive period of his entire scientific career, the Paduan years from 1592 to 1610. He had been offered the chair of mathematics there after a brief, three-year teaching stint at the University of Pisa, his first academic post, where almost from the beginning he had offended his Aristotelian colleagues over explaining matters of motion, all the while currying the favors of friends and patrons, networking to move on. Pisa was the stepping-stone. Padua would prove to be the ideal position for his breakthrough into analysis (notwithstanding strained relations there too with some of his Aristotelian colleagues, whom he referred to as "those goat-turds").[13] The situation united a university of a long-standing and highly respected tradition with the bustling energy of a commercial empire and the practical-minded men who ran it. No Nostradamus could have forecast it, but Galileo's arrival would come to signify a singular union of theory and practice, a marriage with outstanding issue: the birth of modern science.

The university at Padua was the central, state-supported university of the Most Serene Republic of Venice (often referred to with affection simply as La Serenissima). Over the years Il Bo ("the Bull," as students called it after a nearby inn and tavern) had gained a widespread reputation for its school of medicine and especially for its anatomical theater. Andreas Vesalius, "father of modern anatomy," had taught there; William Harvey (1578–1657), "father of modern physiology," had studied there. Moreover, it boasted highly regarded faculties in jurisprudence, theology, and the liberal arts, which at the time included what we would call the sciences. Originally created in

1222 by students and faculty who broke away from the University of Bologna in search of greater academic freedom, Padua had evolved from the student-administered institution of its earliest history to the state sponsorship of Galileo's employment. During the sixteenth century, the university had also become a center for the study of mathematics and sported a lively arena of debates and interest in Archimedes, mechanics, statics, dynamics, and the relations between logic and mathematics. At Padua Galileo entered into a flourishing center of academic learning that furthered and reinforced the studies of his youth and early, Pisan career.[14]

Equally significant were the practical responsibilities and engagements Galileo owed his Venetian overlords because of his position. Although by the seventeenth century the preeminence of Venice was on the wane, with the shifting of commercial and political focus to the states of the north Atlantic and away from Italy, it was still a polity to be reckoned with, possessing enormous wealth and stately power. As such, the doge, the Senate, and their Great Council of advisers continually engaged in naval, military, and civil projects, especially in the late sixteenth and early seventeenth centuries, when the republic essayed to retain its hold on Crete and the remaining portions of a once sprawling Adriatic and Mediterranean empire. Working with officials from the Arsenal, the vast shipyard home of the Venetian navy, Galileo addressed a number of practical subjects: the length of oars and mechanics of rowing Venetian warships; a horse-driven pump; fort placement and design; cannon ballistics; height and distance measurement through triangulation; measurement of longitude; and various applications of Archimedean simple machines (lever, pulley, screw). In this context he developed the military compass itself and, later, telescopes far superior to any others available. Along with everybody else, of course, Venetians grasped at once the benefits of spying ships on the horizon with the telescope some two hours ahead of seeing them with the naked eye. Pleased, they rewarded Galileo. For his innovations he was offered a significantly increased salary and a lifetime appointment at Padua, which he subsequently turned down in favor of returning to Florence, much to the displeasure of his Venetian benefactors.

Into the Paduan frame Galileo brought his own, Florentine upbringing and Renaissance sensibilities. Though born in Pisa, he spent most of his formative years in the ruling town of Tuscany, where his freethinking father had cultivated close ties to the world of musicians and composers, to cultural and intellectual circles, and to literary and artistic traditions (there is some suggestion that among his childhood dreams Galileo once considered becoming a painter; he was certainly quite adept at drawing). Unpretentious and constantly straitened financially, Vincenzo oversaw a modest living from the patronage of Count Giovanni de' Bardi (1534–1612), who headed the Florentine Camerata, a group of musicians, poets, humanists, and intellectuals devoted to discussing and guiding trends in the arts, especially music and drama (the group contributed significantly to the development of early opera). Even with rather limited means, the Galilei family connections reached into the upper tier of cultural life and politics in Tuscany. The young Galileo grew up immersed in this world. He lectured publicly on the structure of Dante's *Inferno* at the Academy of Florence in 1588; he built an organ for Jacopo Corsi, an aristocrat and member of the Camerata who later assumed the patronage of the group after Bardi left; he debated the relative merits of the poets Torquato Tasso and Ludovico Ariosto, the day's literary hot topic; and at his father's knee and insistence, he became proficient with the lute himself. In brief, Galileo acquired and exhibited the qualities of a cultured man of the Renaissance, the likes of a Leonardo da Vinci or an Alberti, someone who approximated the talents of Castiglione's "courtier" (without the athletic or military prowess in Galileo's case).[15]

At Padua, the young academic state employee initially passed his free time mostly nurturing cultural and social interests, all the while fulfilling required teaching and service obligations. Then, after a few rather carefree years he settled into his true vocation in mathematics and science, concentrating his intellectual energies on critical questions surrounding the physics of motion and matter. Uniting the theoretical tradition of university mathematics with the trial and error of practice, Galileo brought what we can recognize as an engineer's mentality to scientific investigation.[16] Bodies in free fall, or following the parabolic arc of projectiles, or swinging on a pendulum cord all succumbed to the reverse engineering of his—and henceforth modern—analysis. Although he would not publish his results until 1638, nearly at the end of his life, the dozen years or so between 1598 and 1610 witnessed Galileo's greatest achievements as a scientist, his subsequent findings with the telescope notwithstanding. In these same years Francis Bacon, Galileo's decidedly nonmathematical, though comparably important English contemporary, was urging followers to examine nature more closely and torture it into revealing its secrets. Bacon told us where to look. But Galileo told us what to look for . . . and how.

* * *

"Men make their own history," wrote Karl Marx, "but not of their own free will; not under circumstances they themselves have chosen but under the given and inherited circumstances with which they are directly confronted."[17] The truth of Marx's comment about the collective "men" (or in our gender-conscious age, humans, for he was referring to the entire species) extends as well, unbroken, to singular individuals. Galileo's inherited circumstances included the historical accumulation of the practical developments we have chronicled leading up to his lifetime. They engrossed him in the emerging, relational numeracy, and its use of empty, abstract, and functional symbols. These were his tools of discovery, as the merchant's Hindu-Arabic numerals, the musician's abstract time, the artist's grids and perspectives, and the astronomer's universal clock all became instinctively part of his thought processes.

Nor was Galileo alone in this regard. Other mathematicians and natural philosophers too had at their disposal the same array of symbols and abstractions, and often employed many of the same mathematical tools. In this world of "mixed mathematics," Tartaglia, for example, had introduced a "new science" in his analysis of ballistics, while Federico Commandino had translated numerous classical mathematic texts and wrote about their utility for military purposes. His pupil and one of Galileo's mentors, Guidobaldo del Monte, published a book in 1577 titled *Mechanicorum liber* (*The Book of Mechanics*), which conjoined treatment of simple machines and Archimedean statics. The Venetian Giovanni Battista Benedetti studied sundials, music, mechanics, and physics, along with mathematics. And there were others too, outside Italy, engaging in comparable investigations—the Dutchman Simon Stevin (1548–1620); the Englishmen John Dee, William Gilbert (1544–1603), John Napier (1550–1617), and Francis Bacon; the German Johannes Kepler (1571–1630); the young Frenchman René Descartes—and many more besides.

But Galileo always seemed to travel a step in front of the others, to inject a bit more distance than most of his contemporaries between himself and the world of qualitative, philosophical, and allegorical mathematics. He did flirt with astrology, his biographers have noted, but mostly to create astrological charts as a means of supplementing income or bolstering patronage.[18] Like father, like son, family obligations left him constantly strapped financially during these years. Yet also like father, like son, he gave not a fig for the symbolic and metaphorical uses of numbers and mathematics, those "sonorous" and comparable qualities of mathematics his father had so pooh-poohed. And like the independent-minded Vincenzo, Galileo disdained authority. Both were often heard to call fools those who relied on authority to win arguments. Thus armed with the new technology's empty, instrumental symbols and abstractions, Galileo's analytical temper and antiauthority temperament together confronted space and time and the bodies that moved through them. Not since Aristotle defined nature as "a principle of motion and change" had enough elements coalesced to make thinkable an alternative to definitions as a means of capturing and conveying the patterns we discern in nature. Galileo's métier found its expression in securing our sensory-laden ephemera in a new, relational mathematics. And with that, the moment of modern science had arrived.[19]

A passage, famous and familiar, merits reiteration: "Philosophy is written in this grand book, the universe, which stands continually open to our gaze. But the book cannot be understood unless one first learns to comprehend the language and read the letters in which it is composed. It is written in the language of mathematics, and its characters are triangles, circles, and other geometric figures without which it is humanly impossible to understand a single word of it; without these one wanders about in a dark labyrinth."[20] Shortly after these words Galileo quoted his friend and former student Mario Guiducci (1585–1646) to reinforce the point, the theme of this part of our story, that "nature takes no delight in poetry." No linguistic or, in modern terms, hermeneutical revelations would lead one to discover the truths of nature; only "demonstrations" in the "language of mathematics" could suffice.[21] Likewise with logic, it was an "excellent instrument to govern our reasoning," explicating what one already knows, but one must rely on "the sharpness of geometry in awakening the mind to discovery."[22] With these and many other similar remarks Galileo began hammering the thin edge of a wedge between the languages separating what we believe and what we know, between the information technologies of conventional, alphabetic literacy and relational numeracy. Henceforth the analytical temper of science, embodied in the emerging, relational mathematics, would steadily depart from the conventional language of religion.

Few contemporary observers saw this. John Donne might well comment on the "doubt" occasioned by the "new philosophy," but neither he nor others had much notion of what the new philosophy would eventually entail. Even Galileo himself, for that matter, continued to wander much of the time in a dark labyrinth. (Ever since Arthur Koestler's influential, though flawed book, "sleepwalker" has become the standard metaphor aptly applied to Galileo, as well as to other early modern natural philosophers.)[23] Still, despite the numerous wrong turns, including one down a huge rabbit hole of trying, incorrectly, to explain the tides, Galileo's analytical temper led him through the labyrinth during these years, bringing science across the threshold into modernity and establishing with it the fault line separating science and religion.

In chapters 9, 10, and 11 we shall follow Galileo as he grappled with and discovered the laws of free fall, pendulum, and projectile motion, as he grappled with and partially broached the mathematization of matter, and as he sought a new account of the relation between science and religion, one growing out of his analytical achievements. Our focus will necessarily fall on developments in early science, less so on religion, for the innovations in the study of nature lay prominently behind the rift between the two. To appreciate the depth and extent of this separation, we need to venture "inside" the emerging, analytical temper, so to speak, to enter as best we can into the process of scientific discovery, exemplified for us by Galileo. Although the Renaissance context of Galileo's investigations may seem a bit foreign to us, and so too his investigative practices, we should bear in mind that the science he came up with is now commonplace and quite elementary, generally introduced in high school. The challenge here lies with figuring out at least some of the mystery involved in his creation of the two new sciences of matter and motion, even while recognizing that the genius of innovation makes its own rules and refuses to bend itself entirely to the historian's narrative will.

Behind his labors lay a palpable, growing confidence in analytical methods that this remarkable son of Florence came to possess. His intellectual surety and assertion were not merely the presumption and moxie of a Renaissance courtier or the Quixotic jousting at Aristotelian windmills, although his oversized ego paraded elements of each.[24] At bottom Galileo's scientific certitudes rested on the abstractions of the new information technology, which we moderns have long since assimilated. These abstractions fed a new understanding of numbers as relations and of how the language of mathematics, freed from allegory, correlated with the phenomena of the natural world. Within them lay embedded, as well, a new understanding of mental operations, to be teased into philosophical awareness by subsequent generations beginning with Descartes. All these factors would in due course bear a great deal on Galileo's conflict with the Church, a conflict that first exposed modernity's fault line, the rift between science and religion. But even before its earliest cracks appeared, the intellectual fault lines were being etched in his self-assurance. And ultimately it would be far less the content of Galileo's scientific findings (including the new worlds he opened up with his telescope) than the functions of the analytical temper—how the scientific mind operates, not what it discovers—that would prove so troubling for religious belief, then and ever since.

CHAPTER 9

The Birth of Analysis

. . . beginning with physical and mathematical arguments.

—Galileo, "On the Copernican Opinion" (1615–1616)

STORIES ABOUND, many of them apocryphal, about Galileo's youth and early scientific career: of his watching a chandelier swing while in church and wondering about the physics of the pendulum, of his worming an entrance into Ostilio Ricci's (1540–1603) lectures on Euclid, of his dropping different weights from the leaning tower in Pisa to disprove Aristotle's theory of freely falling bodies. Our biographical information for these formative years is sparse—often coming from a blind Galileo's own remembrances, dictated in his last years to his then teenaged secretary and later first biographer, Vincenzo Viviani (1622–1703)—and the interweaving of fact and fable seems fated from the outset of Galileo's storied life. Even so, at our remove his early career appears more or less normal for a curious, young Florentine of the late sixteenth century. Galileo's father had wanted him to pursue medicine, then as now a stable, sufficiently lucrative, and socially respectable career. To that end Vincenzo marshaled the family's modest resources and saw his first-born son matriculate at the University of Pisa, noted in Tuscany for its medical faculty. But Galileo's concerns lay elsewhere, in mathematics and science, the latter perhaps inspired somewhat by Vincenzo's own musical profession and his experimental investigations into the mechanics of sound. At Pisa, Galileo likely furthered his interests by attending some classes taught by Filippo Fantoni (1530–1591), who offered courses on Euclid in 1582 and again in 1584. But the major mathematical influence on Galileo was the aforementioned Ricci, court mathematician at Florence. As Galileo's instructor and mentor, Ricci stirred the young man's interest not only in Euclid but, of equal importance, in Archimedes.

Ricci also probably intervened between Galileo and his father, persuading Vincenzo to allow his son pursuit of his mathematical and scientific muse, for Galileo left school in 1585 and began tutoring students privately in Florence and Siena. As his reputation grew in these formative years, Ricci helped him gain an audience in Rome with the distinguished Jesuit mathematician Christoper Clavius (of calendar fame), to whom Galileo had submitted a paper dealing with centers of gravity and by whom he was favorably received.[1] Other Renaissance mathematicians, notably Guidobaldo del Monte and Giuseppe Moletti (1531–1588), much appreciated this early work and likewise encouraged Galileo. In the small world of patronage, courtly behavior, and aristocratic connections, both worked on his behalf when a lectureship in mathematics became vacant at the University of Pisa. Although Galileo never graduated, Ricci and his other supporters painted such a glowing picture of his promise and talents that in 1589 he succeeded his own mathematics teacher, Fantoni, launching his academic career.

As Galileo assumed his duties at Pisa, he did so amid a cultural and intellectual milieu teeming with the empty symbols and abstractions of a new information technology, which we have amply detailed in the practical mathematics of business, music, art, and time reckoning. A child of the Renaissance, he absorbed not only its culture in general but more pointedly for our purposes, those available symbols and abstractions. Over the next two decades, first at Pisa, then at the University of Padua, Galileo constructed a new relation-mathematics from this technology and its core of modern numeracy. With it he captured key features of motion and developed the technical means of breaking down its complexities into simpler components. He worked entirely in Hindu-Arabic numerals, using them to express ratios and proportionalities that functioned like equations, much in the manner demonstrated by his Pisan forbear Leonardo, with the rule of three and other manipulations of the new numbers. He brought the silent music of the spheres down to earth, relying on the regularity of audible musical beat to capture and analyze equal units of abstract time. He mapped motions onto visual grids, whose crisscrossed lines yielded the reference points for dissecting any given moment of a moving body's trajectory and coordinating it with two axes, vertical and horizontal, thus converting the artist's subjective perspective into the scientist's objective reality. And in perhaps the most imaginative leap of all, he made abstract clock time into a coordinate of analysis itself, "squaring" time as one might a unit of spatial length (foot, *braccio,* meter, and the like). Squaring space had long made sense in figuring areas, but squaring time was completely nonsensical and unheard of before Galileo. He himself resisted the move for several years until numbers from his experiments forced him to recognize its place in his discoveries. It was a stunning and original abstraction, an early progeny of modern numeracy.

Using the new technology, Galileo redirected Euclidean geometry from a logical structure based on definitions, postulates, and propositions, which had served to rationalize and consolidate what one could know about fixed space, into an instrument of exploration and discovery pertaining to bodies in motion. A recent Italian translation of Euclid's *Elements* proved to be uniquely significant in this endeavor, for it had included an unadulterated version of the theory of proportion found in book 5, which Galileo mastered in his youth and which became a key analytical instrument throughout his investigations.[2] In particular, a procedure for calculating a relation between ratios known as a "mean proportional" ("geometric mean" in modern terms) allowed Galileo to use in his analyses those wily, irrational magnitudes ("incommensurables") that had proved so intractable for the Greeks. (For a brief explanation of Galileo's mathematical techniques, see Appendix A.10.) Articulated in the symbols of the new numeracy, Galileo's techniques for manipulating proportions established the most rigorous connection between mathematics and physical events until the publication of Descartes's analytical geometry in 1637, nearly three decades after Galileo's major work had been completed.

Evidence of Galileo's use of the new technology in achieving a breakthrough into analysis may be discerned in the three major areas of investigation pertaining to bodies in motion that occupied him throughout his early teaching career and especially during the first decade of the seventeenth century. These investigations led to what are now generally called the "laws" of free fall, pendulums, and projectiles, and in all three areas Galileo exhibited a combination of experimental and mathematical innovations. At the beginning of his career he primarily dissected concepts in the realm of thought, breaking apart definitions to show their absurdity, and with it the irrationality of Aristotle's accounts of falling bodies and motion in general.[3] But early on he also came to see the need for breaking apart complex motions in the reality of experiment, hectoring out of motion its constituents:

abstract units of time, distance, and speed. These components he then correlated with one another through ratios and proportionalities, largely arrayed in the familiar pattern of the merchant's rule of three.

Scholarly consensus has long conceded Galileo's time in Padua (1592–1610) as by far the most decisive in the development of his revolutionary brand of mathematical physics, known technically as kinematics (the study of motion).[4] But for decades in the twentieth century Galilean studies fell under the influence of the French historian of science Alexandre Koyré, who held that Galileo's scientific "discoveries" from this and later periods never went beyond thought problems, such as those introduced in *De Motu* (*On Motion*), an early work written during his Pisa years but never published in his lifetime. From the 1960s forward, however, another generation of Galilean scholars has dispelled Koyré's image of an armchair philosopher and replaced it with that of an active scientific practitioner. Pioneered by Thomas B. Settle and Stillman Drake, researchers have been reproducing Galileo's experiments with motion.[5] They have been guided in these endeavors by a remarkable set of documents, now generally referred to as Codex 72 of the Collezione galileiana.[6] The codex contains over two hundred folio sides describing the experiments Galileo worked on during these critical years. Many of them culminated in his mature work on the physics of motion and matter, *Discourses on the Two New Sciences,* published much later, in 1638. In effect, the folios are Galileo's lab notes. These working papers have established beyond any reasonable doubt that he conducted hands-on experiments in his investigations of nature, and they have offered a fascinating entry into the mind-set and thought processes behind his most important discoveries. The worksheets reveal, in short, the emergence of Galileo's analytical temper.

Even so, the often highly cryptic folios suggest but dark clues to his thinking, for Galileo left no road map with which to guide other readers who have tried to follow his lines of investigation, no notes explaining how to read his notes. Consequently, modern scholars seeking to reconstruct his pathways to discovery have produced widely divergent, and fiercely defended, interpretations of the intellectual events contained in the documents.[7] Snippets of information disclosed in his letters have helped matters somewhat. From a letter dated October 16, 1604 to his Venetian friend Paolo

Sarpi (1552–1623), for instance, it's clear that by then Galileo possessed the law of free fall; from another, to Antonio de' Medici, dated February 11, 1609, that he knew the correct mathematical decomposition of parabolic trajectory into its horizontal and vertical components or axes. These two dates are commonly given as the crowning moments of Galileo's discoveries. But detailed reconstructions of what he knew and when he knew it remain uncertain and highly contested.[8]

Such scholarly uncertainties themselves tender a clue of their own. For Galileo wasn't directly seeking to make scientific discoveries in the sense we moderns mean the term, at least in the early phases of his investigations. That assumes he would have known what was involved in a "scientific discovery" and could recognize one when he had it. Rather, he faced problems and paradoxes posed by the traditional accounts of motion inherited from Aristotle and scholastic thinkers, problems he explored both with a small circle of friends and with growing confidence in the new relational mathematics of his making. Thus we should not be surprised to see him struggling within the linguistic and conceptual framework of his intellectual birthright, to see him showing flashes of deep insight and vision intermingled with large blind spots, to see him unaware when he had actually made a discovery, or to see him puzzling over visualizing, experimenting, deriving, and proving the evidence before him. In the end, he did make the discoveries attributed to him, but they remain less significant than a far deeper achievement underlying his labors, the central focus of this chapter: the invention of scientific discovery itself.[9]

It wasn't war, this drama, or high-level political intrigue, or a love story in the fashion of Abelard and Héloïse, or even a tale of social ambition. But in strictly intellectual terms it was, nonetheless, gripping theater, one in whose audience we still stand awestruck. For at center stage stood the daughter of time, chanting in discrete and equal intervals the integer values of metronomic measurement, and sounding in mellifluous and seductive continuity the flow of change.[10] It was the time of time's singing, and to Galileo's ear both its voices merged into a single harmony. Musical beat measured time; time, pirouetting on itself in squares, measured distance. Both figured centrally in the analytical process of reverse engineering bodies in motion. More prosaically put, Galileo abstracted and severed time from the traditional, hoary definition of motion, only to reconnect it to motion through the new mathematics, a mathematics that shattered the categorical incommensurability between discreteness and continuity, between arithmetic and geometry.

As we explore Galileo's investigations into the patterned behaviors of free fall, pendulums, and projectiles, into the birth of analysis, we need to bear in mind the place of his innovations in the larger context of our central theme, the emerging rift between science and religion. In working through the technical difficulties of devising, interpreting, and correlating his numbers with actual phenomena, Galileo was paying virtually no attention to religion. Nor shall we as we follow his labors and explore his emerging analytical temper. He was also distancing science from its time-honored grounding in the linguistic assumptions and practices of alphabetic literacy and classification, the same assumptions and practices that gave voice to religious tenets and belief. In his own terms he was creating a "language of mathematics" with which to read and understand the book of nature. The new relation-mathematics was not just another conventional language, however, one among many, a language of nature in the same fashion that French, say, is the language of love, or German of metaphysics, or English of practical affairs. It departed radically from the conventional languages in which both traditional (Aristotelian) science and the narrations of religion were historically cast. It became, in the words of a modern physicist, "a system of *connections,* created by the tools of logic."[11] Historically, these logical tools were those of the new 'techno-logic' of information. This was Galileo's doing. On the basis of new and different abstractions, formed with numeracy's symbols, he distilled relations from the ephemera of moving bodies. It was nothing short of a transformation of analysis from its two-thousand-year history of establishing deductions on first principles or concepts—on definitions—into the reverse engineering of modern science, a metamorphosis signaling science's uncoupling from the conventional language not only of philosophy but eventually of religion as well.

* * *

Galileo's budding analytical sensibilities grew in the Pisan years (1589–1592), during which time he launched his scientific career. Like many a young academic, brash and confident, he began by criticizing key features of the intellectual universe he had inherited.[12] In his context that meant confronting the central tenets of Aristotle's physics, especially bodies in free fall, which he did in his own, early manuscript *On Motion.* This thin volume challenged many of the "childish arguments," "errors," and "inept attempts at subtleties" that Galileo believed typified Aristotle's "qualitative" accounts of heavy and light bodies and their motions in various media. Using as his analytical scalpels Euclidean proportion and the hydrostatic principles of Archimedes, he devised numerous thought experiments to show that many common terms and definitions used by philosophers, such as "absolute heaviness," were absurd; that the relative weight between a body and a medium in which it moved should be the proper focus of attention; that mathematical demonstrations allowed the truths of nature to "shine brightly."[13]

Consider, Galileo proposed, the case of falling bodies. According to Aristotle's theory, in modern paraphrase, the speed of a body in motion is directly proportional to the force that moves it and inversely proportional to the resistance impeding it. A chair remains at rest until it is pushed; how fast it moves depends on how much force is pushing it, assuming the resistance remains the same, such as the friction between the chair and a carpeted floor. The same force pushing the same chair on ice will move it much faster as resistance decreases.[14] For bodies in free fall, the force pushing the body would be replaced by the body's weight because the body's heaviness (its *gravitas*) caused it to fall down, toward the center of the earth. Therefore, the heavier the body, the faster it would fall, assuming the resistance, such as the medium of air, remained the same. For instance, if a lead ball and a wooden ball of the same size but different weights were dropped at the same time, the heavier, lead ball should reach the ground first. Well, argued Galileo, imagine two bodies equal to one another in size and weight being dropped simultaneously. Their speeds would be equal. Now imagine joining them together while moving. In such a case, they should double their speed during the fall, according to the prevailing theory. Or imagine two lead balls, the first one a hundred times heavier than the second, being dropped from the sphere of the moon, which divided earthly from celestial motions. If the larger one reaches earth in one hour, then the smaller must require a hundred hours to reach the earth. Or, if two stones, one twice the size of the other, are thrown simultaneously from the same tower, the larger will reach the ground twice as fast as the smaller. Further, imagine tying them together; then the smaller would actually slow the larger's speed of fall. By such examples and "at all times" employing reasoning, Galileo essayed by means of the philosophical *reductio ad absurdum* to bring about the "collapse of Aristotle's view."[15]

In the same work Galileo developed as well the first hints of a restricted concept of inertia and the thought problems that would lead later to his actual experiments with bodies in motion. He spoke of bodies "free from all external resistance" and was already convinced that mathematical proofs concerning such idealized bodies could be valid on their own, regardless of the physical assumptions involved in applying the proofs. At the same time he recognized that confidence in both the proofs and the assumptions of mathematical demonstrations of natural phenomena would be affected by actual experimental tests and their consequences. If not, as he later wrote, demonstrations unsupported by evidence would belong purely to "a world on paper." The interplay between mathematical abstractions and experimental corroboration was crucial. In the words of Galileo's nonpareil scientific biographer, Stillman Drake, "Whether the test induces us to abandon, modify, or accept the assumptions usually depends on the magnitude of the disparities found [between the mathematics and the experiment]."[16] Thus mathematical demonstrations and some sort of experimental testing lay embedded within Galileo's often hostile, always searching critique of the definition-based, Aristotelian natural philosophy, at least that portion of it we now term physics. Both would emerge as critical in Galileo's principal discoveries concerning motion.

In 1592 Galileo doffed his professor's toga at Pisa and donned a new one at Padua. His career had gotten off to a rocky start at the university in his natal town, where he had roundly criticized not only the Philosopher but his own Aristotelian colleagues as well. His pay was low, and lowered further with fines for missing lectures and not wearing academic regalia, and his contract was unlikely to be renewed. Written at the time, his satirical poem "Against the Wearing of the Gown" revealed his own desire to shake the Pisan dust from his sandals. The steady influence and guiding hand of Guidobaldo del Monte brought Galileo to Padua, a situation far more to his liking, and there he seemed to settle in comfortably to what he would later term the "best eighteen years of my life."[17] During the next decade or so, a period marked by a dearth of biographical information, his attention was largely taken up with practical engineering projects (requests from Venetian authorities), teaching obligations (which he took more seriously than at Pisa), and scrambling for cash (an unremitting worry for years).

Then, sometime around the turn of the seventeenth century, Galileo revisited the questions of motion broached earlier. He began devising experiments with balls released as a pendulum bob and with balls rolling down a slightly inclined plane. As indicated by his lab notes and correspondence, he was seeking ways of addressing the issues raised earlier in *De Motu* of the heaviness and lightness of bodies in both freely falling ("natural") and projectile ("violent") motion. Over the next few years these investigations would lead him to recognize the laws governing free fall and pendulums. The law of free fall states that a body falling from rest will travel a distance equaling the square of the time it falls, while that of the pendulum states that the length of a pendulum is directly proportional to the time squared of its period. Galileo did not explicitly set out to discover any such laws.[18] He sought, rather, to slow down the motion of heavy bodies, so that one could measure their behavior and discern mathematical relations between the various components of movement. Slowly moving pendulums and slightly inclined planes were means to this end. His friend Guidobaldo del Monte was performing similar tests with balls rolling along the insides of barrel hoops. By slowing the motion down, it would become possible to measure weights, distances, times, and speeds involved, and to investigate what happens when a body falls freely on the vertical. As Galileo did so, the abstractions of musical beat and time reckoning came to figure momentously in his results.[19]

Galileo's calculations on folio 107v, one of the worksheets described earlier, provide a window into his early thinking on these matters. (See Figure 9.1a.) In the upper-left corner of the folio (Figure 9.1b), there are three columns of numbers, arrayed as follows (columns labeled for reference):

[A]	[B]	[C]
1	1	33
4	2	130
9	3	298
16	4	526
25	5	824
36	6	1,192
49	7	1,620
64	8	2,104

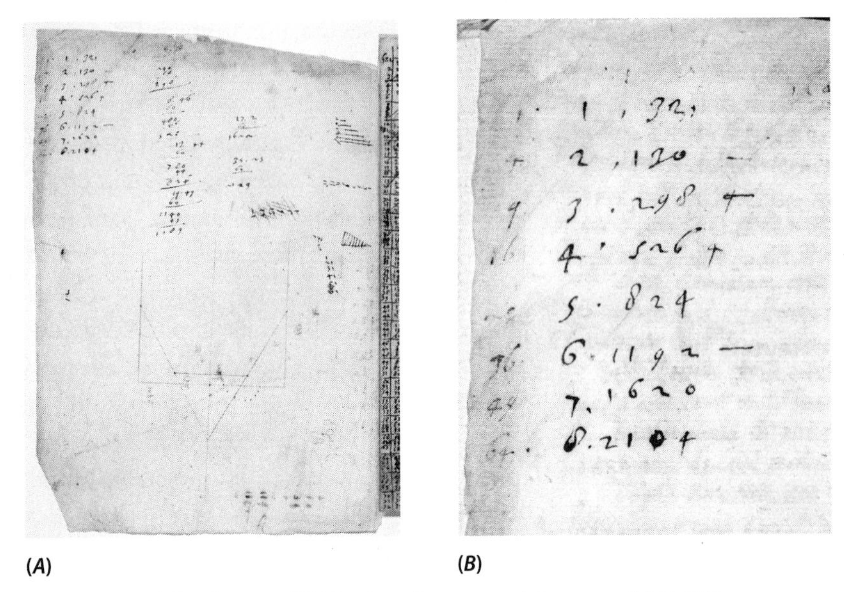

FIGURE 9.1. (*A*) Folio 107v. (*B*) Numbers from upper-left corner of folio 107v.

While the sequence and dating of these entries remain in dispute, it is highly probable that column C represents his first measurements of distance taken from a ball rolling down an inclined plane, measurements stemming directly from musical beats and registered as frets on the plane.[20] The numbers themselves refer to *punti* (points), the name Galileo gave the units he had marked off on a measuring stick. There was no commonly accepted standard of fine measurement, so he created his own, a stick probably made of brass, sixty points in length, with one *punto* (point) equaling 0.94 millimeters in modern, metric reckoning.[21] The inclined plane itself was roughly seven to eight feet long, a ramp slightly elevated to angle of approximately 1.7 degrees with horizontal. Because speed is a function of distance covered in time, a correlation known since antiquity, Galileo next had to devise a way of introducing time measurement into his experiment. That meant equalizing the times the ball traveled through regular intervals in order to discern how fast the ball was traveling. If the ball rolled down a greater distance during the same period of time, it was obviously going faster.

But Galileo had no readily available means of equalizing the times. The newest and most precise clocks could barely measure seconds accurately;

counting fractions of seconds lay far in the future. Likewise with metronomes; their days too lay ahead. And heartbeats, it was well known, were notoriously irregular. There was nothing immediately at hand against which he could measure regular passages of the small amounts of time needed for his purposes. At this juncture he conceived of using regular, musical beat as a pacesetter, likely drawing on his home-schooled, youthful lessons with the lute and his knowledge of musical tempo. Not only was Galileo himself a decent musician, but most people can manage to reproduce a regular beat—for example, by drumming with a pencil on a table. Even average ears can generally manage eighth or sixteenth notes quite easily, assuming a moderate beat, and Galileo's sense of musical time was likely more refined, perhaps good for thirty-second or sixty-fourth notes.[22] To set up the experiment, he devised a means of placing movable "frets" (probably made of gut) across a groove in his inclined plane.[23] The groove kept the ball rolling in a straight line, and when it rolled over a fret, it made a slight ticking sound. By sliding the frets up and down, Galileo could equalize the ticks with the beat in his head, or tapped out, drumlike.[24] Having equalized the times of the ticks, he then measured the distances between the frets, producing the numbers he wrote in column C in the foregoing figure. Just as creating and measuring abstract time as a symbolized reality had been critical to the development of polyphonic tempos and harmonies, made possible by the invention of musical notation, so too did abstract time become a cornerstone of analyzing the physical phenomena of bodies in motion. It was immediately evident that as the ball went farther down the ramp, the spaces between the ticks grew longer, even while the beat remained constant. The ball was clearly increasing its speed during descent, which we know as its uniform acceleration from rest, a conclusion Galileo did not immediately infer.

Now the challenge was to find some sort of ratio or proportion in the measurements. At first, Galileo thought this would be between the speed of the ball and the distance it had traveled, a proportionality of $v \propto d$ in modern symbols, with v designating velocity, d distance, and the symbol $\propto$ proportionality. As a preliminary, he probably tried several numerical sequences to discern a pattern in his distance measurements. We find, for example, the sequence of alternating odd numbers—1, 5, 9, 13, 17, 21—written on the same folio 107v, only to be crossed out, not matching column C. (See Figure 9.2a.) We also find a series of odd numbers on the same page, which is not crossed

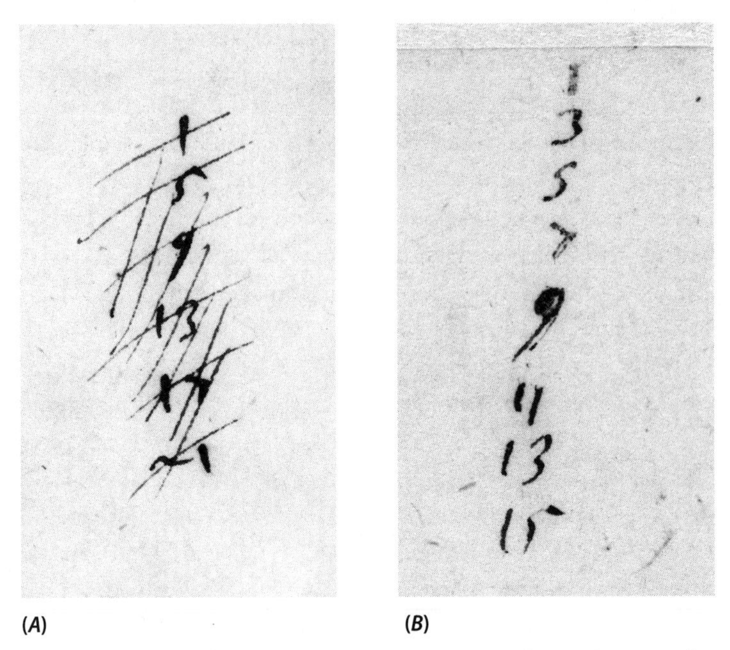

FIGURE 9.2. (*A*) Crossed-out numbers from folio 107v. (*B*) Odd-number series from folio 107v.

out and which at some point Galileo came to see as roughly following the sequence of measured distances between the equal, successive time measures. (See Figure 9.2b.) To see how, let's look again at column C. Each number represents the distance traveled in equal intervals of time as the ball descended the ramp, as follows:

Galileo's distance measurements from F.107v, with intervals added:

[B]	[C]	Interval, (C1 to C2, etc.)	Times
1.	33		Time #1 (.55 sec)
		97	
2.	130		Time #2 (.55 sec, 1.10 total)
		168	
3.	298		Time #3 (.55 sec, 1.65 total)
		228	

4.	526		Time #4 (.55 sec, 2.20 total)
		298	
5.	824		Time #5 (.55 sec, 2.75 total)
		368	
6.	1192		Time #6 (.55 sec, 3.30 total)
		428	
7.	1620		Time #7 (.55 sec, 3.85 total)
		484	
8.	2104		Time #8 (.55 sec, 4.40 total)

Now, taking the first distance of 33 points as the base unit of measure, divide it into the successive distances of the (equal) time intervals. This creates the following table and allows us to recognize the odd-number sequence, which Galileo had entered on his note:

1st Interval, 0 to 1 = 33	÷33 = 1	(or, rounded	1
2nd Interval, 1 to 2 = 97	÷33 = 2.93	off, the odd-	3
3rd Interval, 2–3 = 168	÷33 = 5.09	number	5
4th Interval, 3–4 = 228	÷33 = 6.91	sequence (the	7
5th Interval, 4–5 = 298	÷33 = 9.03	"gnomon" of	9
6th Interval, 5–6 = 368	÷33 = 11.15	the squares in	11
7th Interval, 6–7 = 428	÷33 = 12.96	Greek terms)	13
8th Interval, 7–8 = 484	÷33 = 14.67		15

Although he clearly recognized the odd-number sequence in his measurements, the matter of deciphering it posed a much murkier problem. The numbers by themselves meant nothing; there was no holy *tetractys* or golden ratio lurking here, no Pythagorean harmony or divine finger tracing a path through the symbols. The only question was how to take the sequence of numbers and match it to the physical action of the ball's rolling down the plane. Initially Galileo thought in terms of a direct one-to-one correspondence between units of speed and distance during each of the equalized time intervals. As long as this was his objective he did not think to sum the numbers in the odd-number sequence. Had he done so, the time-squared law might have been readily apparent ($1+3=4$; $1+3+5=9$; $1+3+5+7=16$; and continuing). Still, that would have meant correlating the distance of the ball's cumulative descent with the squaring of time, and, as we mentioned earlier, the squaring of time made no physical sense whatsoever; it was only an empty abstraction, a "derived" magnitude, he termed it. Nearly two years would pass before he could get around this hurdle.[25]

Perhaps the most remarkable breakthrough leading to the law of free fall came in another of Galileo's investigations, this one growing out of his experiments with pendulums, where he likely first saw the correlation of times squared and distances. We have no records here of a Galilean eureka moment analogous to that of his hero Archimedes, certainly no scandalous reports of Galileo's running naked through the streets of Padua.[26] But it's hard not to imagine a singular "rush" moment in his pendulum experiments affecting him comparably, a moment illustrating archetypically how the new technology came to bear on his threshold discoveries. For in some trice he recognized that an abstract number, understood as a ratio or relation, could represent an even greater abstraction, the squaring of time, and that even without a direct, physical correlation or counterpart, time squared could and did play a critical role in deciphering the patterns found in nature.[27] That the abstractions of relational mathematics actually worked would subsequently become perhaps Galileo's greatest insight and discovery, generating a mystery that since then neither he nor any mathematician nor physicist (nor philosopher for that matter) has ever fully resolved. We still don't know why they work, but properly understood and manipulated and applied, empty symbols and their rules of combination can capture nature's most recondite secrets, beginning with motion itself.[28]

In experimenting with pendulums Galileo's general strategy followed that of his approach to the inclined plane: slow down the motion of the ball, now a bob hanging on a string, in order to make accurate measures of its movements and seek ratios between various components (such as length of string, weight and material of bob, time of period, distance of arc). As he conducted his pendulum tests, beginning early in 1602, he was soon able to show that the length of the string was the only variable affecting the period of a pendulum, not the weight or material of the bob or the distance of the arc. This discovery most likely suggested a connection to free fall, for earlier, in *De Motu*, he had argued similarly that the speed of a falling body was independent of its weight.[29]

Whereas with the inclined plane he had equalized discrete units of time and measured the differences in distances the ball rolled, in the pendulum experiments Galileo sought the reverse, to equalize distances and then measure the flow of time during a pendulum's swing. He experimented with a bob's falling through different arcs. This he accomplished with a water clock (or, more precisely, a water stopwatch), a device of his own making, whose operation he described later in the *Two New Sciences.*[30] A water clock was made, simply, from a pail of water hanging from a support, which had a "slender" tube attached to the bottom. Probably using his thumb as a stopper, Galileo collected small amounts of water as it descended through the tube into a beaker. These amounts corresponded to times of motion (say, of a pendulum's period, or a ball's rolling down a ramp). He then weighed the water collected, and this gave him his correlative units, by means of which he could account for the continuous coursing of elapsed time. Recall, any sort of synchronism—correlations of regular occurrences—can be used to record time's passage. He recorded his first water measurements in just this fashion on folio 154v, using grains as his unit of weight. In his reckoning, sixteen grains of water equaled one *tempo,* or unit of time.[31]

As Galileo began his tests using the water clock, he investigated the times of a pendulum's swing through small arcs to the vertical, rather than through an entire period (a full swing and return). Measuring a quarter period allowed him to time the swing with greater precision from the moment of releasing the bob to the sound of its impact against a block fixed at the base of the vertical, which he installed at the side of the bob when it was hanging plumb.[32] From such measurements he was able to correlate the length of a pendulum (measured in *punti*) with the time of its swing to the vertical (measured in grains of water flow).[33] We need not venture very far here into Galileo's actual calculations. In fact, he himself made them in a very roundabout manner, only later eliminating several redundancies. The critical point surfaced when the numbers from his tests led him to recognize that the abstract square of time was directly proportional to the length of the pendulum string and therefore played a direct part in analyzing and explaining the motion of the pendulum.[34] (A compilation of his data and the central correlations he discerned in them may be found in Appendix A.11.) And because in his mind there was a parallel between the motion of a pendulum

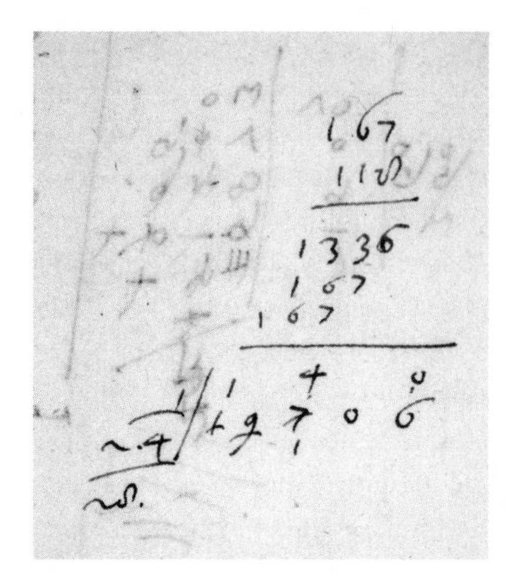

FIGURE 9.3. Calculation from folio 154v, Galileo's calculation of the mean proportional between 118 and 167. Note the alignment of columns in the multiplication.

and that of a falling body, he came to see the place of time squared in explaining free fall as well.

One brief digression can serve to illustrate just how adroitly these numbers and calculations fell into place and gained currency in Galileo's thinking. On folio 154v, where Galileo had recorded his first timings with the water clock, he further confirmed his correlations between pendulum lengths and times in a separate calculation on the same page. There he figured the mean proportional between 118 and 167, two time measurements that corresponded with two pendulum lengths.[35] This calculation gave him 140 *tempi.* (See Figure 9.3.) The mean proportional of the two corresponding pendulum lengths is 9,843, which meant that a pendulum 9,843 *punti* long would swing to the vertical in 140 *tempi.* On the backside of the same page (folio 154r), he wrote that the string is sixteen *braccia* long. From other of his Paduan notes we learn that one *braccio* (arm) equals ± 620 of Galileo's *punti.* Figuring a *braccio* as 615 *punti,* the total length of the pendulum is 9,840 (about thirty feet). Such a pendulum could be hung from a window over the courtyard at the University of Padua, and Drake has calculated that it would pass through a small arc to the vertical in 141 *tempi.* The closeness of these results strongly suggests that Galileo performed an experiment that allowed him to generalize the law of pendulum in its mean proportional form, which is mathematically equivalent to our own law—namely, that the period of a simple pendulum is directly proportional to the square root of its length, or, alternatively expressed, that the time squared of a pendulum's period is directly proportional to its length.[36]

We shall forever remain in the dark as to whether his pendulum findings explain exactly how Galileo correlated the squaring of time with distance or whether some other musing over his data led him to the same conclusion. For centuries writers have sought the holy grail of creation, the philosopher's stone that would account for the creative process and explain the alchemical transformation of our neural networks into intellectual and artistic brilliance—but in vain (at least so far). All one can say here is that at some moment Galileo recorded in the first column of his worksheet folio 107v the squares (1, 4, 9, 16, 25, 36, 49, and 64) next to those distance measurements of a ball's descent down the inclined plane.[37] This confirmed his discovery that the proper account of the speed of a moving body falling from rest depends on the ratio between the distance of the fall and the square of the time required to complete it. There did remain the more traditional task, based on Euclidean and Archimedean examples, of "proving" the relation from known or surmised mathematical axioms, which after false starts and dead-end assumptions, Galileo accomplished later in the *Two New Sciences.* There he utilized the long-standing mean speed theorem as a basis for his deductions.[38] (See Appendix A.12 for a modern reconstruction.) Focused initially on free falls and pendulums, all of Galileo's investigations over the next few years reveal a comparable reverse engineering of his data, a halting mixture of experimentation, mathematical abstractions, derivations from known principles or from newly discovered ones, and corroborating tests.[39]

* * *

By 1604 Galileo had made his initial breakthrough into analysis, using the abstractions of musical rhythm and time reckoning in uncovering the laws of free fall and pendulum. He followed those results by extending his investigations further, employing the visual aids of perspective and coordinate grids as he experimented with compound motions. His most significant findings of the next few years pertained to the third area of his investigations of motion, the parabolic trajectory of projectiles. Earlier, in *De Motu,* Galileo

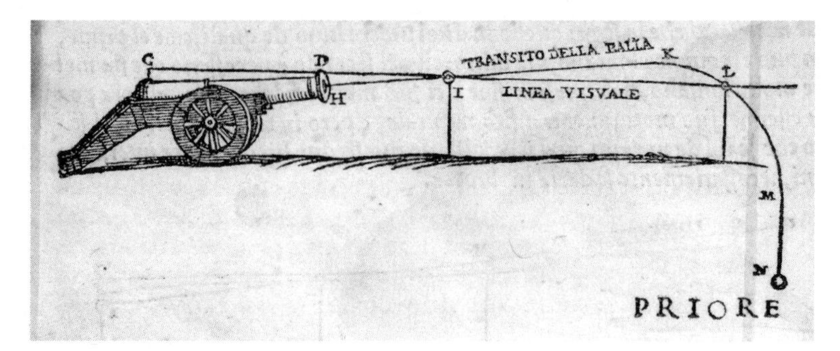

FIGURE 9.4. Tartaglia's depiction of projectile motion, from *Nova Scientia* (1537). The initial, "violent" motion (from H to K) occurred in a straight line, followed by a circular phase (K to M), and concluding with a "natural," straight-line descent to earth (M to N). At first, Galileo followed Tartaglia's lead, not "mixing" violent and natural motions; later, he blended the two motions in his own analysis of projectiles.

had followed the lead of Niccolò Tartaglia in decomposing and depicting projectile motion. Staying within the Aristotelian tradition of definitions, Tartaglia had held that no body could go through "any interval of time or of space with mixed natural and violent motions."[40] From this premise he argued that projectiles passed through three phases from when they were first released—as a ball shot from a cannon for instance. Initially, the "violent" motion occurred in a straight, horizontal line (or elevated), whereupon it passed through the arc of a circle before descending to earth in its "natural," straight-line motion. (See Figure 9.4.) Galileo came to see this picture as deeply flawed both because it relied on the increasingly obsolete incompatibility of categories ("natural" and "violent" motion) and because it failed to address adequately the whole matter of acceleration, which his own experiments and mathematics had more successfully depicted with free fall and pendulums. Further, his new approach was coupled with his recognition, as well, of the "speed law" (sometimes termed "law of descent"), the principle that the speed (v) in free fall is proportional to the square root of the distance fallen ($v \propto \sqrt{d}$, or alternatively, $v^2 \propto d$).

Although scholarly disputes persist over the exact lineage of his thinking that led to these findings, there is considerable agreement that Galileo essentially recorded his experiments on four key folios. On these worksheets he drew sketches of projectile motions, laid out graphically in a grid-like manner, which he then labeled with various numbers, giving the results of

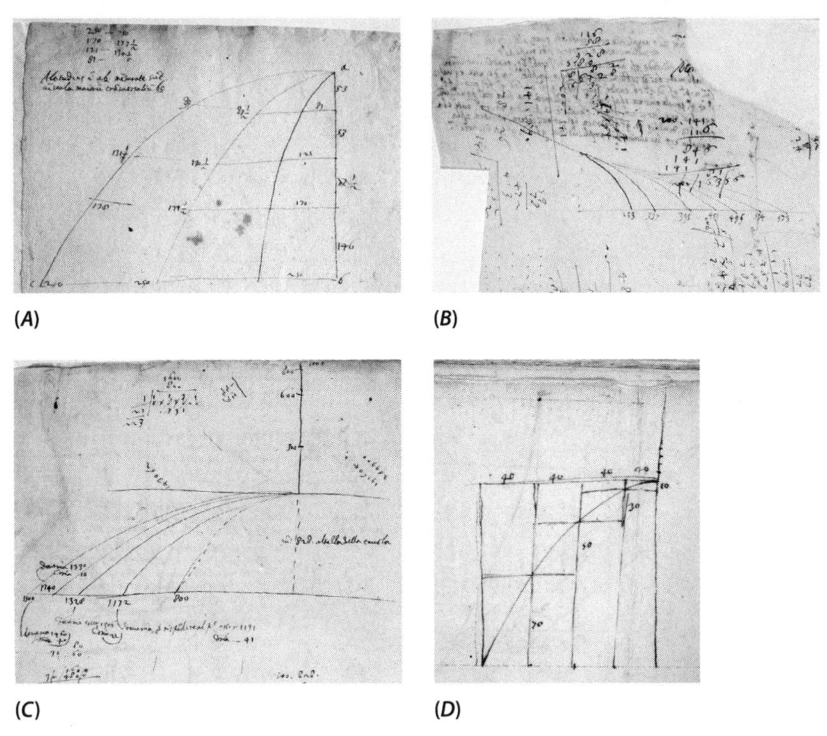

FIGURE 9.5A. From folio 81r.
FIGURE 9.5B. From folio 114v.
FIGURE 9.5C. From folio 116v.
FIGURE 9.5D. From folio 117r.
Galileo's lab notes recording his experiments in grid-like fashion and dealing with accelerated, compound motions and the paths of "movables." All contributed to his discovery of the parabolic trajectory of projectiles.

his tests.[41] (See Figures 9.5a, b, c, and d.) The numbers and grids substantiate how Galileo reverse engineered motion by splitting apart its horizontal and vertical components, measuring each of them, and then recombining them in ratios of speed, time, and distance, probably processing many computations with his compass, as well as performing them on his worksheets.[42] Others before him had used grids for various purposes, as we have noted, but in his hands the grid itself became a tool of analysis, demonstrating a decomposition of motion into its different dimensions.

The diagram Galileo drew on the verso side of folio 116 (Figure 9.5c) serves as a particular case in point, and speaks volumes about the way his mind worked analytically. This was a sketch he made of experiments with a ball rolling down a long, inclined plane from various heights and then launched horizontally from a tabletop as a projectile.[43] The diagram is not drawn to precise scale, but the numbers on the vertical and horizontal lines reveal the visual grid underlying his analysis. (See the schematic illustration in Figure 9.6.) The top horizontal line represents the table, the bottom one the floor. On the dotted vertical line, beneath the table top, there is the number 828 (*punti,* about seventy-eight centimeters), the height of the table from which the launch was made. The numbers on the solid vertical line above the table top also indicate heights recorded in *punti* (300, 600, 800, and 1,000). These were the heights from which a ball was released down a long, inclined ramp (some twenty-three feet in length, he later stipulated in the *Two New Sciences*).[44] The higher the number and point of release were, then, the greater was the speed at the point of the ball's horizontal launch and the farther it would travel. The numbers on the bottom horizontal line represent actual, measured distances the ball traveled after launch. Several of them are accompanied by other numbers and the word *doveria,* which means 'should be' or more precisely, 'would have had to be.' The *doveria* numbers reveal calculations that Galileo made in accordance with his base ratio of 300:800, the height of the first ball roll and launch correlated with the distance it traveled. They were the numbers that ideally should have resulted, given a frictionless universe. He then noted the differences between the mathematical ideal and the actually measured distances, which he designated with the letters 'dria' (his shorthand for the Italian *differentia.*)

The diagram shows clearly Galileo's recognition that projectiles followed a parabolic trajectory, which he had observed through qualitative and visual, vis-à-vis quantitative and analytical, experiments many years before.[45] Moreover, it reveals his awareness of the correlation and interplay between mathematical abstractions and actual test measurements, and his confirmation of parabolic trajectory with the rule of three applied to his tests. By taking the first roll height and travel distance as the base ratio (300:800), he was able to formulate a proportionality of ratios with tests from the other roll heights. This gave him the three numbers he needed to find the fourth, the distance traveled with each new test (thus, $300 : 800 :: 600 : x$). The question in his mind, however, was, just what did the numbers signify? In his discovery of free fall, Galileo initially had sought a direct proportionality between speed and distance

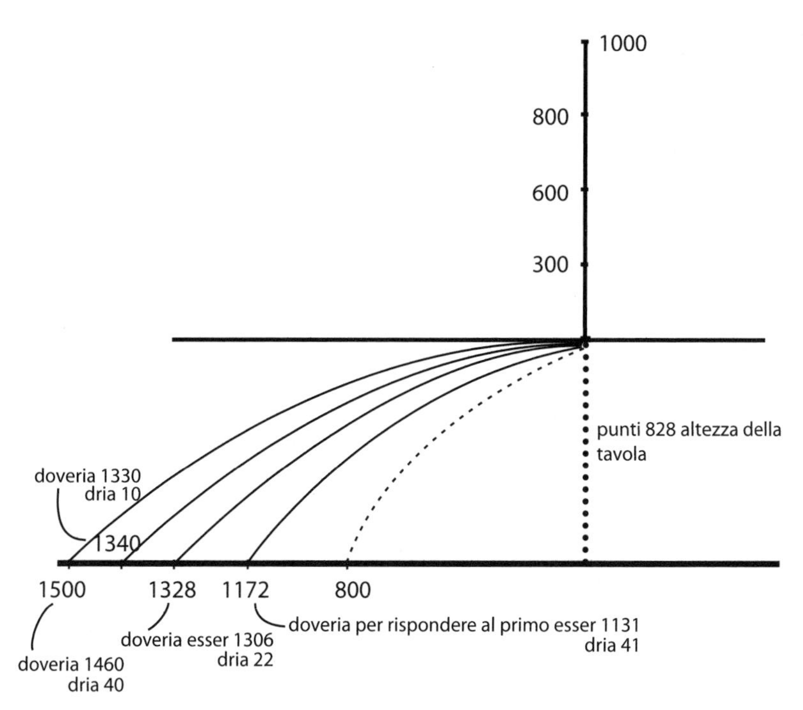

FIGURE 9.6. From folio 116v, schematic illustration of Galileo's analysis of the parabolic trajectory of projectiles. Numbers on the solid vertical line represent the heights from which a ball was rolled down a long, elevated ramp (in *punti,* points). By the vertical dotted line are the words "828 points, table height." Numbers on the bottom horizontal line represent actually measured distances the ball traveled from its tabletop launch point, the intersection of the upper horizontal and vertical lines. Also at the bottom is the expression *doveria,* which translates as "should be" or "would have had to be, given the first result." The *dria* is Galileo's shorthand for *differentia,* or 'difference.' The difference here is between (a) the actual measurements and (b) the distance predicted by calculation, with the "first result" ratio of 300:800 as a base and using the rule of three. The variations derived from friction, resistance, and other external factors.

($v \propto d$), but he eventually realized that the proper correlation was between the square of time and distance ($t^2 \propto d$). He found this confirmed, as well, in the experiments recorded on folio 116v. For if his numbers were direct ratios of speed and distance ($v \propto d$), then the result of the first test, for example, from a ball-roll height of 600 *punti* (*p*), should have been a distance traveled of 1,600 *p* $\left(\frac{300}{800}=\frac{600}{x}; x=1600\right)$, way out of line with his actual

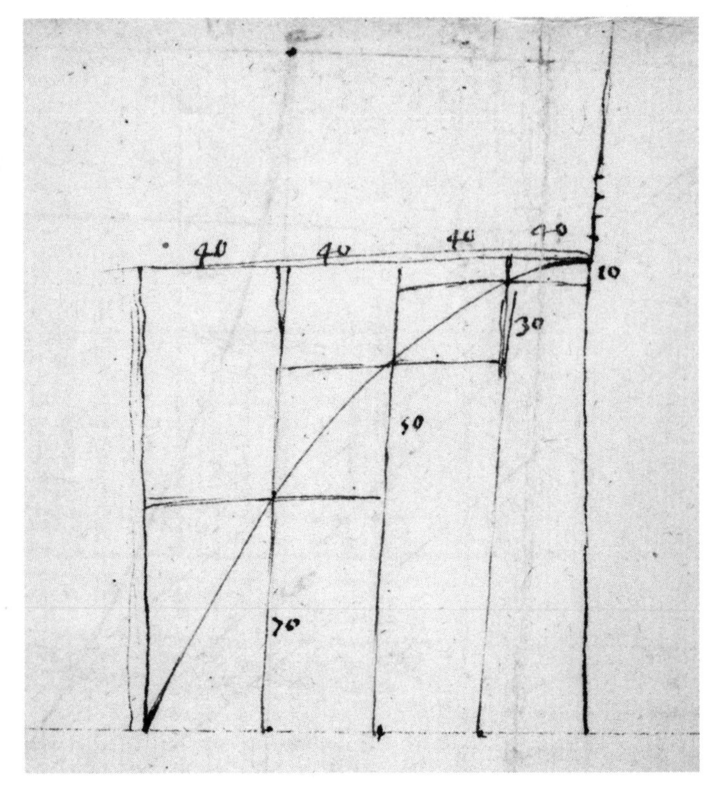

FIGURE 9.7. From folio 117r, Galileo's summary diagram of the parabolic motion of a projectile.

findings of 1172 *p*. But if the proportionality were to follow the speed rule (or speed law), which incorporated time squared in the launch velocity ($v^2 \propto d$) and which he had already discovered on folio 114v, then his actual experimental results would be far closer to the ideal prediction, well within range of what one might anticipate given the external factors of air resistance, friction, the slight wobble of the rolling ball, and so on.[46] (The mathematically inclined can see Appendix A.13 for an algebraic account of the calculations.) With these results Galileo further secured his most fundamental insight into the analysis of motion—namely, that it could be, in our terminology, reverse engineered, broken down into component parts (what we identify as vertical and horizontal axes), and reassembled by means of proportionalities between ratios, later called mathematical formulas. Behind these results, as well, stood the premise that static Euclidean geometry, especially those theorems dealing with proportionalities and ratios, could be employed successfully in an analysis of the phenomena of motion.

An even clearer visual confirmation and summary of the analysis Galileo had invented is revealed on the diagram he drew on folio 117v. (See Figure 9.7.) Here the motion of the projectile is shown to follow regular intervals on both the horizontal and vertical axes. A quick glance at the intervals allows us to ascertain the modern, algebraic formula of a parabola: $y=x^2$. The intervals between vertical lines are marked on the horizontal axis in regular increments of forty units each (*punti*), while the intervals between horizontal lines are noted in regular multiples of ten units each on the y axis. The correlations between the axes yield the odd-number sequence 1, 3, 5, 7, 9, . . . (the "gnomon" of the squares), which Galileo had eventually summed to obtain his time squared and distance correlation in the law of free fall and which he now saw outlined the parabolic path of a projectile. Thus the first point on the curve lies at the junction of interval 40 on the x axis and 10 on the y axis, which Galileo took as the base, 1:1 ratio. The second point lies at 2 intervals of 40 on the x axis and 4 intervals of 10 on the y axis, for a ratio of 2:4. The third point lies at 3 x intervals and 9 y intervals, for a ratio of 3:9, and continuing, with the y axis values always equaling those of the x axis squared. In this fashion Galileo represented squared, derived magnitudes on an axis (not just first-order units), which became one of the coordinates of his analysis. With such a procedure one could in principle take apart the movement of any projectile and then reconstruct it, so to speak, using mathematical proportionalities as instruments. When describing the motion in these contexts, Galileo continued to rely on the term 'impetus' occasionally, but not as an explanatory, causal force. In his reworking, the word referred only to an impressed motion, for throughout these descriptions, his chief concern lay with dissecting, not defining, projectile motion.[47] With this summary example Galileo effectively placed the capstone on an extraordinary, hitherto unprecedented, decade of scientific investigation into the mathematical properties of moving bodies.

Mostly unnoticed, it was nonetheless a breathtaking leap beyond the visual, linear perspective of the Renaissance. Artists had analyzed a point of view, mapping the objects they beheld onto their grids of orthogonal and transversal lines. From the eye to the picture plane to the vanishing points, spatial grids framed objects in proportional relations to one another. The position of the eye determined relations and when the eye moved the perspective changed. In Alberti's immortal phrase, this was the "winged eye." In effect, such a treatment analyzed the motion of sight, breaking it into its components, projecting their three dimensions onto a two-dimensional plane, and then calculating changes on the plane whenever the eye moved. The objects of sight were the variables that depended on one constant, the viewpoint of the eye, and seeing things in space meant coordinating them with that viewpoint. (Recall, for example, the image of the skull in Holbein's painting, *The Ambassadors*.) The illusions of three-dimensionality were a function of what one saw.

Galileo inverted the process, transforming the subjectivity of what one saw into the objects of reality. Rather than breaking down the motion of sight into its coordinate components, he held vision constant, unwinged, and broke apart the movements of the bodies themselves. Once separating motion into two axes, vertical and horizontal, he could then situate a moving body mathematically with a pair of numbers, a ratio. Comparing or combining ratios (with the rule of three) in proportionalities was the (later) mathematical equivalent of devising a formula, creating an equation unifying space and number. Soon, Descartes would build on the algebra of François Viète to formulate explicitly a coordinate scheme and with it the analytical geometry that joined space and number into a single, mathematical system. But Galileo's analyses of "moveables," bodies in motion, had shown the way. He had converted the new information technology and its relational numeracy into a higher-level relation-mathematics and applied it to the world of physical phenomena. With his analytical breakthrough the ancient incommensurability of arithmetic and geometry lay in ruins, soon to become a quaint artifact of "premodern" thinking.

* * *

The abstractions elicited from the new technology had permitted recasting the cardinal and ordinal features of number and had become the catalyst of the metamorphosis from thing-mathematics to relation-mathematics. Surface similarities notwithstanding, this was not old-fashioned, Euclidean geometry. In Galileo's hands the "language of mathematics" began with a first-order abstraction, an empty line, but one that could be filled with the various properties of a moving body in free fall, the swing of a pendulum, or a projectile (natural or violent movements in Aristotelian categories). Motion harbors factors such as distance, speed, acceleration, and, most critically, time. Previously subject to definitions, all these properties now stood as abstract units, or intervals, or magnitudes, or even procedures (for example, squaring a number) in a one-to-one correspondence with points or segments on the line. Similarly with compounded properties like velocity, a ratio of time and distance conjoined. Each point could be analyzed by its relations to axes (x or y), yielding a pair of numbers that stood as a ratio to one another. Numbers themselves became ratios or relations, vis-à-vis a collection of things. Take time again. While a unit of time (such as a day) might correspond with a physical regularity (such as the sun's apparent movement), the square of time—a "derived" magnitude—correlated to nothing beyond other abstract relations, units of weight or mean proportionals for Galileo. Those abstractions eventually linked with phenomena in the law of free fall, enabling one to predict a body's location, speed, acceleration, and later on a host of additional factors associated with motion. Thus did the cardinal correlations join two abstractions—abstract, countable units (or measurements) of, say, distance or time and the visual abstractions of geometry—while the ordinal features of number permitted manipulating these abstractions in accordance with the rules governing the combination of symbols, essentially the rules of arithmetic and the indefinitely repeatable act of counting.

Throughout our narrative we have referred to the twofold nature of abstraction. Recall, etymologically the term 'abstraction' means "to pull," "drag," or "draw away from," and historically it has meant a mental activity that fixes the flux of ephemeral and momentary experience both by distancing ourselves from it so that we are no longer immersed in it and can observe it from some distance or perspective, and by pulling or dragging out of experience some pattern or "mental extract." As a "symbolic species," we engage constantly in practices of abstracting. We do so in manifold ways, two of which have resided at the center of present concerns: (1) the abstractions obtained by means of letter symbols, words, and definitions—the information technology of alphabetic literacy; (2) the abstractions educed by means of number symbols and their algorithmic manipulations—the information technology of numeracy. Our story has followed the transformation of numeracy from its earlier manifestation as thing numeracy, embedded in the classificatory world of words and things, to its modern emergence as "relational numeracy," which has served since then as the foundation of higher relation-mathematics. And just as alphabetic literacy had brought forth the classifying potentiality of speech, so too did modern numeracy awaken and foster the analytical temper.

Abstractions have their counterpart in those features of our experience left behind and deemed irrelevant to the abstracting purposes at hand. And like abstracting, determining irrelevancy harbors a twofold process. Aristotle and scholastics identified irrelevancy as, first, those ephemeral sensations beyond the clutch of words—aspects of sound or time or smell, for examples (words allow us to talk about sounds or odors, not to hear or smell directly). Second, irrelevancies were singled out within the abstractions of words themselves. Words displayed both their "essential," core definitions and their "accidental," supplementary and less relevant features. Thus the essence or *logos* of a falling body's natural motion could be found in its heaviness, its *gravitas;* its accidents lay with the local and transitory components of the body's fall. With modern numeracy irrelevancy emerged as a function of separating a problem into relevant and irrelevant variables, manifest in a pair of operations comparable to those of literacy's abstractions. Initially there was the separation of the mathematized elements of bodies in motion from all their other features, what would later be termed primary versus secondary qualities. In Galileo's words, the shapes, locations, size, movements, and "being one in number, or few, or many," were critical to solving the problems of "moveables," while "tastes, odors, colors, and so on" resided only in our perceptions and therefore were irrelevant to the problems at hand.[48] This has become familiar to us as the conversion of "what" something is (say, *gravitas*) into "how" it works or functions.

Beyond the initial irrelevancy designated by the process of mathematical abstraction was an even more significant means of identifying irrelevant variables, this within the abstractions of relation-mathematics themselves. Earlier, in 1602, Galileo had written to Guidobaldo del Monte of having discovered the chord theorem, which states that the times of descent along all chords of a vertical circle to their lowest points are equal. Because free fall on the vertical of a circle was also a chord, Galileo initially thought that the theorem applied to free fall, whose law would thus be identical to that of a pendulum. His experiments showed otherwise, however. Later he proved that the chord theorem actually did work, but within limits that excluded both a vertical chord and a widely swinging pendulum. Otherwise stated, these behaviors were irrelevant to the chord theorem. The trial and error of developing and applying the abstractions elicited by modern numeracy in their correlations with the phenomena of the natural world thus carried with them likewise their own means of identifying and corroborating irrelevancies. This analytical process has been incorporated into scientific procedures ever since.[49]

Once launched, then, the analytical temper continually reworked the products of its own creation from new vantage points, as abstractions built on abstractions, propelling the mind's critical capacities into an expanding universe of higher mathematics and physics, and eventually into the entire world of nature. As with the plate tectonics slowly shifting on the earth's surface, this would be a gradual, ongoing intellectual process over centuries, one still continuing in our own times (although the pace has quickened considerably). Galileo's innovations were but the start. His own analyses were restricted to motion (kinematics); others after him would extend them to the dynamics of nature, capturing beyond the positional changes of objects in motion the forces and causalities propelling them, and producing what would become known as "classical" or "Newtonian" physics. Once under way, analysis—in all its dimensions—would henceforth steadily supplant the medieval image.

Aristotle had said that puzzlement and awe signified our discontent arising from a lack of knowing, and that contentment replaced our curiosity and wonder once we had acquired a firm grasp of a topic. Of course, he meant once the topic had been properly defined, classified, and demonstrated through syllogistic reasoning. Even with its narrower focus, with qualities stripped away from nature's bare, quantitative patterns, the new analysis heralded a greater measure of confidence and contentment attending the newly won knowledge. In this regard, taking apart a complicated motion and reconstructing it with abstract symbols far outdid the definitions of words; there was something absolutely "right" about the actual measurements and analytical manipulations that came close to those *doveria* ("should be") numbers, about the close correspondence of reality and one's abstractions. Anyone who has tinkered with and disassembled a mechanism of whatever sort—whether an eggbeater or automobile—and put it back together in working order recognizes the feeling. Centuries later, Albert Einstein certainly did, analyzing and reassembling the heavens themselves with his new theory of general relativity. He worried; it was only a "world on paper" until it could be corroborated. But then the numbers generated by Einstein's new mathematics enabled him to calculate and predict precisely a long-recognized anomaly in the orbit of Mercury. The discovery was "by far the strongest emotional experience" in his scientific life, "perhaps in all his life," writes his biographer Abraham Pais. It caused heart palpitations. "Nature had spoken to him. He had to be right."[50] Less far reaching for us lesser mortals, but no less emotionally convincing, we know emphatically how something works if we can reproduce it. So it was with Galileo and the early science of mathematical physics, a confidence born of the analytical temper.[51]

Not only would literacy be diminished in the mathematized discovery processes of science. Gone as well from these new mathematical abstractions were the spirits that had previously animated much of early and medieval numeracy—those things, allegories, mystical realities, metaphysics, celestial harmonies, sonorous qualities, astrological influences, and other enchanting bumps in the night found populating the world of premodern mathematics. The new symbols were empty, nothing more than marks on a page to be correlated with abstract units and manipulated by an increasingly sophisticated array of algorithmic procedures. Over time, discarding the medieval image would come also to mean no longer residing in an animated world of spiritual realities, no longer bathing with Francis in the immediate presence of divinity. Lamentable for some, for others the new numbers were in their own way no less awe inspiring or divine, revealing mysteries of a different sort and commanding assent from the practitioners of the new natural philosophy. Once demonstrated through mathematical patterns and analysis and confirmed through experiment or close observation—forming what would come to be called natural laws—accounts of phenomena would yield a new perception: nature is fecund and wondrous enough without adding human allegory to the world of number. The empty symbols of mathematics tracked, captured, and depicted an intricate and beautiful lacework of moving phenomena. And there was much more to come. If motion could be analyzed in such a fashion, then why not the stuff moving as well? Why not extend the analytical temper to matter itself? As Galileo reflected further on both nature and his own scientific investigations during these Paduan years, he began to imagine widening his analytical forays into what we may call the mathematization of matter, a topic that would even more fully promote the emerging fault line between science and religion, as well as the separation of the two cultures, and to which we now turn.

CHAPTER 10

Toward the Mathematization of Matter

Reefs . . . hard to pass.

—Galileo, *Two New Sciences* (1638)

ENTERING THE MIND OF another and re-creating a process of discovery remains, as we have just seen, forever freighted with uncertainty; even the discoverer does not always know the means whereby his or her own éclat occurred. We are fortunate that in the case of Galileo there is a sense in which it doesn't really matter whether the foregoing reconstruction of his breakthrough into analysis through discovering the laws of free fall, pendulum, and projectiles is entirely correct. It is certainly plausible, highly so. And in this event we are aided by the fact that Galileo did get it right. In others, such as tides and floating bodies, he did not. But, right or wrong, Galileo seldom proved uninteresting. For throughout his investigations he labored to introduce a new, analytical temper into the study of nature, a mind-set in turn wrought from the previous innovations of information technology and its core of modern numeracy. It was a synergy in the making. In taking apart complex motions and reassembling them with the new abstractions, their mathematical proportions and ratios, Galileo was struggling to discern new and different relationships in the phenomena of the moving bodies he was analyzing, peering through a glass darkly. And as some of those relations became increasingly evident and confirmed experimentally, his reliance on the new abstractions and what they could convey became ever more secure.

In paying homage to the haughty Galileo's genius, observers too often infer that he had a thorough grasp of his own thinking. The assuredness, and often audacity, with which he advanced his own views and skewered his many intellectual foes augments this impression all the more. Though brilliant he may have been ("He has five brains and each one is smarter than yours," said of Nobel laureate and quark discoverer, Murray Gell-Mann, but equally applicable to Galileo), Galileo labored in a context foreign to "modern" science. His world was still populated by words and things, in which all natural phenomena possessed qualities; in which science meant finding the right words or definitions and applying them to the things in an organized, taxonomical manner; and in which the notion (from Aristotle) of a real, intrinsic, and total cause residing in the nature of things infused these definitions. This was the intellectual world with which Galileo struggled, and which he eventually helped turn on its ear. The invention of scientific discovery was a halting process.

Two profound insights framed Galileo's overarching vision. First, he realized that the sterility of conventional language could not suffice to expose the most critical and buried truths about the natural world ("Nature takes no delight in poetry"). Second, he recognized that mathematical abstractions could succeed where the abstractions of conventional language could not (without the "language of mathematics . . . one wanders about in a dark labyrinth"). These insights framed Galileo's youthful criticisms of Aristotelian physics, his experimental phase at Padua and beyond, and his more mature reflections on the entire analytical process of investigating nature. The same insights pushed him further into considering what it means to conceive of bodies (matter) mathematically. Imposing an informational screen of mathematical abstraction between the knowing mind and nature created new questions about the material world itself. How could matter, for instance, be analyzed in ways that effectively "mapped" it onto mathematical ratios, calculations, and procedures? What held it together in a continuous whole? What caused it to break apart into discrete pieces? As with his discoveries pertaining to motion, Galileo's footsteps along this path were stumbling and oftentimes confused, not surprising given the context surrounding him and given the inherent thorniness of the problems involved.

For us, the numbers on Galileo's lab notes lay out puzzles to be solved, ones that provide an opening into the mind of their creator. For Galileo, too, the numbers posed puzzles, ones that provided an opening into the mind of the Creator. They also invited more. Mathematizing matter resulted from the mind's further and critical reflection on the abstractions of its own creation, those of modern numeracy. It meant moving into the symbolized universe of relation-mathematics and its connections to the physical world well beyond the conveyances and competence of conventional language, well beyond definitions traditionally conceived. We have already seen a hint of this in our discussion of Galileo's difficulty in accepting the idea of squaring time. Linguistically, as well as physically, 'time squared' simply made no sense; there was no sensory-perceived object or thing called 'time' that could be converted into a 'square,' nothing to which 'time squared' referred (unlike the "man squared" of Da Vinci's drawing, which referred to the geometrical space surrounding the figure). In his efforts to treat matter mathematically, Galileo would confront many such conundrums.

On the surface this may seem a trivial point, for words change their referents all the time as contexts and usages evolve.[1] Earlier we cited and discussed a bit the study of Peter Harrison, who traces the historical "emergence" of the terms 'religion' and 'science' in order to provide insights into their current relations. Both *religio* and *scientia* began as moral terms in antiquity, referring to virtues and mental habits, and only in the early modern period evolved into the more cognitive terms of current usage, changing their referents and meanings along the way. With a different objective, David Wootton (also cited earlier) addresses changes in key terms of conventional language—'facts,' 'evidence,' 'experiments,' 'laws,' 'hypotheses,' 'theories,' 'judgment,' and the like—whose "modern" usage, emerging in the sixteenth and seventeenth centuries, collectively revealed the Scientific Revolution and the arrival of modern science.[2] But in sticking to words, their definitions in context, and developments in conventional language use, even the words of science and religion, both of these fine scholars overlook just how the mathematical abstractions born of the new information technology and modern numeracy supplied an alternative to literacy as a means for discerning patterns in nature.

In mathematics a word's semantic reference would be no longer to a different thing or collection of things or even a process but to a procedure of manipulating symbols in accordance with rules. 'Times squared' refers to naught but the multiplying of an abstract quantity a certain number of repetitions, the same number as the quantity itself, which modern convenience simply denotes with a symbol, t^2. Multiplying furnishes a simplified manner for making repeated additions, and addition is the basic procedure one follows in counting, a manifestation of the ordinal feature of number. So too with other mathematical functions and properties. In fact, abstract number's cardinal and ordinal properties themselves denote the basic procedural rules one can follow in making one-to-one correspondences or counting indefinitely. Mathematician and philosopher Friedrich Waismann summarized this deeper point. "Calculating rules," he wrote, "govern . . . concept formation" from the most foundational of concepts ('equality,' 'greater' and 'smaller,' 'sum,' 'difference,' and so on) all the way skyward into the more complicated, higher reaches of mathematics, an ascent made possible once the relations of mathematics became symbolized.[3] In the context of mathematics, words become a kind of shorthand, pointers indicating procedures or operations one should follow, a function quite different from that found in literacy.[4]

Galileo was on the ground floor of the process, but even early on, as he came to see and as we shall explore in this chapter, the procedures allowed him to carry out operations with symbols and to produce results that words, with all their logic and rhetoric, could not express. Because of the virtual impossibility of translating key components of mathematical analysis of the material world directly into conventional language, therefore, we can appreciate just how difficult were Galileo's attempts to reformulate the nature of matter along mathematical lines. These features of his scientific labors surfaced as he explored the embryonic stages of what scholarly consensus now identifies as an early modern form of atomism, and as he grappled explicitly with the related mathematical question of continuity and discreteness.[5] It was one thing to eliminate this latter distinction in the course of his discoveries pertaining to moving bodies, which we saw in Chapter 9. But it was quite another to address its elimination explicitly, which mathematizing matter demanded. This entire line of thought would require his coming to terms with infinity, procedurally, in a way the Greeks had never envisioned. Like Aristotle, Galileo too found himself responding to Zeno and his paradoxes. Unlike Aristotle, Galileo had a different information technology at his disposal . . . and with it a different temper.

As he pushed deeper into the thicket of relation-mathematics, into the world of analysis, Galileo also moved steadily away from the linguistic underpinnings of the presumptive medieval harmony between science and religion in two critical ways. First and most immediate, Galileo's atomism would challenge key tenets of Aristotelian natural philosophy and by extension, those of Thomistic theology.[6] In philosophy Aristotle and scholastics alike had relied on defining terms (including matter) with reference to their essential and accidental features. At the Council of Trent, the Roman Catholic Church had reaffirmed use of such defining characteristics in the doctrine of transubstantiation, its theological interpretation of the Eucharist, which we noted earlier. For Catholics, belief in the efficacy of the sacrament, and therefore God's grace, turned on such definitions and their substantiating, institutional authority.[7] (Many Protestants in the Calvinist tradition, by contrast, interpreted the Eucharist as a symbolic reenactment of the Last Supper, while Lutherans negotiated a middle ground with their doctrine of "consubstantiation," or Christ's "real presence" in the substance of the bread.) In undercutting the traditional language of essence and accident (or comparable variants, such as 'genus' and 'species' or 'substance' and 'modification'), atomism thus threatened to turn into flummery the central ecclesiastical event of the Catholic Church, the reenactment of Christ's sacrifice.[8]

About the same time as Galileo began thinking more immediately and seriously about the constituents of matter (in 1611 or thereabouts), he was also drawn into a public debate over the Copernican hypothesis of a heliocentric universe (that the earth revolves around the sun) and its attack on the Aristotelian-Ptolemaic cosmology. This engagement followed on the heels of Galileo's own telescopic discoveries of the previous year and his subsequent return to Florence. Based on his understanding of "modern science" and in the context of the Copernican dispute, he would produce a second major challenge to the medieval concord of science and religion during the next few years. This became the earliest version of what Steven Jay Gould in our day has called NOMA (the "non-overlapping magisteria" of science and religion). In Chapter 11 we shall address Galileo's "two truths" rendering of the doctrine and the quarrels over "demonstration" accompanying it. For now we need only acknowledge that his insistence on separating science from religion would be predicated on his perception of the different mental operations proper to each. Science, he often said, was based on sensory experience and necessary demonstrations, the latter combining mathematics and experiment. Religion drew its revelatory "narrations" from the Word of God itself, a Word whose reading required the words of "wise interpreters." Galileo's attempts to mathematize matter drove deeper the wedge between science and religion, especially as he explicated the symbolic and procedural nature of "infinity" or (more accurately) the plural "infinities" and their demonstrable role in his conception of atoms as matter's basic constituents. A science based on such operations and abstractions lay well beyond the linguistic pale of interpreters, no matter how wise. It could have no truck with a religion tied to the conventions of language.

Surfacing notably in his attempts to mathematize matter, then, Galileo's challenges to the medieval harmony between science and religion contributed directly to the uncoupling of numeracy-based science from literacy-based religion and philosophy alike. As the decades and centuries rolled by after Galileo's lifetime, the actual practices of scientists (especially those in mathematical physics) and the reflections of divines, philosophers, and other citizens of the republic of letters would steadily go their different directions, widening the fault line between science and religion and prefiguring as well our own divide between the "two cultures." Until well into the nineteenth century, science would still be called "natural philosophy," with its counterpart in religion carrying the label of "natural theology." But by then both were severed from the actual practices of mathematicians and the investigations of experimental scientists, from the universe of analysis.[9]

Our focus on the emerging fault line between science and religion should not blind us to the obvious facts that most early modern natural philosophers were also believers; that they most often saw natural order, increasingly described through mathematically determined 'laws,' as a manifestation of God's creation (a rereading of the "book of nature"); that they were often motivated by religion to discover as much as could be known about God's handiwork; and that they did not perceive such an emerging and sharp division as we can see in retrospection. Our story has followed the historical unfolding of two different information technologies, literacy and numeracy, which underlay two tempers, classifying and analytical. Their initial, barely detected separation became far more pronounced only with the further passing of years.

* * *

Well known among a small group of mathematically oriented scientists during his Padua years (1592–1610), Galileo burst like a supernova into the awareness of a broader European public with his telescopic discoveries, published in 1610 as *Sidereus Nuncius*. In May of the previous year, he and other Venetians heard rumors of a "spyglass" (*occhiale*) made in the Netherlands, and at the instigation of his friend Paolo Sarpi Galileo soon set about creating one of his own. Available lenses did not allow for anything beyond a power of five or six, so he turned his own talents to lens grinding in order to fashion a much stronger, twenty-power instrument. Then in a dazzling stroke he pointed it skyward. He was not the first to do so, but he was certainly the first to exploit the new telescope, using it to expand beyond normal eyesight and peer into the vastness of space. For nearly four months from November 1609 into the following March, he spent night upon night recording his observations: the moons of Jupiter (which he named "Medicean Stars"); the craggy, apparently mountainous surface of the earth's moon; the millions of stars that made up the Milky Way; and, a bit later, the phases of Venus. Even as he concluded his observations, the last dated March 2, he was preparing the manuscript with his findings for publication. Anxious to scoop others who might be making comparable discoveries (especially of the moon's surface, which lay within grasp of less powerful instruments), he supervised a print run of 550 copies, which appeared mid-March and sold out immediately.[10]

As accolades began pouring in, Galileo parlayed his sudden, newfound fame into a desirable position that allowed him return in September of 1610 to his beloved Florence (and avoid teaching). After a series of negotiations, actually begun several years prior, the young, recently crowned Grand Duke of Tuscany Cosimo II de' Medici succumbed to the enticement of having the newly discovered Medicean Stars named in his honor, and appointed Galileo court mathematician and philosopher. Galileo insisted on the latter title because he wanted to be taken seriously as someone who discovered truths about the natural world, the purview of "natural philosophers," and not just a mathematician who made his calculations hypothetically or suppositionally (*ex suppositione*) as it was then termed. As a part of the discussions leading to his appointment, he had promised to write a small library of scientific works (at least ten books) on "local motion," "mechanics," the "system and constitution of the universe," acoustics, vision and colors, tides, the continuum, the "motions of animals," and "still more."[11] When he arrived in Florence he set about with a new round of investigations, only a few of which, however, had much to do with his promises. For the next decade and a half the topics of floating bodies, sunspots, comets, and tides commanded Galileo's attention. Unlike most of his earlier work on motion, which was known only to a few (and mostly not published until the *Two New Sciences,* in 1638), these topics proved to be highly controversial, challenging even more directly the conventional canons of Aristotelian, classificatory science. And underlying the various disputes of this period was the pressing issue of which world system—Copernican or Aristotelian (or Tychonic)—would best account for new telescopic and experimental data.

Embarking on this new phase of investigations, Galileo devoted part of his attention to an early modern form of mathematically inspired atomism. In scattered observations and comments about the ultimate constituents of matter we find attempts to extend his mathematical and analytical techniques from problems of motion to the composition of matter itself. His initial foray into this uncharted terrain first surfaced in an improbable arena, a controversy over floating bodies. In June 1611, Galileo had returned from a triumphal tour in Rome, where he was feted by the Jesuits, among many others, as the famous author of *Sidereus Nuncius.* At the Collegio Romano, famed mathematician Christopher Clavius himself (whom, recall, Galileo had met earlier) had praised the Medicean court philosopher for his remarkable discoveries. Later in summer at the Florentine villa of his friend Filippo Salviati, Galileo participated in a series of informal, philosophical discussions with several Aristotelian professors from the University of Pisa, along with some of his own followers and former pupils. Among the topics of conversation were the qualities of heat and cold and their role in the condensation and rarefaction of matter. The case of ice, suggested an Aristotelian philosophy professor, Vincenzio di Grazia, demonstrated that cold condenses, whereupon Galileo pounced, replying that because ice floats on water it must be rarefied, not condensed water. To "save the appearances," Di Grazia then offered a fallback explanation that the colder, condensed ice floats because of its shape, a claim Galileo quickly dismissed, declaring blanketly that shape had nothing to do with whether bodies floated or sank, and that only the differences between the densities of matter and the liquid medium in which it was placed determined whether a body would stay afloat.

At this juncture an amateur Peripatetic, Lodovico delle Colombe, stepped forward with a suggestion that the issue be decided by "means of reasons and experiments."[12] Galileo took up the cudgels of the experiment agon and the contest was arranged, but for reasons still somewhat obscure, delle Colombe and he never appeared together. Instead, at various public places delle Colombe performed the one experiment he believed would vanquish Galileo. Taking a thin board of ebony and an ebony sphere of identical weight, he showed that while the sphere sank, the board floated. Voilà—shape affects buoyancy. Galileo remained unconvinced, and the controversy continued to escalate both intellectually and rhetorically. In Italian *colombo* means 'pigeon,' and playing on delle Colombe's name, Galileo referred to his Aristotelian opponents as the "pigeon league," effectively calling them bird-brained. This contrasted with his own, recent membership in the Rome-based Accademia dei Lincei at the invitation of its founder, Federico Cesi (1585–1630). One of the earliest scientific societies, its name roughly translates as the "academy of the (sharp-eyed) lynx." No pigeons here. In an effort to lower the temperature of the debate and to get his former tutor, now court philosopher back on intellectual point, the young Cosimo II rebuked Galileo and urged him to settle the dispute with his pen. Galileo complied with an essay of some fifteen pages that sketched out his main position.[13] Then in the spring of 1612, he expanded his essay into a small book, the *Discourse on Floating Bodies,* which was a more general attack on Aristotelian physics. Published in late May, it immediately sold out, requiring a second edition before the year's end.[14]

The book marked Galileo's first public efforts to articulate ideas about the atomic constituents of matter, which he formulated in his explanation of floating bodies. His starting point, and one of the trickiest issues confronting him, was how to account for delle Colombe's experiment. Galileo laid out a geometrical line of attack. Confident that the major problems of hydrostatics could be solved by considering only the specific gravity of the solid relative to that of the medium, he sought ratios between the volumes of the water raised by a body floating in a container and the portion of the body left submerged. As he did so, he began pushing traditional terms and definitions toward a more procedural, mathematical rendering in keeping with his expanding analysis. He described these ratios as a function of *momento,* a term that falls somewhere between the 'moment' and 'momentum' of modern English. As did mathematicians before him, Galileo used the term in Archimedean fashion to denote the downward tendency resulting from the product of a weight and the length of a lever arm, which was exactly compensated by the upward tendency of the lever's arm and weight on the opposite side of the fulcrum.

To this static notion—a 'moment' of balance between the two—Galileo added the idea of a very small or virtual movement or motion, which he designated with the term *velocitas* (speed), a departure that foreshadowed the concept of 'momentum' as the product of mass and velocity, a vector quantity in modern physics (having both magnitude and direction). As we noted in Chapter 9, in keeping with his kinematic approach Galileo also transformed the term "impetus" from its medieval, qualitative use as an "impressed force" to a term designating only an "impressed motion." The addition of motion to both terms changed their meanings somewhat, nudging their referents toward the calculating procedures of comparing ratios and proportionalities as he had done with moving bodies. In this manner he could equate a *momento* of the solid with that of the liquid medium (water) in a manner analogous to the functioning of a lever or steelyard. When the *momento* of the solid exceeded that of the liquid, the body sank; otherwise it remained floating. Such procedures explained both how bodies floated statically, so to speak, and how they behaved when dropped into a liquid, bobbing up and down ever more slowly until eventually achieving stasis. As with earlier investigations into motion, Galileo here again sought to convert static, Archimedean balances into analyses of dynamic phenomena, among which he included floating bodies.

Working with ratios of *momento,* Galileo showed that for floating bodies the volume of water raised is always less than the submerged portion of the solid. This consideration bore on his close observation that when placed in water the lamina of ebony actually floated slightly below the surface. What kept it there? The "floating body" in delle Colombe's demonstration, he hypothesized, was in reality an "aggregate and composition of an ebony chip and almost equal amount of air."[15] The "adherent air" in effect "joined with the chip" to make it a different body from the ebony alone, one "less heavy than water." (See Figure 10.1.)[16] Just how this sticky air actually attached itself to the ebony chip he had no idea, although later, in *Two New Sciences,* Galileo also speculated that in a number of cases the "introduction of some sticky, viscous, or gluey substance" would be required to explain the cohesion of "the particles of which [a] body is composed."[17] Nor did Galileo know about surface tension, which actually explains the phenomenon. His reasoning along these dead-end pathways, nonetheless, led him to consider

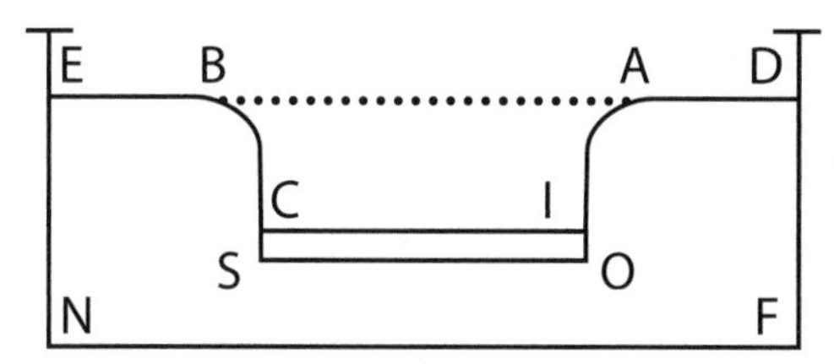

FIGURE 10.1. Adapted from Galileo's drawing of an ebony chip floating in water. The rectangle CSOI represents a cross section of the lamina, floating slightly below the surface of the water, shown by the line EBAD. The area BCIA is the volume of air that has adhered to the chip.

some or even all types of matter as composed of particles capable of adhering to or separating from other particles.

In the terminology of the day, the issue centered on whether matter should be understood as continuous or contiguous. Generally speaking, Aristotelians believed in the continuity of matter, a belief nested in a series of definitions. "Nature," the Stagirite had written, denoted a "principle of motion and change," and motion in turn belonged to a "class" of "continuous" things, which included material bodies that occupied place (local motion was simply a body's change of place). Regardless of the particular composition of a material body (that is, its particular blend of the elements, earth, air, fire, and water), it remained the same type of matter unless it underwent a series of essential internal or external changes.[18] Break a piece of wood into thousands of pieces and each would remain wood. Likewise with water—separate it into thousands of droplets and each is still water. Continuous matter could be cut or divided.

Galileo disagreed, rejecting delle Colombe's Aristotelian references to water's continuity as nothing but a "useless multiplication of accidents." By contrast, he suggested that water be seen as analogous to a collection of separate, contiguous entities, similar to a crowd of people as we walk through it, moving "aside persons already separated and not conjoined," or to a heap of sand as we thrust a stick into it, merely separating the grains of sand already divided. Galileo précised the point, writing of "two manners" of representing "penetration": the one in continuous bodies, where division is necessary; the other in "aggregates of noncontinuous but only contiguous parts," where "there is no need of division, but only of moving." Water and other fluids, he was "rather inclined to believe" at that moment, were of the latter sort, aggregates of contiguous parts.[19]

Fire, another of the four fundamental, earthly elements in the Aristotelian view of nature, also offered evidence of aggregate matter composed of contiguous parts. Here Galileo took his initial cue from Democritus (ca. 460–370 BCE), the atomist from antiquity, who had claimed that in water "ascending igneous atoms sustain heavy bodies of broad shape," while rounder, more narrow bodies of the same material sink. Even though Galileo believed Democritus fundamentally erred in his conviction that shape influenced a body's buoyancy, he praised him for trying to account for the specific differences between "gravity and levity" with the appeal to fire atoms. Moreover, into this generally philosophical discussion, Galileo proposed an experimental test of heating water and then observing whether a plate would ascend from the bottom of the vessel to the surface. His conclusion? Only in cases involving "exceedingly thin" plates of "material little heavier than water" would the fire atoms actually raise up and sustain a body.[20] This was not sufficient evidence to challenge the central principle that only the specific gravities of the body and medium were needed to explain buoyancy, but it did, nonetheless, strongly suggest the particulate nature of fire as well as water. Some years later, in *The Assayer* (*Il Saggiatore*), Galileo expanded the motion of "fire-atoms" to account more generally for the heating of bodies.[21]

Yet even as he began to imagine material bodies composed of atom-like particles, Galileo thought about matter's composition quite differently from Democritus and other atomists of antiquity. They had worked within the Greek philosophical world of words and things to define stability and change in response to an argument proposed by Parmenides of Elea. Parmenides, we recall from earlier, had argued that change of any sort could only be an illusion because nothing comes from nothing (the scholastic *Ex nihilo, nihil fit*). "Being," the most capacious of all abstract categories and definitions, characterized "everything that is" and was accordingly eternal and unchanging. Seeing change as purely an illusion struck Democritus as twaddle, for if all change were illusory, then we would have no means for sorting out illusion from nonillusion, and the definition itself would be meaningless. So he sought alternative nouns and adjectives to describe the changing and unchanging features of our experience. Pursuing a different tack from that later taken by Aristotle, he divided 'being' into a number of 'beings,' carving or cutting being into the smallest possible pieces, each incapable of further division. These he called "atoms" (the Greek *atomos* literally means "uncuttable"). Like the universal 'being' of Parmenides, Democritus's atoms were indivisible, eternal, and unchanging. Infinite in number and distinguished only by shape and size, they were all qualitatively identical. In order to operate, they needed space, and Democritus then gave a twist to the definition of "not-being" (or nothing), calling it instead the "void," making a physical presence of being's absence. In sum, atoms and void composed all existence. The union and separation of atoms explained the changes in the world we perceive, although just how remained quite the mystery; the void was required by definition to give them the space to do so; their unity revealed the unchanging substratum.[22]

Dense, conceptual poetry this, which offered little allure to Galileo except for its mathematical implications. For him, thinking about atoms meant devising operational definitions that allowed him to proceed with the ratios and proportionalities of his analyses. As he reflected on these matters (and engaged in others), he eventually posed the next logical question regarding "particulate" matter: What holds it together? Liquids clearly took their shape primarily from the containers that held them (with notable exceptions, such as water beading on cabbage leaves), but how did particles of solid bodies cohere? In the decades after the controversy on floating bodies he voiced his dilemma. On the one hand, if a dense body were composed of continuous matter, how could one explain rarefaction and contraction without violating its impenetrability or continuity? On the other hand, if one assumes voids or spaces between matter's distinct particles (which might well explain rarefaction and contraction), then how can the unity and solidity of gross bodies be explained?[23] Ice floated because of rarefaction, yes, but as he wrote in the *Assayer,* the question of "how it is possible to rarefy bodies without any separation of their parts" remained "one of the most recondite and difficult questions of all physics."[24] These were the "reefs . . . hard to pass" that he spoke of in the *Two New Sciences.*

* * *

In "Day One" of this last and most "scientific" of his works, Galileo tackled the problem of cohesion, which he raised with the question of how solid bodies break, or conversely what prevents them from coming apart. In keeping with his program of seeking a mathematical analysis of matter, he initially addressed the central features of discreteness and continuity within mathematics itself and then correlated these features with those of the material world. In so doing he accomplished two objectives. First, he was able to demonstrate practically, with a series of one-to-one correspondences, how the long-standing separation of arithmetic and geometry could be overcome, thereby taking a giant step into relation-mathematics. Second, he produced definitions that were in effect procedural or algorithmic in nature, much like the earlier 'time squared,' helping inaugurate a long-standing tradition of mathematical definitions that very often make little sense in conventional language. Thus did he move further away from the world of words and things.

The context for this discussion was provided by the paradox of Aristotle's wheel, a mathematical conundrum of long standing in Galileo's day, although seldom discussed in our own.[25] Aristotle had maintained the categorical separation of arithmetic and geometry, the former deriving from the discreteness of numbers, the latter from the continuity of lines and spaces. This separation entailed a dictum, that nothing continuous (and infinitely divisible) could be composed of discrete, finite indivisibles (such as points or numbers), a dictum Galileo challenged directly as he analyzed the problem of the wheel: "Let us see how it might be demonstrated that in a finite continuous extension it is not impossible for infinitely many voids to be found."[26] The paradox itself derives from rather straightforward thinking about the action of a circle or wheel. If we roll a circle along a tangent, the path traced out in one revolution equals the circle's circumference, and for two circles of unequal dimension their respective paths are, accordingly, unequal. For example, in Figure 10.2, circle A rolls along line XY, and the larger circle A' along the longer line X'Y'.

But if the two circles are concentric and combined in a fixed fashion, say as with a wagon wheel and its interior hub, then something weird seems to happen, as illustrated in Figure 10.3. When the larger circle revolves about the center A, it traces out a line BF along the tangent, while the smaller circle, also revolving about the center A, rolls along line CE, which is identical in length to BF. And even the center point A covers the same distance (AD), without rolling at all. How could this be? Pseudo-Aristotle had addressed the problem as an instance of one body moving another, such as the larger wheel

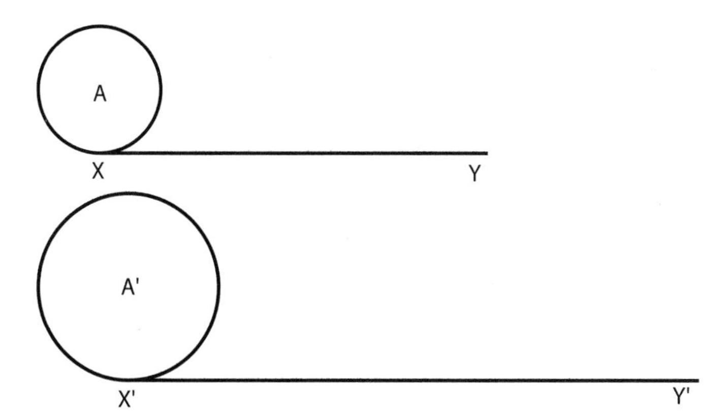

FIGURE 10.2. Aristotle's wheel.

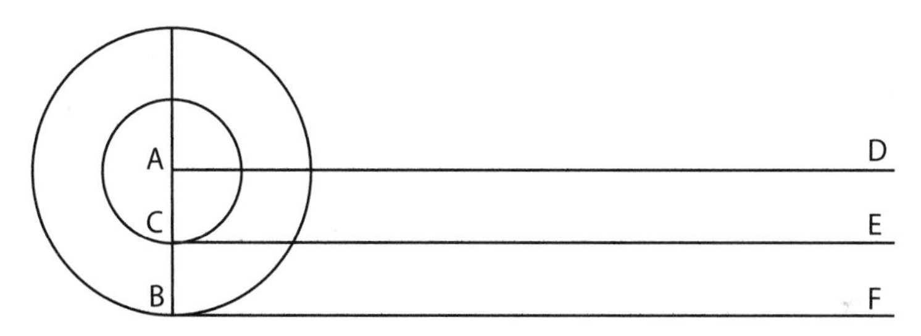

FIGURE 10.3. The paradox of the wheel. As the wheel turns, it traces out the lines CE and BF, which are equal, despite having been created by different circumferences.

moving the smaller. Because forced motion requires the continual contact of a mover and moved, the larger circle in effect forces the smaller to move the same distance as it does. This approach, however, sidestepped the central problem posed by the point-to-point correspondence of two circumferences of unequal length rather than solving it.[27]

Galileo's tactic differed. He began by breaking down, analyzing, the "rolling" motion of concentric polygons. Take for instance a hexagon (ABCDEF in Figure 10.4) containing another (HIKLMN) with the same center G. Now rotate the larger, ABCDEF hexagon. As it tips to the right, side BC will drop down to BQ on line AS, with point C following the arc CQ as it does so and the center G following the arc GC (both arcs represented by

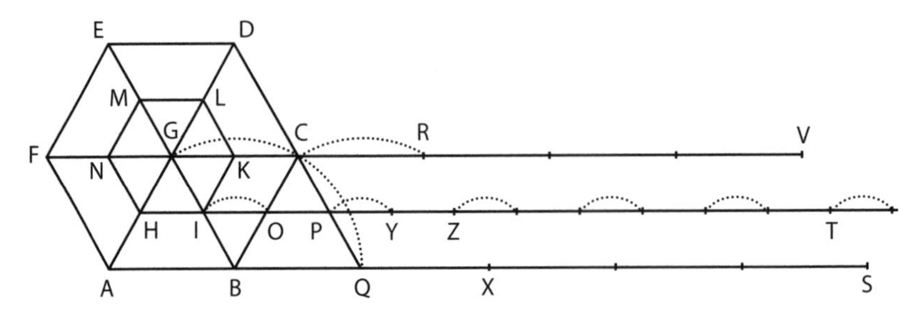

FIGURE 10.4. Galileo's analysis of the wheel paradox.

dotted lines). But note what happens to the smaller, HIKLMN hexagon. Side IK drops to OP, passing over the equal space IO (indicated on the diagram with a smaller dotted arc). Subsequent rotations duplicate the action, making the smaller hexagon travel nearly as far as the larger, but only by "jumping" over alternating, equal segments of line HT. As the polygons get larger—Galileo conceived of them with a thousand sides or even with "one hundred thousand sides"—the "void spaces" jumped over by the smaller, inner polygon become accordingly smaller. This suggested a solution for Galileo, which was to consider circles as polygons "of infinitely many sides." Thus the line traveled by the larger circle equals that traversed by the smaller, but with the difference that the smaller has "the interposition of . . . [infinitely] many voids between . . . the [infinitely many] sides." Galileo elaborated: "Just as the 'sides' [of circles] are not quantified, but are infinitely many, so the interposed voids are not quantified, but are infinitely many." The larger circle is thus crafted, as it were, of infinitely many points, "all filled," while the smaller contains "infinitely many points, part of them filled points and part voids." The two could then be correlated one to one between their points, filled or empty.[28]

To buttress the argument, Galileo also noted that there existed a one-to-one correspondence between numbers and their squares even though it initially appeared as though "all the numbers were many *more* than all the squares." That is, an infinity of integers seems larger than an infinity of squares or cubes because between each square or cube there exist many other integers. Here Galileo was expressing in the realm of pure mathematics the same correlations between distances and times squared he had discovered earlier with his law of free fall, and then extending them to other powers, roots, or numerical sequences in general. The following table illustrates his conception, with the columns representing one-to-one correspondence between members of the different collections (numbers, squares, and cubes), collections at once infinite and countable:

Numbers:	1	2	3	4	5	6	. . .	n	. . .	*ad infinitum*
Squares:	1	4	9	16	25	36	. . .	n	. . .	*ad infinitum*
Cubes:	1	8	27	64	125	216	. . .	n	. . .	*ad infinitum*

The payoff with this analysis lay in further correlating numbers and points on a line, other infinities. Responding to the question of how there could not be more points in a longer line than in a shorter one, Galileo declared, "There are neither more, nor less, nor the same number, but in each [line] there are infinitely many. . . . [The] points in one are as many as the square numbers; in another and greater line, as many as all numbers; and in some tiny little [line], only as many as the cube numbers." While the technical means of correlating numbers and points awaited Descartes and analytical geometry, Galileo was venturing intuitively into the realm of real numbers and anticipating later developments in the mathematical treatment of infinity.[29] With this formulation, his deep intuition lay in recognizing that infinity, "the inexhaustibility of the counting process," stood as a basic assumption of all mathematical thought, arithmetic as well as geometry.[30] For his own purposes and without further technical means at his disposal, he concluded simply that despite initial appearances "the attributes of equal, greater, and less have no place in infinite, but only in bounded quantities."[31] Infinite collections are thus nonquantifiable, although their members can be counted and ordered, and as a consequence different collections of infinities (for example, lines of varying lengths; numerical series of varying powers) could be effectively correlated with one another.

Through these observations and analyses, Galileo had reached the threshold of comprehension and logic in ordinary, conventional language, in effect Aristotle's threshold. But whereas the definition-driven Aristotle inferred that the finite and the infinite could not be correlated or even compared with one another, leading to the separation of arithmetic and geometry, Galileo envisioned the two coming together in a mathematical unity, an infinite and countable aggregate. He could not define this unity with precision but only characterize it suggestively: "The infinite . . . seems to end at unity; from indivisibles is born the ever-divisible; the void seems to exist only by being indivisibly mixed into the plenum; in a word the nature of each of these things alters from our common understanding of it."[32] Indeed, so too for us. This and comparable critical features of the world of mathematical analysis lie well beyond our common understanding in words, tendering, as Galileo so aptly phrased, "marvels that surpass the bounds of our imagination" (well, at least for some).[33] Try, for example, imagining a "filled" point, with no dimensions, then contrast it with a "void" or empty point, likewise with no dimensions. Rooted in the senses, our image-laden, conventional language fails to capture such terms as other than oxymorons or distinctions without difference. Yet, while we may not be able to imagine such entities, we can conceive of procedures and calculations that lead us to them.[34]

The mathematical point next extended to matter as the analysis of "simple lines" held as well for surfaces and solid bodies, which Galileo accordingly claimed were composed of "infinitely many unquantifiable atoms."[35] Thus did he continue to the world of physical atoms the one-to-one mapping, or correspondence, between aggregates (today we would call them sets) of numbers and points. This analysis allowed him to explain the cohesion of solid matter as composed of an infinite number of atoms separated by an infinite number of spaces. He then projected into this microcosm of matter the "celebrated repugnance that nature has against allowing a void to exist," a longstanding, Aristotelian principle, which Galileo had already demonstrated with the macrocosmic case of two slabs of marble, gold, or other dense materials that adhered to one another.[36] For dense matter, the horror of the (unquantifiable) vacua held together the unquantifiable atoms, while less dense bodies had other types of matter (the sticky air, for instance) or mathematically empty space between their parts. And while infinity per se was not quantifiable, it nonetheless gave rise to a quantifiable analysis of matter's parts, providing the mathematical basis for an analysis and explanation of rarefaction and condensation.

In conventional language, describing this world of early, quasi-atomic physics was as problematic as trying to capture in definitions the linguistically quirky world of pure mathematics. Taken as a definition, Galileo's term *corpicelli minimi* ("minimum particles"), for example, comes to us as either a redundancy or a self-contradiction. "Particles" (or "atoms") were defined as the smallest, "uncuttable" pieces of matter. But pairing the noun with the adjective 'minimum' suggests that the uncuttable pieces could be cut in different ways, larger or smaller (maximally or minimally), which makes no sense if the atom is already cut into the smallest possible bits of stuff. Further, if atoms are infinitely small and unbounded, they cannot be 'cut' at all, but if they are bounded and hence quantifiable, then they can be 'cut' further, a procedure without end, and thus are not really atoms. Definitionally we are still at the threshold between the finite and the infinite. Stating that they merge into a unity does suggest something else going on, but with little in the way of common sense to allow its comprehension. The atomic (or ultimately subatomic) world simply does not accord with the words and things of ordinary, conventional language and experience. We are reminded of the famous remarks of Bertrand Russell when he described atoms:

> It is chiefly through ideas derived from sight that physicists have been led to the modern conception of the atom as a centre from which radiations travel. We do not know what happens in the centre. The idea that there is a little hard lump there, which *is* the electron or proton, is an illegitimate intrusion of common-sense notions derived from touch. For aught we know, the atom may consist entirely of the radiations which come out of it. It is useless to argue that radiations cannot come out of nothing. We know that they come, and they do not become any more really intelligible by being supposed to come out of a little lump.[37]

Long before our own atomic age Galileo's nascent atomism had resolved analytically into nonexistent little "lumps" too, mathematico-physical points (some filled, some empty). They made no linguistic sense to him (or to us), but they opened the door to the mathematical analysis of matter.[38]

Thus with the analytical procedures and techniques of a new, relational mathematics did Galileo begin framing a novel understanding of matter: a mathematized, vis-à-vis definitional atomism. The informational symbols of relational numeracy made this possible, having given rise to a greater and more simplified form of abstraction than could be achieved with the technology of literacy. The abstractions of literacy had led to simplifications centering on "essences," those parts of a definition without which a term would not be a term. Often these involved a captured image, such as with the categories 'vertebrate' or 'bipedal,' for example, or words representing natural kinds, such as 'terrestrial,' 'aquatic,' or 'flying,' all species of the more abstract but still image-laden 'animal.' Through the process of abstraction, the most general and intangible of terms, dense categories such as 'being' or 'potentiality' or even 'atom' ('uncuttable being'), had divested themselves largely of imagery's content but nonetheless were still canopied over image-laden words that corresponded to things.

Numeracy's abstractions were simpler and, being so, all the greater and more robust: pure symbols, marks devoid of images and bound only by an algorithmic syntax, the rules of counting.[39] Following the rules led one into an orderly universe, but one beyond words, as when Galileo compared the infinities generated by various number series—integers, squares, cubes. The total of "all the numbers," it would seem, had to be greater than "all the squares." The square of two is four, that of three is nine, and so on to infinity. With each procedure of squaring, one "leaps" over numbers just as Galileo had jumped over the empty voids on Aristotle's wheel. Add in the numbers excluded or leapt over in a sequence and, of course, there had to be many more integers than squares or cubes or numbers of whatever powers. Yet if all these sequences could be extrapolated to infinity, the results would tally up to more than one infinity or one kind of infinity. And how could one talk about one infinity with more numbers than another infinity with fewer? The brief answer is one couldn't. One could perform all these and myriad other operations with numbers that simply could not be captured in words. Where numbers went, matter followed, at least at this point in their respective histories. But where matter went, Aristotelian definitions (or, indeed, any others) could not. No linguistic thread of Ariadne could extricate one from such a labyrinth. But using the same techniques of correspondence and recurrence that attended the history of mathematics from its birth, comparing infinities through the procedural manipulations of symbols would eventually become commonplace, a linguistic mystery converted to mathematical problem.

* * *

Galileo's long and celebrated life as a scientist culminated in his final work, *Two New Sciences*, the manuscript smuggled out of Italy and published in

Holland in 1638.[40] He was then seventy-four years old, blind, and under the Inquisition's sentence of permanent house arrest at his residence in Arcetri, situated in the hills above Florence. Yet the book and the career behind it had definitively launched the central topic of the volume's title: the two new sciences of motion and matter. These sciences became the twin focal points of what scholars centuries later would refer to as the "Scientific Revolution," a series of events that "outshines everything since the rise of Christianity" for it changed the character of "men's habitual mental operations." So remarked one of our earliest contemporary historians of science, Herbert Butterfield.[41] Galileo was by no means alone in this series of events, but his was nonetheless a transformative scientific career, unlike any other between antiquity and the seventeenth century (and perhaps none since), and its implications reached far beyond science itself. On the threshold of modernity, by introducing the analytical temper to students of nature, he contributed mightily to the fracturing of the idealized medieval harmony between science and religion and thus to opening the rift between the two in our modern era. Galileo and his fellow progenitors of modern science were circumscribing an arena of knowledge in which the truths of nature would be sought and captured through analysis, through the abstractions of relation-mathematics, and corroborated by means of experiments. It was an approach to knowledge that at once narrowed our opening onto the natural world and expanded our command over it. Through that narrower opening nature would pour more information and reveal more secrets than could be found in all the words hitherto uttered or written.

In his *Dialogue* Galileo expressed both the fecundity of nature and the incisive role of the new, analytical learning with a "philosophical distinction" between two modes of human understanding, which he identified as "extensive" and "intensive," prefiguring the modern use of these terms in logic. The distinction bore two implications germane to this part of our story. The first dealt with the difference between the infinite understanding of the divine and the finite capabilities of humans. Extensively, with "regard to the [infinite] multitude of intelligible things," Galileo observed, the human understanding is as nothing, "like a zero." God's natural handiwork is infinitely more vast and richer than our restricted learning. Yet, intensively, within our own bailiwick, we understand some propositions "perfectly" and with as much "absolute certainty as nature herself." This is the case for the "pure mathematical sciences, namely geometry and arithmetic," whose truths match those of "the Divine in regard to objective certainty." The second implication was that in approaching the fecundity of nature with those pure mathematical sciences, combined with experiment in the "mixed mathematics" of the day, humans had discovered the key to unlocking nature's truths, truths objectively "having nothing to do with the human will." These were the truths we could conceive, could calculate, could quantify . . . could discover through analysis.[42]

The rest was "poetry."

CHAPTER 11

Demonstrations and Narrations

The Doctrine of Two Truths

> The intention of the Holy Spirit is to teach us how one goes to heaven and not how heaven goes.
>
> —Galileo, quoting Cardinal Baronius, in "Galileo's Letter to the Grand Duchess Christina (1615)"

TUCKED AWAY IN SUNNY ITALY, Galileo lived an academic's relatively ordinary early and middle age. His days were filled with teaching and scientific interests, Venetian duties, domestic demands, and the pleasures of La Serenissima. After his father's death in 1591, and being the eldest son, he had assumed the mantle of the Galilei family, including its financial burdens and the challenge of managing his troublesome, "bullying and devious" mother, the stubborn Giulia Ammannati (1538–1620), who outlived her husband by nearly three decades.[1] Until the financial relief provided by his appointment at the Florentine court in 1610, Galileo found himself constantly cash poor in trying to maintain his motley household. On top of substantial dowries for his sisters Virginia (1573–1623) and Livia (1578–?) and sporadic demands from his musician brother Michelangelo (1575–1631), he had fathered his own hostages to fortune: three children (Virginia, Livia, and Vincenzo) born by his Venetian mistress, Marina Gamba, an "honest courtesan" about whom little is known.[2]

All the while, and mostly beyond the quotidian cares of family, friends, colleagues, and associates, the major events of the day lay elsewhere, with the religious tumult, politics, and warfare of the late sixteenth and early seventeenth centuries. In these years, notwithstanding the importance of the papacy as the ecclesiastical head of Roman Catholicism, Italy was becoming steadily a geopolitical sideshow to northern Europe. Its own Renaissance tourbillon had been swallowed up considerably by the greater Reformation tempests: the French Wars of Religion (1562–1598); the Spanish struggle in the Netherlands and attempted invasion of England (the famous Armada of 1588); the uneasy political settlements in Elizabethan England and the German principalities; the growing and competitive overseas expansion of the Atlantic states; and, somewhat later, the Thirty Years' War (1618–1648). Measured by involvement with the larger occurrences of the age, Galileo's was indeed a quiet and mostly private life, despite an occasionally sizeable crowd attending one of his public lectures (as on the supernova of 1604), or the interest generated by his inventions (sector, telescope), or his academic controversies. Shakespeare might have drawn regularly upwards of 1,500 souls to see one of his celebrated plays at the Globe, but academic careers, like that of Galileo, generally occupied little attention on the broader stage of events.

That changed in the wake of his telescopic discoveries; the censure of Copernicanism in 1616, issued by the Church's Congregation of the Index; and Galileo's subsequent heresy trial of 1633, conducted by the Congregation of the Holy Office (the Inquisition).[3] With these events Galileo became a household name, his case a cause célèbre. The factual outline of the "affair" (as it is now called) still stands as it did then, in haut-relief. And at nearly four hundred years remove from the events, one finds it hard to imagine any documentary evidence or detail yet to be exhumed that would seriously alter it, setting the record straight, as it were.[4] In 1633 Galileo was tried and found guilty of "vehement suspicion of heresy" by the Inquisition for disobeying an earlier injunction against advocating (or even discussing?) Copernicanism.[5] His disobedience was judged manifest in his book on the tides, published in 1632 and titled simply *Dialogue*.[6] The trial made virtually no reference to the science itself, only to the matter of his disobedience and to the degree of intentionality behind it.[7] The science had been adjudicated in the earlier censure of 1616, when certain propositions from Copernicus's work pertaining to its heliocentrism were declared "as altogether contrary to the Holy Scripture" and when Galileo was warned against holding Copernican views.[8] And Galileo was a natural philosopher who at the time of the censure in 1616 had publicly advocated the Copernican theory of the heavens. Thus the central facts. Behind the factual scrim there gathered an imbroglio of personalities and politics, court favorites and ecclesiastical rivalries, theology and philosophy, scientific claims and counterclaims, competing interests and animosities worthy of the Bard's finest inspiration. And at the center of this stage, fanning the flames of controversy on occasion, stood Galileo.

For centuries observers and scholars have narrated these happenings through a variety of perspectives, pigeonholing the blameworthy and praiseworthy and offering numerous interpretations of the censure, the trial, and the aftermath. (A summary of the main accounts may be found in Appendix A.14.) The details of Galileo's case can support most of these views, as well as numerous, frequently subtle permutations of them. And collectively the interpretations themselves more often complement and overlap than rival one another in mutual exclusion. But whatever the interpretation, and whatever adventitious causalities and motives drove the affair, its lasting significance in the present context lay in exposing the emerging fault line between science and religion in the modern world. As should be clear by now, this was not a long-standing rift or warfare between scientific knowledge and religious belief but rather a slowly widening fissure between the two, as though the tectonic plates on which they rested were gradually drifting apart under the weight of separate information technologies: a traditional, alphabetic literacy and a modern, relational numeracy.

Often called the first "modern" scientist, Galileo himself promulgated the first modern version of Steven Jay Gould's "non-overlapping magisteria" of science and religion by explicitly rending asunder scientific "demonstrations" and religious "narrations." As he put it in a well-known epigram, the former tell one "how heaven goes," the latter "how one goes to heaven."[9] Separate in Galileo's mind, the truths of science and religion, nonetheless, were quite compatible with one another because both "derive equally from the Godhead," the books of nature and revelation authored by the same hand. And "two truths," he often stressed, "cannot contradict one another."[10] Still, though equally true, science and religion lent themselves to dramatically different forms of human understanding. The techniques of tempering our knowledge of nature lay across a divide from those of interpreting scripture.

Different information technologies framed the different truths and kept them separate. The truths of science, in Galileo's frequently exercised phrase, derived solely from "sensory experience and necessary demonstrations" written in the language of mathematics and resulting from the analysis of natural phenomena. Such truths bore witness to the divine's operations in nature, that "most obedient executrix of God's orders." Supporting the truths of religion and theology, and cast in the conventional language of words, things, and meanings, the "dictation of the Holy Spirit" revealed God in the "sacred words of Scripture." These revelatory "narrations" required "wise interpreters," for while God could never lie or err, flawed humans surely could and did in their interpretations of Holy Writ. Furthermore, in either field of endeavor, whether scientific demonstrations or religious narrations, mistakes and misinterpretations could only be corrected by trained specialists in the respective areas, be they scientists or theologians, and neither the experts nor the epigones of one should venture into the territory of the other.[11]

Although deeply implicit in the experimental and analytical work of Galileo's Paduan years, the rift between these dimensions of our experience, between science and religion, opened up only during the two phases of the affair as disputes materialized among the participants. In the first phase, culminating with the censure of Copernicanism in 1616, arguments over what counted as a scientific "demonstration" rose to the surface in letters written and circulated by Galileo and Cardinal Robert Bellarmine (1542–1621), the Church's premier theologian at the time. Bellarmine and most of his fellow theologians continued to see demonstrations in traditional terms, in the form of either logical syllogisms or Euclidean proofs. Reason was the same, after all, whether in theology or natural philosophy, and formal demonstrations applied equally to both. They rested on definitions abstracted from sensation and were ultimately grounded in conventional language and literacy. A product of his age, Galileo likewise shared this view, but deviated from it in critical ways as he also insisted on seeing demonstrations in a new guise, one tied to a unique feature of modern numeracy and its abstract connection with reality. We recognize this feature, somewhat technically, as the "functional dependency" intrinsic to relational mathematics (to be explained later). Galileo did not have benefit of our hindsight and terminology. But in his scientific practices he relied on functional dependency constantly. It stood at the core of his experiments and observations and separated the demonstrations of science from those of philosophy and religion. In contrast to Bellarmine's virtually "absolute" standard of deduction from commonly accepted facts and principles, the mathematical demonstrations expressing Galileo's discoveries displayed a far more provisional sort of proof based on evidence.

Beyond the verbal and theoretical dispute over demonstrations, and critical in the affair's second phase, leading to the trial in 1633, lay the entire pitch of "narrations," the expansive territory of conventional language. Nature might take no delight in poetry, but Galileo always did . . . and with relish. We should understand 'poetry' here in the broadly sweeping manner both Galileo and Guiducci had intended, as a synecdoche for the varied conveyances of speech and writing. These included not only the literary arts per se but also narratives and rhetoric in general, as well as the more specific "narrations" of religious interpretations. Throughout his life Galileo revealed himself a virtuoso of his own rhetoric, a "master of persuasion" in the words of one scholar, as well as of the analytical temper.[12] He never ceased to promote the importance of rhetoric for convincing his followers and critics of the legitimacy of his new science, all the while ceding to religion its own narrations, at least on the surface.[13] A Renaissance humanist, as well as forward-looking natural philosopher, he traversed comfortably an intellectual terrain characterized by traditional modes of rhetorical persuasion, philosophical interpretation, and logical demonstration, all of which bore on both science and religion. Toward the end of his life, Galileo reiterated his enduring refusal "to compress philosophical doctrines into the most narrow kind of space . . . and to adopt that stiff, concise and graceless manner . . . bare of any adornment which pure geometricians call their own." To the contrary, he insisted, with its internal content beyond the reach of words, the new science would nonetheless wrap its results in titivations that gave "grandeur, nobility, and excellence to our deeds and inventions."[14]

With his own demonstrations and narrations, then, Galileo stood astride the emerging rift between science and religion, the fault line whose opening he himself had recognized and widened. Yet, by the time of the trial in 1633 the occasion for demonstrations and arguments about them had largely passed. The disagreement between the Galilean and Church camps had shifted mainly to the arena of rhetoric and competing narrations. Galileo composed his narration around the separation of science from religion; the Church essayed to keep science within its theological fold, within its own magisterium. And at this juncture, it was not the science but the rhetoric of persuasion and the power of the Church that would be decisive. In the days long before the popularity of "spin control," both sides in the affair sought to narrate or spin their version of scientific knowledge and religious belief.

Galileo might well have been a virtuoso rhetorician, but even virtuosi have off days, or off performances, misreading their audience. For if his intention were to persuade the Holy Fathers, among others, of the separation of science from religion, and with it the virtues of Copernicanism as science and not faith, he surely bombed in his *Dialogue*.[15] He overreached. His narration, his spin—his poetry if you will—was too persuasive and ran afoul of Pope Urban VIII, the pontiff's advisers, and some well-placed clerics (mostly Jesuits and Dominicans), exposing all the more the crevasse opened up in the discussions about demonstrations. The new demonstrations had clearly separated science from religion, as Galileo well perceived, but his own rhetorical narration played a critical role in exposing and promoting the rift between them. It also netted him a conviction for heresy and a sentence of house arrest for the remainder of his life.

After 1610, relying on the new numeracy, its emerging relational mathematics, and the confidence analysis inspired, Galileo had extended his investigations from the earthly to the celestial spheres. The telescope had provided the pivot point, not in creating the fault line between science and religion per se but in furthering the circumstances that allowed it to be exposed, whether in Galileo's career, in the collapse of the Aristotelian separation of celestial and terrestrial science, in the growing acceptance of a heliocentric cosmology, in the incipient formulation of a scientific method, or in the sleepwalking awareness that the abstractions of mathematics diverged markedly from those of conventional language. In the following pages we shall revisit the two phases of the affair, taking up in turn the dispute over demonstrations and the rhetoric of competing narrations. These topics will provide a fitting conclusion to our own story, for in their aftermath the techniques and practices of the modern, physical sciences had clearly embarked on their own course, establishing a trajectory that would increasingly take the demonstrations of science away from the narrations of religion and make ever wider and deeper the rift between them.[16] Once separated, the two truths would not be rejoined.

* * *

Galileo first developed his doctrine of the "two truths" and the distinction between demonstrations and narrations in a pair of letters written during the years when the "pending controversy" (his term for it) over Copernicanism was gathering storm. The letter to Benedetto Castelli (1578–1643) of December 1613 and the letter to the Grand Duchess Christina of 1615 both came from the period after his discoveries with the telescope and after his career shift in 1610 from Paduan academician to Florentine courtier.[17] Before then, at least from 1597 and probably earlier, Galileo had become increasingly an advocate of the Copernican "hypothesis," joining the sprinkling of astronomers and natural philosophers who could be counted among the Polish astronomer's supporters. At first, controversies stayed largely among the specialists and out of the public eye.[18]

But the new findings publicized in the *Sidereus Nuncius* brought the instantly famous Galileo and his discoveries to the breakfast table, both figuratively and literally, as they did in December of 1613 at a Medici court and family breakfast in Pisa. Present were the pious Duchess Christina, her son Cosimo II (now Duke of Florence), and their entourage, along with various courtiers, local dignitaries, and professors from the university. The latter included Castelli, Galileo's former student, subsequent collaborator and friend, who had been newly appointed to his mentor's old chair in mathematics. At the royal repast he had spoken highly of Galileo's findings with the telescope, especially the "Medicean planets" (moons of Jupiter). Afterward, prompted by a Pisan philosophy professor, Cosimo Boscaglia (ca. 1550–1621), "Her Most Serene Ladyship" questioned Castelli about the compatibility of the Copernican theory and scripture. Castelli wrote to Galileo that he acquitted himself well in response, playing "the theologian with such finesse and authority" as to please both Her Highness and his former teacher. Less certain, Galileo replied to Castelli a week later in a letter drafting his own thoughts on the proper relation between science and religion. As the controversy over Copernicanism and the Bible mushroomed, the letter reached well beyond its recipient, as intended, and into its "primary audience in the shadows," the religious authorities in Rome who were considering the Church's stance on the Copernican matter. Two years later, he elaborated his thoughts further in the letter to Christina, also intending to influence Church officials. But by then the situation in Rome had grown more tense and problematic, leading Galileo prudently to withhold it. Addressed formally to the duchess, the lengthy missive (thirty-nine pages in the printed version of the complete *Opere di Galileo Galilei*) circulated in manuscript only among his closest associates. A copy of it surfaced in 1632, on the eve of his trial.[19]

In the two letters, as well as several other writings, including an untitled essay from the same period, now referred to as "Galileo's Considerations on the Copernican Opinion (1615)," Galileo heralded the conceptual marriage of mathematics and natural philosophy with his often-repeated pet expression, "sensory experience and necessary demonstrations" (he used 'demonstration' or its variants over fifty times in the Christina letter alone). The word 'demonstration' boasted a lengthy history in Galileo's day (along with related terms, such as 'analysis,' 'hypothesis,' and 'science'), and both longstanding and novel meanings found their way into his expression. Unsurprisingly, for as with any early seventeenth-century natural philosopher, Galileo thought and worked within the context of an established terminology, even as he exploded it from within.[20] Thus he continued to employ the traditional senses of 'analysis' and 'demonstration' in devising proofs for his discoveries, as illustrated earlier by the derivation of the law of free fall from the mean speed theorem, or in purely mathematical terms by his confirmation of the chord theorem. When a "conclusion is true," he stated expressly, "by using the method of resolution [analysis] one easily finds some proposition already demonstrated, or some self-evident principle" from which it can be derived.[21] One imagines Euclid or Aristotle or Aquinas or a host of others nodding in agreement.

Nodding in agreement with tradition also was Cardinal Bellarmine. His nod came in the form of a letter to Paolo Antonio Foscarini (1565–1616), a Carmelite friar who had written and distributed a letter of his own in support of Galileo and the Copernican opinion. In his letter-essay (actually booklet, subsequently published) he devoted himself primarily to the exegetical work of interpreting scripture in light of a sun-centered and mobile-earth cosmology, which he and many of his contemporaries also believed had been held by Pythagoras.[22] Foscarini had sent Bellarmine a copy, somewhat naively anticipating a favorable reception.[23] In answer, the highly respected and powerful cardinal initially praised both "the Very Reverend Father" and Galileo (whom he referred to specifically by name) for "prudently limiting [themselves] to speaking suppositionally [*ex suppositione*, 'hypothetically'] and not absolutely" about the Copernican opinion and its alleged agreement with scripture. He went on to say that "if there were a true demonstration" of the earth's movement about the sun, then the Church would have to take great care in explaining passages from scripture that seemed contrary. In such case one would say not that scripture is "false" but only that it had not been properly understood. Galileo, in fact, held the same position.

Yet, as "custodian of the ark" who had spent a lifetime defending the Catholic faith from threats, Bellarmine remained convinced that no such "demonstration" existed. By 'demonstration' he meant the application of Aristotelian, formal logic in explaining matters of theology and natural philosophy alike. This was the rigorous deduction of syllogistic reasoning from accepted principles and facts about the "real" world, not a "suppositional" account of the sort that mathematicians frequently employed to "save the appearances."[24] To eliminate the last scintillas of doubt, and tender an unmistakable tone of admonition, Bellarmine concluded with a direct appeal to the demonstrable common sense that fortified scholastic science. As regards the "sun and the earth," he pronounced, the "scientist . . . clearly experiences that the earth stands still and that the eye is not in error when it judges that the moon and the stars move."[25] This was the basic fact, a cornerstone of natural philosophy and theology alike, and demonstrations followed from it.[26] What could be more direct, more a matter of common sense? What you see is what you get.

Rooted in the common sense of tradition and Aristotelian logic, Bellarmine's standard of proof set an exceedingly high benchmark, and by it the earth's movement about the sun had not been demonstrated in either his or the Church's view. Galileo largely agreed with the standard, but took exception to what he perceived as Bellarmine's narrow interpretation, for it meant effectively that such a proof was well-nigh unreachable by provisional, experimental, or empirical measures.[27] Galileo exploited the term's ambiguity to develop an alternative, opposing version of its application in light of his own analytical breakthrough, to pour new wine into the old wineskin. The method of investigation in the "demonstrative sciences," he noted, utilized "senses, experiments, and observations," all of which were then made "demonstrable" through "analytical methods."[28] The trick was to bring the constantly changing and ephemeral data from the senses into strictly mathematical demonstrations. Galileo was keenly aware of the discrepancies between pure mathematics and the natural phenomena under investigation. To be sure, proofs themselves could be laced into the Spanish boot of rigorous mathematical deduction, but their correlation with actual data was never exact. Recall from our discussion of free fall that the measurements he had taken from the ball rolling down the inclined plane merely suggested, after a considerable while, the Greek gnomon of squares, the odd-number sequence of 1, 3, 5, 7, 9, 11, 13, and 15. They did not correspond to it precisely but were rounded off from the actually measured sequence of 1, 2.93, 5.09, 6.91, 9.03, 11.15, 12.96, and 14.67 points of distance between the frets on the ramp. So it went with all experiential or experimental investigations. They focused on "the sensible world and not . . . a world on paper."[29] Invariably, slippage occurred between the two.

How, then, did one get from the sensible, provisional world of investigation to the mathematical exactitude of the paper proof, from "sensory experience" to "necessary demonstration"? The brief answer, whose lengthier response we have been chronicling for several chapters now, was through the new techniques of abstracting information. Galileo summarized: science depended on "extremely delicate observations and subtle demonstrations . . . supported by abstractions whose understanding requires a very vivid imagination."[30] In place of an abstract honeycomb of words and definitions, nouns and adjectives that correlated to the things in the world outside, the analytical mind abstracted with number symbols an alternative array of patterns, one considerably narrower and more focused in scope. As Galileo had learned from the artist's perspective, what you see is still what you get, but you see it in different ways. In the shift from thing-mathematics to relation-mathematics a new element appears, a functional dependency, which, we saw in Chapter 5, was introduced explicitly with the Hindu-Arabic numeral system in its place holding, zero, and positional counting. The functional dependency manifest in place holding (determining the value of an integer as a function of its columnar position) extended to rule-of-three proportionalities (determining the fourth, unknown quantity as a function of the three known). From these beginnings, numbers themselves eventually came to be understood as functional, or dependent relations, rather than simply collections of things or units.

As the linchpin between sensory data and abstraction, functional dependency bears some elaboration, for although lurking in the emerging information technology of modern numeracy, it was relatively new and unfamiliar in Galileo's day. And outside the sciences it is still not always recognized in our own.[31] Its presence in Galileo's experimental science and practical mathematics marks a new phase in the long evolution of the familiar cardinal and ordinal features of number. In his hands the initial, one-to-one correspondence lay between two abstractions: number symbols and geometrical points, lines, and shapes. The latter, in turn, were visual abstractions derived from matter and motion, thus capturing their "primary qualities," those units of mass, time, distance, position, change of location, and the like.[32] Now often referred to as "mapping" (a synonym for "function"), the central correlation thus united abstract number and visual information as members of two corresponding "sets," combining in a single mental operation the discreteness of arithmetic, the continuity of geometry, and the data or givens of "sensory experience." Galileo invoked business practices to explain: "It would be novel indeed if computations and ratios made in abstract numbers should not thereafter correspond to concrete gold and silver coins and merchandise." And as with the merchant, so too with the "mathematical scientist [*filosofo geometra*]," who seeks the same correspondences between the "abstract" and the "concrete." When there are problems with investigations, they lie with the "calculator who does not know how to make a true accounting"—that is, correlate the world of mathematics with that of sensory experience in a cardinal, one-to-one fashion.[33]

Mathematics historian Edna E. Kramer describes Descartes's "method of analytic geometry" as a "trail-blazing idea" in formulating the techniques of coordinate geometry, which unified algebra and geometry.[34] True enough, but Galileo had already blazed the trail intuitively in his physics, for once number and space had been combined with sensory data in his analytical temper, the capacity of mathematics to graph, map, and explain physical phenomena was secured. Alberti's "logic of likeness" had metamorphosed from art to science.

Further, as Galileo also intuitively grasped, numbers in motion capture phenomena in motion. When understood as a relation, an abstract number itself embodies a rule-governed process or procedure, the comparative interaction or ratio between two separate number symbols. So 7 really signifies 7:1 or 7 / 1 or 14 / 2, the activity of dividing 7 by 1 or 14 by 2, and so on, an activity originating in the ordinal recurrence of number. 'Recurrence,' we recall, denotes the indefinite act of counting, of which division is an abbreviated function. Here, the relation between the dividend (numerator) and divisor (denominator) typifies the dependency in number, a relation such that operations performed on one must be countered by the same operations on the other—such as doubling in the case of going from 7 / 1 to 14 / 2. In this sense, conventional fractions, as well as integers themselves, are somewhat like little equations, for the same procedures must be applied to both the numerator and denominator. Dependency in mathematical functions can be described as a relation of equality, greater, or lesser—what philosophers from Galileo to now generally term a matter of degree, not kind—and the "value" or meaning of a mathematical expression rests on this dependent relation, on counting, not on an association of the number symbol and some collection of things. In fact, collections themselves, identified specifically by number, are but dependent relations. Seven anythings comes to mean seven anythings in relation to one anything. Following the leads of Galileo and Descartes, subsequent natural philosophers and mathematicians would explicitly redefine number to mean relation as just described.

From its earliest appearance in Europe with Hindu-Arabic numerals, the feature of functional dependency reached steadily into more-intricate areas of numerical expressions as the mapping of inputs and outputs gradually grew more explicit and familiar (though obviously not termed as such).[35] The new symbols brought to the surface procedures and operations that had long lain buried and hidden in premodern mathematics; now on paper they could be readily seen. Proportionalities, those ratios of ratios, displayed functional dependency from the outset in Fibonacci's work and in the incipient practices of Renaissance merchants. Businessmen used it in the rule of three (the x factor in $\frac{a}{b}=\frac{c}{x}$) to calculate returns on investment in an ever-changing world. Their livelihoods depended on it. So too were the labors and products of composers, artists, and timekeepers shot through with functional dependency in their manipulations of the abstract symbols of the new information technology. Whether hearing and recording melodies as a function of beats in regular time, seeing and depicting objects as a function of perspective in space, or devising and calculating dates as a function of an abstract numerical continuum, Renaissance purveyors of the mixed mathematical arts constantly relied on this singular feature of mathematical abstraction to express the instructions issued by empty symbols. Stunning and remarkable in all this was the "nothing" on which these practices themselves depended: the place-holding zero in the number system, the rest (absence of sound) in musical notation, the vanishing point in perspective, and the instant (point of no time lapse) between future and past in time reckoning.

Galileo applied these practices and their functional dependency to the study of nature. Using the technology of modern, relational numeracy, he employed abstract numbers to elicit processes from ephemeral sensations, identifying specific features that were dependent on or a function of other features. In the law of free fall, to wit, the distance a body travels from rest is a function of the time of its fall, assuming gravity as a uniform (or constant) rate of acceleration; in the law of the pendulum, the time of a period depends on the length of its string. This was more than just a new way of seeing. The functional component was vital here, for it opened the door to the entire process of reverse engineering that lay at the heart of Galilean, modern analysis. One teased apart complex motions or phenomena, sought the functional relations between the components once isolated, then expressed these relations by means of equations or formulas, or their equivalent. Galileo had neither the algebraic symbols nor the equations we generally employ, as we noted earlier. But he had mastered the functional dependency of analytical abstraction through using ratios and proportionalities in his application of Euclid to experimental investigations.[36] After Galileo, continuing with the breakthroughs into the analytical geometry of Descartes and Fermat and into the calculus of Newton and Leibniz, the functional dependency of abstract number would eventually extend to the highest reaches of mathematics and its deepest connections to the processes of nature.

Awareness of the foregoing was tentative in Galileo's intellectual development, just as were his hesitant forays into the mathematization of matter.[37] But the immediate payoff was both evident and reliable. In contrast to Bellarmine's tradition-based understanding of demonstration, Galileo's experimental or empirical "demonstrations" were tied to showing this functional dependency in ascertaining a "truth of fact and of nature."[38] This he knew well. Throughout his investigations into motion—free fall, pendulum, and projectile—he regularly provided such demonstrations, boosting greatly his confidence as a scientist. Once making his discoveries, he knew he was right and could show it. These were not hidden or occult arts; they were not dependent on fancy philosophical footwork or definitions or the accumulation of textual passages from ancient authorities. The results were in plain sight for all to see, the reasoning there to replicate. In turning his attention to celestial matter and motion, he fully expected to produce comparable, earthly solutions to heavenly problems. And some of his telescopic findings did at least partially validate his expectations. With careful observation and record keeping he was able to demonstrate that the changing positions of the "stars" near Jupiter were a function of their circling the giant planet. Similarly, he showed that the phases of Venus could be understood as a function of the planet's orbit around the sun when viewed from the perspective of an observer on earth.

The clearest and most convincing demonstration of functional dependency regarding celestial phenomena came with Galileo's "necessary proof" that sunspots were on the surface of the sun, or separated from it only "by an imperceptible distance," and that they probably followed the sun's axial rotation.[39] As with many of Galileo's assertions in these years, the proof arose in the context of a dispute, this one beginning in the fall of 1611. Christopher Scheiner (1573–1650), a Jesuit professor of mathematics at the University of Ingolstadt in southern Germany writing under the pseudonym Apelles, had distributed through a friend three brief letters intended for Kepler, Galileo, and some members of the Lincean Academy. In them Scheiner noted that he had been scanning the skies with the new telescope, and he claimed to have been first to discover the dark spots on the sun, a claim Galileo quickly contested.[40] Of more serious issue than who had the prior discovery was the question of understanding the spots, disagreement over which occasioned further exchanges between them throughout 1612. Galileo's letters were subsequently printed in early 1613 and revealed his first published support of Copernicus. For his part, Scheiner continued to work within the general framework of an Aristotelian-Ptolemaic cosmology, seeking to save the immutability of the heavens by fitting new data into an old theory.[41]

Basically, Scheiner thought the spots were stars or planets, or the "denser parts of some heaven" (not clouds), which passed between the sun and the earth. This posed no particular problem for traditional cosmology because, as he argued, before the telescope these could not be seen with the naked eye. The new spyglass had simply made even more wondrous the fixed and immutable heavens. But if, to the contrary, the spots were on the actual surface of the sun, then their constantly changing shape and position would show the heavens themselves subject to "corruption." That wouldn't do.

Committed to Copernicanism, Galileo argued the opposite. The spots were contiguous with the sun's surface, and their movements were most likely a function of the sun's rotating on its own axis, which carried the spots from edge to edge. The question was how to prove, or "demonstrate" the claim.

The demonstration began with gathering more accurate data than Scheiner initially had been able to produce. In the spring of 1612, Castelli had discovered that one could see sunspots by projecting a solar image through a telescope onto a sheet of paper, allowing for much greater accuracy in their measurements and investigation, and without potential eye damage. Galileo was soon using the new method to plot the movements of spots through grid lines he could superimpose on the circle cast by the sun's image.[42] In the second letter of the exchanges with Scheiner, Galileo had appealed to the laws of artistic "perspective" to explain the "maximum foreshortening" exhibited by the spots "near the extremities of the visible hemisphere."[43] Now, having determined that sunspots could be treated as two-dimensional phenomena, projected onto paper, it became a fairly straightforward exercise to measure the effects of foreshortening, using his standby rule-of-three proportionalities and following the progression of the spots on a grid. This he explained in the third letter.

Galileo pursued a twofold comparison strategy. First, he assumed the spots were contiguous with the sun, then calculated the percentage of elongation occurring as they traversed its surface, moving toward the middle of the disc. He showed this in the diagram he drew to accompany his calculation. (See Figure 11.1.)[44] In it he designated the circle, ABD, as the circumference of the sun, on which the spots appeared to move. The line AM represented its diameter, and the vertical lines perpendicular to the diameter were rays of vision from a view at such an "immense distance" (on earth) that they could be considered "parallel." Here he had his grid. The sunspot, μ, was first measured as the space between A and B, and then again as the spaces from B to D, D to L, and so on, continuing to the midpoint of the circumference, where the spot's size was indicated on the diagram by the short line above the letter μ. The first measurement showed the spot to be "very thin" at AB, but the measurement of the spot taken at the middle of the disc indicated that here it was "six times" greater. These reckonings actually accorded closely with the illustrations of Scheiner, who in the meantime had improved his own data and measuring techniques. (In fact, Galileo used

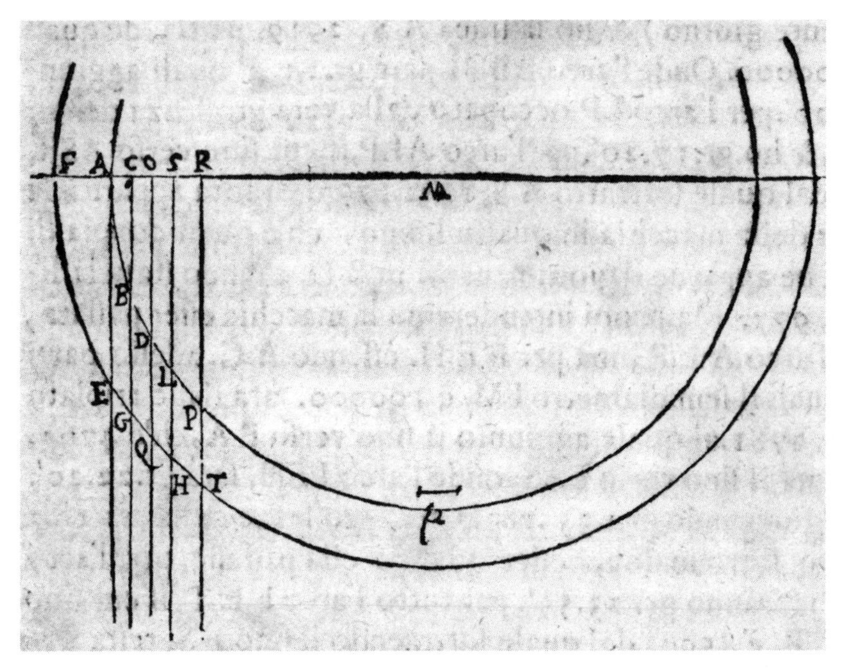

FIGURE 11.1. Galileo's diagram, used to demonstrate the contiguity of sunspots on the sun's surface.

Scheiner's drawings in order to avoid the charge that he, Galileo, had adjusted his sketches to conform to his own conclusions.)

Next, Galileo proposed assuming the spots to be at some distance from the sun's surface in accordance with Scheiner's claim. This he showed with a second circle, FEG, and a second diameter, FM, which he stipulated as one-twentieth (5 percent) longer than the diameter AM of the first measurements. In the second illustration his calculations proved that the middle spot could "never appear enlarged by more than three times" from its location and size in the first space. Further computations based on the same assumptions revealed that if the spot really were away from the sun's surface, even by as little as the proposed 5 percent, it would have doubled in size from the actual, measured appearances.[45] These results ran counter to the data both he and Scheiner had at hand. The conclusion was clear-cut: the spot's proportional elongation in the middle of the disc was a function of its resting on the surface of the sun. Moreover, that being so, sunspot movement was in all probability a function of the sun's axial rotation, although this part of the claim could not be corroborated in the same fashion. The sunspot example illustrated an exciting, new kind of demonstration of heavenly motions, one of "compelling power" born of analysis and reverse engineering, which differed from both the exacting proofs of Euclid and the logical syllogisms of Aristotle.[46] It was contingent on evidence assimilated by the senses, abstracted into numbers through measurements, then subject to the manipulations of mathematical analysis. It was how one got from sensory experience to necessary demonstrations. And it could be replicated by anyone willing to gather data carefully and do the calculations, even Scheiner.[47]

The analysis of sunspots provided a demonstration that explained features of a celestial, physical phenomenon. It was a relatively small and local demonstration, only a piece of the heavenly puzzle. By itself it did not demonstrate the veracity, or even high probability, of the entire heliocentric cosmology. Bellarmine's standard had yet to be reached.[48] Galileo acknowledged as much in the letter to Christina, and conveyed optimism that it was only a matter of time before the cumulative demonstrations of celestial phenomena would provide convincing proof of Copernicanism. In May 1615, writing to his well-connected Florentine friend Piero Dini (1570–1625), who at the time was living in Rome, he explained that the "surest and swiftest way to prove . . . Copernicus . . . would be to give a host of proofs."[49] Other, local demonstrations showing the functional dependency of phenomena seemed to be pointing in the same direction. The moon's craggy surface could be demonstrably compared to the earth's by measuring the shadows cast by its mountains; its glow a function of the sun's reflection off the moving earth. Jupiter's moons could be demonstrated as orbiting the giant planet; the phases of Venus, likewise, showed its solar orbit. To buttress these demonstrations Galileo was working on the one he believed would tip the balance and persuade perhaps even Bellarmine: his theory that the tides were a function of the earth's motions on its axis and around the sun.

There matters stood in March 1616 when the Catholic Church's Congregation of the Index lowered the boom on the Copernican theory, identifying it as "formally heretical" and censuring it—most likely at Bellarmine's instigation, certainly with his active compliance and encouragement.[50] In defending religion, he asserted that the earth's immobility as implied by scripture was essentially a matter of faith, not science, unless it could be convincingly shown otherwise. But his view of demonstration effectively ruled out the latter option, ruled out the possible veracity of the Copernican theory. (A further, and menacing implication: even trying to demonstrate the theory could be construed as a heretical act, one that contradicted a tenet of faith.)[51] Galileo saw heliocentricity and other "natural conclusions" differently, "not [as] matters of faith" but purely as topics of science or natural philosophy on which the new demonstrations of analysis would have a direct bearing.[52] In hindsight, of course, Bellarmine was fated to lose the battle. We know this because we have long accepted Galileo's version of demonstrations in science, together with the abstractions that connect mathematics and natural phenomena. We have also long accepted his insistence that scientific claims be amended or overturned only by other scientific investigations. Above all, we have long accepted Galileo's underlying separation of scientific demonstrations from those of religion or philosophy.[53]

* * *

Throughout the winter of 1616 Galileo had stayed in Rome, trying to persuade Church authorities either to accept the theory of Copernicus or, minimally, not to proscribe it. After the censure he returned to Florence having failed in these objectives. The Congregation of the Index had decreed the "Pythagorean doctrine" to be "altogether contrary to the Holy Scripture." It had "prohibited and condemned" Foscarini's book and "suspended until corrected" the books of Copernicus and Diego de Zuñiga, the latter of whom thirty years earlier had written a small *Commentary on Job* in which he had exhibited the now condemned doctrine.[54] Further, at the request of Pope Paul V, Bellarmine had met privately with Galileo to warn him specifically against holding or teaching Copernicus's theory. Still, Galileo didn't arrive home entirely empty-handed. The decree itself made no mention of him or his writings, even the publication on sunspots. Plus, a bit later, in order to squelch rumors of Galileo's abjuration and to clear his name, Bellarmine had given him a personal certificate exonerating him from all involvement, stating that he had been "notified" of the Church's ruling, nothing more.[55] Henceforth, in Rome's view, the case was closed and the matter resolved.

Hardly so, however. Bellarmine's certificate appeared to give Galileo a bit of wiggle room, for it stated only that the Copernican doctrine could not be "defended or held." It didn't say one couldn't discuss it or, for that matter, other topics pertaining to celestial physics.[56] This was a view held also by others, including the future pope Urban VIII, Maffeo Barberini, a Galileo supporter in 1616 who endorsed speaking of Copernicanism as a hypothesis to save the appearances. Ever mindful of the winds of ecclesiastical opinion, over the next decade and a half Galileo cautiously reentered the arena of debates surrounding the new cosmology. But the debates themselves refocused their attention from the demonstrations of science to competing narrations.[57] In the period culminating with the *Dialogue*, when eyed as a tapestry of the whole, Galileo's central purpose lay in promoting a greater acceptance of the new way of thinking about nature through the analysis generated by numeracy and relational mathematics. A strategy gave direction to the purpose. For if the methods of investigation were to be accepted, then so too eventually might be the discoveries and the content they had generated, including the Copernican theory.[58] This meant discussing the methods when possible (as in the dispute over comets) and, more pointedly, shaping the narrative framework of the new science and its independence from religion and philosophy alike.

So from numbers to words the arena shifted, competing narrations converging less on the content of science than on the question of where science would fit in the bigger picture of life and death on the earthly coil. Amid all the points of interest and controversy generated by the *Dialogue* and surrounding writings, then and over the years, three leitmotifs of Galileo's story put his narrative stamp on the separation of science from religion and philosophy: the indifference of nature; the cumulative effects of demonstrations and evidence; the rejection of any authority, save nature itself, when investigating natural phenomena. The last in particular brought him before the Church's Holy Office and the pope. These were not separate chapters or major episodes of the work. Rather, they subsisted as less pronounced themes coursing throughout it, implications of the new analysis, which Galileo perceived with varying levels of awareness and clarity. Though he often utilized numerous traditional concepts, Galileo also transformed them sufficiently so as to promote through these themes a coherent, innovative, and rival account (to that of Aristotle) of investigating nature.

A key premise of Galileo's narrative of scientific independence was the presumed indifference of the natural world. Nature, he wrote to Castelli, "does not care at all whether or not her recondite reasons and modes of operation are revealed to human understanding."[59] The 'there' was simply there. We figure out nature's doings with the abstractions of numeracy and higher mathematics, with the analytical temper. Following in his father's footsteps, Galileo continued with the expurgation of sonorous and other qualities from abstract numbers. The language of mathematics only described natural processes; it did not ascribe any allegorical or philosophical meanings to them. No qualities in numbers meant no purposes in them either, and the various specific patterns of natural phenomena they captured and conveyed were equally purposeless. Possessing no *telos,* these empty informational symbols allowed one to reach "conclusions" in the "natural sciences" that "have nothing to do with human will."[60]

Galileo's perception here challenged teleological thinking and its final causes at both the general and specific levels. As discussed earlier, Aristotle had argued that the realms of thought and nature shared essential similarities, and that accordingly one employed the same vocabulary of willful activity in explaining "the 'why' of," or causation, in both human affairs and natural philosophy. Terms such as 'agency,' 'patiency,' 'potentiality,' and 'fulfillment' all invoked the intentionality of the human will in explanations. So too did 'purpose' and catchphrases such as "for the sake of" and the watchwords Galileo placed in the speech of Simplicio, his Aristotelian interlocutor: "Nothing has been created in vain, or is idle, in the universe."[61] But the hypothesis of divine purposes or causes was simply too general to be usefully employed in accounting for specific events. Appeals to God wound up explaining everything, and therefore nothing at all. "Whatever begins with a Divine miracle or an angelic operation," he laid out with the rhetorician's stock-in-trade, a double negative, "is not unlikely to do everything else by means of the same principle." Moreover, when scientists include divine purposes in specific explanations, they arrogate to themselves the perspective of divinity. We can certainly say that the sun shines, that its rays produce light and warmth, and that we benefit from them. But, Galileo continued, we cannot say, without being "guilty of pride," that it does so with the "goal" of making our lives better, or any other specific goal. It would be as though a single grape (were it conscious) could say that the "sun's rays should be employed upon itself alone."[62] There was irony here, not lost on most of his readers, in light of his earlier comment that "the truth of knowledge . . . given by mathematical proofs . . . is the same that Divine wisdom recognizes." Regardless, teleological purposes, large or small, were purged from the abstractions of the new relational mathematics and their correlations with nature. Eventually, as satirized by the testimony of Voltaire's Dr. Pangloss, they would be purged from scientific explanation altogether.

At the same time, the independence of science and the indifference of nature to human understanding did not rule out divine purpose in the general scope of human concerns for Galileo. As he put it, writ large "God and Nature are . . . occupied with the government of human affairs." This was a judgment about the benignity of "Divine Providence" as a whole and not about its role in explaining specific natural patterns or phenomena, unlike the final causes found in Aristotle's version of "the 'why' of."[63] Galileo did not explicate this line of thought any further, offering instead his many local demonstrations as illustrations of scientific explanation. Combining an affirmation of God's general purpose in human affairs with a denial of general and specific purposes in explaining natural phenomena would have meant encroaching even more on the realm of theology. And he had learned his lesson, or so he imagined, from his earlier attempt to interpret along Copernican lines the passage in scripture in which Joshua, assuming the sun's orbiting of the immobile earth, commands it to stand still. Galileo was content just to state the point. Humans still had purposes, and those purposes could still reflect divine concern for human life and its ultimate redemption, but they resided outside the pale of scientific explanation.

Complementing the indifference of nature and its accompanying eclipse of final causes was a second motif in the *Dialogue,* Galileo's belief that the cumulative effects of many local proofs would corroborate more general theories, especially those of Copernicus. Like most of his contemporaries Galileo sought certainty in the claims of natural philosophy, those "necessary demonstrations." More than most, he also appreciated the provisional, less than absolutely certain, and progressive nature of the new "science itself, [which] can only improve." In fact, it was already doing so. "Because of many scholars' contributions . . . ," he wrote to the duchess, "one is discovering daily that Copernicus's position is truer and truer." In the preface to the *Dialogue,* he told his "discerning" readers what to expect about the accumulation of evidence. He would appear "openly in the theater of the world as a witness of the sober truth," and he would strive "by every artifice to represent" the "Copernican side" of the debate, "proceeding as with a pure mathematical hypothesis." His readers could then judge for themselves whether or not the smaller and local arguments added up. Then and since, most thought they did just that, with the larger Copernican picture presented by Galileo's mouthpiece, Salviati, being more persuasive than the science and cosmology of Ptolemy and Aristotle, defended by Simplicio. One should always be careful in one's wishes, however, for his readers included the Inquisitors. They dismissed Galileo's shallow claim that his arguments summed to a "mere mathematical caprice," and with unintended irony used the persuasiveness of his own case as evidence against him, of his having disobeyed the censure of 1616.[64]

Throughout the *Dialogue* Galileo marshaled his evidence with the twofold objective of clearing away Aristotelian debris and amassing the local proofs that collectively pointed to the new cosmology, conveniently leaving to the side the theories of Tycho Brahe. He displayed his finely honed polemical skills in criticizing traditional terminology, concepts, perspectives, and claims; he drew on his experiments in the motions of free fall, pendulums, and projectiles; he explained many of the findings of the new telescope; he summarized the demonstration of sunspots and reviewed the debates over comets and over the configuration of the heavens. All this was a prelude to his central, tour-de-force demonstration, that the tides were functionally dependent on a combination of the earth's diurnal rotation and its circumnavigation of the sun. In short, two of the three motions attributed to the earth in the Copernican system explained the "flow and ebb of the ocean waters."[65]

Galileo's theory, of course, was subsequently proved wrong by Newton, who correctly identified tides as a function of the gravitational forces exercised on the earth's motion by the sun and moon. Without a theory of gravity at his disposal, Galileo sought to attribute the flow and ebb entirely to the "mobility of the container" (the earth). He probably acquired the idea from riding in Venetian gondolas and watching the water sloshing in the bottom, for he illustrated his point with a description of water's action in a barge. When the barged stopped its forward movement, the water's "impetus" carried it to the prow, from which it then receded toward the stern of the boat. The analogy was clear. Tides were like the water in the barge, ebbing and flowing with the earth's motions. It was a decent theory, given the context, but overwhelmed by the complexities and variability of data too often at odds with it—not to mention the fact that the earth's motion is continuous; unlike the barge, it doesn't stop. For us the theory's significance resides in Galileo's effort to explain the tides along purely naturalistic lines, assuming as he did the indifference of nature and accumulating mathematicized data to seek corroboration or proof. Along with this he expunged other qualitative and animistic explanations, ones ranging from the "temperate heat of the moon," to the earth's breathing motion, to various astral influences.[66] Equally significant was its place in the overall, provisional picture of the new cosmology. Galileo thought it would be a capstone proof in the accumulating evidence for Copernicus. Unlike many of his other demonstrations, however, this was indirect, inferential. One assumed earthly mobility, then drew out its consequences. Enough of such consequences would eventually corroborate heliocentricity, but the logic of the claim meant it could only be provisional.[67]

Interwoven in Galileo's local demonstrations, partial proofs, and even his incorrect theory of the tides, a third theme threaded through his narrative—namely, that nature itself was the only authority that mattered when doing science. It was the touchstone against which all scientific claims should be evaluated. Two sorts of established authority claimed his attention as he developed this thought. First was the authority of science, which ought to be open always to criticism from scientists, for only ongoing scientific investigation could detect and correct the errors of previous generations. In his lifetime this meant challenging the natural philosophy of Aristotle, which Galileo pursued relentlessly from an early age, not dismissing all opinions out of hand but rather assessing each of them critically. Second, the authority of religion in "matters of faith and morals" simply had no place in scientific investigation, and thus should in no way interfere in it: "Officials and experts of theology should not arrogate to themselves the authority to issue decrees in the professions they neither exercise nor study," he proclaimed, including and especially matters of scientific investigation. Even when found in scripture, matters of fact should be investigated by natural philosophers.[68]

As cardinal, later pope, Maffeo Barberini was not unalterably opposed to this latter claim, as had been Bellarmine. A Florentine noble, patron of the arts and sciences, he had early on supported Galileo in particular and scientific investigation in general, as long they produced suppositional claims about appearances, not absolute claims about reality. After his papal election, he and Galileo even had a half dozen conversations in Rome, during which he encouraged Galileo to continue with his work along these lines. (Pope Urban had also been quite amused by the wit and sarcasm of the *Assayer*.)[69] Some eight years later, working in the favorable circumstances of Urban's early tenure, Galileo published the *Dialogue*. But then Urban reacted swiftly and unexpectedly. As reported by Florence's ambassador Francesco Niccolini (1584–1650), "His Holiness exploded into great anger" upon learning of Galileo's work, proclaiming that "Galilei had dared entering where he should not have, into the most serious and dangerous subjects which could be stirred up at this time."[70] The pope clearly felt betrayed by Galileo, but his reaction far transcended personal feeling. From the perspective of Urban's own, ex cathedra religious narration, Galileo had crossed the line from suppositional science into absolute theology, committing formal heresy in the process.

Scholars long wondered why Urban felt the sense of betrayal so strongly and suddenly. The answer derives from his adherence to a well-established and respected theological tradition, a religiously grounded skepticism and the vanity of human reason it stressed. Within the framework of Catholic theology, this tradition gave rise to a concern much different from Bellarmine's preoccupation, for Urban worried far less about natural facts than about the claims made regarding them. He could support scientific investigation as long as its propositions pertained to hypothetical or provisional knowledge, but he drew the line around the "absolute" nature of its assertions. His position has been termed by one modern scholar the "divine omnipotence objection" to Copernicanism. The argument averred, in a nutshell, that God's infinite power and wisdom trump any human claims to natural knowledge. It became official policy with the papal bull *Inscrutabilis*, issued in 1631, which placed the omniscience and omnipotence of the divine above anything that weak, sinful humans could know or accomplish.[71] Conversely, "absolute" claims about the processes of the natural world put restrictions on God's actions by tying them to the "mathematical sciences" and necessary demonstrations, thus limiting divine omnipotence and in effect denying one of God's essential attributes. So Urban related to Niccolini: "There is an argument no one has been able to answer: that is, God is omnipotent and can do anything; but if He is omnipotent, why do we want to bind him?"[72]

Urban himself had advanced this argument to Galileo in their Roman conversations of 1624, and Galileo had agreed with the pope's position. Further, he had also agreed to respect it and include it in his writings. But in "Day 1" of the *Dialogue* he explicitly asserted that the "intensive" part of (his own?) mathematical understanding "equals the Divine in objective certainty." And then, undaunted and perhaps thinking that the same Urban who had delighted at the sarcasm in *The Assayer* might be amused, Galileo placed the pope's argument in the mouth of Simplicio (Mr. Simpleton) on the last page of the book, a supposed "medicine of the end" that would cure whatever ills had come before.[73] For Urban, whose theology highlighted the frailty and limits of human reason, it was perhaps doubly outrageous and ironic to discover Galileo had placed this central theological point in the mouth of the frailest intellectual of the play. Depending on one's point of view, the rhetorical placement of the argument made not only God appear foolish, for having giving humans the ability to discern natural truths, then making them irrelevant, but even more so his vicar, the pope. This was personal . . . and well beyond. Through his righteous wrath Urban saw in the *Dialogue* a heretical challenge not just to His Holiness but to the supremacy of the divine. It could not be allowed to stand.[74]

In his reaction Urban disregarded the less-than-certain, provisional thrust of Galileo's Copernicanism that generally persisted in the local demonstrations throughout the *Dialogue*. Rather, he was incensed at what he saw as Galileo's "assertively" and "conclusively" claiming its superiority, or "asserting" his opinions about it "absolutely," instead of admitting only its "hypothetical truth."[75] This pitted Galileo against an essential doctrine of the Church, and the pontiff was convinced that he had committed "formal" heresy by denying or challenging a chief attribute of God. As the trial neared, Galileo hoped for a verdict of "slight suspicion of heresy" and a slap on the knuckles by way of penance. To that end, in his second deposition he acknowledged his "error," confessing that through "vain ambition, pure ignorance, and inadvertence" he had gone too far in arguing on behalf of the Copernican theory. But the admission wasn't enough in the face of the pope's insistent theological concern. In the end, after depositions, plea bargaining, behind-the-scenes negotiating, and internal struggles among the prosecuting authorities, the compromise "vehement suspicion of heresy" was agreed on.[76] After his abjuration and sentence, the popular story has it, Galileo reportedly muttered beneath his breath "Eppur si muove" (And yet it moves) in reference to his continuing belief in the moving earth and Copernican heliocentricity. Though assuredly fictitious, the report rings true not only for its defiance but because it underscored Galileo's own narration, his fidelity to the independence of science and to nature as its sole acceptable authority.[77]

* * *

With his doctrine of two truths Galileo etched modernity's imprimatur on the separation of science and religion. By the end of his life, different information technologies had evolved historically to frame each of these dimensions of human experience, giving rise to different forms of abstraction and netting different claims about the outside world and our interactions with it. The two technologies promoted separate and distinct mentalities as well, contrasting tempers of mind: the classifying impulse of literacy; the analytical impulse of numeracy. From its beginnings in antiquity alphabetic literacy had placed a premium on using the data symbols of letters to construct words, which correlated with the objects of the world and brought sensory-laden information to the mind in the form of definitions. Organizing words and things into general and specific categories fostered classification as the predominant means of knowing, while the syntactical and logical structures of definition-based propositions tendered explanations of natural, as well as divine things, insofar as they could be known. These early developments crested in the medieval magisterium of knowing and believing, an idealized harmony of *scientia* and *religio.*

Empty, abstract marks corresponding to sensory data, new information symbols steadily made their appearance throughout the Renaissance, bit by bit scrubbing the medieval image of its animistic, allegorical, and philosophical allusions. In various walks of practical life, and extending as well to cultural and intellectual matters, information was introduced to the mind by means of such symbols in the Hindu-Arabic counting system, in the musical notation of sounds and silence, in the geometrical grids of perspective, and in the abstract continuum of linear clock time. In all these and associated areas, symbols were manipulated according to rules governing their operations and assembled in what we have termed modern, relational numeracy. Forged from numeracy's new symbols, the knowledge of "all dark things" made its transformation from thing-mathematics to relation-mathematics, whose functional relations now mapped the phenomena of nature. This was the new scientific world Galileo engineered with his "sensory experience and necessary demonstrations." It was the world of "number, the language of science." To be sure, the language of words continued to cast its spells in meaningful narrations, including those of religion, but henceforth these would serve primarily to situate us in nature, not to explain it.

After Galileo, the principal question involving science and religion would center on their compatibility, on whether demonstrations and narrations, now split apart by different data-harvesting techniques and information foundations, could accord with one another. This was then, and continues to be now, a narrative matter. Galileo believed, on the surface at least, that the truths of science and religion were compatible; they could not contradict one another he often reiterated. (One wonders what else he could have said with the image of a burning Giordano Bruno at the hands of the Inquisition still freshly looming in many minds, including his own.) Still, though both truths were compatible, he was nonetheless crystal clear about where his own predilections lay: science first, then religion.

Nearly two decades before his conviction, Galileo had gone on notice. Writing to Christina he had insisted that one started with investigations of natural philosophy and only then proceeded to religious interpretation. He supported this narrative spin with a piece of theology taken from one of the Church's fathers, Tertullian: "We postulate that God ought first to be known by nature, and afterward further known by doctrine—by nature through His works, by doctrine through official teaching." (Aquinas too had advanced the same point with his claim, quoted earlier, that "faith presupposes natural knowledge.") The direction of argument here ran from science to religion, ascertaining "the facts first," then using the facts to "guide us in finding the true meaning of Scripture."[78] This heralded the pattern for the decades and centuries to come after Galileo, as believers conversant with scientific learning often sought to interpret scripture in light of science's latest findings, seeking (and achieving?) the sort of compatibility Galileo had suggested. On the premise that nature's designs could be understood as God's handiwork, many churches and individual believers alike eventually made their peace with Copernicanism, and with scientific discoveries and theories since then (although because of the role played by chance in Darwin's account of the descent of species, the theory of evolution has continued to be much trickier and far more problematic than other theories among the sciences, a topic for elsewhere).

Yet a far deeper and more significant pattern in the flow of Galileo's argument from science to religion centered not on specific facts and findings but on the mind-set, or temper that produced them. Noteworthy rhetorical abilities aside, his own temper, the seat of his intellectual confidence, was predominantly and profoundly analytical. And regardless of whether he paid either lip service or true belief to the Catholic Church, or to religion in general, he could never for long put it to bed. His last and most scientifically impressive book, the *Discourses on the Two New Sciences,* appeared in 1638, some five years after his conviction, when he was living in the restricted circumstances of permanent house arrest, when blindness had overtaken him, and when he knew he was deep into the winter of his life's year. The finale of Galileo's scientific career spoke volumes about the future relation between science and religion. With its abstractions, with its methods of investigation, and with its disregard for the claims of religion—with, in sum, its analytical temper—scientific knowing would inexorably distance itself from religious belief. From its origins in the scarcely noticed fault line that Galileo had demonstrated with his science and exposed with his narrative, the Great Rift has descended, ever widening to our own day.

Epilogue

The Great Rift Today

> This way of philosophizing [mathematical science] tends to subvert all natural philosophy.
>
> —Galileo, *Dialogue concerning the Two Chief World Systems* (1632)

WHY DID HUMAN BRAINS—so small in volume, so brief in span—learn to think about and master their world with numbers, musical notes, geometrical grids, and other completely indeterminate, empty symbols? Despite the extent of our contemporary scholarship and intellectual sophistication, at bottom we don't really know. We have just narrated a story of how this happened, but as suggested at the outset of our tale, no one knows the why of things. In particular, no one can explain just why mathematics actually works in allowing us to be the "servant and interpreter of Nature," stealing a phrase from the prescient, albeit nonmathematical, Francis Bacon, writing on the eve of the Scientific Revolution.[1] Our command over empirical phenomena always bears a load of uncertainty; though highly improbable in many instances, new evidence can always overturn our most cherished mathematized natural laws, the so-termed problem of induction.[2] More to the point, it's unreasonable, this mastery, almost unfathomable. Thus Nobel laureate in physics Eugene Wigner: "The enormous usefulness of mathematics in the natural sciences is something bordering on the mysterious . . . and there is no rational explanation for it." Uncertainty and mystery have always permeated our lives and history; they also permeate our science.[3]

The fact of mystery at the heart of modern science has long tendered an opening for believers of religious persuasions and narrations, traditional and otherwise. But the uncertainties and mysteries of science and religion are as day to night, and from the moment of modern science's inception, the gulf between them has inexorably grown wider and deeper. By now, at the end of our tale, this should not surprise us, for the products of each stem from the "principles and parameters" of their underlying information technologies.[4] As we have seen, the invention of writing created information, most probably in response to business and accounting demands as settled societies became more complex. Writing allowed folks to fix information in symbols, whose syntactical and semantic *techné* became increasingly systematic and logical over the centuries, largely because of our ability to take a step back from what we have written, to look at the symbols critically, to create new ones, and to match the symbols with the things of experience. The phonetic alphabet marked the apogee of early script evolution, bringing to its high point the classification potential inherent in speech. Writing and information thus gave voice to the human compulsion to organize experience, to understand the world and to situate oneself in it. From a substrate in abstract letter symbols, to definitions of words, to classifying words and things, and to the reaches of abstract logic and poetic narration, alphabetic literacy supplied the information technology for both science and religion for millennia. Its dominance in life and thought remains the most singular and monumental accomplishment of the Greeks.[5]

An accomplishment long hailed but not always understood. "Classification," wrote Alfred North Whitehead nearly a century ago, stands as a "halfway house between the immediate concreteness of the individual thing and the complete abstraction of mathematical notions."[6] Since then his remark has spawned a host of comparable declarations from scholars, references often made to the "abstract" nature of scientific investigation, including that of Galileo.[7] But the issue here turns not on whether the medieval schoolmen were only partially abstract in their classificatory thinking while Galileo and other early modern scientists took the next step toward full abstractions with their mathematics. Rather, the quandary turns on the sorts of abstractions Galileo and others employed. Reflect for a moment. How could any expression be more abstract, more distant from the concreteness of individual things, than the terms 'being' or 'becoming,' or 'actuality,' 'potentiality,' 'essence,' or 'accident'—the whole gamut of philosophical terminology in the Middle Ages? Medieval linguistic abstractions ascended like Gothic cathedrals, merging their light and space with the vastness of the heavens. In many medieval minds the world may well have been filled and animated by ghosts, goblins, and various enchantments, divine or devilish, but regardless, it was abstract—and soaringly so.

Early logical and mathematical abstractions lay embedded within this technology, whose acme was the deductive reasoning of both Aristotle and Euclid. But abstracting the things of the world through the definitions of words, classifying them in a hierarchy of categories, and eliciting the rules governing their logic tendered only one means of distancing oneself from the fleeting and mercurial phenomena of our experience, of dragging from it mental extracts. Other means of abstraction became available during the late medieval and Renaissance periods. A different substrate of symbols, even emptier than sound-laden letters, crept into human practice and became the building blocks of a new information technology, relational numeracy, interposed between the mind and nature. In summary fashion, we have told this story. Born of this new technological screen, the abstractions of Galilean science were not greater or less than those of Aristotle and the scholastics. They were radically different—a difference of kind, not degree. And they channeled their way into the core of the natural sciences, where they have stayed ever since.

After Galileo words would have less and less to do with the means, technically speaking, of scientific discovery and explanation. Countless phenomena pertaining to gravity, vision, sound, air, combustion, magnetism, electricity, and myriad other topics, which had been once defined and understood qualitatively by words, became steadily subject to mathematical, quantitative analysis. Reaching ever more abstractly from their foundation in relational numeracy, mathematics and physics extended their breadth, depth, and mastery over natural phenomena from the seventeenth century forward. A hierarchy of scientific disciplines emerged in their wake, leading to what nowadays is oftentimes referred to as "reductionism." Some scientists use the term to express in general the understanding of "higher-level" phenomena by explaining the workings of its "lower-level" constituents. In biology, for example, the makeup and function of organisms can be explained in part by giving an account of the cells that compose them. Cellular biology, in turn, can be partially explained by the bonding mechanisms of biochemistry, which themselves derive from the mathematized description of the properties of atomic and subatomic particles and their interactions.

"Reductionism," as Richard Dawkins puts it, "is just another name for an honest desire to understand how things work."[8] And at the most basic, reductionist levels of explaining how natural phenomena work, words fail; mathematical symbols, formulas, and the reverse engineering of analysis take over. Without them one "wanders about in a dark labyrinth."

Not all natural phenomena can be explained reductively in the way just sketched, but abstract, relational mathematics (including statistical analysis) remains nevertheless the key to scientific discovery and explanation. In recent decades scientists have paid a great deal of attention to "emergent" phenomena as one proceeds from lower to higher levels of investigation, or to the consequences of "feedback" loops between levels, or to the sciences of complexity, the mathematics of chaos, and a host of newer topics for investigation. All of these require highly technical, mathematical descriptions.[9] Many involve the use of an additional information technology, that of the electronic computer, whose speed permits mathematical and algorithmic calculations of previously undreamt magnitude.[10] Logical symbols and their rules reside at the electronic heart of computer processing. And since the nineteenth century, understanding the "science of patterns" (the modern definition of mathematics) and its computer-modeling applications and analyses of nature has required delving ever more deeply into the world of symbolic logic and algorithms.[11] The move to a deeper, logical grounding of Galileo's "language of mathematics" and its connections to the physical world affirms, in brief, that any "dreams of a final theory" one may harbor for explicating the processes of nature are ultimately mathematical, not linguistic.[12]

To be sure, words and conventional language would remain important for the sciences. Gathering and classifying information—fact-finding, as it were—became a key component of the empirical tradition inspired by Francis Bacon and carried on ever since by many subsequent natural philosophers.[13] From travelogues to curiosity cabinets to observing cells through a microscope and much, much more, linguistic descriptions of objects and phenomena fed expanding interests in natural philosophy. Using conventional language (vis-à-vis a symbolic or mathematical language), the great eighteenth-century biologists Buffon and Linnaeus launched their own descriptive and synthetic systems of information about natural organisms and brought biological taxonomy to its height, where it remained until the challenges of evolutionary cladistics in recent years. Linnaeus's binomial nomenclature permitted the placement of organisms into fixed categories: 'kingdoms,' 'phyla,' 'classes,' 'genera,' and 'species.' But both Buffon and Linnaeus were clear about their compilations; these were arbitrary and useful ways of organizing information, not mirrors of nature, and certainly not explanations of phenomena.

Throughout the history of science from Galileo's time to the present, there has grown a critical distinction we need to highlight. On the one hand is the virtual impossibility of translating the analytical content of relational mathematics and physics directly into the conventional language of words and things. (The wave-particle "weirdness" of quantum mechanics provides a current illustration of the point.) This was the Aristotelian threshold of incommensurability between discreteness and continuity that Galileo addressed. On the other hand, we still need to discuss in commonsense ways of speaking something of the results and procedures that issue from analysis, as well as the truths it captures. Otherwise stated, thinking analytically and mathematically about natural phenomena stands quite apart from describing in conventional language the procedures, results, and significance of analytical thinking, as Galileo also instinctively knew. For the most part he negotiated these conceptual brambles quite comfortably, even brilliantly on occasion, within his own intellectual environment.

Words still matter in the sciences, then, but as they did for Galileo, principally as a means of making sense in conventional language of the results of empirical and theoretical investigations conducted in the language(s) of mathematics. "Nature does not care what we call it," Richard Feynman once wrote, précising the idea in our own times, "she just keeps on doing it."[14] We figure out nature's "doing" with the reverse engineering of analysis, our modern analytical temper, driven by the empty symbols and logical rules of the information technology of numeracy and its processing. These symbols are now reduced even further and stripped bare, down to the zeros and ones of computers, but their historical cornerstones were laid in the Renaissance. Later, the good bishop Berkeley termed empty symbols the "ghosts of departed quantities" (his expression for the then-new infinitesimals); French encyclopedist Jean Le Rond d'Alembert simply called them "phantoms."

Today's scientific thought, our own phantom world, has been built on these cornerstones. Figuratively and literally . . . and convincingly . . . they've taken us to the moon and beyond.

Even as the tools of alphabetic literacy, words and conventional language, became less and less relevant in scientific discovery and explanation, they continued to express the presence of mystery in religious narrative. Until well into the nineteenth century, many, if not most natural philosophers could at least nominally be counted among the folds of established religious congregations. New discoveries in the processes of nature, the regularities of divine craftsmanship, reinforced belief in the "parts of one stupendous whole, / Whose body nature is and God the soul," intoned poet Alexander Pope (1688–1744), as arguments from design breathed new life into natural theology.[15] All the same, for many others, including those who continued to believe throughout these years, the realm of mystery shrank steadily, pared away by the analytical temper and its grounding in the information technology of numeracy. In the wake of Newton's prodigious creation of classical physics with his *Mathematical Principles of Natural Philosophy* (1687), for instance, John Locke (1632–1704) wrote of the "reasonableness of Christianity," while John Toland (1670–1722) went further into a kind of pantheism with his own version of *Christianity Not Mysterious.* Writers following Voltaire (François-Marie Arouet, 1694–1778) espoused a sparse "deism," which clung to the idea of God as creator of natural laws and moral sensibilities but little beyond. Philosopher David Hume wrote that the only remaining miracle was the "continued miracle" that some people unreasonably believe in the Christian story, which "subverts all the principles of . . . understanding."[16] By the early nineteenth century, that stalwart of the Enlightenment Thomas Jefferson (1743–1826) had excised from the Gospels all those passages he thought smacked of miracles or the supernatural in order to arrive at the "life and morals of Jesus." The result, now known as the *Jefferson Bible,* was a "wee-little book."[17]

Viewed in this light, the sprawling Romanticist movement can be seen as an attempt to reclaim and even enlarge the realm of mystery in people's lives. Some artists, poets, and authors sidestepped the sciences altogether, focusing on intuition, emotion, and a holistic embracing of mysteries in all their variety, including a revival of interest in "fairy tales" (to wit, the brothers Grimm) and the enchantments of the Middle Ages (thus the French author

François-René de Chateaubriand, 1768–1848). Others, taking their lead from William Wordsworth (1770–1850), challenged science's "murder to dissect" propensities head-on and promoted a more fulsome and aesthetically, emotionally charged response to nature.[18] Natural philosophers themselves frequently sought to expand the mystery they found in their investigations. For a while William Paley's *Natural Theology* (1802) reinvigorated the design arguments and the "watchmaker God" by infusing an affective, narrative content into scientific analysis, a theme highlighted by the first "modern" philosopher of science, John F. W. Herschel (1792–1871). A polymath of nineteenth-century science, Herschel, one of the founders of the short-lived but highly influential Analytical Society during his student years at Cambridge, wrote a widely read, popular work, *A Preliminary Discourse on the Study of Natural Philosophy* (1830). In it he combined the "habit of abstraction" and the power of "analysing" with the "contemplation of general laws which powerfully persuades us to merge individual feeling" into a belief in divine providence.[19] Still, despite Herschel's urging, along with that of many contemporaries (William Whewell, 1794–1866, for example), by the mid-nineteenth century scientists like Michael Faraday (1791–1867), while still affirming that "the book of nature . . . is written by the finger of God," were insisting equally on "an absolute distinction between religious and ordinary [science-based] belief."[20]

The dénouement of this sketchily lined course of development lay with the Darwinian revolution, which brought the biological sciences into the analytical temper and laid open their future to the information technology of numeracy. (Darwin himself claimed that eventually the "terms used by naturalists . . . will cease to be metaphorical," but the story remains to be written of Darwin's own analytical temper and the effects of numeracy on his and subsequent biological thought.)[21] In the wake of these and comparable developments in the march of the analytical sciences, the breadth and depth of the fault line between the two information technologies has become even more visible and pronounced. For from the late nineteenth century to our own times we have reached the point where observers and participants alike (scientists or religionists) have come to view the widening separation between science and religion as an impasse, or even war zone. The incompatibility between them is simply too great to overcome either through scientific techniques or by means of conventional discourse framing religious beliefs.

We live in a variegated age, one whose collective cultures are torn by the rift between science and religion.[22] Many lament the loss of traditional religions and their enchantments and as a way to grapple with the ongoing presence of uncertainty and mystery in our lives, urge a return to or continuation of religious doctrines carried forward from the past. (As noted at the outset, Huston Smith seeks a common thread in all historically grounded, mainstream religions, the distinction between "*this-world* and an *Other-world*.") Many still cling to a belief in the "heavenly city of the eighteenth-century philosophers" and the Enlightenment ideals of progress.[23] The employment of science, technology, and industry in the service of improving the human estate, and universally extending their benefits politically and socially, continues to inspire thought, political action, and sacrifice. (To be sure, the impending, human-generated climate-change crisis clearly casts a pall over these possibilities.) Many too have sought and seek solace in newer, modern enchantments, such as nationalism and other 'isms,' often called "secular" and ideological forms of religion, or in any among an array of New Age spiritualties. We could continue with the list of our own age's deep, often desperate quests in search of believable enchantments, but the foregoing suffices to establish our central point, for none of these "beliefs" is scientific. Mystery still resides with us, but science stands mute in its presence, while traditional narratives of religious beliefs remain incapable of commanding universal assent or of even securing widespread agreement (much like the fracturing Christianity underwent during the Reformation). In short, neither science nor religion can bridge the gap between science and religion.

But where science and religion cannot go, perhaps history can. In the end, even as religion shrinks from science, we continue to tell stories, without which not only would our lives be bleak and bereft, "the last rose of summer" as the Irish song has it, but more to the point we would have little or no footing for understanding how it came to pass that we are who we are, where we are. "Narrative makes meaning," observes Charles Taylor. "Through story . . . we find or devise ways of living bearably in time."[24] Though a poignant and suggestive observation, Taylor's remark requires a critical amendment. In order for narrative to possess the power of bridging meaningfully the divide between science and religion it must tell the story, with evidence, of how the divide came about. It must be historical. Our own historical narrative has focused on the deep incompatibility of religion and science that arose because of a major shift in their respective underlying information technologies, literacy and numeracy, which separated them on the matter of understanding nature. A profound irony pervades our account and indeed our times. For our story is written by means of the traditional technology of literacy and the narrations of conventional language we derive from it (storytelling with facts, we called it earlier). Only through these means can we appreciate how, based on numeracy, modern analysis of the natural world arose and subverted traditional religious beliefs. Likewise, only through these linguistic means can we discern how analysis continues to corrode our linguistically formulated beliefs and "realities" with its virtual, phantomlike presence.

Appreciating historical narratives might not assuage everyone's existential discontent or nervousness. But it does allow us to satisfy our curiosity, as well as to live "bearably in time." And even though on occasion it may appear tinctured with nostalgia for an unattainable past, there is nonetheless continuing grandeur in this view of history, for it supplies our most salutary means of embracing modern mystery and living in the presence of the Great Rift.

Appendixes

A.1 Logical Demonstration and the Syllogism

Codified by Aristotle in his logical works (*The Organon*), the syllogism became the common form of depicting a logical demonstration during the Middle Ages. It entailed three statements or propositions—a major premise, a minor premise, and a conclusion (preceded by "therefore" or *ergo*—indicated here by the modern symbol ∴)—and was often arrayed as in the following examples:

	Example A		**Example B**
Major Premise:	All men are mortal	*Major Premise:*	All Martians are green
Minor Premise:	Socrates is a man	*Minor Premise:*	Xantippe is a Martian
Conclusion:	∴ Socrates is mortal	*Conclusion:*	∴ Xantippe is green

The syllogism functioned exclusively *within* the realm of thought, connecting premises that were either commonly accepted or previously established through dialectical inquiry. The truth or falsity of the premises themselves had no bearing on the formal validity of the argument. Thus, example B is as valid as example A, even though we can safely say that Xantippe (Socrates's argumentative wife) was neither a Martian nor green.

Each statement of the syllogism contained a subject (S) and predicate (P), and the two premises themselves shared a common, or "middle" term (M). Statements could be universal ('all') or particular ('some' and its variants, including 'one') and either affirmative or negative, which led to four types of "categorical propositions," labeled A, E, I, and O: A = All S is P; E = No S is P; I = Some S is P; O = Some S is not P. (The letters come from the Latin *AffIrmo* and *nEgO,* meaning "I affirm" and "I deny.") Different combinations of the statements produced the "mood" of a syllogism (AAA, OAO, EIO, and so on), while the positions of the middle terms led to four different logical "figures" or patterns of argument. Combined, all these variables produced some 256 possible types of syllogism, of which 24 yielded valid, logical inferences, some stronger than others. As a mnemonic device, scholastic logicians gave each of these 24 arguments names containing the letters identifying the syllogism's mood and figure. For example, the pattern Barbara included three universal statements of type A (B*A*rb*A*r*A*) and was written as mood and figure AAA-1, as illustrated with the following:

Major Premise (A):	All men are mortal (All M is P)
Minor Premise (A):	All Greeks are men (All S is M)
Conclusion (A):	∴ All Greeks are mortal (All S is P)

From this skeletal structure, countless variations, nuances, and interpretations occupied logicians for centuries, until the invention of modern, symbolic logic.

A.2 Calculating the Beast of Revelation

> Revelation 13:18 (Revised Standard Version): "Let him who has understanding reckon the number of the beast, for it is a human number, its number is six hundred and sixty six."

The Catholic numerologist Peter Bungus devised a system based on gematria whereby the letters correlated with numbers, as follows (*I* and *J* were the same letter; so too *U* and *V*; there was no *W*):

A	B	C	D	E	F	G	H	I or J	K	L	M	N	O	P	Q	R	S	T	U or V	X	Y	Z
1	2	3	4	5	6	7	8	9	10	20	30	40	50	60	70	80	90	100	200	300	400	500

With this code he summed the numbers corresponding to the letters in Martin Luther's name:

M	A	R	T	I	N	L	U	T	H	E	R		
30	1	80	100	9	40	20	200	100	8	5	80	=	673

Together, 'Martin' and 'Luther' totaled 673, in excess of 666, the number given the beast (devil) in Revelation. Bungus then fiddled with his numbers, Latinizing Luther's surname (LUTERA), but not his given name. The results, more to Bungus's liking, demonstrated that Luther was indeed the devil:

M	A	R	T	I	N	L	U	T	E	R	A		
30	1	80	100	9	40	20	200	100	5	80	1	=	666

Not to be outdone, Michael Stifel, monk, mathematician, and early supporter of Luther, directed comparable numerical legerdemain against Pope Leo X. He began with the Latin form of the name, LEO DECIMUS, and treated it using the same code Bungus had employed, with the following result:

L	E	O	D	E	C	I	M	U	S		
20	5	50	4	5	3	9	30	200	90	=	416

The tally here was far short of 666, so Stifel too toyed with his numbers. Departing from Bungus's schema, he considered only the letters that are also Roman numerals: *L, D, C, I, M*, and *V* (the last because *U* = *V*). He then subtracted *M* because it symbolized *mysterium* and thus could be discarded as unknowable. As a capstone to his computations, he added an *X*, which stood for the "ten" in 'Leo the Tenth.' Finally, placing the remaining Roman numerals in descending numerical order and summing their values, he achieved his desired goal:

DCLXVI = 666

Voilà: the pope was the devil incarnate.

A.3 Medieval Poetic Meter and Rhythmic Modes

The six modes of poetic meter used by the Notre Dame creators of the rhythmic modes and notation are as follows:

Mode	Rhythmic Pattern (– = long, or stressed; ᴗ = short, or unstressed)	Metrical Terms	Modern Musical Notation
I	– ᴗ	Trochee (Trochaic)	♩ ♪
II	ᴗ –	Iambus (Iambic)	♪ ♩
III	– ᴗ ᴗ	Dactyl (Dactylic)	♩ ♪ ♪
IV	ᴗ ᴗ –	Anapaest (Anapaestic)	♪ ♪ ♩
V	-- ---	Spondee (Spondiac)	♩ ♩ ♩ ♩ ♩
VI	ᴗ ᴗ ᴗ ᴗ ᴗ	Tribrach (Tribrachic)	♪ ♪ ♪ ♪ ♪

As an illustration of poetic meter, recall the common expression "Ta DAH," which is actually an iambic foot (mode II), with the stress on the second of two syllables. Five "Ta DAHs" create a line of iambic pentameter, the most common meter in English-language poetry. This can be seen in William Shakespeare's Sonnet 73:

∪	—		∪	—		∪	—		∪	—		∪	—	
That	**time**	|	**of**	**year**	|	**thou**	**mayst**	|	**in**	**me**	|	**be**	**hold**	|
∪	—		∪	—		∪	—		∪	—		∪	—	
When	**yel**	|	**low**	**leaves,**	|	**or**	**none,**	|	**or**	**few,**	|	**do**	**hang**	|
∪	—		∪	—		∪	—		∪	—		∪	—	
Up	**on**	|	**those**	**boughs**	|	**which**	**shake**	|	**a**	**gainst**	|	**the**	**cold,**	|
∪	—		∪	—		∪	—		∪	—		∪	—	
Bare	**ru**	|	**in'd**	**choirs**	|	**where**	**late**	|	**the**	**sweet**	|	**birds**	**sang**	|

A.4 The Rule of Three and Metrical Rhythm

Constructed from Johannes Tinctoris, the following schema is technically a *sesquialtera* proportion, written in perfect time with a major prolation:

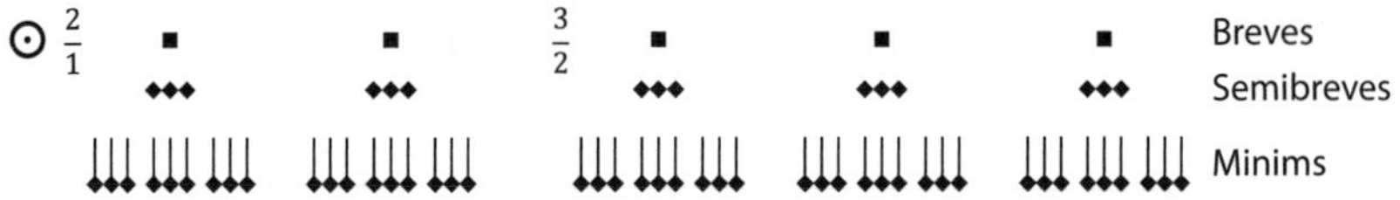

⊙ $\frac{2}{1}$ ■ ■ $\frac{3}{2}$ ■ ■ ■ Breves

◆◆◆ ◆◆◆ ◆◆◆ ◆◆◆ ◆◆◆ Semibreves

Minims

The ʘ at the beginning indicates that the passages here are written in perfect time, major prolation—ternary divisions, with each note subdivided into three of the next-smaller unit. The first section is governed by a *dupla* (double) proportion ($\frac{2}{1}$). This section (we would say the first two measures) is thus comprised of two breves (double whole notes in modern notation) and their corresponding divisions. Each breve can be broken down into three semibreves (whole notes), and each of the semibreves in turn can subdivide into three minims (loosely, modern half notes, but in a ternary system there are three minims to a whole). Any combination of breves, semibreves, and minims equaling the duration of two breves may be used in this part of the composition. Then in the second section (the next three measures, so to speak), Tinctoris shifted to a new proportion ($\frac{3}{2}$), representing a new time value. Now three breves replace the two of the first section, with their corresponding prolations or divisions. The challenge was to find out how many minims could be used in the second section. To solve that problem, the rule of three was used as noted.

A.5 Creating the Pythagorean Scale

Beginning with the tone made by any plucked, struck, or bowed string, we can produce others related to it by whole-number ratios. Let t designate the frequency of the tone's vibrations. Doubling the string's frequency gives a ratio of $2t{:}t$, or simply 2:1. This is the octave of the original tone. (On a piano keyboard, count the white keys from middle C to the next octave higher: C–D–E–F–G–A–B–C′. Here, C = t and C′ = $2t$.) Next, triple the vibrations of the string; this yields a frequency three times the original, or $3t$. The interval between the frequencies of third and second tones, $3t$ and $2t$, is $3t{:}2t$, or 3:2, a ratio producing an additional tone, whose frequency is $\frac{3}{2}t$. This tone lies between t and $2t$ and gives us a scale of three notes: t, $\frac{3}{2}t$, and $2t$. (On the piano, $\frac{3}{2}t$ is the note five tones above C, or G, called the "perfect fifth.") With this additional note, we can fill in other intervals of the scale from C to C′. First, subtract from C′ a perfect fifth, which nets the interval of $2t : \frac{3}{2}t = \frac{4}{3}t$. The result, 4:3, is the ratio for the perfect fourth (F on the piano keyboard).[1] Now we have a scale of four notes: $t, \frac{4}{3}t, \frac{3}{2}t, 2t$ (C–F–G–C′). These are the "pure," Pythagorean tones—with whole-ratio intervals based on the numerals one through four (the *tetractys*). Further intervals result from iterating the same procedure. Having the fourth and the fifth (F and G on the piano), we can establish the interval between them, and thus between two successive tones (called the 'major second' or 'whole tone'): $\frac{3}{2}t : \frac{4}{3}t, = \frac{9}{8}$. With the whole tone interval of $\frac{9}{8}$, the remaining gaps may be filled:

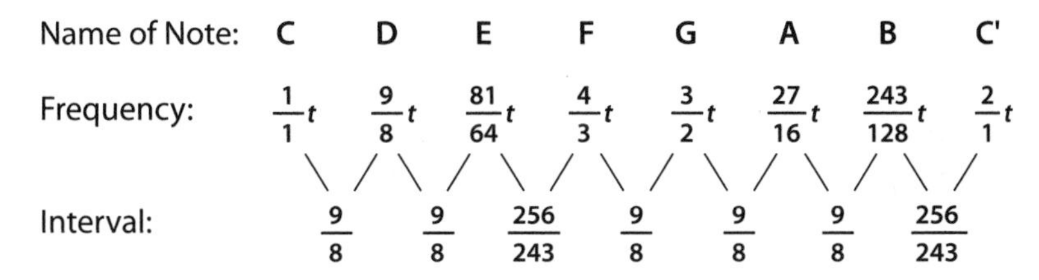

Note the narrower intervals between E and F, and between B and C′; these are generally termed a 'minor second' or 'semitone.' Their ratios are $\frac{4}{3} \div \frac{81}{64} = \frac{256}{243}$ and $\frac{2}{1} \div \frac{243}{128} = \frac{256}{243}$, which Plato had observed.

A.6 The Spiral of Fifths, Dissonance, and the Tuning Problem

We can see the mathematical source of dissonance if we follow the so-called spiral of fifths in the Pythagorean scale. Start with a root note of any frequency, a string of any length. With the ratio for the fifth at 3:2, this means that a string two-thirds the length of the original or root tone will sound a harmonious pitch five steps higher. Now take the fifth as the new root and create a second fifth from it. The second fifth, in turn, becomes another new root, another fifth is created, and continuing. Were we to do this twelve times, we would expect to return to our initial tone or starting point, seven octaves higher, because twelve fifths equal seven octaves. In modern figuring, an octave equals twelve half steps or semitones, while a fifth is composed of seven half steps or semitones. So twelve cycles of fifths equals seven cycles of octaves. Or it should. (It does on a modern piano tuned with equal temperament, but the tone intervals on a piano have all been fudged a bit.) What actually occurs acoustically can be shown by straightforward arithmetic, comparing twelve fifths, with their ratios of 3:2, and seven octaves, with their ratios of 2:1 (recall from Appendix A.5 that to add or combine intervals, we multiply their ratios):

$$\frac{3}{2}\times\frac{3}{2}\times\frac{3}{2}\times\frac{3}{2}\times\frac{3}{2}\times\frac{3}{2}\times\frac{3}{2}\times\frac{3}{2}\times\frac{3}{2}\times\frac{3}{2}\times\frac{3}{2}\times\frac{3}{2}=129.746$$

$$\frac{2}{1}\times\frac{2}{1}\times\frac{2}{1}\times\frac{2}{1}\times\frac{2}{1}\times\frac{2}{1}\times\frac{2}{1}=128.0$$

The spiral of fifths goes beyond the cycle of octaves, resulting in a ratio of 129.746:128, or 1.014. (The ancients calculated a similar ratio of 81:80, the difference between a major tone of 9:8 and a minor tone of 10:9: $\frac{9}{8}\div\frac{10}{9}=\frac{81}{80}$. Its decimal equivalent is 1.013). This may seem insignificant, for the last fifth is only about a quarter of a semitone high. But it creates a troublesome discrepancy, termed a "comma," that makes the last note completely unusable. Try as one might, the numbers just don't add up. Thus irrationality permeates the scale; we perceive this irrationality as dissonance. The tuning problem centers on spreading the comma or discrepancy throughout the octave, or figuring out where best to place it.

A.7 The Roman Calendar and the Celestial Sphere

On the eve of Julius Caesar's reform, the Roman calendar was luni-solar, tied to the months and their imprecise correlation with the sun-governed seasons. A lunation, or lunar cycle, averaged roughly 29.5 days, which led to an initial division between "hollow" months of 29 days and "full" months of 30 days. By the first century BCE, the year had grown to twelve months, four of which—March, May, Quinctilis (later July), and October—had 31 days, with all the rest at 29, except for February, which had 28. This produced a lunar year of 354, or sometimes 355 days. High priests periodically curtailed February after 23 or 24 days and inserted (intercalated) an additional month to bring lunar cycles into rough conformity with the solar year and its seasons, actions often fraught with political interests and decisions. Much as we count the minutes leading up to an hour (for example, 'ten minutes to six'), the days in a month were counted backward from three fixed points: Ides, Nones, and Kalends. Each month was split in half by the Ides (from the Etruscan term *idus* for 'divide,' originally at the full moon). Nine days before the Ides fell another marker day, Nonus ('ninth,' originally at the half moon), while the first day of the month was the Kalends (*kalendae,* at the new moon). Counting included both ends of the sequence; thus the Nones fell on what we would call the seventh day of the month, nine inclusive days before the fifteenth (Ides).

Caesar supplanted this calendar with one based entirely on the solar, or tropical year, the time of the sun's complete passage through the zodiac from the vernal equinox until its return to the point of origin. Beginning with the Babylonians, early astronomers saw the heavens as a large sphere surrounding the earth and planets, still a useful convention for plotting the locations and movements of the fixed stars, sun, moon, and planets. (See Figure A.7.1.) The north celestial pole was understood as an extension of the earth's North Pole, with the polar star (Polaris) providing the fixed point around which the other stars, including the sun and planets, revolved every twenty-four hours, east to west, at a rate of fifteen degrees per hour. This "daily" rotation constituted one of two major movements perceived among the heavenly bodies. The ancients also observed a second dominant motion in the skies, an "annual" motion associated with the seasons and with the wandering stars—the sun, the moon, and all the other planets. As the year progressed, they saw (as can we) the sun follow a path around the celestial

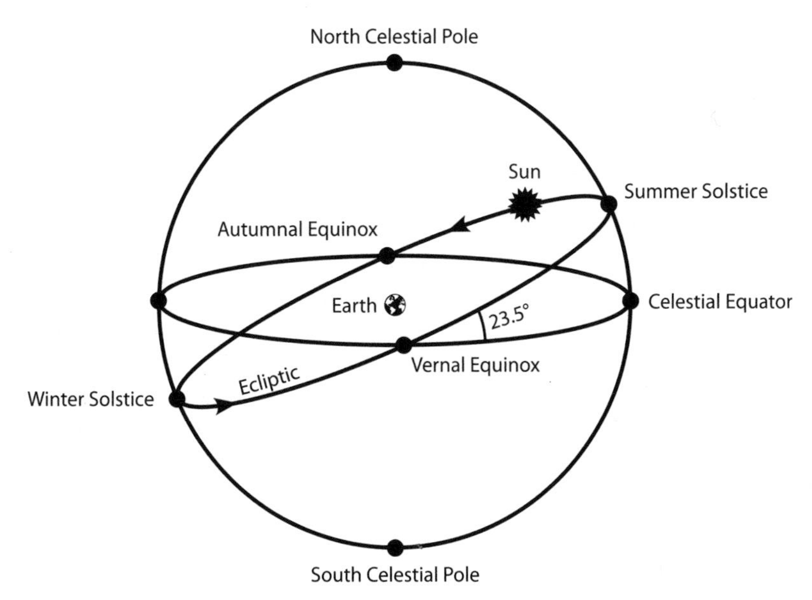

FIGURE A.7.1. The celestial sphere.

sphere along a line called the ecliptic, moving from west to east, season to season, at approximately one degree per day. (This path, we know, actually results from the tilt of the earth's axis, at roughly 23.5° from the plane of its orbit around the sun, an inference not available to pre-Copernican, geocentric astronomers.)

As the sun travels north of the equator, the days lengthen in the northern latitudes until the June (summer) solstice, the longest day of the year, after which the sun travels southward for the next six months. The sun's path crosses the celestial equator at two points, the days of the vernal and autumnal equinoxes, each with an equal number of daylight and nighttime hours. The time required for the sun to travel around the ecliptic through the cycle of seasons, from one vernal equinox to the next vernal equinox, is the solar or tropical year. The planets follow the sun's path along a band of sky several degrees on each side of the ecliptic; this band is the zodiac. In Figure A.7.2, the line in the center of the ring represents the ecliptic. As sun and planets proceed along the ecliptic in their own orbits, they pass in succession in front of the background of the zodiac constellations on the sphere of fixed stars.

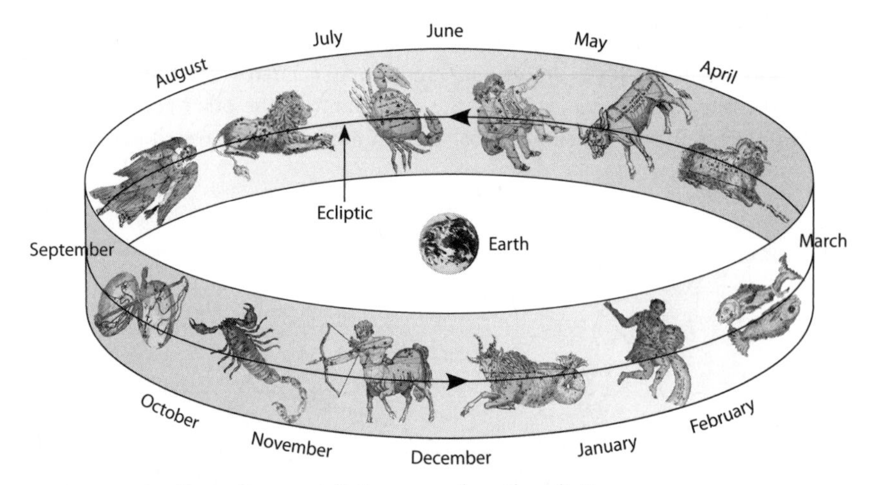

FIGURE A.7.2. The zodiac constellations seen along the ecliptic.

By Caesar's day, astronomers had figured the period of the solar year as 365.25 days, and, minus the quarter day, this became a common year. To account for the deleted time, every fourth year one day was added to the 28 days of February to make a 'leap' or 'bissextile' year, so called because of Caesar's practice of intercalating by repeating the "sixth day before the kalends of March" (February 24). The extra day was known as "twice sixth." The months continued in use as civil divisions, but lunar cycles had nothing to do with determining calendar labels. New moons and so on were noted simply by their relations to the days and months of the solar system.

A.8 The Precession of the Equinoxes

The term 'precession of the equinoxes' refers to the slight changes in the vernal and autumnal equinoxes, the moments when the sun, traveling on its ecliptic path, crosses the celestial equator and from which solar calendars generally proceed. (See Figure A.8.1.) The precession derives from the difference between the solar and the sidereal years. The former marks the sun's annual return to the equinoctial point as seen from the earth, while the latter refers to the measurement of the equinox, and hence a sun's year, from a distant star (like figuring the sidereal month). The sidereal year is twenty minutes and twenty-three seconds longer than the solar year.

Now known to be caused by the earth's slight wobble on its own axis, in turn caused by the gravitational pulls of the sun and moon, the equinoxes drift behind the fixed-star constellations in the zodiac at a rate of roughly one degree (or day) every seventy-two years, or twenty-eight degrees after two thousand years, nearly

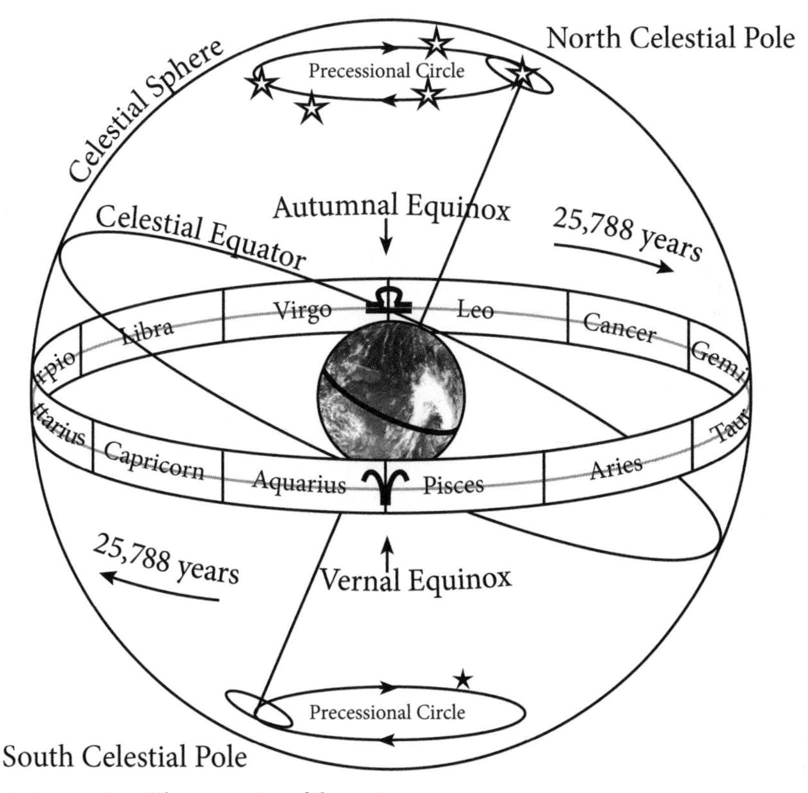

FIGURE A.8.1. The precession of the equinoxes.

an entire thirty-degree zodiac sign. In Caesar's era the vernal equinox appeared on the zodiac between constellations of Aries and Pisces. Now it occurs at the border of Pisces and Aquarius (hence the dawning of the "Age of Aquarius," so called by New Age spiritualists). The overall precession period is about 25,788 years, the time it takes the vernal equinox to travel around the ecliptic. Associated with the precession and also caused by the wobble are slight variations in the obliquity of the ecliptic, the 23.5° angle between the celestial equator and the ecliptic plane now known as the earth's axial tilt.

A.9 The Astrolabe

Figure A.9.1a shows the so-called Chaucer astrolabe, similar to the one Geoffrey Chaucer described. Created in 1326, it is the oldest dated astrolabe in Europe and measures about five inches in diameter, a typical size. Figure A.9.1b shows the components of a comparable device.

FIGURE A.9.1A. The Chaucer astrolabe.

The astrolabe's parts consisted of a circular "mater" ('mother'—"moder" in Chaucer's English) or base, around which was cast or welded a raised ring, called the "limb," which displayed time and degree scales. Inserted into the recessed mater were removable plates, sometimes called 'climates'; etched on these were local skies projected from a specific latitude. The plates were interchangeable so as to permit using the astrolabe in different locations. Astrolabes were generally crafted and sold with three plates, each etched on both sides, making it possible to use the instrument in six different locales. Overlaying the plate was another, often very ornate and carved out circular piece, the "rete" (Latin for 'net'), whose etchings and points showed a number of stars and the projection of the ecliptic. Over the rete was placed a "rule," used to line up scales and take readings. The whole instrument was held together by a pin, which permitted the turning of rete and rule. On the backside of the astrolabe, along with zodiac scales and a calendar, was generally a movable sighting guide called the "alidade" (from the Arabic *al-'iḍāda,* or 'ruler'), which allowed the user to find the altitudes of the sun or stars and doubled as a surveying instrument for taking earthly elevations.

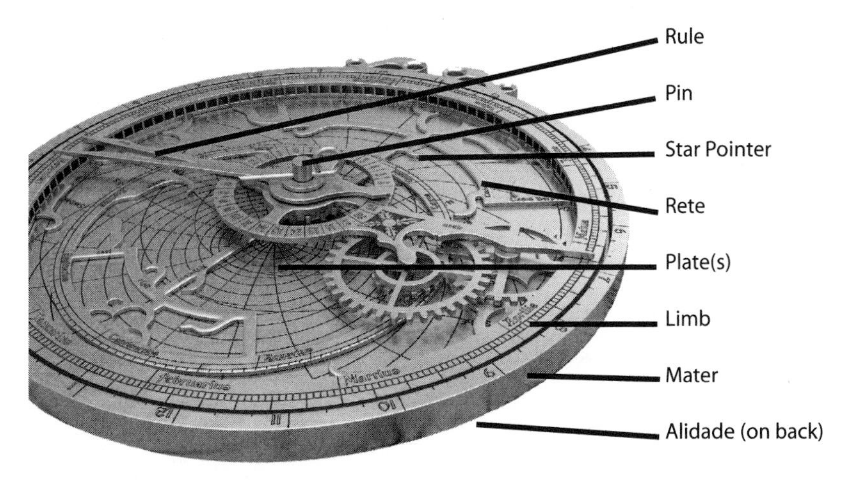

FIGURE A.9.1B. The astrolabe's parts.

A.10 Galileo's Mathematical Techniques

Calculation with ratios and proportions constituted a central feature of Galileo's mathematical analysis. The actual units of measure he utilized were arbitrary, for he worked in proportionalities of the general form *a*:*b* :: *c*:*d* (*a* is to *b* as *c* is to *d*), nowadays conventionally written as an equation of fractions: $\frac{a}{b} = \frac{c}{d}$. This was the old and familiar form of the "rule of three" or "merchant's rule" that Leonardo of Pisa had so thoroughly explained and whose use had spread far and wide throughout the Renaissance, the same rule that Galileo's geometric compass could manipulate with mechanical facility.

The form of the rule of three also underlay a calculating technique known as the "mean proportional," which Galileo had taken from Euclid's *Elements*, book 5. In modern terms, the mean proportional is the positive number x that serves as the mean between two extremes, such that *a*:*x* :: *x*:*b*, or algebraically, $\frac{a}{x} = \frac{x}{b}$. Often termed the "geometric mean," the equation of a mean proportional can be expressed as $x^2 = ab$ or $x = \sqrt{ab}$, a notation not available to Galileo. For example, in the simple equation $\frac{4}{6} = \frac{6}{9}$, 6 is the geometric mean or mean proportional between 4 and 9, which Galileo would have written rhetorically with the expression, '4 is to 6 as 6 is to 9' (4:6 :: 6:9).

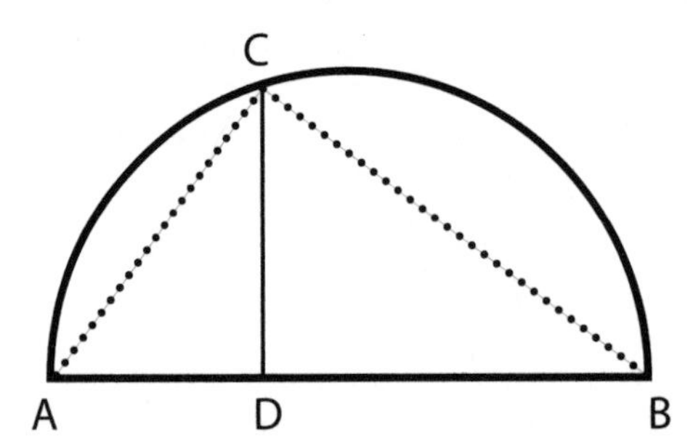

In Euclidean geometry mean proportionals may be constructed with a ruler and compass and proved rigorously by the similarity of triangles. Thus, in the diagram the altitude CD to the hypotenuse of a right triangle ABC is the mean proportional between the segments into which it divides the hypotenuse, AD and DB, or $\frac{AD}{CD} = \frac{CD}{DB}$. This is sometimes called the "altitude rule" (part of hypotenuse ÷ altitude = altitude ÷ other part of hypotenuse). Further, each leg of a right triangle is the mean proportional between the hypotenuse and the projection of the leg on the hypotenuse. That is, $\frac{AB}{AC}=\frac{AC}{AD}$ and $\frac{AB}{CB}=\frac{CB}{DB}$, sometimes called the "leg rule" (hypotenuse ÷ leg = leg ÷ projection).

Expressed in the Hindu-Arabic numerals of the new numeracy, this mathematical equivalent of an algebraic equation allowed Galileo to use and transform static, geometric construction into an analysis of motion. Not only could one correlate ratios between fixed spaces, but with mean proportionals one could apply the ratios to "movables," as Galileo termed bodies in motion. It became possible to analyze the irrational magnitudes of motion into equalized components of space and time, then address the challenge of connecting them.

A.11 Galileo's Analysis of Pendulum Motion

The following table represents Galileo's data, compiled by Stillman Drake, pertaining to the relations between the lengths and times of pendulum swings. From them one can derive the law of pendulum in a restricted form. The first two numbers in each column are actual measurements found in Galileo's notes. They permit the extension of both columns of numbers, up through the last number in the length column, which is also found, significantly, in one of the folios. Column [A] was formed by successive doubling, indicated by the numbers in the brackets, here added for clarity. Column [C] was formed by alternate doublings, again noted here by the bracketed numbers.

	[A] Length of pendulum in punti		[C] Fall time to the vertical in grains of flow	
1.	870	[1]	668½	[1]
2.	1,740	[2]	942	
3.	3,480	[4]	1,337	[2]
4.	6,960	[8]	1884	
5.	13,920	[16]	2,674	[4]
6.	27,840	[32]	3,768	

From these data Galileo created his own, standardized unit of time (a *tempo*—plural, *tempi*) and demonstrated the correlation between a pendulum's length and the time squared of its quarter period. He did so by connecting the columns with two computations. First, he calculated the mean proportional between two and the lengths of the pendulum (he likely used two as one extreme because of doubling the lengths). Second, he divided the weights in grains by sixteen, a procedure he performed on folio 189v. These two operations produced a direct correspondence between the length of the pendulum and the square of its fall time (a quarter period), as shown in the virtual equivalencies of columns [B] and [D] of the following table:

	[A] Pendulum Length (punti)		[B] Mean Proportional between Length and 2	[C] Fall Time to Vertical (grains)		[D] Fall Time ÷ 16 (16 grains = 1 tempo)
1.	870	[1]	41.71	668½	[1]	41.78
2.	1,740	[2]	58.99	942		58.88
3.	3,480	[4]	83.42	1,337	[2]	83.56
4.	6,960	[8]	117.98	1884		117.75
5.	13,920	[16]	166.85	2,674	[4]	167.13
6.	27,840	[32]	235.96	3768		235.50

Thus column [A], pendulum lengths, correlates directly to column [C], the squares of the times. Other measurements and calculations from additional folios tie directly to these data and confirm Galileo's discovery of the law of pendulum, expressed in mean proportional terms. Its equivalent, the modern formula for the period of a simple pendulum, is $T = 2\pi\sqrt{\frac{l}{g}}$, where T = the pendulum's period (the time required for one complete oscillation), l = the length of the pendulum's string, and g = the force of gravity. Because π and g are both constants, the square root of the length is the only variable on which the period depends. Squaring both sides of the equation yields $T^2 \propto l$, time squared is proportional to length.

A.12 Mean Speed Theorem and Free Fall

As discussed earlier (Chapter 4), Nicole Oresme of Paris in the fourteenth century devised the mean speed theorem (or rule), which states that a body uniformly accelerated from an initial velocity (v_1) to a final velocity (v_2) would travel the same distance (d) in a time interval (t) as if it had moved with a constant, mean speed ($\bar{v}$) between v_1 and v_2 from start to end. Whereas Oresme had expressed and verified this geometrically, we may demonstrate it algebraically with the formula $\bar{v} = \frac{v_1 + v_2}{2}$. From this only a few short steps allow us to see the derivation of the familiar equation for Galileo's times-squared law found in science textbooks:

(1) $\bar{v} = \frac{v_1 + v_2}{2}$ Mean speed rule

(2) $\overline{v}=\frac{1}{2}(0+v)$	From (1), average velocity from rest (0) to final speed (v) (or $\overline{v}=\frac{1}{2}v$)
(3) $d=\overline{v}t$	Distance traveled (d) = average velocity multiplied by amount of time (t) traveled
(4) $d=\frac{1}{2}(v)t$	From (2), (3)
(5) $v=at$	Where a = uniform acceleration; final velocity v results from multiplying the acceleration of a body by the amount of time (t) traveled
(6) $d=\frac{1}{2}(at)t$	From (4), (5)
(7) $d=\frac{1}{2}at^2$	From (6), the standard formula for free fall
(8) $d=\frac{1}{2}gt^2$	Substituting in (7) the acceleration of gravity, g (32.174 ft / s², read as feet per second per second, or 9.81 m / s², read as meters per second per second), as a constant for a

It is now known that Galileo's formula is generally true for motion in any circumstances in which the acceleration is constant.

A.13 Parabolic Trajectory and the Speed Law

Galileo's method of calculating ideal predictions of the tests on folio 116v is for us somewhat cumbersome and rather counterintuitive. We may demonstrate their equivalents algebraically, working with the speed law: $v \propto \sqrt{d}$ or $v^2 \propto d$. With the latter expression one can figure velocities of moving bodies at the moment of their horizontal launch and their corresponding projected distances. (Higher velocities of projectiles generate greater distances traveled.) The basic proportionality Galileo worked with, using the rule of three, was: $a{:}b :: c{:}d$, $\left(\frac{a}{b}=\frac{c}{d}\right)$ In these tests a corresponds to the projected distance produced by the first velocity squared ($v_1{}^2$), when the ball was dropped from b, the first height (h_1); c represents each of the subsequent test velocities and distances (v_2^2), the unknown distance sought, and d the subsequent test heights (h_2). Combined, the proportionality is $v_1^2 : h_1 :: v_2^2 : h_2$. This in turn can be written as

(a) $\frac{v_1^2}{h_1} = \frac{v_2^2}{h_2}$ and solved for v_2:

(b) $\frac{v_1^2(h_2)}{h_1} = v_2^2$

(c) $v_2 = \sqrt{\frac{v_1^2(h_2)}{h_1}}$

Taking the last as a base formula, we can substitute the values in folio 116v to see the calculations, using the constants of $h_1 = 300$ *punti* (*p*) and $v_1 = 800$ *p*. The first test, from a drop of $h_2 = 600$ *p*, should be $v_2 = \sqrt{\frac{800^2(600)}{300}}$, or $v_2 = 1131.37$, which Galileo rounded to 1131 *p*. This was his *doveria* number, what the distance "should have been" in a frictionless universe. His actual measurement was 1172 *p*, a difference, he noted, of 41 *p*. Likewise for the second test, from a drop of $h_2 = 800$ *p*: $v_2 = \sqrt{\frac{800^2(800)}{300}}$, or $v_2 = 1306.39$. This predicted result (rounded) differed from the test (1328 *p*) by 22 *p*, which he entered on his worksheet. The same calculations were made with the other tests, with the differences between the ideal and the actual occurrence indicated in each case.

A.14 Major Interpretations of the Galileo Affair

The Warfare Thesis. The trial was a major episode in the long-standing warfare between science and religion. The Roman Catholic Church condemned a free-thinking scientist, demanded a recantation of his claims, and declared certain scientific propositions as heresies, the holding of which would be punishable by the Inquisition.[2]

The Hubris Hypothesis. The Church was willing to adjust its views and, if necessary, to reinterpret passages of scripture when they conflicted with the science of the day in order to accommodate the scientists' understanding of nature, once a "true demonstration" of their positions had been formulated. But Galileo had not met the standard and with ambiguous proofs went too far in claiming more than the evidence would warrant.[3]

The Godfather Scenario. Because of his aggressive, abrasive, and arrogant personality, Galileo had created numerous personal and political enemies within the Church, who utilized the available mechanisms of the Inquisition to exact their revenge when the opportunity presented itself, "an intrigue engineered by a group of obscure and disparate characters in strange collusion."[4]

The Courtier's Demise. Galileo's rise and fall mirrored that of other typical courtiers who acquired, placated, and then alienated their patrons and supporters.

The dénouement of this particular drama followed the pope's personal conviction that with the *Dialogue* Galileo "had betrayed" Urban's expectations and "contravened the promises made."[5]

The following accounts generally appeal to a broader historical context.

The Geopolitical Web. Galileo's case was ensnared in a wider political struggle within the Church as the papacy tried to address geopolitical issues pertaining to the (now-termed) Thirty Years' War, struggles involving the Spanish Hapsburgs, their influences at the papal court, the French Bourbon dynasty, Bavarians, and a host of others, including the Protestant Swedes, with whom the papacy had formed a secret alliance. In the turmoil and crossfire Galileo lost his patronage from key figures in the Vatican and for political reasons from Urban VIII himself.[6]

The Scholastic Shuffle. Various Church officials (especially Jesuits and Dominicans) objected not to "science" writ large but rather to Galileo's attack on "Aristotelian science." Such natural philosophy was woven into the broader fabric of scholasticism, which the Church had reviewed and reaffirmed at the Council of Trent (1545–1563), and to whose defense it remained largely committed. Centrally, for example, its own doctrine of the Eucharist, transubstantiation, relied on the long-held Aristotelian distinction between substance and appearance, or essence and accident. Abandon Aristotle (or, more to the point, Thomas Aquinas), went the argument, and the Eucharist goes, and with it an entire host of theological doctrines embodied in the authority of Rome.[7]

The Cast of the Die (Bad Timing, Bad Luck). The Roman Catholic Church found itself in a major retrenchment mode because of the religious controversies of the Reformation, pursued an increasingly literalist strategy in scriptural exegesis, and was thus incapable of being as tolerant in 1616 and again in 1633 as it has been in other, both pre- and post-Tridentine, ages. This made Galileo "the victim of extraordinarily bad luck in the timing of the Copernican debate."[8]

Illustration Credits

Note: Every effort has been made to identify copyright holders and obtain their permission for the use of copyrighted material. Notification of any additions or corrections that should be incorporated in future reprints or editions of this book would be greatly appreciated.

Figure 1.1: Evolution of the alphabet. Table creation and permission courtesy Orly Goldwasser, Professor of Egyptology, Hebrew University. All rights reserved.

Figure 2.1: The Aristotelian cosmos. Courtesy the Smithsonian Libraries.

Figure 3.1: Greek figurate numbers. Illustration by Sally Sheedy.

Figure 4.1a: Gerbert's equilateral triangle. Courtesy Lawrence J. Schoenberg Collection, Kislak Center for Special Collections, Rare Books and Manuscripts.

Figure 4.1b: Gerbert's equilateral triangle, reconstruction. Diagram by Sally Sheedy.

Figure 4.2a: Oresme, latitude of qualities. From Oresme's *Treatise on the Configuration of Qualities;* adaptation by Sally Sheedy.

Figure 4.2b: Oresme, mean speed theorem. From Oresme's *Treatise on the Configuration of Qualities.*

Figure 4.3: Oresme's geometric proof. Diagram by Sally Sheedy.

Figure 5.1: The evolution of Hindu-Arabic numerals. "Numerals, a time travel from India to Europe," © 2003–2017 by Gianni A. Sarcone, giannisarcone.com. All rights reserved.

Figure 5.2: Finger counting, or dactylonomy. Photo credit: NYPL / Science Source.

Figure 6.1: Neumes, written in *campo aperto.* Public domain, Wikimedia Commons.

Figure 6.2: Gregorian chant, square-note notation. Public domain, Wikimedia Commons; adaptation by Sally Sheedy.

Figure, p. 169: Line A, musical notation. Mensural notation by Sally Sheedy.

Figure, p. 169: Lines A and B, musical notation. Mensural notation by Sally Sheedy.

Figure 6.3: The harmonic series. Illustration by Sally Sheedy.

Figure 7.1: Leonardo da Vinci, *Vitruvian Man.* Photo credit: Scala / Art Resource, NY.

Figure 7.2a: Donatello, *Saint Mark,* original. Photo credit: Scala / Ministero per i Beni e le Attività culturali / Art Resource, NY

Figure 7.2b: Donatello, *Saint Mark,* copy in alcove. Photo courtesy John H. Ludas.

Figure 7.3: Brunelleschi's mirror experiment. Illustration by Sally Sheedy, adapted from "Brunelleschi's Perspective," by Jim Anderson.

Figure 7.4: Alberti's perspective construction. Illustration by Sally Sheedy, adapted from "The Westologist: Cultural Insight," website by Pierre Assier, https://thewestologist.wordpress.com/author/passier/page/2/.

Figure 7.5a: Holbein the Younger, *The Ambassadors* (1533). © The National Gallery, London. Bought 1890.

Figure 7.5b: Skull detail from *The Ambassadors.* © The National Gallery, London. Bought 1890.

Figure 7.6: Uccello, perspective chalice (1450). Photo credit: Gallerie degli Uffizi, Gabinetto Fotografico.

Figure 7.7: Perspective diagram. From Piero della Francesca, *De prospectiva pingendi;* adaptation by Sally Sheedy.

Figure 7.8a–c: Sketches of a head. From Piero della Francesca, *De prospectiva pingendi,* Reggio-Emilia copy; photos and permission, Biblioteca Panizzi, Reggio-Emilia. All rights reserved.

Figure 8.1: Verge and foliot escapement. Reprinted with permission from Encyclopædia Britannica, © 2007 by Encyclopædia Britannica, Inc.

Figure III.1: Galileo's geometric and military compass. Photograph © 2010 Museo Galileo, Florence—Photographic Archives.

Figure 9.1a: Folio 107v. Permission, Ministero dei beni e delle attività culturali e del turismo / Biblioteca Nazionale Centrale, Firenze. All rights reserved.

Figure 9.1b: Numbers from upper left corner of folio 107v. Permission, Ministero dei beni e delle attività culturali e del turismo / Biblioteca Nazionale Centrale, Firenze. All rights reserved.

Figure 9.2a: Numbers from folio 107v. Permission, Ministero dei beni e delle attività culturali e del turismo / Biblioteca Nazionale Centrale, Firenze. All rights reserved.

Figure 9.2b: Numbers from folio 107v. Permission, Ministero dei beni e delle attività culturali e del turismo / Biblioteca Nazionale Centrale, Firenze. All rights reserved.

Figure 9.3: Calculation from folio 154v. Permission, Ministero dei beni e delle attività culturali e del turismo / Biblioteca Nazionale Centrale, Firenze. All rights reserved.

Figure 9.4: Tartaglia's depiction of projectile motion. Courtesy Max Planck Institute for the History of Science, Berlin.

Figure 9.5a: From folio 81r. Permission, Ministero dei beni e delle attività culturali e del turismo / Biblioteca Nazionale Centrale, Firenze. All rights reserved.

Figure 9.5b: From folio 114v. Permission, Ministero dei beni e delle attività culturali e del turismo / Biblioteca Nazionale Centrale, Firenze. All rights reserved.

Figure 9.5c: From folio 116v. Permission, Ministero dei beni e delle attività culturali e del turismo / Biblioteca Nazionale Centrale, Firenze. All rights reserved.

Figure 9.5d: From folio 117r. Permission, Ministero dei beni e delle attività culturali e del turismo / Biblioteca Nazionale Centrale, Firenze. All rights reserved.

Figure 9.6: Schematic drawing from folio 116v. Adaptation from folio 116v by Sally Sheedy; permission, Ministero dei beni e delle attività culturali e del turismo / Biblioteca Nazionale Centrale, Firenze. All rights reserved.

Figure 9.7: From folio 117r. Permission, Ministero dei beni e delle attività culturali e del turismo / Biblioteca Nazionale Centrale, Firenze. All rights reserved.

Figure 10.1: Galileo's drawing of an ebony chip floating in water. From Galileo, *Discourse on Floating Bodies* (1663); adaptation by Sally Sheedy.

Figure 10.2: Aristotle's wheel. Illustration by Sally Sheedy.

Figure 10.3: The paradox of the wheel. From Galileo, *Two New Sciences;* adaptation by Sally Sheedy.

Figure 10.4: Galileo's analysis of the wheel paradox. From Galileo, *Two New Sciences;* adaptation by Sally Sheedy.

Figure 11.1: Galileo's diagram of sunspots on the sun's surface. Courtesy Library of Congress.

Figure, p. 330 (Appendix A.4) : Musical notation—mensural. Mensural notation by Sally Sheedy.

Figure, p. 331: Pythagorean scale intervals. Rendering by Sally Sheedy.

Figure A.7.1: The celestial sphere. Illustration by Sally Sheedy.

Figure A.7.2: The zodiac. Illustration by Sally Sheedy.

Figure A.8.1: The precession of the equinoxes. Illustration by Sally Sheedy.

Figure A.9.1a: Chaucer's astrolabe. Photograph © Trustees of the British Museum.

Figure A.9.1b: The astrolabe's parts. Photograph iStockphoto.com; adaptation by Sally Sheedy.

Figure, p. 338: Mean proportional diagram. Diagram by Sally Sheedy.

Figure, note 14, p. 412: Cuneiform numerals. Public domain, Wikimedia Commons; adaptation by Sally Sheedy.

Figure, note 37, p. 423: Franco's principal notes. Mensural notation by Sally Sheedy.

Figure, note 30, p. 434: Euclid's visual cone. Diagram by Sally Sheedy.

Figure, note 9, p. 456: Diagram of compass operation. Diagram by Sally Sheedy.

Notes

Introduction

1. Peter Harrison narrates the evolution of the concepts 'science' and 'religion' in *The Territories of Science and Religion* (Chicago: University of Chicago Press, 2015), 1–54 and throughout.

2. There has emerged in recent decades an industrial-strength body of both popular and scholarly works seeking to define science and religion and the encounter between them. Among them, for difficulties in "defining science," see, for example, John Lennox, *God's Undertaker: Has Science Buried God?* (Oxford: Lion Books, 2007), 31–46. For a hostile look at religion and the "thankless task" of defining it (which leads him to rely on the *Oxford English Dictionary*), see Jerry A. Coyne, *Faith vs. Fact: Why Science and Religion Are Incompatible* (New York: Viking, 2015), 41–63, and throughout. In their efforts to delineate science and religion, and to argue for and against their compatibility, Lennox and Coyne make manifest most clearly the contemporary rift between the two. Other examples abound. Among well-known, contemporary scientists, mathematicians, and philosophers, for instance, a number of avowed atheists (for example, Coyne, Richard Dawkins, Steven Weinberg, Lawrence Krauss, and Daniel Dennett) engage in continual jousting with those who retain a theistic commitment to religion (for example, John Polkinghorne, Kenneth R. Miller, Lennox, Russell Stannard, and Antony Flew) in tournaments of scholarly and scientific argument, philosophical reflection, and polemic.

3. Houston Smith, *Why Religion Matters: The Fate of the Human Spirit in an Age of Disbelief* (New York: HarperCollins, 2001), xiv, 1–5, 19–20, 160–161, 214, and throughout. Although he finds some solace in the optimism and hope of counterculture New Agers and praises their refusal to acquiesce in scientism, Smith also finds their thinking "conceptually . . . pretty much a mess," for their "enthusiasms jostle one another promiscuously" (161). See too *The Huston Smith Reader,* ed. Jeffery Paine (Berkeley: University of California Press, 2012), 151–218, for Smith's extended account of "the Big Picture."

4. For Smith, these include primarily Hinduism, Buddhism, Confucianism, Taoism, Islam, Judaism, and Christianity. See Huston Smith, *The World's Religions,* 50th anniversary ed. (1958; New York: HarperCollins, 1991). In a similar vein, philosopher Thomas Nagel finds behind the different mainstream religions a feature they hold in "common"—namely, "the idea that there is some kind of all-encompassing mind or spiritual principle . . . and that this mind or spirit is the foundation of the existence of the universe, of the natural order, of value, and of our existence, nature, and purpose." Thomas Nagel, *Secular Philosophy and the Religious Temperament: Essays, 2002–2008* (Oxford: Oxford University Press, 2010), 4–5 and more generally his opening essay, 3–17.

5. Smith, *Why Religion Matters,* 214, ital. original. Some current statistics seem to belie Smith's observations: roughly 90 percent of Americans profess a belief in God or some sort of religious persuasion; over 70 percent of Americans claim to be Christians of one denomination or another; in the first decade of the twenty-first century, Americans bought more religious books than they did books on science (including textbooks); and, globally, some two-thirds of the world's population counts itself among the adherents of one of the traditional religions. The United States appears to be more religious than Europe, with over 50 percent of Americans claiming that religion is "very important" to them, while in most European nations the percentage is far smaller (for examples: France, 13 percent; Britain, 17 percent; Germany, 21 percent, Spain, 22 percent). Pew Research Center, "The American-Western European Values Gap, Survey Report," February 29, 2012, http://www.pewglobal.org/2011/11/17/the-american-western-european-values-gap/. Gallup and Pew Forum surveys also reveal that the number of active adherents in nonindustrialized regions of the globe far exceeds that of western Europe and even the United States. Overall, according to a 2012 WIN-Gallup Survey, 59 percent of the world's people deem themselves religious (WIN-Gallup International, "Global Index of Religiosity and Atheism, 2012," http://www.wingia.com/web/files/news/14/file/14.pdf), while a 2012 Pew Research Project reports only 16.3 percent of them as "unaffiliated" (Pew Research Center, "The Global Religious Landscape," December 18, 2012, http://www.pewforum.org/2012/12/18/global-religious-landscape-exec/). Comparably, a Spiegel Online "Atlas of Faiths" sets the net "adherents" figure at roughly two-thirds of global population. "Atlas of Faiths," Spiegel Online [January 22, 2007 date of posting], http://www.spiegel.de/img/0,1020,775969,00.gif. All such data as the foregoing should be spoon-fed with a good dose of skepticism. Nevertheless, the overall conclusion seems to be that the majority of folks in the world are believers and participants in one of the traditional religions, which makes one wonder about Smith's "spiritual crisis."

6. For example, see his early book, Steven Weinberg, *The First Three Minutes: A Modern View of the Origin of the Universe* (New York: Basic Books, 1977). Weinberg won the Nobel Prize for his work unifying weak and electromagnetic inter-

actions. On atheism, see his article, "Without God?," *New York Review of Books,* September 25, 2008.

7. Many of Weinberg's essays have been collected in Steven Weinberg, *Facing Up: Science and Its Cultural Adversaries* (Cambridge, MA: Harvard University Press, 2001); for the quotations, see 242, 234. See too his recent collection, *Lake Views: This World and the Universe* (Cambridge, MA: Harvard University Press, 2009), esp. 6–27, 210–217.

8. For many believers, Weinberg's reliance on "explanation" misses the point of religion entirely. Matters of faith and morals, the traditional markers of religion, fall instead within a narrative account of human experience, which does not "explain" so much as situate one in a world to which is ascribed a special kind of significance or meaning. For Christians, the story is of sin and redemption; for Muslims, peaceful submission to the will of God (Allah); for Zoroastrians, the eternal struggle between good and evil; for Buddhists, the path to enlightenment (nirvana). This is not "explanation" in the ordinary (that is, modern, scientifically derived) sense of the term, for which see Peter Achinstein's influential work *The Nature of Explanation* (Oxford: Oxford University Press, 1983). The "why" question is sui generis, asked and answered. A loving God whose wisdom transcends mere mortals bestows the why or meaning of existence in the mere fact of existence. Further explanation is neither needed nor part of the religious experience. Weinberg's placement of both science and religion in the category of "explanation" pits them together in a field circumscribed by scientific, epistemic parameters, one casting to the outside the narrative components central to religion.

9. Stephen Jay Gould, *Rocks of Ages: Science and Religion in the Fullness of Life* (New York: Ballantine Books, 1999), 3, 5–6.

10. Dean Hamer, *The God Gene: How Faith Is Hardwired into Our Genes* (New York: Anchor Books, 2004); John Calvin, *Institutes of the Christian Religion,* trans. Henry Beveridge (London, 1599), bk. 3, ch. 23, sec. 7. A discrepancy permeates even Gould's reasonable account. On the one hand, he insists that we always debate and discuss matters "under a magisterium." Gould, *Rocks of Ages,* 6. But on the other hand, that being so, how can we debate and have dialogues between the magisteria, between science and religion, as he does? The latter consideration, it seems, assumes a superior vantage point or magisterium, a "higher" level of engagement, some sort of Venn diagram that encircles them both.

11. Michael E. Hobart and Zachary S. Schiffman, *Information Ages: Literacy, Numeracy, and the Computer Revolution* (Baltimore: Johns Hopkins University Press, 1998).

12. Hans Christian von Baeyer, *Information: The New Language of Science* (2003; Cambridge, MA: Harvard University Press, 2005), 4. In his study, von Baeyer notes that "information is . . . a vague, ill-defined concept" (10), which he seeks to clarify and expand into an answer to one of the "really big questions" in science, especially

physics—namely, how is the material world constructed from information, or, how does one get "it from bit" (ix–xiv)? As will become readily apparent, I am using the term 'information' in a more general, qualitative sense with references to specific, historical ways of providing the "connection between matter and mind" (17). This is a study of intellectual and cultural history, not a work of science, in which 'information' tenders a means for describing how the human mind functions in its cognitive interactions with the outside world.

13. The context for Burke's distinction was a discussion of the rights of man, which he held were "in a sort of *middle* [between being "metaphysically true" and "morally and politically false"], incapable of definition, but not impossible to be discerned." Edmund Burke, *Reflections on the Revolution in France,* ed. Thomas H. D. Mahoney (New York: Bobbs-Merrill, 1955), 71.

14. Science writer James Gleick calls the alphabet the "founding technology of information." James Gleick, *The Information: A History, a Theory, a Flood* (New York: Vintage Books, 2011), 11. But the alphabet actually came later in the evolution of information and literacy, as we shall see in Chapter 1. Information per se was created in its earliest forms with the initial abstraction of writing from speech. Its antecedents lay in tokens, "emblem slotting," and accounting practices, all of which coalesced in the first writing script, proto-cuneiform, dating roughly from between 3200 and 3100 BCE. From there, literacy and information developed apace as pictographic, ideographic, and syllabic forms of writing increasingly elicited the classifying potential of speech and oral communication. Sometime around the first millennium BCE, the Greeks created the first phonetic alphabet, using the signs of Phoenician syllabic writing to capture phonetically the pronunciation of their own language, whose sounds bore no relation to Phoenician symbols. See Hobart and Schiffman, *Information Ages,* introduction, pt. 1, and the bibliographical essay, 279–294.

15. The striking image of "mind-forg'd manacles" comes to us via a different context, William Blake's response to human-created repression in "London," from his *Songs of Experience.*

16. Ann M. Blair, *Too Much to Know: Managing Scholarly Information before the Modern Age* (New Haven, CT: Yale University Press, 2010), 1–61, and throughout, chronicles the early modern information overload and the traditional classifying attempts to master it. See too Hobart and Schiffman, *Information Ages,* ch. 4.

17. Amos Funkenstein, *Theology and the Scientific Imagination: From the Middle Ages to the Seventeenth Century* (Princeton, NJ: Princeton University Press, 1986), 297.

18. Galileo Galilei, *The Assayer,* in *Discoveries and Opinions of Galileo,* trans. Stillman Drake (New York: Anchor Books, 1957), 237–238.

19. Jagjit Singh, *Great Ideas of Modern Mathematics: Their Nature and Use* (New York: Dover, 1959), 8. A note on terminology: throughout this book I shall refer to 'numeracy' in its broad, contemporary sense as the ability "to reason with num-

bers and other mathematical concepts," a wide-reaching description drawn from many sources and articulated by the United Kingdom's National Numeracy organization. It includes both arithmetic and geometry, as well the abilities to handle information and computations. For the general definition, see "What Is Numeracy?," National Numeracy, accessed October, 2017, https://www.nationalnumeracy.org.uk/what-numeracy; also, for a more detailed account of the recent history and use of the term, see W. Alex Neill, "The Essentials of Numeracy" (paper presented at the New Zealand Association of Researchers in Education Conference, Christchurch, New Zealand, December 6–9, 2001), http://www.nzcer.org.nz/system/files/10604.pdf. In "Counter Culture: Toward a History of Greek Numeracy," *History of Science* 40 (2002): 321–352, Reviel Netz looks at numeracy's "cognitive history" in its "cultural, political, and economic" context, especially as applied to ancient Greece. Additionally, I shall be developing the historical distinction between "thing numeracy" and "relational numeracy" as the information technologies underlying Singh's contrast between premodern "thing-mathematics" and modern "relation-mathematics," respectively.

20. Funkenstein, *Theology and the Scientific Imagination,* 34–35, writes that in the seventeenth century, "mathematics . . . was to change, in substance and ideal, from an inventory of ideal entities and their properties into a *language.*" He notes that the capstone of these developments was put into place by Gottfried Wilhelm Leibnitz and his "general science of relations" (*scientia generalis de relationibus*), "abstractions . . . indispensable for the ordering of phenomena" (316). Further, he incisively argues that the new language was the basis of a "constructive theory of knowledge," as opposed to the ancient and medieval "*contemplative* ideal," in which "knowledge or truth is found, not constructed" (298–299). Despite the richness of his account, however, he offers no explanation of the origins of the new language, nor of its underlying technology, the subject of the present narrative.

21. Going back at least to Immanuel Kant, philosophers, scientists, and scholars have described in various ways the idea of a mental screen between the knowing mind and the outside world. For a succinct contemporary statement, see, for example, Peter Dear, *Discipline and Experience: The Mathematical Way in the Scientific Revolution* (Chicago: University of Chicago Press, 1995), 13: "How we apprehend the world through experience depends on the ways in which we conceptually formulate experience. How we see things is strongly conditioned by the mental categories we bring to our perceptions." Also see Norwood Russell Hanson's modern classic, *Patterns of Discovery* (Cambridge: Cambridge University Press, 1958). The philosophical perspective of "linguistic relativism," which articulates the role of language in shaping consciousness, addresses many of the same issues. It derives its contemporary label from the names of two linguists, Edward Sapir and his student Benjamin Whorf, who first developed the Sapir-Whorf hypothesis. In the strong version, the hypothesis claims that the language a person uses determines how he or she perceives the world; in the weaker version

(favored here), 'influences' replaces 'determines.' For an illustration of how the hypothesis applies to a comparison of Western languages and Chinese, see Bryan W. Van Norden, *Introduction to Classical Chinese Philosophy* (Indianapolis: Hackett, 2011), 244–246. Chris Swoyer discusses lucidly the philosophical dimensions of Sapir and Whorf's claims in his article "Relativism," in *The Stanford Encyclopedia of Philosophy,* Winter 2010 ed., ed. Edward N. Zalta, http://plato.stanford.edu/archives/win2010/entries/relativism/. None of these perspectives addresses directly the present topic of how information technology frames experience by interposing itself between knower and known. Our concern here lies not with consciousness writ large but rather with how, as information technologies, literacy and numeracy set parameters to the scientific knowledge of nature.

22. We owe to Max Weber (and, before him, the Romantic poet, playwright, and philosopher Friedrich Schiller) the trenchant phrase "disenchantment of the world." For Weber it referred to "intellectualization and rationalization," the advance of scientific reasoning and progress that characterizes the "fate of our times" and in whose wake were swallowed those "mysterious incalculable forces" of yesteryear. Ever since Weber's original essay, the theme of disenchantment has figured centrally in discussions of "modernization" and "secularization," producing a huge and controversial literature that vastly exceeds present concerns. Here we shall restrict our focus to one dimension of disenchantment that accompanied the emerging information technology: the expulsion of intrinsic, qualitative content from mathematical symbols. See Max Weber, "Science as a Vocation," in *From Max Weber: Essays in Sociology,* trans. and ed. H. H. Gerth and C. Wright Mills (1922; New York: Oxford University Press, 1946), 129–156. For a magisterial treatment of disenchantment and secularization, many of whose central themes are complemented by the current narrative, see Charles Taylor, *A Secular Age* (Cambridge, MA: Harvard University Press, 2007), esp. 1–295.

23. See Neal Ward Gilbert, *Renaissance Concepts of Method* (New York: Columbia University Press, 1960), 3–115 and throughout. The writings of these three classical authors provided the textual arena of Renaissance debates. Aristotle's *Analytics* (both *Prior* and *Posterior*), plus his *Topics* and other works, proffered analyses of arguments, definitions, and demonstrations as regards both their form and content. Second only to Aristotle in influence, the well-known Greek physician-philosopher Galen (Claudius Galenus) had scattered throughout his medical treatises many descriptions of analysis and synthesis, which Renaissance thinkers often termed the "resolutive-compositive" method. Although known mostly as the guiding progenitor of medicine from the ancient world, Galen had also authored numerous works on logic and scientific "method," most of which were lost in antiquity. Finally, to a far lesser degree, the thirteen books of Euclid's *Elements* invoked analysis in the geometrical method of "linear demonstration," as it was formerly termed. As mathematics gained currency throughout the sixteenth century, Euclid's mode of analysis became increasingly prominent.

24. The locution comes from François Viète, *The Analytic Art,* trans. T. Richard Witmer (Kent OH: Kent State University Press, 1983). In the Renaissance, scholastics and humanists alike debated the nuances of key words such as *ars* (art), *techné* (skill, craft), *methodus* (method), *doctrina* ("habit of mind"), *inventio* ("methodological guidance"), and *via* (way, sometimes "shortcut")—fluid terms for fluid times. Their efforts in these debates generally attempted to classify the procedures that involved orderly habits of mind and invent new ones. See Gilbert, *Renaissance Concepts of Method,* 3–115, esp. 5, 12, 16, 42, 55, 65, 119. Viète's contribution to the debates helped galvanize a new understanding of the "whole analytic art" as the "science of correct discovery in mathematics." Viète, *Analytic Art,* 12. For a fine survey of "Viète's analytic program," see Michael Sean Mahoney, *The Mathematical Career of Pierre de Fermat, 1601–1665,* 2nd ed. (1973; Princeton, NJ: Princeton University Press, 1994), 26–48; see also Jacob Klein, *Greek Mathematical Thought and the Origin of Algebra,* trans. Eva Brann (1968; New York: Dover, 1992), 150–185.

25. Viète, *Analytic Art,* 32. Technically speaking, Viète used the symbol = not for "equality" but rather for "difference," or subtraction. Instead he designated equality rhetorically with the Latin *aequabitur.* The equals sign itself was a novelty introduced by Robert Recorde in his algebra, titled *The Whetstone of Witte* (London, 1557), where he also employed the + and − signs for the first time in English. After Recorde's introduction of = into the vocabulary, however, it did not appear in print again until 1618, some sixty-one years later. See the still timely and magisterial work by Florian Cajori, *A History of Mathematical Notations: Two Volumes Bound as One,* vol. 1, *Notations in Elementary Mathematics;* vol. 2, *Notations Mainly in Higher Mathematics* (LaSalle, IL: Open Court, 1928, 1929; New York: Dover, 1993), 164–167, 181–187, 297–307. Citations refer to the Dover edition.

26. Among mathematicians, in the terminology of the day, 'analysis' generally referred to algebraic analysis as launched by Viète; continued in the work of Fermat, Descartes, and others; and described by Jean Le Rond d'Alembert (1717–1783) in the eighteenth-century French *Encyclopédie* (entries titled "Analytique" and "Analyse," vol. 1). In the nineteenth century the term became more restricted, referring to developments in calculus, especially its being placed on the rigorous, logical foundation provided by Augustin-Louis Cauchy (1789–1857) and others. Our historical use of the term extends 'analysis' to a much broader sweep of developments in the history of science and scientific culture. As an example, see my earlier account of analysis and its role in knowledge organization: Michael E. Hobart, "The Analytical Vision and Organization of Knowledge in the *Encyclopédie,*" *Studies on Voltaire and the Eighteenth Century* 327 (1995): 153–181. See, too, Tobias Dantzig, *Number: The Language of Science: A Critical Survey Written for the Cultured Non-mathematician,* 4th ed. (1930; Garden City, NY: Doubleday, 1954).

27. Italian examples alone include Niccolò Tartaglia (ca. 1499–1557), Giovanni Battista Benedetti (1530–1590), Guidobaldo del Monte (1545–1607), Federico

Commandino (1509–1575), and Bernardino Baldi (1553–1617). In recent decades scholars have made great strides in situating Galileo ever more thoroughly in the context of sixteenth-century developments in mathematics, of burgeoning engineering practices, of Renaissance culture, and of the practical realities of university and courtly life, materials we shall be examining in due course.

28. See David Wootton, *The Invention of Science: A New History of the Scientific Revolution* (New York: HarperCollins, 2015). Wootton devotes an entire early chapter to "inventing discovery," concluding that "it was the idea of discovery that made the new science, and the new set of intellectual values that underpinned it, possible" (108, more generally 57–109). In this captivating study, Wootton (who also has written a fine biography of Galileo) challenges the current historical consensus that no overarching story or "master narrative" can be told about the Scientific Revolution or about the early modern relations between science and religion. Although here I am emphasizing developments in physics and mathematics over those in astronomy, the latter of which Wootton stresses, I have relied heavily on his overall interpretation, which is complemented by the present thesis: Galileo's contributions to the idea of discovery and especially its technical manifestations in analyzing the physics of motion were central to the invention of modern science and its eventual separation from religion.

29. "Decree of the Index (5 March 1616)," in Maurice A. Finocchiaro, ed., *The Galileo Affair: A Documentary History* (Berkeley: University of California Press, 1989), 149. Finocchiaro offers a thorough documentary account of the affair's essential features: Galileo's support of Copernicanism; his jousting with officials, clerics, and academics; his warning at the hands of Cardinal Bellarmine in 1616; his reemergence into public prominence in the 1620s after the elevation of Urban VIII to the papacy; the publication of the *Dialogue on the Two Chief World Systems;* his trial of 1633; and his conviction of "vehement suspicion of heresy," subjugation, and lifetime sentence of house arrest. We shall take up these matters in Chapter 11.

30. See Taylor, *Secular Age,* 12–13 and more generally 1–22. His opening chapter, "The Bulwarks of Belief" (25–89), provides an especially winsome introduction to the premodern intellectual world.

31. The following provide a noteworthy introduction to the prominence of early modern religion in bringing about science and modernity: Eugene M. Klaaren, *Religious Origins of Modern Science: Belief in Creation in Seventeenth-Century Thought* (Grand Rapids, MI: William B. Eerdmans, 1977); Michael Allen Gillespie, *The Theological Origins of Modernity* (Chicago: University of Chicago Press, 2008); Stephen Gaukroger, *The Emergence of a Scientific Culture: Science and the Shaping of Modernity, 1210–1685* (Oxford: Oxford University Press, 2006); Peter Harrison, *The Fall of Man and the Foundations of Science* (Cambridge: Cambridge University Press, 2007); James J. Bono, *The Word of God and the Languages of Man: Interpreting Nature in Early Modern Science and Medicine,* vol. 1, *Ficino to Descartes*

(Madison: University of Wisconsin Press, 1995); Funkenstein, *Theology and the Scientific Imagination;* and Taylor, *Secular Age*. For much of this literature, it helps to have recently looked up the word 'hermeneutics.'

32. Gaukroger, *Emergence of a Scientific Culture,* 77–83 and more generally 46–86.

33. Some historians refer to these challenges as the "theological crisis of late medieval thought." See, for example, Gillespie, *Theological Origins of Modernity,* 19–43.

34. Taylor, *Secular Age,* 90–99; see too Brian P. Copenhaver, "Astrology and Magic," in *The Cambridge History of Renaissance Philosophy,* ed. Charles B. Schmitt et al. (Cambridge: Cambridge University Press, 1988), 264–300.

35. Harrison, *Fall of Man,* 89–138.

36. Michel de Montaigne, "Apology for Raymond Sebond," in *The Complete Essays of Montaigne,* trans. and ed. Donald Frame (1957; Stanford, CA: Stanford University Press, 1965), 318–457; John Donne, "An Anatomy of the World: The First Anniversary," in *John Donne's Poetry,* ed. A. L. Clements (New York: W. W. Norton, 1966), 73. See the influential work of Richard H. Popkin, *The History of Scepticism from Erasmus to Spinoza,* rev. ed. (Berkeley: University of California Press, 1979).

37. Galileo Galilei, "Letter to the Grand Duchess Christina (1615)," in Finocchiaro, *Galileo Affair,* 92–94.

38. C. S. Lewis, *The Discarded Image: An Introduction to Medieval and Renaissance Literature* (Cambridge: Cambridge University Press, 1964). While Lewis focused primarily on literary evidence and themes, I am enlarging the scope of the "medieval image" and its eventual "discarding" to embrace as well intellectual attitudes and patterns of thought pertaining to science and religion, and their underlying information technologies. Harrison too has adopted the term to denote the disintegration of the "symbolic and causal orders" of the Middle Ages. See Harrison, *Territories of Science and Religion,* 74–78.

39. Some historians of science and religion insist that no "master narrative" can explain either the emergence of science or the interactions between the two. John Brooke and Geoffrey Cantor, for example, caution against projecting backward onto history "essentialist" definitions drawn from our own times, force-feeding information anachronistically into the predetermined categories ('science' and 'religion') of our narratives and squeezing "events into a preconceived mold." John Brooke and Geoffrey Cantor, *Reconstructing Nature: The Engagement of Science and Religion,* Glasgow Gifford Lectures (New York: Oxford University Press, 1998), 107, 109. See too John Hedley Brooke, "Religious Belief and the Natural Sciences: Mapping the Historical Landscape," in *Historiography and Modes of Interaction,* vol. 1 of *Facets of Faith and Science,* ed. Jitse van der Meer (Lanham, MD: Pascal Centre for Advanced Studies in Faith and Science / University Press of America, 1996), 1–26; John Hedley Brooke, *Science and Religion: Some Historical Perspectives* (Cambridge: Cambridge University Press, 1991), esp. 1–51; David C.

Lindberg and Ronald L. Numbers, eds., *God and Nature: Historical Essays on the Encounter between Christianity and Science* (Berkeley: University of California Press, 1986), 1–18; and Gary B. Ferngren, ed., *Science and Religion: A Historical Introduction* (Baltimore: Johns Hopkins University Press, 2002), ix–xiv, 3–29. One scholar, David Wilson, has even suggested that we forsake the terms 'science' and 'religion' altogether in order to write about science and religion because these modern terms are suffused with "several" different meanings and are misleading "when applied to past thought." David Wilson, "On the Importance of Eliminating 'Science' and 'Religion' from the History of Science and Religion: The Cases of Oliver Lodge, J. H. Jeans, and A. S. Eddington," in van der Meer, *Historiography and Modes of Interaction,* 27. Well, nothing like committing suicide to eliminate the fear of death. We might just as nonsensically forgo using the term 'atom' in reference to Greek atomists who coined it because our term means something quite different from theirs. Still, these warnings from serious scholars need to be taken seriously. No one would gainsay that history is complex, that nineteenth-century "warfare between science and religion" readings have been superseded, and that many detailed and local histories published in recent years are rich and rewarding to read and instructive to ponder. But that history—the past—is bound by context doesn't mean that history—the account—need (or can) avoid our own words. To do so would mean abandoning all interpretation and consigning written history to a matter of reading and copying documents, the labor of scriveners. We create narratives, even master narratives, with the language we have. And, indeed, the insights and language of our own day, such as recognizing the place of information technology in our thought and lives, can and should open new doors onto the past.

Religio and *Scientia*

1. For Catharism as a heresy and the role of the newly created mendicant orders (Franciscans and Dominicans) in combatting it, see Roger French and Andrew Cunningham, *Before Science: The Invention of the Friars' Natural Philosophy* (Brookfield, VT: Ashgate, 1996), 99–173, 202–230, and throughout. In their intellectual offensives against the "Manichean" heresies of the Cathars, both orders enlisted reason and the study of nature's goodness as a manifestation of God's creation, although in different fashions. While the Franciscans emphasized the harmony of science and religion through the furthering of an "inner spiritual life" (209) and comparable themes of Neoplatonism that accompanied the study of nature, the Dominicans, as we shall explore, relied more explicitly on an Aristotelian approach to understanding the natural world and its implications for theology.

2. Saint Bonaventura, *The Life of St. Francis* [ca. 1263], trans. Reverend Fathers of the College of Bonaventura (London: J. M. Dent, 1904), 110–112.

3. In this instance "little flowers" makes no specific reference to the saintly beauty and innocence of Francis. *Fioretti* is the Italian translation of the Latin *florilegium,* which in the Middle Ages referred to a compilation of excerpts from other writings. The word derives from the Latin *flos* (flower) and *legere* (to gather)—literally a gathering of flowers—and in turn was adapted from the Greek *anthologia,* of comparable meaning and from which our own 'anthology.'

4. Pope John Paul II declared Francis the patron saint of ecology in 1979.

5. In his modern, meditative immersion into the medieval spirit and mind, Henry Adams noted that Francis, "the ideal mystic saint," had fulminated against the science of his contemporaries. He did this, however, not because of its methods or content but because of the conceit exhibited by its scholastic proponents. Francis sought true poverty (*domina nostra paupertas*), which meant primarily the "poverty of pride," the emptying of the self into the fulsomeness of the divine. See Henry Adams, *Mont-Saint Michel & Chartres* (1905; New York: Anchor Books, 1959), 370–384. To be sure, contradictions tend to be swallowed by mystical visions. Still, one wonders whether the pride manifest in fulminating against pride doesn't somehow defeat the purpose, sort of oxymoronically, like a self-conscious humility. These matters are probably best left to theologians.

6. For a worthwhile and sympathetic treatment of Francis in the context of medieval Catholic reform, see Lawrence S. Cunningham, *Francis of Assisi: Performing the Gospel Life* (Grand Rapids, MI: William B. Eerdmans, 2004), esp. chaps. 1 and 2.

7. See, for example, John of Salisbury, *The Metalogicon of John of Salisbury: A Twelfth-Century Defense of the Verbal and Logical Arts of the Trivium,* trans. Daniel D. McGarry (1955; Berkeley: University of California Press, 1962), bk. 4, ch. 17, 228–229: "Nothing that agrees with reason is out of harmony with God's plan. In obedience to the Divine mind, one will move through his allotted span of life making happy progress."

8. Peter Harrison, *The Territories of Science and Religion* (Chicago: University of Chicago Press, 2015), 7–19. Jerome had translated the Greek *thrēskeia* in James 1:27 into the Latin *religio,* which referred less to a body of beliefs than to a form of worship. *Scientia* derived from the Latin *scire,* which meant "to know" and after Augustine was often used interchangeably with "knowledge." *Sapientia* (from which our 'sapience') was Cicero's Latin for the Greek *sophia,* or "wisdom." Saint Augustine, *De Trinitate,* bk. 12, sec. 15, 25, cited in Eileen Serene, "Demonstrative Science," in *The Cambridge History of Later Medieval Philosophy: From the Rediscovery of Aristotle to the Disintegration of Scholasticism, 1100–1600,* ed. Norman Kretzmann, Anthony Kenny, and Jan Pinborg (Cambridge: Cambridge University Press, 1982), 499–500. See too the remark of Honorius of Autun (1080–1154): "*Scientia* deals with earthly matters, while *sapientia* deals with divine matters," cited in French and Cunningham, *Before Science,* 56.

9. See as well John of Salisbury, *Metalogicon,* bk. 4, ch. 13, 222, and bk. 4, ch. 19, 231. Technically, for Aristotle and his followers, *scientia* involved the search for

the middle terms of syllogisms, which allowed one to establish the causal demonstrations that explained empirical phenomena. See "*Scientia*," in *Medieval Science, Technology, and Medicine: An Encyclopedia,* ed. Thomas F. Glick, Steven J. Livesey, and Faith Wallis, Routledge Encyclopedias of the Middle Ages, vol. 11 (New York: Routledge / Taylor and Francis, 2005), 455–458. Also see Aristotle, *Metaphysics,* 1025b5–30, in *The Complete Works of Aristotle: The Revised Oxford Translation,* 2 vols., ed. Jonathan Barnes (Princeton, NJ: Princeton University Press, 1984): "Science . . . involves reasoning . . . with causes and principles" and all "sciences mark off some particular being—some genus, and inquire into this." Hereafter all Aristotle citations are from this edition. The pagination refers to Immanuel Bekker's standard scholarly Greek edition of Aristotle, first published in 1831, with the references being to the page, column, and line of the text. Modern scholars still employ these locator references when writing about Aristotle.

10. Harrison, *Territories of Science and Religion,* 7–19; Thomas Aquinas, *Summa Theologica,* pt. 2, qu. 57, art. 2, in *Basic Writings of Saint Thomas Aquinas (The Summa Theologica; The Summa Contra Gentiles),* 2 vols., English Dominican trans., trans. Laurence Shapcote, ed. Anton C. Pegis (New York: Random House, 1945), 2:431–432; Thomas Aquinas, *Commentary on the Posterior Analytics of Aristotle,* trans. Fabian R. Larcher, reedited and HTML formatted by Joseph Kenny, bk. 2, lec. 1, http://dhspriory.org/thomas/english/PostAnalytica.htm#201. Most of Aquinas's texts may be found in English translation online at http://dhspriory.org/thomas/. See too Tobias Hoffman, "The Intellectual Virtues," in *The Oxford Handbook of Aquinas,* ed. Brian Davies and Eleonore Stump (Oxford: Oxford University Press, 2012), 327–336.

11. Aquinas, *Summa Theologica,* pt. 2, qu. 52, art. 2, in *Basic Writings,* 2:397–398. Conversely, for Aquinas, and another contrast to our present day, "carnal vices, namely gluttony and lust, . . . cause man's attention to be very firmly fixed on corporeal things, so that in consequence man's operation in regard to intelligible things is weakened." That is, too much food and sex dull the mind. Thomas Aquinas, *Summa Theologica,* pt. 2 of pt. 2, qu. 15, art. 3, Benziger Bros. ed. (1947), trans. Fathers of the English Dominican Province, http://dhspriory.org/thomas/english/summa/SS/SS015.html#SSQ15OUTP1.

12. Joseph Wawrykow, "The Theological Virtues," in Davies and Stump, *Oxford Handbook of Aquinas,* 287–307.

13. Amos Funkenstein contrasts the medieval, contemplative ideal of knowledge and its inner connections to the divine with that of what he terms "ergetic knowledge," or "knowledge by doing," which marked a later, seventeenth-century "new ideal of knowing," one that assumed a greater separation not only of knower from known but also of humans from God. Amos Funkenstein, *Theology and the Scientific Imagination: From the Middle Ages to the Seventeenth Century* (Princeton, NJ: Princeton University Press, 1986), 290–299.

14. Aquinas, *Summa Theologica,* pt. 1, qu. 14, art. 1, and pt. 1, qu. 44, art. 1, in *Basic Writings,* 1:136, 427.

15. The smithing imagery is medieval: "After he [Aristotle] had procured the instruments of invention and mastered their use, he set himself . . . to the forge and worked away at hammering out a crucible to serve in his scientific analysis of reasoning." John of Salisbury, *Metalogicon,* bk. 4, ch. 1, 204. For Dante, see Dante, *The Inferno,* trans. John Ciardi (New York: New American Library, 1982), canto 4, line 131.

16. Plato's works too were mostly lost to western Europe until the Renaissance, but many of his ideas survived and were handed down over the centuries, filtered through the writings of authors influenced by Neoplatonism. An exception to the absence of translations from Greek was supplied by John Scotus Eriugena (ca. 800–ca. 877), who rendered from Greek into Latin the works of Pseudo-Dionysius, as well as several other writings from the Greek Christian theological tradition.

17. Toby E. Huff, *The Rise of Early Modern Science: Islam, China, and the West,* 2nd ed. (Cambridge: Cambridge University Press, 2003), 63.

18. Ibid., 179–189. See also French and Cunningham, *Before Science,* 33–69; Edward Grant, *The Foundations of Modern Science in the Middle Ages: Their Religious, Institutional, and Intellectual Contexts* (Cambridge: Cambridge University Press, 1996), 33–53; Pearl Kibre and Nancy G. Siraisi, "The Institutional Setting: The Universities," in *Science in the Middle Ages,* ed. David C. Lindberg (Chicago: University of Chicago Press, 1978), 120–144; "Universities," in Glick, Livesey, and Wallis, *Medieval Science,* 495–499; Nancy G. Siraisi, *Medieval and Early Renaissance Medicine: An Introduction to Knowledge and Practice* (Chicago: University of Chicago Press, 1990), 55–64; and Norman Zacour, *An Introduction to Medieval Institutions,* 2nd ed. (New York: St. Martin's, 1976), 196–219. For a general overview of the twelfth century, see the still valuable and highly readable work of Charles Homer Haskins, *The Renaissance of the Twelfth Century* (1927; Cambridge, MA: Harvard University Press, 1955).

19. Initially, the term *universitas* meant literally a self-governing corporation, a legal entity with its own governing structure and various rights and privileges, which applied to merchant or craft guilds of different types. Early on, faculties themselves (of arts, theology, law, or medicine) were called universities, while the *studium generale* comprised a loose association of different faculties and stood for what we would today term a university. The *studium generale* gained legal status in the thirteenth century, and toward the end of Middle Ages the term "university" began to replace *studium generale* in common parlance.

20. "Universities," 497.

21. Grant, *Foundations,* 23 and, more generally, 18–32. For the diffusion of Greek and Arabic works in western Europe, see also Edward Grant, *A History of Natural Philosophy: From the Ancient World to the Nineteenth Century*

(Cambridge: Cambridge University Press, 2007), 130–142; David C. Lindberg, "The Transmission of Greek and Arabic Learning to the West," in Lindberg, *Science in the Middle Ages,* 52–90; James Hannam, *The Genesis of Science: How the Christian Middle Ages Launched the Scientific Revolution* (Washington, DC: Regnery, 2011), 61–66; Thomas Goldstein, *Dawn of Modern Science: From the Ancient Greeks to the Renaissance* (New York: Da Capo, 1995), 92–129; and Howard R. Turner, *Science in Medieval Islam: An Illustrated Introduction* (Austin: University of Texas Press, 1995), 201–216.

22. See Grant, *Foundations,* 18–32, and Lindberg, "Transmission." For a list of Gerard's translations, along with a eulogy compiled by his students, see "A List of Translations Made from Arabic into Latin in the Twelfth Century," trans. Michael McVaugh, in *A Source Book in Medieval Science,* ed. Edward Grant (Cambridge, MA: Harvard University Press, 1974), 35–38. In his article on the reception of Aristotle in medieval Europe, Bernard G. Dod includes a useful tabulation of Latin translations of Aristotle's works, along with Greek and Arabic commentaries. See Bernard G. Dod, "Aristoteles latinus," in Kretzmann, Kenny, and Pinborg, *Cambridge History of Later Medieval Philosophy,* 45–79.

23. David C. Lindberg, *The Beginnings of Western Science: The European Scientific Tradition in Philosophical, Religious, and Institutional Context, 600 B.C. to A.D. 1450* (Chicago: University of Chicago Press, 1992), 218–240; Grant, *History of Natural Philosophy,* 68–94.

24. By midcentury a dispute had emerged between the more radical Aristotelians, Averroists, among whom Siger of Brabant (ca. 1240–1280s) and Boethius of Dacia (d. after 1283), and the more conservative, Aristotelian-rooted theologians, including Thomas Aquinas, over the correct interpretation and uses of Aristotle's works, especially in their theological implications—for example, on the issues of the earth's eternity and the soul's survival after death. The conflict between the two parties led to the Church's intervention with the "Condemnation of 1277," an inquisitorial declaration condemning 219 propositions, many dealing with Aristotelian natural philosophy. Included in this syllabus of errors were some twenty propositions held by Aquinas himself; later on, adjustments were made in time for Aquinas's elevation to sainthood in 1323. See Edward Grant, "Science and Theology in the Middle Ages," in *God and Nature: Historical Essays on the Encounter between Christianity and Science,* ed. David C. Lindberg and Ronald L. Numbers (Berkeley: University of California Press, 1986), 49–75; Stephen Gaukroger, *The Emergence of a Scientific Culture: Science and the Shaping of Modernity, 1210–1685* (Oxford: Oxford University Press, 2006), 47–49, 59–77; and Hans Thijssen, "Condemnation of 1277," in *The Stanford Encyclopedia of Philosophy,* Fall 2008 ed., ed. Edward N. Zalta, http://plato.stanford.edu/archives/fall2008/entries/condem nation/.

25. Earlier prohibited at the University of Paris (1210), Aristotle's writings became officially part of the arts curriculum there in 1255, with cautions that the

masters of arts teaching them "shall be required to finish the texts which they shall have begun." Then as now, Aristotle was not an easy read, or listen. See the statute "The Natural Books of Aristotle in the Arts Curriculum at the University of Paris, 1255," trans. Lynn Thorndike, in Grant, *Source Book in Medieval Science,* 43–44.

26. See Grant's fine chapter "What the Middle Ages Did with Its Aristotelian Legacy," in *Foundations,* 86–126; and James Doig, "Aquinas and Aristotle," in Davies and Stump, *Oxford Handbook of Aquinas,* 33–44.

27. *Fides quaerens intellectum* was the motto of Saint Anselm (1033–1109). See Adams, *Mont-Saint Michel & Chartres,* 387, for the connections between the "Church Intellectual" and the "Church Administrative . . . , both expressing—and expressed by—the Church Architectural."

28. Charles Taylor reminds us to bear in mind that various debates and disputes in the High Middle Ages should "be understood against the background of the harmonious order." Charles Taylor, *A Secular Age* (Cambridge, MA: Harvard University Press, 2007), 92. In speaking of a presumptive "harmonious ideal" or order, we are invoking the terms somewhat loosely in the Weberian sense of an "ideal type" or of a Wittgensteinian "family resemblance." Both notions capture the image of a cluster of ideas held cohesively by a community, even though each member of the community may not subscribe to all the particulars of the ideal or family resemblance. Certainly, not every scholastic thinker endorsed the harmonious ideal in the same manner. Nonetheless, when we read their works and compare them with those of later intellectuals, say, of the seventeenth and eighteenth centuries or, more noticeably, of our own day, the thrust of scholastic thought lay with integrating science and religion, nature and God, in ways that sharply contrast with the thought of later ages. For a brief description of Weber's "ideal type," see H. H. Gerth and C. Wright Mills, introduction to *From Max Weber: Essays in Sociology,* trans. and ed. H. H. Gerth and C. Wright Mills (1922; New York: Oxford University Press, 1946), 57–61. Wittgenstein's notion of "family resemblance" may be found in Ludwig Wittgenstein, *Philosophical Investigations,* rev. 4th ed., trans. G. E. M. Anscombe, P. M. S. Hacker, and Joachim Schulte, ed. P. M. S. Hacker and Joachim Schulte (1953; Oxford: Blackwell, 2009), secs. 67–77.

29. Thus Aquinas argued that error arises when definitions are either malformed "of mutually repugnant parts" or misapplied, as when joining the definition of a "circle to a man." Aquinas, *Summa Theologica,* pt. 1, qu. 17, art. 3, in *Basic Writings,* 1:184.

30. Ibid., pt. 1, qu. 85, art. 1, in *Basic Writings,* 1:813. Also see pt. 2 of pt. 2, qu. 15, art. 3, Benziger Bros. ed., http://dhspriory.org/thomas/english/summa/SS/SS015.html#SSQ15A3THEP1.

31. Ibid., pt. 1, qu. 2, art. 2, in *Basic Writings,* 1:20–21.

32. Ibid., pt. 1, qu. 1, arts. 1, 2, in *Basic Writings,* 1:5–7.

33. "Names are said of God and creatures in an *analogous* sense, . . . according to proportion." Ibid., pt. 1, qu. 13, art. 5, in *Basic Writings,* 1:118–121. See Gyula

Klima, "Theory of Language," in Davies and Stump, *Oxford Handbook of Aquinas,* 371–389; and Brian Davies, "The Limits of Language and the Notion of Analogy," in ibid., 390–397.

34. "There is a twofold mode of truth in what we profess about God. Some truths about God exceed all the ability of the human reason. Such is the truth that God is triune. But there are some truths which the natural reason also is able to reach. Such are that God exists, that He is one, and the like. In fact, such truths about God have been proved demonstratively by the philosophers, guided by the light of the natural reason." Thomas Aquinas, *Summa Contra Gentiles,* trans. Anton C. Pegis, bk. 1, ch. 3, http://dhspriory.org/thomas/english/ContraGentiles1.htm#.

35. Thus Aristotle: "Though one perceives the particular, perception is of the universal—e.g. of man but not of Callias the man." Aristotle, *Posterior Analytics,* 100a15–100b5. Thus Ockham: "For every universal cognition is a cognition of a singular." William of Ockham, *Quodlibetal Questions, Vols. 1 and 2: Quodlibets 1–7,* trans. Alfred J. Freddoso and Francis E. Kelley (New Haven, CT: Yale University Press, 1991), quod. 1, qu. 13, 64.

36. See Philotheus Boehner, introduction to *Philosophical Writings: A Selection,* by William of Ockham, rev. ed., trans. Philotheus Boehner, ed. Stephen F. Brown (1955; Indianapolis: Hackett, 1990), ix–li; John E. Murdoch, "William of Ockham and the Logic of Infinity and Continuity," in *Infinity and Continuity in Ancient and Medieval Thought,* ed. Norman Kretzmann (Ithaca, NY: Cornell University Press, 1982), 165–206; Paul Vincent Spade and Claude Panacio, "William of Ockham," in Zalta, *Stanford Encyclopedia of Philosophy,* Win 2016 ed., http://plato.stanford.edu/archives/win2016/entries/Ockham/; and Alfred J. Freddoso, "Ockham on Faith and Reason," accessed October 13, 2017, http://www3.nd.edu/~afreddos/papers/f&rcam.htm. The term "Ockham's razor" was coined by the mathematician William Rowan Hamilton in 1866: "There exists a primary presumption of philosophy. This is the law of Parcimony: which prohibits, without a proven necessity, the multiplication of entities, powers, principles or causes; above all, the postulation of an unknown force where a known [force] can account for the phaenomenon. We are, therefore, entitled to apply 'Occam's Razor' to this theory of causality." Quoted in Toni Vogel Carey, "Parsimony (in as Few Words as Possible)," *Philosophy Now* 81 (October/November 2010), https://philosophynow.org/issues/81/Parsimony_In_as_few_words_as_possible.

37. William of Ockham, *Quodlibetal Questions,* quod. 2, qu. 3, 103–104.

38. Ibid., quod. 4, qu. 12, 292–298. Ockham actually introduced another type of term (or "predication"), which he called "denominative" and which meant something like what we would call a connotative definition. But when explored more deeply, such terms were shown to harbor either a univocal or an equivocal sense, and his dilemma returned. It scarcely needs adding at this juncture that both Aquinas and Ockham created highly complex systems of thought and logic, to whose details we can only make brief, passing reference.

39. Scholastic thinkers were frequently given nicknames by their peers, as we have seen earlier with the "Angelic Doctor." Ockham had two of them, "Venerable Inceptor," which referred to his never having finished his degree in theology (an "inceptor" was a student who stood at the threshold of earning a degree), and "More than Subtle Doctor," a reference to John Duns Scotus's nickname of "Subtle Doctor," reflecting the widely held belief that Ockham had surpassed Scotus's linguistic and logical subtleties.

40. Taylor, *Secular Age,* 94, 97, 284; Gaukroger, *Emergence of a Scientific Culture,* 81–83. Michael Allen Gillespie sees the roots of modernity in these disputes, claiming that "'atheistic' materialism . . . has a theological origin in the nominalist revolution." Michael Allen Gillespie, *The Theological Origins of Modernity* (Chicago: University of Chicago Press, 2008), 36.

41. Other scholastics advanced a theory of "divine illumination" as the principal source of our knowledge, or at least one of them. This tradition traced back from Saint Bonaventure to Anselm to Augustine and to its roots in Neoplatonism. In the generation following Aquinas, Henry of Ghent (ca. 1217–1293) wrote his own *Summa* and other works in an extended effort to synthesize both the Platonic and Aristotelian strains of medieval thought. He too thought science extended to theology, but only when supplemented by the "special illumination of a supernatural light," divinely placed ideas ("exemplars," he called them) in our minds. See Henry of Ghent, *Summa of Ordinary Questions: Articles Six to Ten on Theology,* trans. Roland J. Teske (Milwaukee: Marquette University Press, 2011), art. 6, qu. 1, 23–27. Steven P. Marrone, *Truth and Scientific Knowledge in the Thought of Henry of Ghent* (Cambridge, MA: Medieval Academy of America, 1985), offers a worthwhile introduction not only to the thought of Henry of Ghent but also to the whole issue of science and theology.

42. For Aquinas's theory of adequation, see the selections from *Quaestiones Disputatae* in Richard McKeon, ed. and trans., *Selections from Medieval Philosophers* (New York: Charles Scribner's Sons, 1930), 2:159–170.

43. Thomas Aquinas, *Questiones Disputatae de Veritate: Questions 1–9,* trans. Robert W. Mulligan (Chicago: Henry Regnery, 1952), qu. 2, art. 3, arg. 19, http://dhspriory.org/thomas/QDdeVer.htm. In John of Salisbury's comparable expression, scientific "reasoning is dependent on the data provided by sense-experience." *Metalogicon,* bk. 4, ch. 9, 216.

1. A World of Words and Things

1. A good summary of John's life and works may be found in Kevin Guilfoy, "John of Salisbury," in *The Stanford Encyclopedia of Philosophy,* Fall 2008 ed., ed. Edward N. Zalta, http://plato.stanford.edu/archives/fall2008/entries/john-salisbury/.

2. John of Salisbury, *The Metalogicon of John of Salisbury: A Twelfth-Century Defense of the Verbal and Logical Arts of the Trivium,* trans. Daniel D. McGarry

(1955; Berkeley: University of California Press, 1962), bk. 3, prologue, 142–143, bk. 4, prologue, 203. Though offering a sanctuary from worldly cares, John's intellectual life itself, like that of most scholars, was anything but calm. In *Metalogicon,* he sought to out-joust his black-knight foe, one "Cornificius," a personification of those who had misused and mistaught Aristotle, who had sought to use their sophistry for material gain, and who had ill-used their God-given gifts of rhetorical ability and intellectual acumen. In this context, his defense of the properly taught trivium became an exemplar of medieval educational theory and practice.

3. Ibid., bk. 1, chs. 13 and 14, 38–39, ch. 20, 58, ch. 23, 65.

4. In our day, Walter J. Ong, one of the most erudite students of literacy as a technology (dubbed the "thinking man's McLuhan"), has examined the differences between "primary oral cultures" and those shaped by varying degrees of literacy, exploring in depth how writing systems "affect the word" and stand as not only "mere exterior aids but also interior transformations of consciousness." See in particular his now-classic study, *Orality and Literacy: The Technologizing of the Word* (1982; London: Routledge, 1988), 1, 81, more generally 77–114, and throughout.

5. John of Salisbury, *Metalogicon,* bk. 4, ch. 18, 230. One thinks here of T. S. Eliot's lament for the discarded image, which evokes the same medieval hierarchy (*Choruses from the Rock,* 1934): "Where is the wisdom we have lost in knowledge? / Where is the knowledge we have lost in information?"

6. As an outgrowth of analysis, the modern study of language, either as an information technology (which informs the present narrative) or as a science (linguistics, with its subdisciplines of syntax, semantics, pragmatics, phonetics, and many others), has deepened and broadened greatly our understanding of the ways words function. This is neither to gainsay nor to denigrate philosophical accounts of language, which in many ways remind one of scholastic endeavors. The difference between (1) historical, technological, or linguistic approaches; and (2) those of philosophy lies with their objectives, perhaps for us more a matter of emphasis than of sharp distinction, degree rather than kind. The former concentrate on how language evolves over time or how it functions in context. The latter target language (along with logic) as the screen between the mind and the outside world, primarily in order to determine what sorts of possible claims can be made about knowledge or reality. As a reflexive move, runs the argument, one must first study the instrument making the claims before studying the claims themselves. The enduring allegory of Plato's cave lies behind such perspectives.

7. In *The Birth of the Past* (Baltimore: Johns Hopkins University Press, 2011), Zachary Sayre Schiffman incisively chronicles the emergence of our awareness of the historical past from antiquity through the Enlightenment. For the Middle Ages, see esp. 77–134.

8. John of Salisbury, *Metalogicon,* bk. 3, ch. 8, 183.

9. Ibid., bk. 4, ch. 19, 231.

10. For the idea that, as a cultural product, information was created with the invention of writing, see Michael E. Hobart and Zachary S. Schiffman, *Information Ages: Literacy, Numeracy, and the Computer Revolution* (Baltimore: Johns Hopkins University Press, 1998), introduction and pt. 1; and James Gleick, *The Information: A History, a Theory, a Flood* (New York: Vintage Books, 2011), 28–41.

11. John of Salisbury, *Metalogicon,* bk. 3, ch. 5, 175, ch. 8, 185.

12. Aristotle, *Topics,* bk. 1, 101b1–5, bk. 8, 155b7–15. The *Organon* ("instrument") comprises Aristotle's treatises on logic, which ancient commentators had grouped together. These included *Categories, On Interpretation, Prior Analytics, Posterior Analytics, Topics,* and *On Sophistical Refutations.* See Robin Smith, "Aristotle's Logic," in Zalta, *Stanford Encyclopedia of Philosophy,* Spring 2009 ed., http://plato.stanford.edu/archives/spr2009/entries/aristotle-logic/. A good introduction to scholastic logic may be found in William Kneale and Martha Kneale, *The Development of Logic* (1962; Oxford: Oxford University Press, 1984), ch. 4.

13. Paleolithic rock art, drawings and paintings dating at least as far back as forty thousand years, suggest the early, well-developed, symbolic capabilities of *Homo sapiens,* probably associated with "spirituality" or "religion," the latter perhaps a form of "shamanism." Even earlier still, the "conditions are present to detect an elementary form of symbolic thought and a certain detachment from a purely material reality." See Jean Clottes, *What Is Paleolithic Art? Cave Paintings and the Dawn of Human Creativity,* trans. Oliver Y. Martin and Robert D. Martin (Chicago: University of Chicago Press, 2016), 15, 27, 32, and more generally 5–38. Robert C. Berwick and Noam Chomsky, in *Why Only Us: Language and Evolution* (Boston: MIT Press, 2016), 53–88, argue that human language facility and its capacity for symbolic thinking emerged rather suddenly around sixty thousand years ago and have remained basically unchanged since then.

14. Some observers have noted that the human capability of abstracting momentary ephemera in language, spoken or written, at least to the degree we can accomplish it, delineates our chief difference from other animals. We can abstract the lessons, patterns, objects, and occurrences from one situation and then apply them to another. We can create and capture "rules of conduct, beliefs about our histories, and hopes and fears about imagined futures." We are a "symbolic species." See Terrence W. Deacon, *The Symbolic Species: The Co-evolution of Language and Brain* (New York: W. W. Norton, 1997), 423, chs. 1–3 and throughout. Aristotle, too, believed that language distinguished human societies from those of other social animals—bees, for example. See Aristotle, *Politics,* bk. 1, 1253a7–18.

15. José Ortega y Gasset, *The Idea of Principle in Leibnitz and the Evolution of Deductive Theory,* trans. Mildred Adams (New York: W. W. Norton, 1972), 58–62; Hobart and Schiffman, *Information Ages,* 3–4, 13–14.

16. Eric A. Havelock, *The Muse Learns to Write* (New Haven, CT: Yale University Press, 1986), 54 and throughout.

17. Bryan W. Van Norden, *Introduction to Classical Chinese Philosophy* (Indianapolis: Hackett, 2011), 236–242. Modern Chinese writing employs some fifty thousand symbols for different words, six thousand of which are commonly known and used by literate Chinese.

18. An early manifestation of the urge to classify can be found in the Sumerian practice of list making, most notably in the Sumerian King List (ca. 2100 BCE). Scribes compiled the King List to commemorate and memorialize the national revival under the leadership of the Third Dynasty of Ur, when the descendants of the Sumerians regained control of Mesopotamia. The scribal compilations listed specific kings and their heirs, and related their deeds, much as can be found in earlier records. But, more importantly, they also devised categories of organization, such as "kingship," as rubrics to facilitate and preclude somewhat the exhaustive practice of naming. Theirs was not yet full-fledged classifying. Rather, the subordinate use of some categories aided the sequential decoding and recalling of names for narrative, commemorative purposes. See Hobart and Schiffman, *Information Ages,* 51 and generally 32–56.

19. Quoted in I. J. Gelb, *A Study of Writing,* 2nd ed. (Chicago: University of Chicago Press, 1963), 13.

20. Gleick, *Information,* 33.

21. Gelb, *Study of Writing,* 69.

22. In scripts from Mesopotamia, for instance, some signs designated vowels, not consonants, such that one sign stood for the syllables *ga, ka,* and *qa;* to interpret the sign properly, the reader had to infer which consonant should be paired with the vowel. Similarly, West Semitic scripts, stemming from Egyptian hieroglyphs, employed consonantal signs to which were added vowel sounds. This also required considerable interpretive skill on the part of the reader, for words written without vowels are not readily pronounceable, as with the following illustration: "Wrds wrttn wtht vwls r nt rdl prnncbl." Hobart and Schiffman, *Information Ages,* 65–68.

23. As with many subjects, Aristotle observed the basics: "Language is the articulation of voice by the tongue. Thus, the voice and larynx can emit vowel sounds; consonantal sounds are made by the tongue and the lips; and out of these language is composed." Aristotle, *History of Animals,* bk. 4, 534b33–535a1.

24. John of Salisbury, *Metalogicon,* bk. 1, ch. 14, 39.

25. Ibid., bk. 1, ch. 15, 41–42.

26. Much of the debate revolves around the path-breaking work of Eric A. Havelock, especially his *Preface to Plato* (Cambridge, MA: Harvard University Press, 1963), but also see *Prologue to Greek Literacy* (Cincinnati: University of Cincinnati, 1971, from lectures delivered in 1970), *Muse Learns to Write,* and *The Literate Revolution in Greece and Its Cultural Consequences* (Princeton, NJ: Princeton University Press, 1982). For an introduction to the debates, the biblio-

graphies in Ong, *Orality and Literacy,* and Hobart and Schiffman, *Information Ages,* complement those of Havelock.

27. Aristotle, *Parts of Animals,* bk. 2, 660a4; Aristotle, *De Interpretatione,* 16a3–5. Beyond his extensive writings about rhetoric and poetry, Aristotle had virtually nothing to say explicitly about the syntax or semantics of language per se, further indication that he could simply take alphabetic literacy for granted.

28. Aristotle, *Posterior Analytics,* bk. 1, 75b30–33: "A definition is either a principle of demonstration or a demonstration differing in position or a sort of conclusion of a demonstration." In *Metaphysics,* bk. 7, 1034b20–22, he wrote that "a definition is a formula [*logos,* also rendered as "account"], and every formula has parts, and as the formula is to the thing, so is the part of the formula to the part of the thing." Recently, Marguerite Deslauriers has constructed a quite technical and elaborate case for the "unity of definitions and of their objects" in *Aristotle on Definition,* Philosophia Antiqua: A Series of Studies on Ancient Philosophy, vol. 109, ed. K. A. Algra et al. (Boston: Brill, 2007), 8–10 and throughout. My account owes much to her work.

29. Aristotle, *Topics,* bk. 1, 101b36–37. In *Posterior Analytics,* bk. 2, 90b30–31 and 93b29, Aristotle added variations on the definition of definition: "Definition is of what a thing is and of substance," and "Definition is . . . an account [*logos*] of what a thing is."

30. Aristotle, *Physics,* bk. 2, 194b24; Thomas Aquinas, *Commentary on the Gospel of St. John, Part 1: Chapters 1–7,* trans. James A. Weisheipl (Albany, NY: Magi Books, 1998), ch. 1, lec. 1, sec. 26.

31. Basically, Aristotle recognized these features (semantics and syntax) as the form of language. See Aristotle, *On the Soul,* bk. 1, 403b1–7, where, in discussing the soul, he breaks down definitions into their "material conditions," their "form or account [*logos*]," and their "purpose or end."

32. Aristotle, *On the Heavens,* bk. 1, 270b16–25; Aristotle, *Meteorology,* bk. 1, 339b20–28. In early, Homeric Greek, "ether" (*aithēr*) meant "pure, fresh air" or "clear sky." It was personified in Greek mythology, according to Hesiod, as the deity Aether (god of "light and day"), son of Erebus (god of deep darkness) and Nyx (goddess of night), and characterized as the pure essence in which the gods lived and which they breathed, analogous to the air breathed by mortals. See Hesiod, *Theogeny,* trans. Norman Brown (Indianapolis: Bobbs-Merrill, 1953), 56–57. It is related to the Greek term *aithō,* an affinity that led Anaxagoras to identify ether with fire, a mistake according to Aristotle (*On the Heavens,* bk. 3, 302b2–4).

33. Terrence Irwin, *Aristotle's First Principles* (Oxford: Oxford University Press, 1990), 26–50. In *Topics,* bk. 1, 101b3–4, Aristotle defined "dialectic" as "a process of criticism wherein lies the path to the principles of all inquiries."

34. Both *horos* and *terminus* originally signified piles of stones, and later markers, which separated fields, delimiting property. The Greeks made Hermes a

god of boundary markers, among other things, and he stood at crossroads to differentiate roadways. In Greek the correct path or journey was called *methodos,* the word derived from *meta* ("beyond") and *hodós* ("way," "road," "journey," or "path"). Because he oversaw the right way to go, Hermes was also considered a god of salvation. From such storied beginnings came eventually Aristotelian, logical abstractions. Like boundaries, terms face two directions. Looking outward on reality, they pretend to tell us the veracity of what they see; they correspond to things. Looking inwardly, they are concerned with their own classifying logic and order. Today in logic we identify these features as the extension and intension of a concept. Ortega y Gasset, *Idea of Principle,* 58–62.

35. In addition to its more familiar translation as "word," the etymology of *logos* reveals a long and storied life span, with various meanings depending on context, including the following list from Charles H. Kahn: "saying, speech, discourse, statement, report, account, explanation, reason, principle, esteem, reputation, collection, enumeration, ratio, proportion." Charles H. Kahn, *The Art and Thought of Heraclitus: An Edition of the Fragments with Translation and Commentary* (Cambridge: Cambridge University Press, 1981), 29. Kahn translates *logos* as "account" in Heraclitus's fragments 1 and 3: "Although this account [*logos*] holds forever, men ever fail to comprehend, both before hearing it and once they have heard" (frag. 1); "Although the account [*logos*] is shared, most men live as though their thinking were a private possession" (frag. 3).

36. Aristotle, *Metaphysics,* bk. 7, 1037b25–27.

37. Immediately following his often-quoted opening, Aristotle continued: "An indication of this is the delight we take in our senses; for even apart from their usefulness they are loved for themselves; and above all others the sense of sight." Aristotle, *Metaphysics,* bk. 1, 980a22–25.

38. See Jonathan Lear, *Aristotle: The Desire to Understand* (Cambridge: Cambridge University Press, 1988), 1–14.

39. Thomas Aquinas, *Summa Contra Gentiles,* bk. 3, ch. 63, in *Basic Writings of Saint Thomas Aquinas (The Summa Theologica; The Summa Contra Gentiles),* 2 vols., English Dominican trans., trans. Laurence Shapcote, ed. Anton C. Pegis (New York: Random House, 1945), 2:112. See too ibid., bk. 3, ch. 50, in *Basic Writings,* 2:90: "The desire for knowledge naturally implanted in all intellectual substances does not rest unless, having acquired the knowledge of the substance of the effects, they know also the substance of their cause. Consequently, since separate substances know that God is the cause of all the things whose substances they see, their natural desire does not rest, unless they see God's substance also."

40. Aristotle deviated from Plato's dichotomous "method of division" for defining, which divided categories into twos. Thus, to establish a definition of X, first ascertain the largest class G into which it falls. Then divide G into its two largest parts—G_1 and G_2, for example—and see under which of those smaller parts X falls. Suppose that X comes under G_1, then further subdivide G_1 into $G_{1,1}$

and $G_{1,2}$, and locate X in one of these yet smaller groupings. Continue the procedure until a subdivision becomes identical with X. The entire process of division thus determines the nature and scope of X. Each division in this process requires subdividing some "kind," the Greek term for which is *genos* (genus), on the "basis of a relevant difference." Aristotle disagreed with Plato, believing that by dividing any category into only two parts, one of the parts always remains negative and therefore does not correlate positively with any objects or phenomena. For "defining" in Plato and Aristotle, see Kenneth Dorter, *Form and Good in Plato's Eleatic Dialogues: The Parmenides, Theaetetus, Sophist, and Statesman* (Berkeley: University of California Press, 1994), 13–17, 166; Robin Smith, " Logic," in *The Cambridge Companion to Aristotle,* ed. Jonathan Barnes (Cambridge: Cambridge University Press, 1995), 27–65; Pierre Pellegrin, *Aristotle's Classification of Animals: Biology and the Conceptual Unity of the Aristotelian Corpus,* trans. Anthony Preus (Berkeley: University of California Press, 1986), 1–49; Deslauriers, *Aristotle on Definition,* 27–30. Also see Plato, *Sophist,* 235c, trans. Nicholas P. White, and *Phaedrus,* 265d–266d, trans. A. Nehamas and P. Woodruff, in *Plato: Complete Works,* ed. John M. Cooper (Indianapolis: Hackett, 1997). All citations of Plato are from this collection. As with Aristotle, the locator numbers refer to a standard scholarly Greek edition of Plato, this one first appearing in Paris (1578) and edited by the French scholar Henri Estienne (Stephanus in Latin). The references are thus to the "Stephanus numbers"—page and section—still utilized by many modern Plato scholars.

41. John of Salisbury, *Metalogicon,* bk. 3, ch. 8, 182. Later, Aquinas reiterated the same idea: "The sciences are divided off in the same manner as things are." See Thomas Aquinas, *Commentary on Aristotle's Generation and Corruption,* trans. Pierre Conway and R. F. Larcher, prologue, http://dhspriory.org/thomas/english/GenCorrup.htm#0.

42. Smith, "Logic," 51–52; Aristotle, *Topics,* bk. 6, 143b20: "The genus and differentia constitute the account of the species." The theory of natural kinds has many logical permutations, some of which Aristotle addressed. For example, although one might subsume the differences 'terrestrial,' 'aquatic,' or 'flying' under the genus 'animal,' one cannot say all 'animals are terrestrial' (or 'aquatic,' and so on). Thus, technically, 'animal' is "predicable of each of the terms by which it is divided," and not the converse. So, even though the genus (here, 'animal') is more general than the differences, one cannot convert it into the subject of a universal statement whose predicates are the differences. Membership in categories flows upward, so to speak, through predicates, toward greater levels of generality. See Deslauriers, *Aristotle on Definition,* 30.

43. Aristotle, *Generation of Animals,* bk. 2, 732a25–732b15.

44. J. L. Ackrill, *Aristotle the Philosopher* (Oxford: Oxford University Press, 1981), 31, summarizes: "Aristotle . . . took for granted that Greek linguistic usage and habits of thought accurately and finally reflected objective reality."

45. Aristotle, *Categories,* 1a1–15. Often rendered as "univocal" (one voice), J. L. Ackrill translates the Greek *sunōnumon* more directly as "synonymous" ("with name"): "When things have the name in common and the definition of being which corresponds to the name is the same, they are called *synonymous.*" See too Aristotle, *Topics,* bk. 4, 123a33–37 ("genus is always predicated of its species in its literal sense") and Aristotle's discussion of epithets and metaphors in *Rhetoric,* bk. 3, 1404b27–1405b33. As noted earlier, both Aquinas and Ockham adopted Aristotle's account of univocity as the primary way to understand words and language, but they drew different inferences regarding secondary, metaphorical and analogical, extensions.

46. John of Salisbury, *Metalogicon,* bk. 1, ch. 10, 32, bk. 1, ch. 13, 37, bk. 2, ch. 3, 78–79, bk. 2, ch. 6, 84. As we shall see later, Galileo will turn the scholastic notion of demonstration on its ear, with far-reaching consequences for modern analysis.

47. For an introduction to this brilliant logician, philosopher, theologian, exegete; teacher, debater, and tragic figure, see Peter King, "Peter Abelard," in Zalta, *Stanford Encyclopedia of Philosophy,* Summer 2015 ed., http://plato.stanford.edu/archives/sum2015/entries/abelard/. Abelard's relationship with Heloise still resonates as one of history's most intriguing love stories, best absorbed through reading the letters between them, *The Letters of Abelard and Heloise,* rev. ed., trans. Betty Radice, ed. M. T. Clanchy (New York: Penguin Books, 1974), which includes Abelard's own account of his turbulent life, *Historia calamitatum* (*The Story of My Misfortunes*).

48. We should note that harmony did not always mean agreement any more than a contrapuntal, descant line in music matches one for one the notes of a melody. A legal anecdote may help clarify. In 1148 or thereabouts, a canon lawyer from Bologna named Gratian compiled and published a textbook for the study of canon law under the title *Concordia discordantium canonum* (more commonly known as *Decretum Gratiani* [*Gratian's Decree*]). The original title translates as "The harmony of discordant canons," and Gratian's work exploited an early version of scholastic method in order to reconcile apparently contradictory canons from previous centuries. The text stood among the earliest of efforts to bring ecclesiastical rules and practices together in a single, coherent body of canon law, eventually comprising part of the *Corpus Juris Canonici* of the Catholic Church until its replacement in 1917 by the *Codex Juris Canonici.* Much like Gratian, many medieval scholars sought a *concordia discordantium,* a presumptive harmony, between reason and faith, science and religion, using the basic tools of alphabetic literacy, common sense, definitions, and logic to reconcile seemingly contradictory positions.

49. Among the Greeks before Aristotle, Plato and a few others had been explicitly concerned with questions of truth and falsity, with valid inferences, and with definitions, but Aristotle's *Prior Analytics,* write the Kneales, was "undoubtedly the first systematic treatise on formal logic." Kneale and Kneale, *Development of*

Logic, 16, and more generally 1–22. See also Jonathan Barnes, "Life and Work," in Barnes, *Cambridge Companion to Aristotle*, 25; Smith, "Logic," 47; David Keyt, "Deductive Logic," in *A Companion to Aristotle*, ed. Georgios Anagnostopoulos (West Sussex: Wiley-Blackwell, 2013), 31; Robin Smith, "Aristotle's Theory of Demonstration," in ibid., 51–65; and G. E. R. Lloyd, "What Was Mathematics in the Ancient World? Greek and Chinese Perspectives," in *The Oxford Handbook of the History of Mathematics*, ed. Eleanor Robson and Jacqueline Stedall (Oxford: Oxford University Press, 2009), 7–12.

50. Aristotle, *Prior Analytics*, bk. 1, 24a20–25.

51. There was also the mathematical use of the term "demonstration," found at the end of each of Euclid's proofs in the *Elements* with the phrase *quod erat demonstrandum*, more generally known by its initials, QED. It means literally "what was to have been demonstrated." The Latin QED, in turn, was a later rendering of the Greek phrase *oper edei deixai*, abbreviated as OEΔ in Euclid, which literally translates as "what was required to be proved." The *Elements* is a "synthetic treatise," beginning with definitions, postulates, and common notions, and then proceeding step by step toward the construction of more complicated geometrical propositions. Its proofs move from the known (or accepted) to the unknown, from the simple and particular to the more complex and general. A rigorous series of deductions makes this movement possible, with each step forward being justified by existing definitions, postulates, or previously proven propositions. To say "QED" at the end of a proof, then, meant the proposition in question had been properly analyzed and the series of deductions leading to it had been rigorously and clearly demonstrated. See Thomas L. Heath, *A History of Greek Mathematics*, vol. 1, *From Thales to Euclid* (1921; New York: Dover, 1981), 371.

52. Modern discussions of the logic of syllogism can be found in most introductory textbooks, including any of the fourteen editions of Irving M. Copi, *Introduction to Logic*, first published in 1953 (New York: Macmillan). For an account of "categorical propositions," summarized in Appendix A.1, including the A, E, I, and O statements, see the second edition (1961), 133–152.

53. Kneale and Kneale, *Development of Logic*, 67–81, 198–246; Keyt, "Deductive Logic."

54. Aristotle, *Posterior Analytics*, 71b17–22, 78a23–78b3.

55. The division between the "fact" and the "reason why" played a crucial role in medieval demonstrations. Causal explanations were described as demonstrations, the passage from premises to conclusion in an argument, and finding the cause of something therefore required determining the appropriate premises. The premises themselves were frequently the results of previous logical argument and demonstrations, and the whole procedure continued back to the first principles or basic elements of any science. Thus Aristotle, *Physics*, bk. 1, 184a14–15: we know a thing when "we are acquainted with its primary causes or first principles, and have carried our analysis as far as its elements."

56. In this context, one further technical distinction deserves note, for it helped guide the schoolmen into the knowledge of natural things and their causes. This was the distinction between the *demonstratio propter quid* and the *demonstratio quia. Propter quid* demonstrations were based on the nature of the thing being investigated, the essence of an object (what it actually is—understood as its "thingness" or "quiddity," from the Latin *quidditas*), while those designated as *quia* focused on the thing's effects. See Thomas Aquinas, *Summa Theologica,* pt. 1, qu. 2, art. 2, in *Basic Writings,* 1:20–21.Today's philosophers employ "a priori" and "a posteriori" as rough equivalents to capture the same basic distinction between conceptual and empirical matters. "A priori" (meaning "from earlier") refers to arguments developed purely within the realm of thought—that is, before experience. "A posteriori" (literally, "from later") depends on experience or empirical evidence brought to the mind by the senses. The scholastic distinction was comparable. "A bachelor is an unmarried male" is a conceptual truth, based on the essence or definition of the thing (bachelor, in this case), a matter of the *demonstratio propter quid.* "All bachelors are happy (or sad)" is an empirical, a posteriori generalization, falling within a *demonstratio quia* for scholastics.

57. The term 'analysis' derives from the Greek *analusis* (rendered in Latin as *resolutio*), which means a "loosening up" or "dissolution", and in philosophy it came to mean solving or dissolving a problem. Scholars generally note three intellectual contexts of the term's application: philosophical (the dialogues of Plato), logical (the *Organon* of Aristotle), and geometrical (Greek geometry, especially Euclid). An entire industry of books, articles, and commentary is devoted to the term and idea of analysis. Michael Beaney provides a good philosophical introduction in "Analysis," in Zalta, *Stanford Encyclopedia of Philosophy,* Winter 2017 ed., http://plato.stanford.edu/archives/win2017/entries/analysis/. Thomas L. Heath explains the movement from analysis to synthesis in Euclid's geometry in Euclid, *The Thirteen Books of the Elements,* 2nd ed., 3 vols., trans. Thomas L. Heath (New York: Dover, 1956), 1:137–151. See too Norman Gulley, "Greek Geometrical Analysis," *Phronesis* 3, no. 1 (1958): 1–14. Especially valuable, the study of Jaakko Hintikka and Unto Remes calls attention to the role of "auxiliary constructions" in framing deductive inferences, discussed by Pappus (ca. 290–ca. 350), whom the authors describe as presenting analysis as an "upward" movement in search of "premises rather than as a sequence of conclusions." Jaakko Hintikka and Unto Remes, *The Method of Analysis: Its Geometrical Origin and Its General Significance,* Boston Studies in the Philosophy of Science, vol. 25, ed. Robert S. Cohen and Marx W. Wartofsky (Dordrecht, Holland: D. Reidel, 1974), xiv-xv and throughout, esp. chs. 2, 3, 8, and 9.

58. Aristotle, *Posterior Analytics,* bk. 1, 70b9–24. See also Smith, "Aristotle's Theory of Demonstration," 55–59.

59. Thus analysis can be "proved by deduction." Aristotle, *Prior Analytics,* bk. 1, 50a30–31; John of Salisbury, *Metalogicon,* bk. 2, ch. 3, 79.

60. Thomas Aquinas, *Commentary on John,* ch. 1, lec. 1, sec. 25.

61. Thomas Aquinas, *De Anima,* lec. 1, and *De Trinitate,* bk. 4, 2, in *Saint Thomas Aquinas: Philosophical Texts,* trans. Thomas Gilby (New York: Oxford University Press, 1960), 19–22.

62. Aquinas, *Commentary on John,* ch. 1, lec. 1, sec. 25.

63. Aquinas, *Summa Theologica,* pt. 1, qu. 34, art. 3, in *Basic Writings,* 1:337–338.

64. Aquinas, *Commentary on John,* ch. 1, lec. 1, sec. 61 and ch. 1, lec. 2, sec. 74.

65. Skeptical conventions as regards classifying were countered until well into the eighteenth century by another metaphysical tradition of long-standing duration, that of the "great chain of being," whose history Arthur O. Lovejoy traced in his modern classic, *The Great Chain of Being: A Study of the History of an Idea* (Cambridge, MA: Harvard University Press, 1936).

66. As noted, Aristotle too was quite aware of the difficulties of forming proper categories. Lotze is cited by Ernst Cassirer in *Substance and Function and Einstein's Theory of Relativity,* trans. William Curtis Swabey and Marie Collins Swabey (1923; New York: Dover, 1953), 7. A worthy introduction to various types of classifying, as well as the whole field of semiotics and its methods of understanding early texts, may be found in the rich, scholarly, and imaginative works of Umberto Eco, especially his recent collection, *From the Tree to the Labyrinth: Historical Studies in the Sign and Interpretation,* trans. Anthony Oldcorn (Cambridge, MA: Harvard University Press, 2014); see above all the title essay, 3–94, on the dictionary and the encyclopedia. As a discipline, semiotics is generally regarded as devoted to the widely ranging study of "meanings" conveyed through signs and symbols—their syntactic structures, semantic references, and pragmatics. It also, by extension, looks at the ways different societies classify or organize cultural and natural phenomena. By contrast, looking at early developments in scientific and philosophical thought from the standpoint of their underlying information technologies gives us greater purchase on the differences between the symbolic characteristics of conventional language(s) and those of the emerging and relational mathematics integral to the rise of modern science.

67. C. S. Lewis, *The Discarded Image: An Introduction to Medieval and Renaissance Literature* (Cambridge: Cambridge University Press, 1964), 10.

2. Demonstrable Common Sense

1. A good introduction to the notion of common knowledge in the Middle Ages may be found in Peter Dear, *Revolutionizing the Sciences: European Knowledge and Its Ambitions, 1500–1700* (Princeton, NJ: Princeton University Press, 2001), 3–8, 10–29.

2. Beyond popular images, many seventeenth-century philosophers had great fun at Aristotle's expense. Thus Descartes found completely unintelligible the

Aristotelians' definition of motion as "the act of being which is in potency, in so far as it is in potency" (Descartes, *Le Monde,* ch. 7), to which Pierre Gassendi (1592–1655) boomed in agreement: "Great God! Is there a stomach in the world strong enough to digest that?" Joseph Glanvill (1636–1680) snorted that "manifest qualities" (such as gravity) "teacheth nothing . . . beyond the empty signification of their names, . . . an account . . . we might expect from a Rustick." Quoted in Stephen Gaukroger, *The Emergence of a Scientific Culture: Science and the Shaping of Modernity, 1210–1685* (Oxford: Oxford University Press, 2006), 458.

3. Over the centuries Tertullian's writings have served as a fount of pithy and recognizable sayings, sort of an early, Christian version of *Bartlett's Quotations.* For the ones quoted here, see Tertullian, *On the Prescription of Heretics,* ch. 7, http://www.tertullian.org/works/de_praescriptione_haereticorum.htm.

4. Thus Clement: "Philosophy . . . was given to the Greeks, as a covenant peculiar to them—being, as it is, a stepping stone to the philosophy which is according to Christ." Clement of Alexandria, *Miscellanies,* bk. 6, ch. 8, https://archive.org/details/writingsofclemen02clem.

5. The stories of Origen are recounted in the *Ecclesiastical History* of Eusebius of Caesarea (ca. 260 / 265–339 / 340), bk. 6, http://www.newadvent.org/fathers/250106.htm. See also Henry Chadwick, "The Early Christian Community," in *The Oxford History of Christianity,* ed. John McManners (Oxford: Oxford University Press, 1990), 60–61 and more generally 21–69. A somewhat different and more sympathetic interpretation of Origen's life and theology is provided by Charles Freeman, *A New History of Early Christianity* (New Haven, CT: Yale University Press, 2009), 176–195.

6. Augustine of Hippo, *The Confessions of St. Augustine,* trans. J. G. Pilkington (New York: Boni and Liveright, 1927), bk. 8, ch. 20, 158. Gaukroger, *Emergence of a Scientific Culture,* 58, and more generally 49–59, argues that Augustine had created a "unity of philosophy and theology in which metaphysics is effectively a science of God," doctrines holding sway over his medieval successors until challenged by the arrival of Aristotle's works. A contrasting view is offered by Charles Freeman in *The Closing of the Western Mind: The Rise of Faith and the Fall of Reason* (New York: Vintage Books, 2002), where he stresses that with the "absorption of Platonism" by Christians, it proved "impossible to find secure axioms from which to start . . . rational argument" (xviii), and thus that "Augustine's theology . . . came to lie in faith rather than reason," its Platonic influences notwithstanding (286). Reason would reenter the Latin West only with the advent of Aristotle and his acolyte Thomas Aquinas. Freeman even suggests that, rather than bringing Aristotle into the Christian fold, the tale commonly told, Aquinas himself was "converted to Aristotelianism" (327).

7. See M.-D. Chenu, *Nature, Man and Society in the Twelfth Century: Essays on New Theological Perspectives in the Latin West,* ed. and trans. Jerome Taylor and Lester K. Little (Toronto: University of Toronto Press, 1997), esp. his essay "The

Platonisms of the Twelfth Century," 49–98. David Albertson, *Mathematical Theologies: Nicholas of Cusa and the Legacy of Thierry of Chartres* (Oxford: Oxford University Press, 2014), 1–89, and throughout, provides a nuanced and interesting account of how "neopythagorianism" interwove Neoplatonic themes in medieval Christian theology, especially that of Thierry of Chartes.

8. Thomas Aquinas, *Summa Theologica,* pt. 1, qu. 46, art. 2, in *Basic Writings of Saint Thomas Aquinas (The Summa Theologica; The Summa Contra Gentiles),* 2 vols., English Dominican trans., trans. Laurence Shapcote, ed. Anton C. Pegis (New York: Random House, 1945), 1:452.

9. On the Incarnation, see, for example, Thomas Aquinas, *The Apostles' Creed,* trans. Joseph B. Collins (New York: Wagner, 1939), art. 2, http://dhspriory.org/thomas/english/Creed.htm#2, where he argued from the Aristotelian principle "that different things have different modes of generation" to the ideas that in a man "the soul by its act of thinking begets the word," and hence, by analogical extension, that with God the act of thinking begets his Word (*Logos*) in the form of Christ. Thus, Aquinas concluded this particular mixture of reason and faith, philosophy and theology: the "Word of God is of the one nature as God and equal to God."

10. The century between, roughly, 1250 and 1350 witnessed the pinnacle of this initial attraction. Not everybody bought in, however; many Franciscans, among others, were less enthralled by the new learning and never abandoned their Augustinian-Neoplatonist heritage. Some scholastics—notably, Henry of Ghent (mentioned earlier)—sought an integration of the "divine illumination" features of Neoplatonism with the more naturalistic teachings of Aristotle, whose works he had studied with Albert the Great. By the fourteenth century, challenges were already being mounted from nominalists, pietists, mystics, and others critical of the excessive rationalism encouraged by Aristotelian thought. Nonetheless, Aristotelian-based scholasticism continued to thrive in university curricula until the eighteenth century, and beyond.

11. For Aristotle's account of "common sensibility," see *On the Soul,* bk. 3, 425a14–427a16; for his treatment of light and heavy, see *On the Heavens,* bk. 4, 307b27–308a33.

12. Aquinas adopted the same account of the elements and their composition from contraries, although he reserved God as the ultimate cause of all reality. See Thomas Aquinas, *Aquinas on Creation: Writings on the "Sentences" of Peter Lombard,* trans. Steven E. Baldner and William E. Carroll (Toronto: Pontifical Institute of Mediaeval Studies, 1997), bk. 2, dist. 1, qu. 1, art. 1, 63–68.

13. Aristotle, *On Generation and Corruption,* bk. 2, 330b1–331a6. Aristotle's account of sensations may be found in *On the Soul,* bk. 2, 416b31–424b19. There existed another way in which sensations were thought to be common—namely, that they were held by each of us in a common location, unified in our consciousness. Aristotle believed we do not possess a separate consciousness for sight, another for

sound, and so forth. Rather, sensations all derive from one central, organic source, which he erroneously supposed to be the heart. Aristotle, *Parts of Animals,* bk. 3, 666a13–15; Jonathan Lear, *Aristotle: The Desire to Understand* (Cambridge: Cambridge University Press, 1988), 51–52.

14. Aristotle, *On the Heavens,* bk. 3, 300a20–300b8.

15. Before the advent of modern physics, two main approaches governed these discussions. In the first, Aristotle explained that a corporeal substance, such as the external medium air, provided the source of continuous movement. Because the violent or unnatural force had to stay in contact with the projectile, it followed that as the original mover put the rock in motion, he also activated the air around it. This in turn caused the air (or other surrounding medium) to propel the projectile farther, but with diminishing motive power. In the second, devised by Jean Buridan (ca. 1300–ca. 1358), an "incorporeal" force called "impetus" was imparted from the initial mover to the body in motion, which in turn kept the body moving, though with decreasing force and speed because of resistance. Buridan's approach lent itself to quantifying various features (distance, weight, velocity) of projectile motion and to investigating the relations between them. Though remaining firmly within the Aristotelian framework, he thus started down one of the pathways that would lead to the emergence of modern analysis. See Anneliese Maier, "The Nature of Motion," in *On the Threshold of Exact Science: Selected Writings of Anneliese Maier on Late Medieval Natural Philosophy,* trans. and ed. Steven D. Sargent (Philadelphia: University of Pennsylvania Press, 1982), 21–39; John E. Murdoch and Edith D. Sylla, "The Science of Motion," in *Science in the Middle Ages,* ed. David C. Lindberg (Chicago: University of Chicago Press, 1978), 206–264; and Edward Grant, *The Foundations of Modern Science in the Middle Ages: Their Religious, Institutional, and Intellectual Contexts* (Cambridge: Cambridge University Press, 1996), 95–98. We shall explore these matters in greater detail in Chapter 4.

16. Aristotle did at times offer descriptions using the terms "ratio" and "proportion." In *On the Heavens,* bk. 1, 273b30–274b18, for example, he wrote, "A given weight moves a given distance in a given time; a weight which is as great and more moves the same distance in a less time, the times being in inverse proportion to the weights." Elsewhere, in *Physics,* bk.7, 249b27–250a25, he used the term "rules of proportion," also sometimes translated as "ratio," in discussing relations involving the forces between moving bodies (a mover and a moved), distance, and time. Although he did believe that mathematics could be applied practically to some questions in physics, his efforts along those lines bore little resemblance to modern, quantitative analysis, notwithstanding R. J. Hankinson's overreaching claim that "Aristotle's principles represent a bold attempt to bring physical phenomena within the grip of generally mathematical ratios." R. J. Hankinson, "Science," in *The Cambridge Companion to Aristotle,* ed. Jonathan Barnes (Cambridge: Cambridge University Press, 1995), 147 and generally 140–167. The Stagirite's physics remained rooted in definitions arising from conventional lan-

guage. Indeed, the Greek term for "proportion" is *analogia,* from which we derive 'analogy.' For Aristotle, making analogies involves a linguistic comparison between the analogue (or source) topic and the target topic, a comparison he conveyed with the expression "is proportionate [analogous] to." Thus, his assertion that time is in "inverse proportion" to "weight" resembled more a qualitative comparison, much closer to analogy, than to mathematical physics.

17. Aristotle, *On Generation and Corruption,* bk. 2, 328b25–330a1. In our own times the differentials between contrary states of energy underlie much of the science of thermodynamics and its famed four laws.

18. C. S. Lewis, *The Discarded Image: An Introduction to Medieval and Renaissance Literature* (Cambridge: Cambridge University Press, 1964), 94–95.

19. The heliocentric interpretation was more clearly and fully formulated in the century after Aristotle by Aristarchus of Samos (313–ca. 230 BCE).

20. Aristotle, *On the Heavens,* bk. 1, 268b23–269b30: "Let us then apply the term 'heavy' to that which naturally moves towards the center, and 'light' to that which moves naturally away from the center." "All simple motion, then, must be motion either away from or towards or about the center." See too Aristotle, *Meteorology,* bk. 2, 365a25–7: "It is absurd to think of up and down otherwise than as meaning that heavy bodies move to the earth from every quarter, and light ones, such as fire, upwards."

21. Aristotle, *On the Heavens,* bk. 2, 286a5–7.

22. Ibid., bk. 1, 270b13–19.

23. "The substance of the heaven and stars we call ether . . . because it is in continual motion, revolving in a circle, being an element other than the four." Aristotle, *On the Universe,* 392a5–9. Ether posed a conundrum for Aristotle and scholastics alike. Change was associated with matter, but the heavens were eternal and thus unchanging. So, was ether matter or not? No conclusive answers were forthcoming, although referring to the ethereal matter of the celestial sphere became commonplace in medieval times. See Grant, *Foundations,* 65.

24. This was a central objective of astronomers after Claudius Ptolemy (ca. 100–ca. 175) and throughout the Middle Ages, to be discussed later.

25. Aristotle was not always clear on the details of this matter, sometimes referring to many "unmoved movers," each associated with a separate crystalline sphere. In general, however, he resorted to a single source of unmoved motion to account for all the movement in the cosmos. See G. E. R. Lloyd, *Aristotle: The Growth and Structure of His Thought* (Cambridge: Cambridge University Press, 1968), 144–153; and Grant, *Foundations,* 67.

26. Aristotle, *Metaphysics,* bk. 12, 1072a19–1073a3. Within this general cosmological picture, natural philosophers of the Middle Ages faced several central difficulties left unexplained by their Philosopher. First, if all celestial matter were composed of the same ether, then how could one account for the apparent differences between the visible ether of stars and planets and the invisible ether filling

the rest of the heavens? Second, how could moving, ethereal matter cause movement in the nonethereal matters of the sublunary realm? Third, and most significantly, how could motion be imparted materially to the spheres themselves from some nonmaterial cause? Grant, *Foundations,* 65–67.

27. Aquinas, *Summa Theologica,* pt. 1, qu. 1, art. 2, in *Basic Writings,* 1:7.

28. Aristotle, *Physics,* bk. 3, 200b10–25. The Greek term *physis* (from which our 'physics') designated generally what we mean by "nature." A good introduction to Aristotle's analysis of change may be found in J. L. Ackrill, *Aristotle the Philosopher* (Oxford: Oxford University Press, 1981), 23–33.

29. Quoted in Jonathan Barnes, "Life and Work," in Barnes, *Cambridge Companion to Aristotle,* 12 and more generally 6–15. Barnes notes that the extant writings of Aristotle are all "esoteric" in the ancient meaning of the term—that is, "tough and technical," intended only for philosophers and for use within the Lyceum. Of his "exoteric" writings, those written with more style and elegance for a broader public, only a few fragments have survived. Aristotle's own advice on writing suggests the incompleteness of his prose: "It is a general rule that a written composition should be easy to read and therefore easy to deliver. This cannot be so where there are many connecting words or clauses, or where punctuation is hard." Aristotle, *Rhetoric,* bk. 3, 1407b12–15.

30. The writings of Aristotle have commanded commentary for well over two millennia, both to interpret his ideas and, even more often, to address problems and inconsistencies within his system. In addition to the excellent Akrill, Lloyd, and Lear volumes already cited, earlier but still quite useful are John Herman Randall Jr., *Aristotle* (New York: Columbia University Press, 1960), and David Ross, *Aristotle,* 6th ed. (1932; New York: Routledge, 2004). The more ambitious will be well rewarded with the earlier-cited Terrence Irwin, *Aristotle's First Principles* (Oxford: Oxford University Press, 1990).

31. Grant, *Foundations,* 127–131.

32. Aristotle, *Physics,* bk. 3, 200b25–201a14; Aristotle, *On Generation and Corruption,* bk. 2, 328b25–35: "For coming-to-be and passing-away occur in naturally constituted substances only given the existence of sensible bodies."

33. G. E. R. Lloyd, *Demystifying Mentalities* (Cambridge: Cambridge University Press, 1990), 81–82. The Latin *ex nihilo nihil fit* (nothing comes from nothing) was the common scholastic rendering of Parmenides's dictum. Parmenides, "Fragments and Commentary," in *The First Philosophers of Greece,* ed. and trans. Arthur Fairbanks (London: K. Paul, Trench, Trubner, 1898), 86–135.

34. Aristotle, *Physics,* bk. 1, 186a22–186b13, 191a25–191b26. Aristotle exploited a logical point to challenge Parmenides. Logically, from 'All S is P,' it does not follow that 'All P is S.' From 'All cats are animals' we cannot logically infer 'All animals are cats.' "Whiteness," he explained, cannot exist apart from a white thing, but nonetheless we can say that "whiteness" is. It differs from a white thing ("that which is white") "in definition." So too with "being" or "what just is." In logical

terms, "what just is is not attributed to anything, but other things are attributed to it." Thus "what just is" can also be white, but white is not identical to "what just is" (any more than all animals are cats). But, according to Parmenides's substantive formulation, if "being white is not what just is," it follows that "what is [that is, "being"] white is not" and hence "what just is is not." This is clearly a contradiction, from which Aristotle concluded that the word 'is' possesses "more than one meaning," which Parmenides failed to see. Parmenides had not perceived the distinction between (1) the 'is' in its substantive formulation of "being" ("just what is"), and (2) the 'is' as a copula in a logical statement. As a logical copula connecting the subject and predicate of a proposition, 'is' can perfectly well connect "what just is" and "not what just is." As Ackrill notes, for Aristotle 'is' "does not always assert *identity* . . . [but also] serves to ascribe a characteristic to something." Ackrill, *Aristotle,* 25. Thus we can indeed say of nonbeing that it is, or vice versa, and complement either subject with predicates. Applied to change, we can therefore say a thing can become something else, and, consequently, change is not an illusion. Aristotle's move here further distanced the knower from the known as he shifted from (a) talking about being to (b) talking about how we talk about being—in essence, the philosophical leap.

35. Aristotle, *Physics,* bk. 3, 200b25–28. Without all the "being" language and references, physicists today capture something of Aristotle's meaning with the distinction between potential and kinetic energy, a system's (measurable) ability to do work. The former designates the energy stored in a body or system, available for work, while the latter is the energy of motion, the work actually being done.

36. Ibid., bk. 3, 201a28-33; Aristotle, *Metaphysics,* bk. 9, 1046a10.

37. Aristotle, *Physics,* bk. 2, 192b12–16. This translation is from Lear, *Aristotle,* 15.

38. Ibid., bk. 2, 193b13.

39. Francis Bacon, *The Great Instauration,* in *The Works of Francis Bacon,* 2 vols. (New York: Hurd and Houghton, 1878), 1:42. Akrill notes, similarly, that Aristotle produced "extremely general and abstract formulations backed up by just one or two standard examples." Ackrill, *Aristotle,* 27.

40. Aristotle, *Physics,* bk. 1, 185a13–15. Aristotle's use of "induction" differs considerably from the modern sense. While both start with particular sense impressions, the modern version gleans its abstractions from a far wider array of observed facts, including those derived from experiments. Aristotle, to repeat, sought primarily to explain facts already known, not to come up experimentally with new ones.

41. Ibid., bk. 2, 194b17–195a3. My account here owes much to Lear, *Aristotle,* 15–54.

42. For his summary of the arguments, see David Hume, *An Enquiry concerning Human Understanding* (LaSalle, IL: Open Court, 1956), 49. Immanuel Kant claimed that causality was among the a priori categories—see *Critique of Pure*

Reason, trans. Norman Kemp Smith (1929; New York: St. Martin's, 1965), 113. But regardless of their philosophical assumption—"natural instinct" versus "a priori" categories—modern scientific practices generally comport with Hume's description of the passage from constant conjunction to causal connection.

43. Lear, *Aristotle,* 27.

44. Ibid., 26–33; Aristotle, *Physics,* bk. 2, 194b16–195a3.

45. Aristotle, *Physics,* bk. 3, 202a12–37.

46. "By gradual advance . . . we come to see clearly that in plants too that is produced which is conducive to the end—leaves, e.g. grow to provide shade for the fruit." Aristotle, *Physics,* bk. 2, 199a20–30.

47. "Things that are for the sake of something include whatever may be done as a result of thought or of nature." Ibid., bk. 2, 196b22–24.

48. For an account of Aristotle's "non-intentional teleology," see R. J. Hankinson, "Philosophy of Science," in Barnes, *Cambridge Companion to Aristotle,* 124–135.

49. In the nineteenth century, G. W. F. Hegel and (by some accounts) Karl Marx revived much of the language and style of Aristotle's efficient causation, the former in his descriptive metaphysics, the latter in his social and economic analyses. Even so, neither focused chiefly on the natural world. Aristotle, by contrast, drew his illustrations from social and artistic experience for the express purpose of explaining how we were to understand nature.

50. Among scholastics, a few (Siger of Brabant, Boethius of Dacia) had apparently held with Aristotle that the universe is eternal, uncreated, at least insofar as "prime matter" is eternal and coeternal with God. Bets were hedged, though, as Siger and Boethius alike also maintained that God created both matter and the world (even though both were eternal). This was a tricky conceptual maneuver. In general, when faith conflicted with Aristotle, faith had to prevail. The majority of scholastic Aristotelians believed with Aquinas that the world's eternity could not be established through reason. Therefore, biblical creation could be accepted without "scientific" refutation, with the world's beginning understood as "an object of faith, but not of demonstration or science." For Aristotle on eternity, see *On the Heavens,* bk. 1, 283b6–32; and Grant, *Foundations,* 74–76. Among "radical," as well as conservative, Aristotelians, however, creationist causality was still applied to existing natural processes.

51. Thus Aquinas: "The same effect is from God and from the natural agent." Thomas Aquinas, *Summa Contra Gentiles,* bk. 3, ch. 70, in *Basic Writings,* 2:129.

52. Aristotle, *Physics,* bk. 2, 199a7–8.

53. Aquinas, *Summa Theologica,* pt. 1, qu. 45, art. 7, in *Basic Writings,* 1:444 (italics in the original). See the Gifford Lectures of the neo-Thomist philosopher Étienne Gilson, *The Spirit of Mediaeval Philosophy,* trans. A. H. C. Downes (New York: Charles Scribner's Sons, 1936), for a sympathetic account of Christian medieval philosophy and its treatment of transitive causality, among other topics,

84–107. 'Transitive' now generally refers to a feature in mathematics: if *a* is in Relation (R) to *b* and *b* is in R to *c,* then *a* is in R to *c*. For example, the relations 'greater than,' 'less than,' and 'equal to' are all transitive.

54. Aquinas, *Summa Theologica,* pt. 1, qu. 3, art. 2, in *Basic Writings,* 1:28. The modern Protestant theologian Paul Tillich challenged what he saw as the "theological theism" of the scholastics and others with the notion that God is the "ground of being," not a "Being" per se, even one of the highest imaginable order. The latter, he argued, makes God a thing, something created, and not the Creator. Tillich sought to transcend traditional theism by exploring the content of "absolute faith"—the God above God, who embraces nonbeing into his own Being in the power of self-affirmation. See, for example, Paul Tillich, *The Courage to Be* (New Haven, CT: Yale University Press, 1952), 178–190 and throughout. These are metaphors for an existentialist age, but they are intended to address the dilemma of being, posed by Parmenides and answered in a different way by Aristotle and Aquinas. Still, in the spirit of Francis we should give Saint Thomas his due: "God is the principle of all being, . . . not contained in any genus." We may speak of divine creativity and categorize its manifestations without constraining God by our classification system. For Aquinas divine creativity will always surpass the ingenuity of human classifying (an argument to which we shall return in the "affair" of Galileo). See Aquinas, *Summa Theologica,* pt. 1, qu. 3, art. 5, in *Basic Writings,* 1:31–32. Other scholastic theologians, most notably Meister Eckhart (ca. 1260–ca. 1327), pursued an alternative tradition of negative or "apophatic" theology, in which God could only be described by references to what he is not. "God is neither good, nor better, nor best; hence I speak as incorrectly when I call God good as if I were to call white black." In short, one cannot "chatter about God . . . [without] telling lies and sinning." Meister Eckhart, *Meister Eckhart: The Essential Sermons, Commentaries, Treatises, and Defense,* trans. Edmund Colledge and Bernard McGinn, preface by Huston Smith (Mahwah, NJ: Paulist, 1981), 80, 207.

55. Aquinas, *Summa Theologica,* pt. 1, qu. 45, art. 1, in *Basic Writings,* 1:433 (italics in the original). For Aquinas on emanations generally, see ibid., 433–446.

56. Aristotle, *Metaphysics,* bk. 12, 1072b10–30 (italics in the original).

57. Ibid., bk. 12, 1072b3–4. We still retain the linguistic intimacy between 'motion' and 'being moved,' between 'motion' and 'emotion.'

58. These references are cited by Grant, *Foundations,* 67. Wikipedia currently lists ten popular songs with the title "Love Makes the World Go 'Round."

59. "In God there is love, because love is the first movement of the will and of every appetitive power." Aquinas, *Summa Theologica,* pt. 1, qu. 20, arts. 1–4, in *Basic Writings,* 1:215–228.

60. John Paul II, "Encyclical Letter: *Fides et Ratio*" (1998), http://w2.vatican.va/content/john-paul-ii/en/encyclicals/documents/hf_jp-ii_enc_14091998_fides-et-ratio.html; the quotations are, in order, from secs. 16, 43, 39, 80, 93, 9, 66, 88, 47, 49, 61.

61. See Arthur C. Danto, *Connections to the World: The Basic Concepts of Philosophy* (New York: Harper and Row, 1989), 1–13, 38, 192, and throughout.

62. Galileo Galilei, *The Assayer,* in *Discoveries and Opinions of Galileo,* trans. Stillman Drake (New York: Anchor Books, 1957), 238.

Teeming Things and Empty Relations

1. Mark Thakkar, in "Mathematics in Fourteenth-Century Theology," in *The Oxford Handbook of the History of Mathematics,* ed. Eleanor Robson and Jacqueline Stedall (Oxford: Oxford University Press, 2009), 619, and more generally 619–638, notes that "all would-be historians of medieval mathematics must ask themselves where to look for their subject matter." A few scholastics, of course, did write on mathematics. Thomas Bradwardine (1290–1349), Richard Swineshead (fl. ca. 1340–1354), and Nicole Oresme (ca. 1320–1382) come to mind. But mostly one must look elsewhere for evidence of developments in mathematical thinking. Thakkar's own investigations uncover illustrations of mathematics use in theology.

2. See Toby E. Huff, *The Rise of Early Modern Science: Islam, China, and the West,* 2nd ed. (Cambridge: Cambridge University Press, 2003), 47–117 and throughout. Huff provides a comparative study of three civilizations with an eye toward explaining the social, cultural, and institutional reasons behind the rise of modern science in the Latin West.

3. Reviel Netz has compiled a list of 144 individuals who were known or likely mathematicians in Greek culture during the millennium beginning with the fifth century BCE. Beyond this small number, he estimates there were probably at most some three hundred "Greek mathematicians active in antiquity" by the time of Proclus (414–485). See Netz's incisive work *The Shaping of Deduction in Greek Mathematics: A Study in Cognitive History* (Cambridge: Cambridge University Press, 1999), 282–283.

4. "The mathematical requirements for even the most developed economic structures of antiquity can be satisfied with elementary household arithmetic, which no mathematician would call mathematics." O. Neugebauer, *The Exact Sciences in Antiquity,* 2nd ed. (Princeton, NJ: Princeton University Press, 1952; New York: Dover, 1969), 71–72.

5. Aristotle, *Metaphysics,* bk. 1, 981b18–26.

6. Thus G. E. R. Lloyd, in *Demystifying Mentalities* (Cambridge: Cambridge University Press, 1990), 97, emphasizes the "competitiveness . . . of Greek intellectual life and culture" in the "conception and practice of proof." For an account of "play" in Greek culture, see the classic work of Johan Huizinga, *Homo Ludens: A Study of the Play-Element in Culture* (1950; Boston: Beacon, 1955), 71–75 and throughout. Aristotle's observation has been corroborated in modern times by the meticulous scholarship of Reviel Netz, whose analyses of Greek mathematics, es-

pecially Archimedes, emphasize its ludic character. See Reviel Netz, "The More It Changes . . . Reflections on the World Historical Role of Greek Mathematics," in *Greek Science in the Long Run: Essays on the Greek Scientific Tradition (4th c.* BCE–*17th c.* CE*)*, ed. Paula Olmos (Newcastle upon Tyne: Cambridge Scholars, 2012), 152–168. In addition to Netz, *Shaping of Deduction,* see also, among his other works, Reviel Netz, *Ludic Proof: Greek Mathematics and the Alexandrian Aesthetic* (Cambridge: Cambridge University Press, 2009); Reviel Netz, *The Transformation of Mathematics in the Early Mediterranean World: From Problems to Equations* (Cambridge: Cambridge University Press, 2004); and Reviel Netz and William Noel, *The Archimedes Codex: Revealing the Secrets of the World's Greatest Palimpsest* (2007; London: Orion Books, 2008).

7. Netz, *Shaping of Deduction,* 271–312.

8. John J. Roche, *The Mathematics of Measurement: A Critical History* (London: Athlone, 1998), 1–50 and throughout. In her discussion of Graeco-Roman mathematics, Serafina Cuomo remarks that among some authors "mathematical practices" often carried great "moral and political significance," while others either denied their importance altogether or pursued mathematics as a form of tranquility (*ataraxia*). During the Hellenistic period it appears that occasionally mathematicians applied their labors in the service of practical engineering problems, but the evidence is scanty, to say the least. See Serafina Cuomo, *Ancient Mathematics* (New York: Routledge, 2001), 204, 208, and more generally 201–211. There remains a controversial and ongoing question of where to draw the line between pure and applied mathematics in the ancient world, of how much craftsmanship (say, architecture) owed to theory (say, of proportions). Some of this is a matter of definition. People of all cultures, of course, counted and measured in myriad practical, often sophisticated ways, activities that, in Neugebauer's account in *Exact Sciences in Antiquity,* had little to do with mathematics per se.

9. See *The Didascalicon of Hugh of Saint Victor: A Medieval Guide to the Arts,* trans. Jerome Taylor (New York: Columbia University Press, 1961), 63, 67. A careful classifier and observer of the liberal arts, as well as theologian and mystic, Hugh wrote that the proper concern of mathematics is "abstract quantity," a subject he partitioned into the major categories of "continuous" and "discrete" (magnitude and multitude). He further divided magnitude into "immobile" (geometry) and "mobile" (astronomy) and declared of multitude that arithmetic concerns what "stands in itself," while music regards what stands "in relation to another multitude." It bears mention that Hugh's treatment of the seven liberal arts in the *Didascalicon* predates the rediscovery and new translations of Aristotle; Hugh relied mainly on the commentaries of Boethius and Augustine. For a modern account of the *Didascalicon* and its place in Hugh's overall oeuvre, see Paul Rorem, *Hugh of St. Victor* (Oxford: Oxford University Press, 2009), 15–37.

10. In his popular and quite readable study *The Measure of Reality: Quantification and Western Society, 1250–1600* (Cambridge: Cambridge University Press,

1997), Alfred W. Crosby builds a good case for the growing importance of quantification throughout the Renaissance, and growth in pantometry (universal measurement) certainly fed into the emerging technology of modern, relational numeracy during the period. Still, for most, words continued to matter far more than numbers (except perhaps, as we shall see, for number words and their religious associations in numerology, the cabbala, magic, mysticism, and so forth). Not until the existing, classifying means of information mastery were overwhelmed by too much information, one of the consequences of printing, did alternatives start to appear attractive, especially as the inability of resolving religious conflicts through words shattered the tattered remnants of Christendom's unity during the religious wars of the Reformation. The emergence of quantification, relational numeracy, modern mathematics, and the analytical temper itself needs to be understood against the backdrop of a numberless world.

11. Proper qualification should attend the phrase "literate culture." Certainly what we would call literacy, the basic ability to read and write, was not prevalent among the majority of Greeks, or indeed of people in any society until well into our own era. (Estimates are that during the late days of the Republic perhaps only 10 percent of Romans were literate.) But by the middle of the fourth century BCE, among the political, cultural, and intellectual elites, writing grew to be paramount as a vehicle of expression. "The Greek Muse," concludes Havelock, "had truly learnt to write, and to write in prose—and even to write in philosophical prose." Eric A. Havelock, *The Muse Learns to Write* (New Haven, CT: Yale University Press, 1986), 116 and more generally 98–116.

12. Netz, *Shaping of Deduction,* 306; Netz, "The More It Changes," 161.

13. Netz provides a thoroughly convincing analysis of the place and unity of mathematics in the culture of Greek literacy. See Netz, "The More It Changes," 160–168; and Netz, *Shaping of Deduction,* especially his chapters "The Lettered Diagram" (12–67), "The Mathematical Lexicon" (89–126), and "The Historical Setting" (271–312). The phrase "formulaic language" should not be confused with mathematical equations or formulas; rather, it refers to a "(relatively) rigid way of using groups of words." Netz, *Shaping of Deduction,* 127. See also Netz, *Ludic Proof,* ix–xv.

14. Occasionally astronomy, music, and optics were classified as among the *mathematica media,* along with statics and the science of weights, which referred primarily to measurements.

15. Charles Taylor, *A Secular Age* (Cambridge, MA: Harvard University Press, 2007), 29, 32, and more generally 29–41. Stephen Wilson packs a lot of information pertaining to popular enchantments in his work *The Magical Universe: Everyday Ritual and Magic in Pre-Modern Europe* (London: Hambledon and London, 2000).

16. See Euan Cameron's account of superstition in the Middle Ages, and particularly of Aquinas on "angelogy, demonology, and the relationships between

spirit and matter," in *Enchanted Europe: Superstition, Reason, & Religion, 1250–1750* (Oxford: Oxford University Press, 2010), 89–102 and more generally 29–139.

17. Frederick A. Homann, introduction to *Practical Geometry (Practica Geometriae),* by Hugh of Saint Victor, trans. Frederick A. Homann (Milwaukee: Marquette University Press, 1991), 1–30.

18. See Menso Folkerts, "The Importance of the Latin Middle Ages for the Development of Mathematics," in *Essays on Early Medieval Mathematics: The Latin Tradition* (Burlington, VT: Ashgate, 2003), sec. I.

19. Hugh of Saint Victor, *Didascalicon,* 64–66.

20. Thierry of Chartes, *Tractatus de sex dierum operibus,* quoted in David Albertson, *Mathematical Theologies: Nicholas of Cusa and the Legacy of Thierry of Chartres* (Oxford: Oxford University Press, 2014), 111. Albertson's study follows the thread of theological reflection on mathematics from Pythagoras and Plato through the Neoplatonism of Augustine, Boethius, and other figures from late antiquity into the Middle Ages, where it extended to the work of Thierry of Chartres and, especially, Nicholas of Cusa (1401–1464).

21. John E. Murdoch, "Infinity and Continuity," in *The Cambridge History of Later Medieval Philosophy: From the Rediscovery of Aristotle to the Disintegration of Scholasticism, 1100–1600,* ed. Norman Kretzmann, Anthony Kenny, and Jan Pinborg (Cambridge: Cambridge University Press, 1982), 564; Nicole Oresme, *A Treatise on the Configuration of Qualities and Motions,* trans. Marshall Clagett, in *A Source Book in Medieval Science,* ed. Edward Grant (Cambridge, MA: Harvard University Press, 1974), 243–253.

22. See Marshall Clagett, *The Science of Mechanics in the Middle Ages* (Madison: University of Wisconsin Press, 1959), 331–346. In Chapter 4, we shall examine these developments in greater detail.

23. Charles Freeman, *The Closing of the Western Mind: The Rise of Faith and the Fall of Reason* (New York: Vintage Books, 2002), 323–333.

24. Michael S. Mahoney, "Mathematics," in *Science in the Middle Ages,* ed. David C. Lindberg (Chicago: University of Chicago Press, 1978), 146 and more generally 145–178.

25. See the modern classic of Frances A. Yates, *The Art of Memory* (Chicago: University of Chicago Press, 1966), throughout, but esp. chs. 4–6.

26. Anthony Grafton, *New Worlds, Ancient Texts: The Power of Tradition and the Shock of Discovery,* with April Shelford and Nancy Siraisi (Cambridge, MA: Harvard University Press, 1992); David Wootton, *The Invention of Science: A New History of the Scientific Revolution* (New York: HarperCollins, 2015), 57–109.

27. See the still valuable work of Marie Boas, *The Scientific Renaissance, 1450–1630* (New York: Harper and Row, 1962), 144–153; and Wootton, *Invention of Science,* 182–186.

28. Ann M. Blair, *Too Much to Know: Managing Scholarly Information before the Modern Age* (New Haven, CT: Yale University Press, 2010). For a more detailed

account of commonplace thought and the "rupture of classification" brought on by printing, see Michael E. Hobart and Zachary S. Schiffman, *Information Ages: Literacy, Numeracy, and the Computer Revolution* (Baltimore: Johns Hopkins University Press, 1998), 87–111 and bibliography, 283–285.

29. John Donne, "An Anatomy of the World: The First Anniversary," in *John Donne's Poetry,* ed. A. L. Clements (New York: W. W. Norton, 1966), 73.

30. Scholars generally note a resurgence of classifying in the natural sciences during and after the seventeenth century, born of the influence of Francis Bacon and the empirical turn to fact gathering from nature. See, for example, Peter Dear, *The Intelligibility of Nature: How Science Makes Sense of the World* (Chicago: University of Chicago Press, 2006), 39–66. But behind the new mode of gathering information lay a new mode of making sense of it, our current topic, the analytical temper.

31. For the Greeks, *techné* meant generally "art," "artifice," "skill," or "craft"—the practical arts and the methods that characterized them. In our time Ursula Franklin, metallurgist and physicist, simply calls it "practice," "the way things are done around here." See Ursula M. Franklin, *The Ursula Franklin Reader: Pacifism as a Map,* intro. Michelle Swenarchuk (Toronto: Between the Lines, 2006), 137. See too her Massey Lectures, published as *The Real World of Technology,* rev. ed. (Toronto: House of Anansi, 1999), viii and throughout.

32. See Paul F. Grendler, *Schooling in Renaissance Italy: Literacy and Learning, 1300–1600* (Baltimore: Johns Hopkins University Press, 1989). Michael S. Mahoney observes that with the recovery of ancient mathematical works in the fifteenth and sixteenth centuries, European mathematicians could start thinking in Greek mathematical terms again, but not quite, for a "peculiar combination of medieval, Renaissance, and Greek motifs found in the sixteenth century constitutes the major factor in the emergence of modern mathematics at the turn of the seventeenth." Mahoney, "Mathematics," 146, 170, and in general 145–178. Stated simply, my central argument is that the new information technology of relational numeracy provided the catalyst for amalgamating these motifs and making modern mathematics possible.

33. It bears reiterating that my thesis here pertains only to the disenchantment of mathematics as a critical component of the early modern creation of analysis and mathematical physics, and hence the separation of science and religion. I leave to others the more broadly sweeping and global claims about the contributions of science and scientific rationality to modernization and secularization—Max Weber's "disenchantment of the world."

34. Ernst Cassirer, "Some Remarks on the Question of the Originality of the Renaissance," *Journal of the History of Ideas* 4, no. 1 (1943): 51 and more generally 49–56.

3. Early Numeracy and the Classifying of Mathematics

1. Both literacy and numeracy emerged from tokens and emblems, the primordial signs that predated writing and counting and that gave humankind its first means of record keeping. Michael E. Hobart and Zachary S. Schiffman, *Information Ages: Literacy, Numeracy, and the Computer Revolution* (Baltimore: Johns Hopkins University Press, 1998), 35–41, 116–118. In *The Math Gene: How Mathematical Thinking Evolved and Why Numbers Are like Gossip* (New York: Basic Books, 2000), Keith Devlin summarizes recent studies that show the evolutionary connection between language and "number sense" in human brains. (See esp. 1–70.) Devlin also distinguishes "number sense" from mathematics in a manner comparable to the present distinction between numeracy and higher mathematics. As used here, 'numeracy' does indeed assume an innate "number sense" in humans, but our attention is directed to the cultural form that "number sense" shouldered as a technology of information management. Though both word and number spring jointly and naturally from the brains of *Homo sapiens* (the frontal lobe serves as the "language center," while the left parietal lobe houses our "intuitive number module"), their cultural and historical paths diverged early on in the settled communities of the Fertile Crescent. See also Stanislas Dehaene, *The Number Sense: How the Mind Creates Mathematics,* 2nd ed. (New York: Oxford University Press, 2011), for greater details on many of the experiments that lead to Devlin's claims.

2. "Sesostris divided the country among all the Egyptians, giving each man the same amount of land in the form of a square plot. This was a source of income for him, because he ordered them to pay an annual tax. . . . This was how geometry as a land-surveying technique came to be discovered and then imported into Greece." Herodotus, *The Histories,* trans. Robin Waterfield, intro. Carolyn Dewald (Oxford: Oxford University Press, 1998), bk. 2, sec. 109, 135–136.

3. The expression comes from the opening lines of the Egyptian Rhind Mathematical Papyrus (RMP), the earliest-known mathematics instructional text, which dates from the mid-sixteenth century BCE. Sometimes the hieratic text is referred to as the Ahmose Papyrus, named after its copyist. Ahmose reported that he was copying an even earlier text, which scholars have placed in the second half of the nineteenth century BCE. A more complete rendering of the opening lines reads, "Correct method of reckoning, for grasping the meaning of things and knowing everything that is, obscurities . . . and all secrets." The RMP gets its modern name from a Scottish antiquarian, Alexander Henry Rhind, who while visiting Egypt in 1858 purchased the papyrus, probably from an illegal excavation. At his death it was acquired by the British Museum. A transcription, analysis, and facsimile have been produced by Gay Robins and Charles Shute in *The Rhind Mathematical Papyrus: An Ancient Egyptian Text* (New York: Dover, 1987).

4. Though the line between them is blurred, we need to bear in mind the discernible distinction between numeracy, which provided the basic symbols of

counting and measuring, and higher mathematics, which developed its own world out of these more primitive abstractions.

5. 'Constraint' here doesn't mean conceptual impossibility. Reviel Netz has argued that the Greeks *could* perform many feats found in modern mathematics—compute with fractions, manipulate geometrical relations with arithmetical operations, articulate a written decimal positional system, and the like—but that within the confines of a ludic and literary culture, other matters were far more important. Reviel Netz, "It's Not That They Couldn't," *Revue d'histoire des mathém matiques* 8 (2002): 263–289. He expands this argument in Reviel Netz, *Ludic Proof: Greek Mathematics and the Alexandrian Aesthetic* (Cambridge: Cambridge University Press, 2009), summarized in 230–241.

6. For the separation of practical and theoretical mathematics, see Markus Asper, "The Two Cultures of Mathematics in Ancient Greece," in *The Oxford Handbook of the History of Mathematics,* ed. Eleanor Robson and Jacqueline Stedall (Oxford: Oxford University Press, 2009), 107–132. Netz summarizes (*Ludic Proof,* 8): "In general . . . Greek geometry . . . is abstract rather than concrete, marking it off from the physical sciences, while, inside the theoretical sciences, it is marked by its opposition to arithmetic." See too Reviel Netz, *The Shaping of Deduction in Greek Mathematics: A Study in Cognitive History* (Cambridge: Cambridge University Press, 1999), 271–312.

7. In his highly polemical study, *The Forgotten Revolution: How Science Was Born in 300 BC and Why It Had to Be Reborn,* trans. Silvio Levy (New York: Springer-Verlag, 2004), Lucio Russo overreaches by far with his claim that the third century BCE witnessed a "scientific revolution" in our contemporary sense of the term. Nonetheless, he does provide evidence for a much greater measure of engagement with practical matters and inventions than is usually acknowledged, lending support to the idea of a third-century golden age of Greek science and mathematics. See also Serafina Cuomo, *Ancient Mathematics* (New York: Routledge, 2001), 62–142; and Mark A. Peterson, *Galileo's Muse: Renaissance Mathematics and the Arts* (Cambridge, MA: Harvard University Press, 2011), 40–42, 48–51. Still, we must remain cautious. Of technological devices from the Hellenistic period, the Antikythera mechanism, a complicated, calendric apparatus, essentially an analog computer for calculating the positions of the sun, moon, and planets, is the most famous, dating probably from the late second to early first centuries BCE. Certainly, a host of problems surrounding mechanics (Archimedes), astronomy (Euclid, Apollonius), optics (Euclid), and geometry and conic sections (Archimedes, Euclid, Apollonius), just to skim the surface, contributed mightily to the dramatic growth of Greek mathematics during the third century BCE, even though little direct evidence connects these developments with practical matters of technology or economics.

8. See esp. Netz, *Ludic Proof,* and G. E. R. Lloyd, *Demystifying Mentalities* (Cambridge: Cambridge University Press, 1990).

9. Netz, "It's Not That They Couldn't," 285.

10. Netz, *Ludic Proof,* 2. Elsewhere Netz summarizes, "Greek writing was dominated by verbal, indeed literary values" and exhibited a "certain . . . preference: literature is ranked above science, inside science philosophy is ranked above mathematics; persuasion (to the Greeks, the central verbal art) is ranked above precision and natural language above other symbolic domains." For Archimedes, then, the "value of efficient calculation is important . . . , but, in the context of a literary treatise, it is trumped by an even more important value, that of proximity to natural language." Netz, "It's Not That They Couldn't," 285–287.

11. Since the advent of formal set theory, following the pioneering work of Georg Cantor (1845–1918), the cardinal numbers are often called the "counting numbers" or the "natural number aggregate" (or set), symbolized as {1,2,3,4 . . . }, while 'ordinal number' is defined as the "order type of a well-ordered set"—that is, sets that have a "definite beginning and a next term after any term except the last." See Edna E. Kramer, *The Nature and Growth of Modern Mathematics* (Princeton, NJ: Princeton University Press, 1981), 4–7, 577–597. Here I am using the terms in their broader, historical senses to designate the "matching idea" and the "counting idea," features of numeracy born with the creation of whole number arithmetic. See Lucas N. H. Bunt, Phillip S. Jones, and Jack D. Bedient, *The Historical Roots of Elementary Mathematics* (New York: Dover, 1976), 1–4. For the terms 'cardinal' and 'ordinal,' see also Jagjit Singh, *Great Ideas of Modern Mathematics: Their Nature and Use* (New York: Dover, 1959), 6–7; and Tobias Dantzig, *Number: The Language of Science: A Critical Survey Written for the Cultured Non-mathematician,* 4th ed. (1930; Garden City, NY: Doubleday, 1954), 8–9.

12. For introductions to early counting and numeracy, in addition to Dantzig, *Number,* and O. Neugebauer, *The Exact Sciences in Antiquity,* 2nd ed. (Princeton, NJ: Princeton University Press, 1952; New York: Dover, 1969), the following well repay their reading: Karl Menninger, *Number Words and Number Symbols: A Cultural History of Numbers,* trans. Paul Broneer (1969; New York: Dover, 1992); Graham Flegg, *Numbers: Their History and Meaning* (London: Andre Deutsch, 1983); John D. Barrow, *Pi in the Sky: Counting, Thinking, and Being* (Oxford: Oxford University Press, 1992); Georg Gheverghese Joseph, *The Crest of the Peacock: Non-European Roots of Mathematics,* 3rd ed. (Princeton, NJ: Princeton University Press, 2011); Thomas L. Heath, *A History of Greek Mathematics,* 2 vols. (1921; New York: Dover, 1981); and Cuomo, *Ancient Mathematics.* An earlier, standard account of the development of Hindu-Arabic numerals is David Eugene Smith and Louis Charles Karpinski, *The Hindu-Arabic Numerals* (Boston: Ginn, 1911).

13. Stanislas Dehaene has devised a computer simulation of the neural network that likely exists in animal brains at the heart of their primitive "number sense." Still in a mostly hypothetical phase, described as the "accumulator metaphor," the program maps the "number-detecting neurons" that allow animals (including humans) to identify and register differences in small collections of events or

objects. Some experimental corroboration of the program has already been established, including the discovery of single, number-detecting neurons in the cortex of cats and recent empirical support for the existence of number neurons in monkey brains. More research promises further discoveries along these lines. For our interests, the core of the program and research to date reinforces the "cardinal" principle of number as the basis of the number sense, the correspondence between an event and the firing of number-detecting neuronal cells. Dehaene notes, carefully, that this is not yet our digital counter, not precisely a one-to-one correlation, but rather a "fuzzy" one, an analogical accumulator that operates on a "floating" level. Nonhuman animals may well have a "number sense," a "mental representation of quantities," but not a number symbol or concept or language. See Dehaene, *Number Sense,* 17–29.

14. As regards cardinal correlations, physicist and historian of science John J. Roche, in *The Mathematics of Measurement: A Critical History* (London: Athlone, 1998), 10, and throughout, distinguishes the "target group" (objects to be tallied) and the "auxiliary group or tally" (the marks that represent them, a form of "nonverbal counting"). Roche's study is rich with insights, especially useful for sorting out many ambiguities surrounding numerical abstraction and measurement.

15. We do, of course, possess visual images of large collections of things—a plague of locusts, an army of Persians, a pyramid of rocks—but they are vague and imprecise as images. Artists generally capture these collections with suggestive lines, not by drawing each unit member. Numerical precision means applying to our vague image the abstract number word and symbol of a counted collection. See Devlin, *Math Gene,* 39–70, for the argument that our innate number sense is limited roughly to recognizing 'three,' after which the abstractions of counting must be employed.

16. Dantzig, *Number,* 8.

17. Barrow, *Pi in the Sky,* 51, 92.

18. We can see this process with our own denary (decimal) numbers. Having counted by integers (1, 2, 3, 4, 5, . . .) to the base 10, repetitions were devised (ten plus one, ten plus two, . . .), with some denary systems giving these new number words; in others, number words were formed from a combination of the base 10 and the integers 1–9, such as 'eighteen,' 'nineteen,' and so on. When the base was attained a second time, a new word indicating twice the base ('twenty') was formed, while reaching the base a larger number of times—say, four—led to multiples and multiplying: four times the base or forty. After the multiple of the base, one then added the integers 1–9 in order to obtain the intermediate numbers between multiples of the base—for example, 44 equals four times the base 10, plus the integer 4. When the base 10 was reached the tenth time, it was multiplied by itself (one hundred), or "squared" as the Greeks called it, whereupon the squared base was again multiplied, digits added and so on as before. With a fully developed base system, one could count indefinitely large collections and rank

order all their members. We can schematize symbolically the natural order of increasing magnitude of any additive counting system with a base B as follows: 1, 2, 3, . . . B−1, B, B+1, B+2, . . . 2B, 3B, . . . B(B−1), B^2, $2B^2$, $3B^2$, $B^2(B-1)$, . . . and so on. Barrow, *Pi in the Sky,* 92.

19. Euclid, *The Thirteen Books of the Elements,* 2nd ed., 3 vols., trans. Thomas L. Heath (1926; New York: Dover, 1956), 2:277. Euclid's definition was quite common in Hellenic and Hellenistic antiquity, as Heath notes elsewhere. Heath, *History of Greek Mathematics,* 1:69–70.

20. Roughly, the first nine Greek letters represented the digits 1–9, the next nine represented base multiples 10–90, and the final six represented base-squared multiples 100–600. Three additional ciphers, the archaic letters *vau, koppa,* and *sampi,* actually represented 6, 90, and 900 and were interspersed with the twenty-four-letter Greek alphabet to expand the sequence up to 999. See Thomas L. Heath, *A Manual of Greek Mathematics* (1931; Mineola, NY: Dover, 2003), 11–35.

21. Scholars sometimes distinguish between "empirical geometry" and "deductive geometry" to indicate the process of geometrical abstraction. For example, see Tobias Dantzig, *Mathematics in Ancient Greece* (1954, 1983; Mineola, NY: Dover, 2006), 41. See also Roche, *Mathematics of Measurement,* 18–19, 34.

22. In addition to the aforementioned works of Roche, Netz, and Cuomo, on whom I have relied heavily, see Wilbur Richard Knorr, *The Ancient Tradition of Geometric Problems* (New York: Dover, 1986). Too numerous to cite, substantial accounts of Greek higher mathematics may be found in the writings of Sir Thomas L. Heath, David E. Smith, Morris Kline, Tobias Dantzig, Otto Neugebauer, and Jacob Klein, among others. Their works are generally accessible, but not always for the mathematically shy.

23. One of the seven sages of ancient Greece, Thales of Miletus (ca. 624–ca. 546 BCE) was actually credited in classical sources with numerous firsts: the first philosopher, the first scientist, . . . and the first to use geometry. But, by tradition, Pythagoras was the first to introduce the serious study of mathematics into the Greek world. Sorting out fact from fiction among early Pythagoreans remains an area of classical scholarship where speculation invariably threatens to outpace evidence, of which there is precious little. For contemporary introductions not only to Pythagoras and his followers but also to the scholarly problems classicists face in mining plausible assertions about them, see the fine studies of Charles H. Kahn, *Pythagoras and the Pythagoreans: A Brief History* (Indianapolis: Hackett, 2001), 5–22 and throughout; and Christoph Riedweg, *Pythagoras: His Life, Teaching, and Influence,* trans. Steven Rendall (Ithaca, NY: Cornell University Press, 2005). These may be supplemented by Carl Huffman, "Pythagoras," in *The Stanford Encyclopedia of Philosophy,* Fall 2011 ed., ed. Edward N. Zalta, http://plato.stanford.edu/archives/fall2011/entries/pythagoras/. A valuable collection of ancient sources discussing Pythagoras, compiled by Kenneth Sylvan Guthrie in 1920, has been reissued and updated by David Fideler, ed., *The Pythagorean Sourcebook and*

Library (Grand Rapids, MI: Phanes, 1987, 1988). Historian Serafina Cuomo generalizes the overall point, reminding us that the paucity of "evidence does not allow us to be certain about anything" pertaining to ancient mathematics. Cuomo, *Ancient Mathematics,* 2. All commentary on early numeracy and ancient mathematics, therefore, including the present, should be qualified with a very large 'perhaps.'

24. Riedweg, *Pythagoras,* 51, 53–55, 63–64. Three biographies of Pythagoras date from late antiquity, their authors listed here chronologically: Diogenes Laertius (fl. third century CE); Porphyry (ca. 234–ca. 305); Iamblichus (ca. 245–ca. 325). These are available in Fideler, *Pythagorean Sourcebook,* 141–156, 123–135, and 57–122, respectively. Each of these authors relates more marvels than his predecessors; generally speaking, the further removed from Pythagoras's life, the fuller and more marvelous, one might say "creative," the account. Kahn, *Pythagoras and the Pythagoreans,* 5.

25. All initiates into Pythagorean communities had to swear an oath on the *tetractys:* "By him who gave to our soul the *tetractys,* / The Source and root of ever-flowing nature." Quoted in Kahn, *Pythagoras and the Pythagoreans,* 31–32. This is the earliest mention of the oath and comes from an account of Pythagoras given by the Stoic philosopher Posidonius, as reported by Sextus Empiricus. These writings occurred, respectively, five and seven centuries after the cult's founding, and further encourage us to take a page out of Sextus's skepticism.

26. Asper, "Two Cultures of Mathematics," 124–125, notes that Plato possessed at best an "outsider's interest" in mathematics, a point echoed by Netz's remark in *Shaping of Deduction,* 276, that Plato's "use of mathematics is done at a considerable distance from it."

27. The doorway inscription might be apocryphal, but it remains quintessentially Plato. For his remarks on the role of geometry as it pertains to the philosopher-king, see Plato, bk. 7, *Republic,* 525b–527c.

28. Whitehead's comment comes from his Gifford Lectures: "The safest general characterization of the European philosophical tradition is that it consists of a series of footnotes to Plato." Less often noticed is Whitehead's clarification, immediately following, in which he refers to Plato's thought as an "inexhaustible mine of suggestion." What else might one expect from the first systematic philosopher? Alfred North Whitehead, *Process and Reality: An Essay in Cosmology* (1929; New York: Harper and Row, 1957), 63. Plato's relation to Pythagoras resembles that of his ties to Socrates; we know about Pythagoras and Socrates largely through the Platonic filter.

29. The Greek *máthema* originally meant "learning" or "education," and the linkage between *máthema* and 'mathematics' per se may well be a later association. Thus, "the Pythagorean movement was the place where *mathémata* changed its meaning from 'doctrines,' i.e. matters being learned, to 'knowledge of number and magnitude,' i.e. 'mathematics.'" Jens Høyrup, *Measure, Number, and Weight:*

Studies in Mathematics and Culture (Albany: State University of New York Press, 1994), 10. Sometimes *mathematikoi* were called "students" and the *akousmatikoi* "hearers," while all followers were also referred to as "esoterics." See Iamblichus, "Life of Pythagoras," in Fideler, *Pythagorean Sourcebook,* 74, 76. Plato was acquainted with some of the *mathematikoi* (for example, Eudoxus, Theaetetus, and Archytas), but was also influenced by those who wanted to preserve the community's way of life, the *akousmatikoi,* particularly as regards their moral behavior and their related commitment to the pursuit of eternal verities. It bears reiterating too that Plato's interest in mathematics emphasized as much the ideal, harmonizing, and religious dimensions associated with the *akousmatikoi* as the problem-solving manipulations of the *mathematikoi,* an interest that left a profound imprint on subsequent generations of Platonists and Neoplatonists. Not until mathematics became modern, relational and analytical, did it subvert the spiritual dimension, one of the central themes of this book. Even so, many modern mathematicians still remain in their heart of hearts thoroughgoing Platonists.

30. Riedweg does remark that under the "Platonic surface, older material" pertaining to Pythagoras can sometimes be discerned. Riedweg, *Pythagoras,* 25. Knorr, *Ancient Tradition of Geometric Problems,* 1–99, does a masterful job of recreating a plausible account of pre-Euclidean developments in geometric problem solving. Beyond the studies of Knorr, Riedweg, Kahn, and Cuomo, introductions to Eudoxus are provided by G. E. R. Lloyd, *Early Greek Science: Thales to Aristotle* (New York: W. W. Norton, 1970), 80–98, and by Heath, *History of Greek Mathematics,* 1:322–335, while a good summary of the current state of scholarship surrounding Archytas may be found in Carl Huffman, "Archytas," in Zalta, *Stanford Encyclopedia of Philosophy,* Fall, 2011 ed., http://plato.stanford.edu/archives/fall2011/entries/archytas/.

31. Ptolemy I Soter (ca. 367–ca. 283 BCE), a Macedonian general under Alexander the Great and founder of the Ptolemaic dynasty in Egypt (323 BCE), personally sponsored Euclid, from whom he tried to learn geometry, although to little avail. After a long struggle, he finally asked Euclid if there "was not a shorter road to geometry than through the *Elements,*" whereupon Euclid replied with the line that has haunted arithmephobes ever since: "There is no royal road to geometry." The story was related several centuries later by the Neoplatonic philosopher Proclus, *Proclus: A Commentary on the First Book of Euclid's Elements,* trans. Glenn R. Morrow (1970; Princeton, NJ: Princeton University Press, 1992), 57.

32. Plutarch, among others, recounts the murder of Archimedes, an event metaphorically epitomizing subsequent developments. In 212 BCE during the conquest of Sicily, Marcus Claudius Marcellus (ca. 268–208 BCE), an important military figure during the Second Punic War, had laid siege to Syracuse, whose defenses Archimedes was reputed to have engineered with everything from catapults to fire-generating mirrors and other war machines. With the impending collapse of Syracuse, Marcellus had ordered Archimedes captured alive, the better to learn

from his enemy. Lost in a geometrical problem, however, Archimedes refused to leave immediately with his captor, a Roman soldier who became enraged and dispatched him forthwith, much to the chagrin of his commander. Plutarch, *Lives of the Noble Romans,* trans. John Dryden, rev. Arthur Hugh Clough, ed. Edmund Fuller (New York: Dell, 1959), 104–106. Of course, it wasn't only the Romans who brought about the slow decline of mathematics and science in Alexandria, although they figured significantly in the process. The reasons are many and complex, and quite beyond the pale of present concerns. See G. E. R. Lloyd, *Greek Science after Aristotle* (New York: W. W. Norton, 1973), 33–52, 154–178; Cuomo, *Ancient Mathematics,* 212–262; and Reviel Netz, "The More It Changes . . . Reflections on the World Historical Role of Greek Mathematics," in *Greek Science in the Long Run: Essays on the Greek Scientific Tradition (4th c.* BCE*–17th c.* CE*),* ed. Paula Olmos (Newcastle upon Tyne: Cambridge Scholars, 2012), 152–168.

33. Bunt, Jones, and Bedient, *Historical Roots of Elementary Mathematics,* 75–83. The *tetractys* also provided the foundation of musical harmony, a subject treated in Chapter 6.

34. Thus Nicomachus of Gerasa (ca. 60–ca. 120) noted that while the "sciences are always sciences of limited things," all "multitude and magnitude are by their own nature . . . infinite." Nicomachus of Gerasa, *Introduction to Arithmetic,* trans. Martin L. D'Ooge, in *Great Books of the Western World,* ed. Robert Maynard Hutchins et al. (Chicago: Encyclopaedia Britannica, 1952), 11:812.

35. Kahn dryly notes that this is the stuff more of "legend than history." Kahn, *Pythagoras and the Pythagoreans,* 35. See too Riedweg, *Pythagoras,* 26, 109–110; Heath, *History of Greek Mathematics,* 1:90–91; and Roche, *Mathematics of Measurement,* 19.

36. In the fashion cast by the Greeks, this problem no longer carries much philosophical interest. Their question was, how could an object *be* at the same time one thing and many things? How could a tree *be* both one tree and also many leaves, limbs, pieces of bark, and the like? When we pose the question now, we do not ask whether the tree is identical with many things—an absurdity—but only what does it mean to say the tree is composed of many parts. Yet in a different form the problem resurfaces. Consider a tree *T* composed of a collection of its parts *P* at any given moment in time. Since a tree can survive the loss of some of its parts (through pruning and so on), *T* continues to exist when *P* no longer does. Is *T,* then, something other than *P,* or, more generally, is each thing distinct from the sum of its parts? See R. M. Sainsbury, *Paradoxes* (Cambridge: Cambridge University Press, 1988), 7 and more generally 5–24, for an interesting introduction to Zeno's paradoxes. With his own introduction to a fascinating and challenging collection of essays on the topic, Wesley C. Salmon surveys modern treatments. See Wesley C. Salmon, introduction to *Zeno's Paradoxes,* ed. Wesley C. Salmon (1970; Indianapolis: Hackett, 2001), 5–44; the volume also contains Bertrand Russell's "The Problem of Infinity Considered Historically," 45–58. In *History of*

Greek Mathematics, 1:271–283, Heath showed how the paradoxes led to "banishing the idea of the infinite" among Greek mathematicians.

37. The most famous of Zeno's paradoxes was that of Achilles and the tortoise. The swift-footed Achilles must overtake the plodding tortoise, who is some distance ahead of him. The race starts; both move. He runs to where the tortoise was, only to discover it has moved on a little farther. He runs next to the point where the tortoise had moved, only to find the tortoise has moved again, though not as far this time. Achilles repeats and repeats his dashes in this fashion, but although he runs far faster than the tortoise, he never catches the plodder. G. E. L. Owen has argued that Zeno's paradoxes fit a pattern, with two of them ("Achilles and the Tortoise" and "The Dichotomy") designed to disprove the continuity of space and time and another two ("The Arrow" and "The Stadium") refuting the atomic, discrete structure of space and time. See G. E. L. Owen, "Zeno and the Mathematicians," in Salmon, *Zeno's Paradoxes,* 139–163.

38. Euclid devoted book 5 (and part of book 7) of the *Elements* to a theory of proportion, first perhaps developed by Eudoxus, which on some readings applied to incommensurable, as well as commensurable, quantities and thereby introduced real numbers and thus a treatment of the irrational. For the arguments for and against this interpretation, see Heath's commentary in Euclid, *Elements,* 2:112–129. True or not, during the Middle Ages neither of the Latin versions of Euclid found in the universities (one based on an Arabic text, another on a Greek text) contained an unadulterated account of book 5 of the *Elements.* As we shall see later, Galileo relied on a new translation of Euclid in Italian, which did contain an accurate book 5, and which provided the tools of his analysis. Modern mathematicians also generally credit Eudoxus with having devised the "method of exhaustion," sometimes interpreted as a precursor to integral calculus. Much historical anachronism infiltrates these observations.

39. José Ortega y Gasset, *The Idea of Principle in Leibnitz and the Evolution of Deductive Theory,* trans. Mildred Adams (New York: W. W. Norton, 1972), 58–62; Aristotle, *Categories,* 4b20–5a17; Aristotle, *Physics,* bk. 3, 200b19. Somewhat strange to us at first, the notion of a "common boundary" may be pictured in the framework of Euclidian geometry. If one cuts a line (say, with another perpendicular to it), the cut serves as the boundary, or point between two lines. The point is indivisible and has position, but no magnitude. It stands as the limit (terminus, boundary or end) of a line. A line in turn is defined as the continuous magnitude between two points. From which, Aristotle argued, being continuous and infinitely divisible, a line, strictly speaking, cannot be composed of points, even though neither lines nor points could be conceived without the other. Numbers, however, do not belong to such a continuous realm: "If five is part of ten the two fives do not join together at any common boundary but are separate; nor do the three and the seven join together at any common boundary." Here there is no "cut" that severs the number realm, no indivisible point standing between numbers.

One cannot shade imperceptibly from one number to another, as one can from point to point on a continuous line. (Clearly, we are not here in the arena of real numbers.) Notice too, it followed that numbers could not stand in a direct, one-to-one correspondence with the points on a line. The realms of arithmetic and geometry remained therefore incommensurable. We should mention that Aristotle took Zeno's paradoxes seriously and claimed to have resolved them. Without our going into detail here, Aristotle's counterargument in effect denied Zeno's incommensurable comparison of finite and infinite magnitudes and claimed that the proper correlations of space and time required finite segments of each, one of which could then be employed as the standard for comparing the other. Aristotle, *Physics,* bk. 5, 233a22–233b17. In refuting Zeno, Aristotle thus maintained the separation of continuity and discreteness.

40. In recent memory, Bertrand Russell highlighted the significance of the "problem" of incommensurability, identifying it as "one of the most severe and at the same time most far-reaching problems that have confronted the human intellect in its endeavor to understand the world." Bertrand Russell, *Our Knowledge of the External World* (1929, 1956; New York: Mentor Books, 1960), 129.

41. Aristotle, *Posterior Analytics,* bk. 1, 76b4–5, 74b38–39, and generally 74a38–77a4.

42. Its better-known title, the *Almagest,* derives from a medieval Latin rendition of the Arabic *Al-majisti,* itself a corruption of the Greek name under which the work came to be known in later antiquity, *Ho megale syntaxis* (The great compilation).

43. *Ptolemy's Almagest,* trans. G. J. Toomer (Princeton, NJ: Princeton University Press, 1998), 37. Toomer's translation is a noteworthy piece of scholarship in its own right, with thoroughgoing annotations and commentary. See his introduction (1–26) for a primer on how to read the *Almagest.*

44. Both Apollonius and Hipparchus introduced the use of epicycles and eccentrics into figuring planetary orbits. Hipparchus's numerous accomplishments included the discovery of the precession of the equinoxes, calculation of the mean lunar month, improved estimates of the size of the sun and moon, and a catalog of over 850 stars, as well as the first systematic use of trigonometry. Ptolemy relied heavily on the work of Hipparchus, whom he mentions frequently, and whose star chart he expanded into his own catalog of 1,028 stars. See Heath, *History of Greek Mathematics,* 2:253–259, 273–276; Lloyd, *Greek Science after Aristotle,* 53–74, 113–135.

45. *Ptolemy's Almagest,* 36.

46. Liba Chaia Taub, *Ptolemy's Universe: The Natural Philosophical and Ethical Foundations of Ptolemy's Astronomy* (Chicago: Open Court, 1993), 139 and throughout. In her accessible study Taub develops both the Aristotelian and Platonic strains of Ptolemy's astronomy and his "eclectic" philosophical views.

47. *Ptolemy's Almagest*, 35–37, 40–41. "[Thus one can distinguish] the corruptible from the incorruptible by [whether it undergoes] motion in a straight line or in a circle, and heavy from light, and passive from active, by [whether it moves towards the center or away from the center]." Ibid, 36.

48. Ibid, 36.

49. Ibid, 36–37. See also Taub, *Ptolemy's Universe*, 135–153.

50. Still the best introductions to the "two-sphere" universe of Ptolemy are those of Thomas S. Kuhn, *The Copernican Revolution: Planetary Astronomy in the Development of Western Thought* (New York: Vintage Books, 1957), 1–133, and Neugebauer, *Exact Sciences in Antiquity*, 145–207.

51. The expression is from Plato, *Timaeus*, 37d. This, of course, was a well-known matter of perspective. From the standpoint of the heavens, a solar year, for example, was eternal; from the standpoint of an earthly observer, a solar year betokened the changing seasons and comparable events. Similarly with other astronomical phenomena.

52. The pagan Neoplatonist Simplicius of Cilicia (ca. 490–ca. 560) had copied a long passage from Geminus that explicated the practice of saving the appearances or phenomena: "For it is no part of the business of an astronomer to know what is by nature suited to a position of rest, and what sort of bodies are apt to move, but he introduces hypotheses under which some bodies remain fixed, while others move, and then considers to which hypotheses the phenomena actually observed in the heaven will correspond. But he must go to the physicist for his first principles." Quoted in Thomas L. Heath, *Aristarchus of Samos: The Ancient Copernicus* (1913; New York: Dover, 1981), 276. See *Ptolemy's Almagest*, 420–423, for Ptolemy's discussion of hypotheses. See too Bernard R. Goldstein, "Saving the Phenomena: The Background to Ptolemy's Planetary Theory," *Journal for the History of Astronomy* 28 (1997): 1–12. We shall further discuss the framing of "hypotheses" in later chapters.

53. Aristotle, *Posterior Analytics*,bk. 1, 71a1. Aristotle's scarcely disguised hostility toward mathematics makes one wonder why Ptolemy was such a willing follower of the head Peripatetic, until we realize just how committed Ptolemy was to the premise that mathematics could not discover anything new but only rationalize what was seen or known. This premise underscored the unchanging, eternal nature of mathematics as a tool for explicating and contemplating the unchanging, eternal divine in its visible, spatial realm.

54. Beyond definitions, Netz has identified a number of linguistic "formulae" that "import the properties of natural language into a sub-language" used to create a "structured system" of deductions in Greek mathematics. Netz, *Shaping of Deduction*, 163, 127–167.

55. See the brilliant, though underappreciated, work of physicist Géza Szamosi, *The Twin Dimensions: Inventing Time and Space* (New York: McGraw-Hill, 1986),

67, 72, and more generally 57–87. We shall take up Szamosi's important distinction between measuring and keeping track of time in our treatment of music.

56. Aristotle, *Movement of Animals,* 698a25.

57. See Roche, *Mathematics of Measurement,* 1–50 and throughout. Roche remarks (34) that around the time of Plato "Greek theoretical geometry distanced itself considerably from practical geometry and mensuration." As a consequence of the problem of incommensurables, "theoretical geometry became non-metric."

58. For all their wisdom and brilliance, the Chinese, for example, never developed formal, abstract logic, complete with the rules of reasoning that connected words and things, or geometrical demonstrations. See Bryan W. Van Norden, *Introduction to Classical Chinese Philosophy* (Indianapolis: Hackett, 2011), 116–119.

59. Aristotle, *Physics,* bk. 4, 219b5–9; Aristotle, *Metaphysics,* bk. 14, 1092b19–23.

60. Roche, *Mathematics of Measurement,* 37 and more generally 36–39. Netz puts a premium on the "lettered diagram" as the visual means of expressing such objects. Netz, *Shaping of Deduction,* 12–67. See too Pseudo-Aristotle (probably written by one of his students), *Mechanics,* 847b16–848a37.

4. Thing-Mathematics

1. "Gerbert to Adalbold," ca. February–March 999, in *The Letters of Gerbert, with His Papal Privileges as Sylvester II,* trans. Harriet Pratt Lattin (New York: Columbia University Press, 1961), 299–302. The manuscript sources, to which both had access, consisted of a survey of mathematical literature compiled anonymously in the seventh century, the so-called *Corpus agrimensorum.* It included virtually all the mathematics available in the early Middle Ages: snippets of practical material from the Roman *agrimensores* (land surveyors) and mathematical passages from early medieval authors, such as Boethius, Isidore, and Cassiodorus, as well as a fragment from Euclid's *Elements,* book 1. The first proposition in Euclid's book 1 explained how to construct an equilateral triangle, but not how to find its area. See Harriet Pratt Lattin, introduction to *Letters of Gerbert,* 3–29, for a brief account of Gerbert's life, and also Marco Zuccato, "Gerbert of Aurillac," in *Medieval Science, Technology, and Medicine: An Encyclopedia,* ed. Thomas F. Glick, Steven J. Livesey, and Faith Wallis, Routledge Encyclopedias of the Middle Ages, vol. 11 (New York: Routledge / Taylor and Francis, 2005), 192–194. A charmingly written, but not always reliable, biography of Gerbert is Nancy Marie Brown, *The Abacus and the Cross: The Story of the Pope Who Brought the Light of Science to the Dark Ages* (New York: Basic Books, 2010). Brown accurately recites the facts of Gerbert's life (scarce as they are) but exaggerates considerably his historical significance.

2. His actual figures had produced a recorded height of $25\frac{5}{7}$ feet and an area of $385\frac{5}{7}$ square feet. The modern formula for the area, A, of an equilateral triangle

may be derived directly from the Pythagorean theorem and expressed algebraically as $A = \frac{\sqrt{3}}{4} s^2$, with *s* corresponding to any of its equal sides. Using the numerical measurements of Gerbert, this computes to an area of 389.7114 . . . square feet. Gerbert's figure of $385\frac{5}{7}$ is within ~1 percent of the modern value. See Costantino Sigismondi, "Gerbert of Aurillac: Astronomy and Geometry in Tenth Century Europe," *International Journal of Modern Physics* 23 (2013): 467–471. For Boethius on triangular numbers, see Boethius, *Boethian Number Theory: A Translation of the De Institutione Arithmetica,* trans. Michael Masi (Amsterdam: Editions Rodopi, 1983), 133–135.

3. "Gerbert to Adalbold," 300. Boethius had called the equilateral triangle the "mother of all numbers," because it was "born from unity," the source of all numbers and figures. Boethius, *De Institutione Arithmetica,* 134.

4. In Gerbert's words, the rule is "always give a seventh [more] to the side and allow the remaining six parts for the height." With the side of seven feet he computed the area as twenty-one square feet ($6 \times 7 \div 2$). This is not precisely the area of an equilateral triangle, whose value in modern expression is 21.2176. . . . Nor, by extension, was his triangle precisely equilateral.

5. For an excellent survey of the state of mathematics during the Middle Ages, see Menso Folkerts, "The Importance of the Latin Middle Ages for the Development of Mathematics," in *Essays on Early Medieval Mathematics: The Latin Tradition* (Burlington, VT: Ashgate, 2003), sec. 1.

6. The Latin correspondence is available with French commentary in Paul Tannery, *Mémoires Scientifiques,* vol. 5, *Sciences exactes au Moyen Age* (Toulouse: Édouard Privat; Paris: Gauthier-Villars, 1922), 229–303. The story is also told in the context of a discussion of tenth- and eleventh-century intellectual life in R. W. Southern, *The Making of the Middle Ages* (New Haven, CT: Yale University Press, 1953), 201–202 (more generally, 170–218), and in Stephen C. McCluskey, *Astronomies and Culture in Early Medieval Europe* (Cambridge: Cambridge University Press, 1998), 177–178.

7. Oscar G. Darlington, "Gerbert, the Teacher," *American Historical Review* 52, no. 3 (1947): 456–476, discusses Gerbert's career in nearly hagiographical terms, referring to him as one the "great teachers of all times."

8. Founded by Count, later Saint, Gerald (ca. 855–ca. 909), the monastery at Aurillac became closely allied with its more famous counterpart at Cluny, and like Cluny it promoted monastic reform and independence, along with the advance of learning and moral uprightness of its monks. In northern Spain, Gerbert spent three years (967–970) largely under the supervision of Atto, bishop of Vic, some sixty kilometers north of Barcelona, and at the nearby Monastery of Santa Maria de Ripoll. The extent of his acquisition of Arabic science and mathematics has been debated, but not his exposure to them.

9. Later, the Neoplatonist philosopher Proclus drew the practical / theoretical line between the mathematics that "attends the sensibles" and that part concerned with the "intelligibles." See Proclus, *Proclus: A Commentary on the First Book of Euclid's Elements,* trans. Glenn R. Morrow (Princeton, NJ: Princeton University Press, 1992), 31–33; Thomas L. Heath, *A History of Greek Mathematics,* vol. 1, *From Thales to Euclid* (1921; New York: Dover, 1981), 10–18.

10. Folkerts notes that earlier practical geometries did contain parts that were "likely" used directly in measuring distances and areas and in the making and using of instruments, but that most such practical knowledge was transmitted orally among skilled manual workers and craftsmen and was not a main focus of the classroom. Folkerts, "Importance," sec. 1.

11. Boethius, *De Institutione Arithmetica,* 71; "Gerbert to Otto Caesar," October 25, 997, in *Letters of Gerbert,* 296–297. A few days before, the young emperor Otto III had written Gerbert asking him to promote his own royal "zeal for study" by explaining the "book on arithmetic," a manuscript of Boethius's *De Arithmetica* that Gerbert had gifted him. "Emperor Otto to Gerbert, His Teacher," October 21, 997, in *Letters of Gerbert,* 294–295. In some of his earlier, extant letters, Gerbert wrote to one of his students (Constantine, a monk at Fleury) further details about Boethius's arithmetic and musical harmony (*De Musica*), as well as instructions on constructing a hemisphere for making astronomical observations, on using an abacus, and on categorizing numbers as "simple" or "composite," "digits" or "articles." "Gerbert to Constantine," 978(?), 978–980(?), 979 (?), first half of 980, in *Letters of Gerbert,* 36–47. For an account of the religious dimensions of Gerbert's pursuit of mathematics, see Joseph V. Navari, "The Leitmotiv in the Mathematical Thought of Gerbert of Aurillac," *Journal of Medieval History* 1 (1975): 139–150.

12. "Gerbert to Bishop Wilderode," December 31, 995, in *Letters of Gerbert,* 246–247, 261. This letter was in essence a legal brief concerning the prosecution of Arnulf, the archbishop of Reims, and his struggles with Hugh Capet of France. Precedent for interpreting the number 318 had been established in Genesis 14:14, where the members in Abraham's household are described as totaling 318, which equaled the alphanumerical sum of the name Eliezer, their chief. Eliezer had been designated Abraham's heir before Yahweh promised the patriarch a son. Alphanumerical correlations worked in Hebrew as well as Greek.

13. Nicomachus of Gerasa, *Introduction to Arithmetic,* trans. Martin L. D'Ooge, in *Great Books of the Western World,* ed. Robert Maynard Hutchins et al. (Chicago: Encyclopaedia Britannica, 1952), 11:811; Boethius, *De Institutione Arithmetica,* 76.

14. The term *denominativa* also appeared in discussions of rhetoric and logic, with different nuances in each case. See A. George Molland, "Mathematics," in *The Cambridge History of Science,* vol. 2, *Medieval Science,* ed. David C. Lindberg and Michael H. Shank (Cambridge: Cambridge University Press, 2013), 515 and more generally 512–531. In Boethius, *De Institutione Arithmetica,* 74, Masi's otherwise noteworthy translation sidesteps the Latin cognate along with its more direct

meaning: "All of those things are in the domain of number [denominatives of numbers]."

15. Boethius marveled that "from the beginning, all things whatever" have been "formed by reason of numbers." *De Institutione Arithmetica,* 75.

16. Steven Pinker argues that human language facility, the "language instinct," was born of natural selection and propelled by a "cognitive arms race" as early humans competed with one another for survival. Recognizing "realities" for what they "are" was critical to this process and underlies the deeply held belief in the connection between words and things or, as we shall see, numbers and things. Steven Pinker, *The Language Instinct: How the Mind Creates Language* (New York: William Morrow, 1994), 332–369.

17. See John R. Searle, *Intentionality: An Essay in the Philosophy of Mind* (Cambridge: Cambridge University Press, 1983), ch. 1, and throughout, for a clear and accessible introduction to contemporary philosophical discussions concerning intentionality as a "property" of mental states "directed" toward "objects and states of affairs in the world."

18. William Shakespeare, *The Merry Wives of Windsor,* 5.1.4.

19. Amos Funkenstein, *Theology and the Scientific Imagination: From the Middle Ages to the Seventeenth Century* (Princeton, NJ: Princeton University Press, 1986), 307–312.

20. David Albertson, *Mathematical Theologies: Nicholas of Cusa and the Legacy of Thierry of Chartres* (Oxford: Oxford University Press, 2014), 40–75. I have relied considerably on Albertson's interpretation for the theological dimensions of thing-mathematics in the Middle Ages. For a still valuable account of number mysticism and Augustine's contribution to its Christian version, see Vincent Foster Hopper, *Medieval Number Symbolism: Its Sources, Meaning, and Influence on Thought and Expression* (1938; Mineola, NY: Dover, 2000), throughout, but especially 69–135 for its presence in medieval thought and letters.

21. Augustine of Hippo, *On Order (De Ordine),* trans. Silvano Borruso (South Bend, IN: St. Augustine's, 2007), bk. 2, chs. 14–16, secs. 39–44, 103–109.

22. Ibid., bk. 2, ch. 14, sec. 41, 105.

23. Augustine of Hippo, *On Free Choice of the Will,* trans. Anna S. Benjamin and L. H. Hackstaff (Indianapolis: Bobbs-Merrill, 1964), bk. 2, ch. 8, line 79, 53; Hopper, *Medieval Number Symbolism,* 33–49.

24. Augustine of Hippo, *The City of God against the Pagans,* ed. and trans. R. W. Dyson (Cambridge: Cambridge University Press, 1998), bk. 20, ch. 5, 972.

25. "This number ten signifies perfection; for to the number seven which embraces all created things, is added the trinity of the Creator." Augustine of Hippo, *Against the Fundamental Epistle of Manichaeus,* 10–11, trans. Richard Stothert, from *Nicene and Post-Nicene Fathers, First Series,* vol. 4, ed. Philip Schaff (Buffalo, NY: Christian Literature, 1887), rev. and ed. for New Advent by Kevin Knight, http://www.newadvent.org/fathers/1405.htm.

26. Augustine of Hippo, *De Genesi ad Litteram (The Literal Meaning of Genesis)*, bk. 4, 3–4, quoted in Albertson, *Mathematical Theologies*, 76.

27. Augustine of Hippo, *On Free Choice of the Will*, bk. 2, ch. 11, lines 126–130, 64–65.

28. Augustine of Hippo, *City of God*, bk. 9, ch. 15, 377–379; M.-D. Chenu, *Nature, Man and Society in the Twelfth Century: Essays on New Theological Perspectives in the Latin West*, ed. and trans. Jerome Taylor and Lester K. Little (Toronto: University of Toronto Press, 1997), 64.

29. Albertson, *Mathematical Theologies*, 75–80; Chenu, *Nature, Man and Society*, 60–65.

30. Albertson, *Mathematical Theologies*, 86–89.

31. See Dermot Moran, "John Scottus Eriugena," in *The Stanford Encyclopedia of Philosophy*, Fall 2008 ed., ed. Edward N. Zalta, http://plato.stanford.edu/archives/fall2008/entries/scottus-eriugena/; Albertson, *Mathematical Theologies*, 88–89; Chenu, *Nature, Man and Society*, 60–88.

32. For the various strains of Platonism in the twelfth century, see Chenu, *Nature, Man and Society*, 49–98.

33. Thierry of Chartres, *Tractatus de sex dierum operibus*, quoted in Albertson, *Mathematical Theologies*, 114.

34. Albertson, *Mathematical Theologies*, 110–118; Augustine of Hippo, *On Christian Doctrine*, trans. D. W. Robertson Jr. (Indianapolis: Library of Liberal Arts, 1958), 10.

35. Boethius, *De Institutione Arithmetica*, 73–75.

36. Thierry of Chartres, *Commentum super Boethi librum de Trinitate*, quoted in Albertson, *Mathematical Theologies*, 125–126. See also Albertson's commentary, 132–139. Albertson does a fine job with this material, exposing Thierry's "new hermeneutic" of number through many of its subtleties, twists, and turns. It is revealing to reflect a bit on what the term 'hermeneutic' adds over and above the more straightforward term 'interpretation.' The word itself derives from a Latinized version of the Greek *hermeneuō*, meaning "interpret." Since the seventeenth century hermeneutics has been used to identify and expose hidden meanings in texts, beginning with the Bible but extending to other writings and experience as well. The tradition of hidden meanings in texts or nature or reality dates to antiquity and ever since has carried with it the assumption that words can reveal the ultimate meanings of things, even in negative or so-termed apophatic ways that point to the limits of conventional language. This assumption prevailed in the Middle Ages, reinforcing the medieval attachment to the things of reality, captured by definitions, and hence disclosing the classifying temper. Thus did Thierry and Nicholas of Cusa, later on, interpret number hermeneutically as a divine, linguistic structure underlying the meaning of the cosmos. From the modern, default point of view and its "presumption of unbelief" (Charles Taylor, *A Secular Age* [Cambridge, MA: Harvard University Press, 2007], 13), the foundational assumption of ulti-

mate, hidden meanings unearthed by hermeneutical methods of interpretation (as opposed to historical, contextual, structural, semiotic, or other critical approaches to texts and experience) founders on the matter of corroboration or, in Karl Popper's more incisive and accurate expression, on "falsifiability." Hermeneutically grounded claims are impossible to disprove. This leaves their purveyors firmly planted in what is commonly known as the "hermeneutical circle," what the less kind might call "wishful thinking." Still, these issues are far more complicated and merit further consideration than we can afford here. For a good introduction to the term, its use, and attendant problems, see C. Mantzavinos, "Hermeneutics," in Zalta, *Stanford Encyclopedia of Philosophy,* Fall 2016 ed., http://plato.stanford.edu/archives/fall2016/entries/hermeneutics/.

37. Nicholas of Cusa, *On Learned Ignorance,* in *Nicholas of Cusa: Selected Spiritual Writings,* trans. H. Lawrence Bond (New York: Paulist, 1997), 87–93. Clyde Lee Miller, "Cusanus, Nicolaus [Nicolas of Cusa]," in Zalta, *Stanford Encyclopedia of Philosophy,* Fall 2015 ed., https://plato.stanford.edu/archives/fall2015/entries/cusanus/, provides a substantial overview of Cusanus's thought and works. For a modern, hermeneutical interpretation of Cusanus's apophatic theology and the wisdom of learned ignorance, see Johannes Hoff, *The Analogical Turn: Rethinking Modernity with Nicholas of Cusa* (Grand Rapids, MI: William B. Eerdmans, 2013), 1–24 and throughout.

38. Nicholas of Cusa, *On Learned Ignorance,* 121–127; Albertson, *Mathematical Theologies,* 180–198.

39. Hopper, *Medieval Number Symbolism,* 102–103.

40. See Annemarie Schimmel, *The Mystery of Numbers* (Oxford: Oxford University Press, 1993), 219–221, 273–274, and throughout. Schimmel has composed a "Little Dictionary of Numbers," which covers many numerological enchantments in each of the numbers one through twenty-two successively and describes larger numbers "up to ten thousand." Also see Karl Menninger, *Number Words and Number Symbols: A Cultural History of Numbers,* trans. Paul Broneer (1969; New York: Dover, 1992), 262–267, for an account of early "letters and numbers." The sequence of triangular numbers is 1, 3, 6, 10, 15, 21, 28, 36, 45, 55, 66, 78, 91, 105, 120, 136, 153, ... The number 153 is the seventeenth in the series.

41. For these and other examples, see Stephen Wilson, *The Magical Universe: Everyday Ritual and Magic in Pre-modern Europe* (London: Hambledon and London, 2000), 446–449 and throughout.

42. Edward Grant, *Science and Religion, 400 BC–AD 1550: From Aristotle to Copernicus* (Baltimore: Johns Hopkins University Press, 2004), 101, 131. Many horoscopes were produced in the Middle Ages for political and personal reasons alike. Yet it is not clear how much they actually influenced decisions and behavior. See Richard Kieckhefer, *Magic in the Middle Ages* (Cambridge: Cambridge University Press, 1989), 120–133. For a good overview of astrology in medieval times, see David C. Lindberg, *The Beginnings of Western Science: The European Scientific*

Tradition in Philosophical, Religious, and Institutional Context, 600 B.C. to A.D. 1450 (Chicago: University of Chicago Press, 1992), 274–280.

43. Funkenstein, *Theology and the Scientific Imagination,* 310.

44. James A. Weisheipl, "The Interpretation of Aristotle's *Physics* and the Science of Motion," in *The Cambridge History of Later Medieval Philosophy: From the Rediscovery of Aristotle to the Disintegration of Scholasticism, 1100–1600,* ed. Norman Kretzmann, Anthony Kenny, and Jan Pinborg (Cambridge: Cambridge University Press, 1982), 531 and more generally 521–539. Ockham's attack on the reality of "universals," his insistence on the singularity of beings, and his defense of the univocity of logical terms contributed greatly to emptying ontological or metaphysical content from numbers and calculations. Further, his critique of the existence of "indivisibles," his constructionist version of the infinite divisibility of continua, and his account of the inequality of infinites helped focus attention on the thorniest of mathematical issues. See John E. Murdoch and Edith D. Sylla, "The Science of Motion," in *Science in the Middle Ages,* ed. David C. Lindberg (Chicago: University of Chicago Press, 1978), 216–217. See too John E. Murdoch, "William of Ockham and the Logic of Infinity and Continuity," in *Infinity and Continuity in Ancient and Medieval Thought,* ed. Norman Kretzmann (Ithaca, NY: Cornell University Press, 1982), 165–206.

45. Peter Lombard, *Sentences,* quoted in Edward Grant, *The Foundations of Modern Science in the Middle Ages: Their Religious, Institutional, and Intellectual Contexts* (Cambridge: Cambridge University Press, 1996), 99. For a fine, succinct account of the doctrine of intension and remission of forms, see more generally 93–104.

46. Nicole Oresme, *On the Configuration of Qualities (De configurationibus qualitatum),* trans. Marshall Clagett, in Marshall Clagett, *The Science of Mechanics in the Middle Ages* (Madison: University of Wisconsin Press, 1959), 348–349.

47. Grant, *Foundations,* 99. Ever since the pioneering investigations of the French historian and physicist Pierre Duhem (d. 1916), scholars have come to appreciate both the complexity and novelty of the fourteenth century's "modern way." Duhem's studies in Leonardo da Vinci had led him to unearth and publicize the existence of flourishing schools of natural philosophy in the fourteenth century, including those of Jean Buridan, Nicole Oresme, and others, all of whom grappled with problems they had found in Aristotle's physics. While few have followed Duhem in his claim that the fourteenth century, rather than the seventeenth, revealed the onset of modern science, most students of the period interpret the theory of impetus, the mean speed theorem, and comparable doctrines as remarkable departures from traditional, Aristotelian dynamics and as, minimally, presaging later developments. After Duhem, the research and analyses of Anneliese Maier, A. C. Crombie, and Marshall Clagett, followed by those of John E. Murdoch, Edward Grant, and Edith D. Sylla, among others, have mined a rich vein of scholarly activity whose historical implications for the seventeenth-

century Scientific Revolution continue to provoke lively debate. Good introductions to these studies may be found in Anneliese Maier, *On the Threshold of Exact Science: Selected Writings of Anneliese Maier on Late Medieval Natural Philosophy,* trans. and ed. Steven D. Sargent (Philadelphia: University of Pennsylvania Press, 1982), 3–20; Murdoch and Sylla, "Science of Motion"; and Edith D. Sylla, "The Oxford Calculators," in Kretzmann, Kenny, and Pinborg, *Cambridge History of Later Medieval Philosophy,* 540–563.

48. Lesser figures, either in the college or associated with the array of problems occupying the Calculators, included Richard Kilvington, Walter Burley, Richard Billingham, John Bode, and John Dumbleton.

49. Sylla, "Oxford Calculators," 542. These disputations included the *sophismata,* logical and physical puzzles employed to hone student wits. By the fourteenth century the Merton College instructors were posing *sophismata* involving the quantitative treatment of qualities and the comparison of speeds of bodies in motion. See too Edith D. Sylla, "Mathematical Physics and Imagination in the Work of the Oxford Calculators: Roger Swineshead's *On Natural Motions,*" in *Mathematics and Its Applications to Science and Natural Philosophy in the Middle Ages: Essays in Honor of Marshall Clagett,* ed. Edward Grant and John E. Murdoch (Cambridge: Cambridge University Press, 1987), 69–101.

50. The earliest known statement of the theorem occurred in a work of William Heytesbury titled *Rules for Solving Sophisms (Regulae solvendi sophismata, 1335),* trans. Ernest Moody, in Clagett, *Science of Mechanics,* 270–289. See Clagett's preceding essay, "The Merton Theorem of Uniform Acceleration," 255–269, and also Walter Roy Laird, "Change and Motion," in Lindberg and Shank, *Cambridge History of Science,* 2:404–435.

51. Richard Swineshead, "On Motion (a Fragment)," in Clagett, *Science of Mechanics,* 243; Oresme, *On the Configuration of Qualities,* 347, 351. Note, the medieval Latin term *velocitas* was loosely equated with speed. In modern terms both speed and velocity were scalar, having only magnitude; neither was a vector quantity, which describes magnitude and direction.

52. Heytesbury, *Rules for Solving Sophisms,* 237; Oresme, *On the Configuration of Qualities,* 358. The expression "uniformly difform" referred to regular, uniform changes in acceleration or deceleration, as opposed to "difformly difform qualities," which connoted irregularly varied acceleration and deceleration.

53. For Greek diagrams, see Reviel Netz, *The Shaping of Deduction in Greek Mathematics: A Study in Cognitive History* (Cambridge: Cambridge University Press, 1999), 12–67. Clagett claims that with his diagrams, Oresme made a "first step towards describing functions in terms of 'curves.'" Clagett, *Science of Mechanics,* 364. But if so, it was but a vague hint at best. Clagett's comment misleads for two reasons. First, Oresme had no conception of analyzing a point or any sort of line or curve into a pair of axes, the heart of analytical geometry, and, moreover, no thought at all that his diagrams corresponded to anything in nature. Second,

and by contrast, he sought to define those critical points he did identify on the diagrams, such as the mean point of an acceleration, in the traditional, classifying framework of words and things, a problematic matter we shall take up later.

54. Referring to Figure 4.3, $\angle EFB = \angle CFG$ (vertical angles are equal), $\angle BEF = \angle CGF$ (both are right angles), $EF = FG$ (F bisects GE), and $\Delta EFB = \Delta CFG$. Euclid, *The Thirteen Books of the Elements,* 2nd ed., 3 vols., trans. Thomas L. Heath (1926; New York: Dover, 1956), bk. 1, prop. 26, 1: 301. Adding either of these triangles to the area BFGD produces the triangle BCD and the rectangle EGBD, which are thus equal.

55. Heytesbury, *Rules for Solving Sophisms,* 236.

56. Grant, *Foundations,* 101.

57. Oresme, *On the Configuration of Qualities,* 348.

58. John E. Murdoch, "Infinity and Continuity," in Kretzmann, Kenny, and Pinborg, *Cambridge History of Later Medieval Philosophy,* 564. Scholars still debate whether a divide exists between premodern and modern science, the so-called continuity question. Some have cited the work of the mean speed theorists as corroboration of an evolutionary "continuity" existing from ancient to medieval to modern science. Opponents of this view stress the revolutionary nature of sixteenth- and seventeenth-century developments, especially the creation of the experimental and analytical methods associated with modern science. As will become increasingly apparent, this latter interpretation informs the present narrative, which, we suggested earlier, complements the recent studies by David Wootton, especially his *Invention of Science: A New History of the Scientific Revolution* (New York: HarperCollins, 2015). The earlier historiography of the entire debate is exhaustively surveyed by H. Floris Cohen in *The Scientific Revolution: An Historiographical Inquiry* (Chicago: University of Chicago Press, 1994); part 3 is especially useful for addressing the "slippery concept" of the scientific revolution. See too I. Bernard Cohen's compendious *Revolution in Science* (Cambridge, MA: Harvard University Press, 1985), esp. chs. 1–3.

59. Grant, *Foundations,* 103. Space precludes further treatment of the Calculators, but we should not leave them without acknowledging their ingenious contributions to the early history of mechanics: the role of "impetus" in explaining projectile and free-fall motion, the enchantment-free mathematics of depicting motions, and the opening of lines of inquiry that in Galileo's day—after the new information technology of relational numeracy had gained currency—would prove to be critical.

60. The Greeks could certainly measure things practically and in fine detail, as illustrated by this spoof in Aristophanes's play *The Clouds,* in *Lysistrata and Other Plays,* rev. ed., trans. Alan H. Sommerstein (New York: Penguin Books, 2002), 80: "STUDENT: Socrates, a moment ago, asked Chaerephon how many of its own feet a flea could jump. One of them had just bitten Chaerephon's eyebrow and jumped over on to Socrates' head. STREPSIADES: And how did he measure it? STUDENT: In

an elegant way. He melted some wax and put the flea's feet into it, so that when it set the flea had a stylish pair of slippers on. And then he took the slippers off and used them to measure out the distance [walking heal to toe]." Measurement indeed.

61. See the classic study by Lucien Goldmann, *The Hidden God: A Study of Tragic Vision in the Pensées of Pascal and the Tragedies of Racine,* trans. Philip Thody (1956, trans. 1964; London: Routledge and Kegan Paul, 1964). For some, God was "hidden"; for many natural philosophers, God was manifest in nature.

62. The disenchantment of language only went so far. Even today, words still retain magic, still harbor mysteries and aesthetic qualities, even when applied to nature—one thinks of Lewis Thomas or Loren Eiseley or Rachel Carson. Nonetheless, seeing the beauty of nature through words is not the same as explaining natural phenomena. Poetry stands separate from science. Not so in the Middle Ages.

63. Nicomachus, *Introduction to Arithmetic,* 813.

5. Arithmetic

1. Leonardo Pisano [Fibonacci], *Fibonacci's Liber Abaci,* trans. L. E. Sigler (New York: Springer-Verlag, 2002). To the modern reader the title can appear confusing. *Abaci,* also spelled *abbaci* in Leonardo's book and elsewhere, is the plural of *abbaco,* which translates as "abacus," but did not refer to our image of a mechanical computing device with its wire frame and beads (probably developed in China as far back as the second century BCE). By the thirteenth century, *abbaco* had come to mean "calculation," which was usually performed on a counting board of the same name and which utilized moveable pebbles (the Latin for 'pebble' was *calculus*) or flat, metallic tokens. Leonardo's book explained how to calculate, but without using a counting board, relying instead on Hindu-Arabic numerals. Complicating matters further, those using the traditional counting board were called "abacists," while "algorismists" referred to followers of the new *Liber Abaci.* See Karl Menninger, *Number Words and Number Symbols: A Cultural History of Numbers,* trans. Paul Broneer (1969; New York: Dover, 1992), 332–388, 425–438. Joseph Gies and Frances Gies, in *Leonard of Pisa and the New Mathematics of the Middle Ages* (1969; Gainesville, GA: New Classics Library, 2000), offer a pleasant introduction to Leonardo's life, career, and accomplishments. A good, recent summary of available information may be found in Keith Devlin, *The Man of Numbers: Fibonacci's Arithmetic Revolution* (New York: Walker, 2011). Richard E. Grimm provides a thorough analysis of the autobiographical "Prologue" from the six extant manuscript copies, "The Autobiography of Leonardo Pisano," *Fibonacci Quarterly* 11, no. 1 (1973): 99–104.

2. Leonardo's series begins with one pair of rabbits and the premises that each pair of rabbits will sire another pair in one month and a second in a second month

before becoming stew. The first pair, 1, at the end of one month becomes 2 (the original pair plus the newly sired one, which must wait a month before parenting), after two months 3 (the original, its first offspring, and the original's second pair), after three months 5, after four months 8, and continuing. The series is thus 1, 2, 3, 5, 8, 13, 21, 34, 55, 89, 144, 233, 377 (at the end of one year) . . . , wherein each number is the sum of the previous two. Leonardo, *Fibonacci's Liber Abaci,* 404–405; D. J. Struik, ed., *A Source Book in Mathematics: 1200–1800* (1969; Princeton, NJ: Princeton University Press, 1986), 1–4. For interesting and accessible accounts of many applications of the Fibonacci series, particularly as they pertain to the "golden ratio," phi, see Mario Livio, *The Golden Ratio: The Story of Phi, the World's Most Astonishing Number* (New York: Broadway Books, 2002), 92–115, and H. E. Huntley, *The Divine Proportion: A Study in Mathematical Beauty* (New York: Dover, 1970), 141–163.

3. For a well-drawn description of the institution and evolution of the *fondaco* throughout the Mediterranean region, see Olivia Remie Constable, *Housing the Stranger in the Mediterranean World: Lodging, Trade, and Travel in Late Antiquity and the Middle Ages* (Cambridge: Cambridge University Press, 2003), 306–361. As an aside, Bugia also exported a high-quality wax, well suited for candles, which became named *bougie* in French, after the town of its source.

4. See Leonardo, "Prologue," in *Fibonacci's Liber Abaci,* 15–16 and an example of checking addition by "casting out nines" on 40–41. "Casting out nines" (in Latin, *abjectio novenaria*) is a method of long standing, dating at least to the third century CE and furthered by Avicenna, whereby basic computations may be verified (discussed later).

5. Besides a second edition of the *Liber Abaci,* completed in 1228, Leonardo wrote several other manuscripts treating problems of mathematical roots, practical geometry, and number theory.

6. Little studied before the 1960s, the pioneering bibliographical and manuscript studies of Gino Arrighi and Warren van Egmond, especially the latter's *Practical Mathematics in the Italian Renaissance: A Catalog of Italian Abacus Manuscripts and Printed Books to 1600,* Monografia 4 (Florence: Instituto e Museo di Storia della Scienza, 1980), have opened the abacus tradition to a flurry of scholarship in the last three decades. Some worthwhile introductions to this work include Frank J. Swetz, "Commercial Arithmetic," in *Medieval Science, Technology, and Medicine: An Encyclopedia,* ed. Thomas Glick, Steven J. Livesey, and Faith Wallis (New York: Routledge / Taylor and Francis, 2005), 133–135; Frank J. Swetz, "Perspectives," in *Capitalism and Arithmetic: The New Math of the 15th Century, Including the Full Text of the Treviso Arithmetic,* trans. David Eugene Smith (LaSalle, IL: Open Court, 1987), 1–35; Jane Gleeson-White, *Double Entry: How the Merchants of Venice Created Modern Finance* (New York: W. W. Norton, 2012), 34–48 and throughout; Albrecht Heeffer, "The Abbaco Tradition (1300–1500): Its Role in the Development of European Algebra," *Suuri Kaiseki Kenkyuujo Koukyuuroku* 1625

(2009): 23–33 (available in various online versions); Jens Høyrup, "Leonardo Fibonacci and *Abbaco* Culture: A Proposal to Invert the Roles," *Revue d'histoire des mathématiques* 11 (2005): 23–56; and Jens Høyrup, *Jacopo da Firenze's "Tractatus Algorismi" and Early Italian Abbacus Culture* (Basel: Birkhäuser Verlag, 2007). Paul F. Grendler, *Schooling in Renaissance Italy: Literacy and Learning, 1300–1600* (Baltimore: Johns Hopkins University Press, 1989), describes abacus schools thoroughly in the context of Renaissance education.

7. Devlin, *Man of Numbers,* 108–111; Grendler, *Schooling in Renaissance Italy,* 1–6, 71–83, 306–323.

8. The *Treviso Arithmetic* was the first of the printed *liber abbaci.* Notable publications thereafter included a wide-ranging primer produced by Piero Borghi and printed in Venice in 1484, which became a best seller and was reprinted fifteen times during the next eighty years; a *Trattato d'arismetricha* (from a manuscript written ca. 1463) by Benedetto da Firenze, Leonardo da Vinci's teacher; a mammoth how-to book of some six hundred folio pages by Luca Pacioli, generally considered the "father" of double-entry bookkeeping (*Summa de arithmetica, geometria, proportioni et proportionalitá,* 1494); and a remarkable *Trattato d'Abaco* by painter extraordinaire Piero della Francesca, about which we shall have much more to say later. See Swetz, *Capitalism and Arithmetic,* 1–35; Devlin, *Man of Numbers,* 6–7, 107, 119; and Gleeson-White, *Double Entry,* 66–78.

9. While steadily expanding, the new counting system's adoption faced sizeable opposition in the days before printing, mostly because fraud was easier to commit in the new system than with Roman numerals, and because computing on the traditional counting board could be done quickly and accurately. Nevertheless, however rapid the computations, the abacus could not store numerical information. This fact, combined with the increasing availability of paper, the ease of erasing and correcting mistakes, an expanding demand for more sophisticated calculations, the growing commercial familiarity of the new system, and ultimately the sheer facility of calculating, especially accounts in double-entry bookkeeping, with their straight columns and rows, all helped the new counting techniques attain widespread popularity.

10. Devlin, *Man of Numbers,* 119. Gerbert of Aurillac, who, as we saw, studied in Spain and who later became Pope Sylvester II, is generally credited with the initial introduction of Hindu-Arabic numerals to Europe in the tenth century, while in the early twelfth Adelard of Bath translated from Arabic the arithmetic of Mohammed Al-Khowârîzmî and Robert of Chester translated Al-Khowârîzmî's treatise "The Calculation of Reduction and Confrontation." Both works employed the Hindu-Arabic notation. Recently, Joseph Mazur has summarized the sometimes intense and nationalistic scholarly debates over the arrival of Hindu-Arabic numerals in Europe and their dissemination; see Joseph Mazur, *Enlightening Symbols: A Short History of Mathematical Notation and Its Hidden Powers* (Princeton, NJ: Princeton University Press, 2014), 1–80.

11. In a working paper for the National Board of Economic Research, William N. Goetzmann argues that Leonardo was "the first to develop present value analysis for comparing the economic value of alternative contractual cash flows." William N. Goetzmann, "Fibonacci and the Financial Revolution" (NBER Working Paper No. 10352, National Bureau of Economic Research, Cambridge, MA, March 2004).

12. Leonardo, *Fibonacci's Liber Abaci,* 17. See also Tobias Dantzig, *Number: The Language of Science: A Critical Survey Written for the Cultured Non-mathematician,* 4th ed. (1930; Garden City, NY: Doubleday, 1954), 9: "The operations of arithmetic are based on the tacit assumption that we can always pass from any number to its successor, and this is the essence of the ordinal concept."

13. Leonardo, *Fibonacci's Liber Abaci,* 17, 19, 28.

14. In this system a vertical, cuneiform wedge stroke was used as the basic unit symbol. For the digits 1–9, strokes were added one by one and stacked in rows of three; for the number 10, the same wedge stroke was made, only now turned on its side, horizontally instead of vertically; the multiples 20, 30, 40, and 50 were likewise achieved by stacking horizontal strokes, as illustrated in this sequence from Old Persian cuneiform:

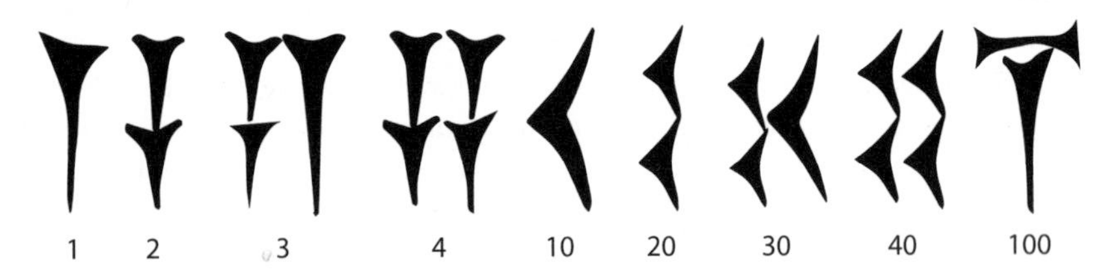

To this stage the system was purely additive. But rather than continue with a growing and unwieldy aggregation of horizontal strokes, numbers above 60 were formed by repeating the unit, vertical stroke on the left, followed by additional strokes (both horizontal and vertical) to the right. The unit stroke on the left thus stood for the sixties column (or 60×1). Further, to the left of it, they placed other strokes in the 60^2 column (60×60 or 3,600) and so on. For example, stylus marks corresponding to the digits 2 8 7 in the Babylonian, base-60 system would read, from left to right, $2 \times 60 \times 60 + 8 \times 60 + 7$ (or $7{,}200 + 480 + 7$—that is, 7,687 in denary notation). John Barrow suggests an easy way of adjusting to this system—namely, by recalling how we calculate time. Thus '2 8 7' would be two hours, eight minutes, and seven seconds, or a total of 7,687 seconds. John Barrow, *Pi in the Sky: Counting, Thinking, and Being* (Oxford: Oxford University Press, 1992), 85. For excellent descriptions of Babylonian mathematics, see O. Neugebauer, *The Exact Sciences in Antiquity,* 2nd ed. (Princeton, NJ: Princeton University Press, 1952; New York: Dover, 1969), 1–70; and Georg Gheverghese Joseph, *The Crest of the Peacock:*

Non-European Roots of Mathematics, 3rd ed. (Princeton, NJ: Princeton University Press, 2011), 125–179.

15. Joseph, *Crest of the Peacock,* 136–139.

16. Historically the place-value principle was developed independently in four distinct, ancient civilizations: the Mesopotamian (Babylonian) in the second millennium BCE; the Chinese, with their rod numerals, early in the Common Era; the Mayan, with a base-20 counting system, between 200 and 900 CE; and the Indian, between 200 and 400 CE, with a decimal-based, place-value system that eventually spread to the entire globe. Besides the Indian development of a symbol for zero, its only other modern use has been found in the Mayan system, which remained culturally isolated from the world beyond Mesoamerica. For the development of Indian mathematics, see Joseph, *Crest of the Peacock,* 22, 338–371. Robert Kaplan, *The Nothing That Is: A Natural History of Zero* (Oxford: Oxford University Press, 2000), 114–115, and throughout, narrates the odyssey of the zero.

17. Barrow, *Pi in the Sky,* 92. Otto Neugebauer called place-value notation "one of the most fertile inventions of humanity . . . properly compared with the invention of the alphabet." Neugebauer, *Exact Sciences in Antiquity,* 5.

18. Gies and Gies, *Leonard of Pisa,* 51.

19. Leonardo, *Fibonacci's Liber Abaci,* 46, 56, and throughout for the first seven chapters, 17–126. Of course, practices and rules for computing accompanied all premodern counting systems. But while earlier "ideograms" could convey general procedures, specific numerical results proved more elusive and difficult to achieve. In Babylonian mathematics, for instance, even with the use of numerical tables, the "lack of a general notation appears to be much more detrimental than in the handling of purely algebraic problems." Accordingly, the "contents of Babylonian mathematics remained profoundly elementary." Neugebauer, *Exact Sciences in Antiquity,* 43–48.

20. Leonardo, *Fibonacci's Liber Abaci,* frontispiece, 20, 50.

21. Ibid., 127–128, 134–135, 148, 179–184, 213–224, 227.

22. Ibid., 127–128. In modern notation, the rule of three is a:b::c:d, or, expressed algebraically, $\frac{a}{b}=\frac{c}{d}$. Assuming d is the unknown sought, it follows that $ad=bc$, and $d=\frac{bc}{a}$. See Joseph, *Crest of the Peacock,* 217–218, 347. Euclid had provided the geometrical equivalent to the rule of three in Euclid, *The Thirteen Books of the Elements,* 2nd ed., 3 vols., trans. Thomas L. Heath (1926; New York: Dover, 1956), bk. 6, prop. 12, 2:215.

23. Ibid., bk. 5, defs. 5 and 6, 2:114; bk. 7, def. 20, 2:278. Much controversy has surrounded these definitions, especially over the question of whether they extend to incommensurable quantities and the entire domain of real numbers. Still, whatever the scope of the definitions, they remain static definitions, a world apart

from the changing functionalities of Leonardo's treatment. That later mathematicians (for example, Augustus De Morgan) would interpret these definitions in light of modern, functional mathematics, and thus capable of treating incommensurables algorithmically, should not blind us anachronistically to this fact. See Heath's discussion in ibid., 2:114, 120–129, 278, 292–293.

24. Leonardo, *Fibonacci's Liber Abaci,* 531 and more generally 531–615.

25. Ibid., 128, 533. Leonardo used the phrase "method of proportion" repeatedly throughout his work. Many scholars in the nineteenth and early twentieth centuries interpreted Euclid's account of proportions as expressing an "algebraic geometry," a view now largely discredited. For a good summary of the debates and the current consensus on this issue, see Ivor Grattan-Guinness, "Numbers, Magnitudes, Ratios, and Proportions in Euclid's Elements: How Did He Handle Them?," *Historia Mathematica* 23 (1996): 355–375. Grattan-Guinness argues that Euclid conceived of quantity in three "ontological" categories: "numbers, geometrical magnitudes, and ratios." Leonardo's "method" for proceeding in his computations with the new numerals not only pointed the way toward algebra but also took a major step toward emptying these categories of their ontological, "thing" content.

26. Thus, instead of countless numerical examples of proportionalities ($\frac{2}{3}=\frac{6}{9}$, $\frac{18}{54}=\frac{1}{3}$, and so on), one could simply express all of them with letters $\left(\frac{a}{b}=\frac{c}{d}\right)$, a modern notation not available to Leonardo.

27. The Arabic title translates literally as "the calculation of reduction and confrontation," with *jabr* and *muqabala* denoting, respectively, the practices of transferring negative terms to the other side of an equation (thus, in modern terms, $a=x-6$ becomes $a+6=x$) and canceling similar terms on both sides of an equation (thus, $a+b-6=x-6$ becomes $a+b=x$). *Jabr* also designates the practice of setting a broken bone, the origin of today's medical use of the term 'reduction.'

28. See Kirsti Andersen and Henk J. M. Bos, "Pure Mathematics," in *The Cambridge History of Science,* vol. 3, *Early Modern Science,* ed. Katherine Park and Lorraine Daston (Cambridge: Cambridge University Press, 2006), 696–723. On 708–709 the authors recount the story of Tartaglia's "discovery" of the solution to cubic equations, actually found initially by Scipione del Ferro (1465–1526), which involved a rivalry, a deception, and a betrayal by Girolamo Cardano (1501–1576), who first published the results.

29. Michael Sean Mahoney, *The Mathematical Career of Pierre de Fermat, 1601–1665,* 2nd ed. (1973; Princeton, NJ: Princeton University Press, 1994), 2–7.

30. The citations are as originally published in Robert Recorde, *The Whetstone of Witte,* facsimile of the sole edition imprinted at London by John Kingston, 1557 (Derby, UK: TGR Renascent Books, 2013), 9–14, 149–153, 285–287. The title of this work carried a pun, for *cosa* in Italian ("thing" or algebraic "unknown") was similar to the Latin *cos,* which meant "whetstone," a stone used for sharpening axes,

knives, and other edged tools. Hence the reference to the sharpening of one's wit through the study of mathematics. Besides introducing the cossic art, Recorde presented numerous practical examples, applying the new numeracy to military formations, bricklaying, and geography, as well as promising a future book on navigation, a study prevented by his death. A brief introduction to Recorde's works may be found in Vera Sanford, "Robert Recorde's *Whetstone of Witte,* 1557," in *From Five Fingers to Infinity: A Journey through the History of Mathematics,* ed. Frank J. Swetz (Chicago: Open Court, 1994), 376–386. For a more extended exposure to Recorde's work and career, see Gareth Roberts and Fenny Smith, eds., *Robert Recorde: The Life and Times of a Tudor Mathematician* (Cardiff: University of Wales Press, 2012); for current purposes, the essay of Ulrich Reich, "*The Whestone of Witte:* Content and Sources," 93–121, is useful, while that of John V. Tucker, "Data, Computation, and the Tudor Knowledge Economy," 165–187, especially complements the current treatment of the new numeracy as an information technology.

31. In the early nineteenth century British logicians labored to show that the Aristotelian and Euclidean systems were logically identical. These efforts came to naught, largely on the question of interpreting the copula 'is.' See Michael E. Hobart and Joan L. Richards, "DeMorgan's Logic," in *Handbook of the History of Logic,* vol. 4, *British Logic in the Nineteenth Century,* ed. Dov M. Gabbay and John Woods (Amsterdam: Elsevier, 2008), 283–329.

32. Perhaps the most famous example of a procedural definition appeared much later, in the early nineteenth century, with the "delta-epsilon" definition of a "limit" advanced by the French mathematician Augustin-Louis Cauchy: "When the successively attributed values of the same variable indefinitely approach a fixed value, so that finally they differ from it by as little as desired, the last is called the *limit* of all others." Cauchy then moved on to articulate the definition in the "precise language of inequalities." "Let δ, ε be two very small numbers; the first is chosen so that for all numerical [that is, absolute] values of h less than δ, and for any value of x included [in the interval of definition], the ratio $(f(x+h)-f(x))/h$ will always be greater than $f'(x)-\varepsilon$ and less than $f'(x)+\varepsilon$." Seen by most historians of mathematics as the rigorous basis of calculus, the definition itself basically described the procedures one could follow in manipulating intervals that, however small, approached an end point but never reached it. Of course, all this lay far in the future during the days of Leonardo, but nonetheless his descriptions of the basic procedures one could follow in much simpler calculations embarked on a trajectory of mathematical definitions that contrasted sharply with those of conventional language—what one can do with a number, a line, or any of their subsequent derivations, not what they are. Cauchy, quoted in Judith V. Grabiner's excellent article "Who Gave You the Epsilon? Cauchy and the Origins of Rigorous Calculus," *American Mathematical Monthly* 90, no. 3 (1983): 185–194.

33. See the classic account of Frances A. Yates, *Giordano Bruno and the Hermetic Tradition* (Chicago: University of Chicago Press, 1964).

34. Leonardo, *Fibonacci's Liber Abaci,* 40–41; see also Swetz, *Treviso Arithmetic,* 46–49, for a like account of casting out nines. Later, in extending mathematical reasoning to other areas of rational discourse, Leibnitz too would rely on casting out nines to correct errors: "By means of this, once a reasoning in morality, physics, medicine or metaphysics is reduced to these terms or characters, one will be able to apply to it at any moment a numerical test [casting out nines], so that it will be impossible to be mistaken if one does not so desire." *Gottfried Wilhelm Leibniz: The Art of Controversies,* trans. and ed. Marcelo Dascal, New Synthese Historical Library (New York: Springer, 2008), 277.

35. Hippolytus of Rome, *Refutation of All Heresies,* trans. J. H. MacMahon, in *Ante-Nicene Fathers,* vol. 5, ed. Alexander Roberts, James Donaldson, and A. Cleveland Coxe (Buffalo, NY: Christian Literature, 1886), bk. 4, ch. 14, rev. and ed. for New Advent by Kevin Knight, http://www.newadvent.org/fathers/050104.htm; Thomas L. Heath, *A History of Greek Mathematics,* vol. 1, *From Thales to Euclid* (1921; New York: Dover, 1981), 115–117.

36. Many manuscripts containing onomantic tables referred to a letter purportedly written by Aristotle to Alexander, titled the *Secret of Secrets,* in which the famous teacher gave advice to his more famous student about impending battles and other matters. See Charles Burnett, "The Eadwine Psalter and the Western Tradition of the Onomancy in Pseudo-Aristotle's *Secret of Secrets,*" in his collection of essays *Magic and Divination in the Middle Ages: Texts and Techniques in the Islamic and Christian Worlds* (Brookfield, VT: Ashgate, 1996), sec. 11.

37. For gematria, see Dantzig, *Number,* 40–41, and Annemarie Schimmel, *The Mystery of Numbers* (Oxford: Oxford University Press, 1993), 34–35, 180–188, 276–277. Schimmel summarizes the Reformation situation succinctly and without flummery (276–277): "With some skill one can find it [666] in almost every name by counting the numerical values of the letters (and often shifting them until the result is achieved)." I. B. Cohen, *The Triumph of Numbers: How Counting Shaped Modern Life* (New York: W. W. Norton, 2005), 60–63, decodes the Bungus-Stifel numerical gerrymandering, as had the nineteenth-century mathematician and logician Augustus De Morgan in his *Budget of Paradoxes* (1872), entry "Bungus on the Mystery of Number." Stifel, it bears mentioning, had prophesied the end of the world on the basis of his mathematical calculations. It was to have occurred in 1533. When it didn't, and after a brief arrest for his failed prediction, Stifel turned more seriously to the study of mathematics, publishing in 1544 a cossist work, *Arithmetica Integra,* that heavily influenced Recorde.

38. For scapulimancy, see Charles Burnett, "Scapulimancy (Divination by the Shoulder-Blades of Sheep)" and "Arabic Divinatory Texts and Celtic Folklore: A Comment on the Theory and Practice of Scapulimancy in Western Europe," in *Magic and Divination in the Middle Ages,* secs. 12, 13. Burnett's account includes many fascinating illustrations and diagrams documenting this strange practice, which, mutatis mutandis, resembles that of nineteenth-century phrenologists.

6. Music

1. *The Didascalicon of Hugh of Saint Victor: A Medieval Guide to the Arts,* trans. Jerome Taylor (New York: Columbia University Press, 1961), 61, 67, 69, 71, 77, 152–153.

2. These associations may be found in the Platonic rather than the Aristotelian tradition. Aristotle had defined harmony as a "composition of contraries" and thus not applicable to the heavens or to the soul per se. See Aristotle, *On the Heavens,* 290b12–17, and Aristotle, *On the Soul,* 407b30–33. All the same, he held "metaphorically" that a "single harmony orders the composition of . . . the whole Universe." Aristotle, *Topics,* 123a33–36; Aristotle, *On the Universe,* 396b25–27.

3. The phrase comes from Cicero's *Dream of Scipio,* book 6 of *De Re Publica.* Boethius's expression was the "silent and noiseless course" of the heavens. While the music of this "celestial spinning" does have a kind of sound, he and others thought, for "many reasons" it "does not reach our ears." See the selections from Boethius's *De institutione musica* in Giulio Cattin, *Music of the Middle Ages I,* trans. Steven Botterill (Cambridge: Cambridge University Press, 1984), 163–166, and in Oliver Strunk, ed., *Source Readings in Music History: Antiquity and the Middle Ages* (1950; New York: W. W. Norton, 1965), 79–86.

4. Plato, *Republic,* bk. 7, 530d–531c, and bk. 10, 616b–618b. Because of the soul's dissonance, arising from its entanglements with the sensory-bound body, Plato also spoke of the practical and ameliorating power of instrumental music (including song) in elevating human morality, improving mental health through both pleasure and solace, and strengthening character, a power often neglected and abused. *Republic,* bk. 3, 398c–402b.

5. The *Timaeus* was the only dialogue of Plato extensively available in Latin to intellectuals of the Middle Ages. Cicero had translated all of it, but it was a fourth-century philosopher, Calcidius, who bequeathed the document (up to section 53c) to later generations. All other knowledge of Plato came indirectly, much of it from the writings of "Neoplatonists," including Augustine, Boethius, and Isidore of Seville, among others.

6. Plato even had observed the ratio for the "semitone," 256:243. See the *Timaeus,* 34b1–37d8, 1238–1241. More on intervals, ratios, and the Pythagorean scale later.

7. The "counter-earth" comes from Aristotle's criticism in *On the Heavens,* bk. 2, 293a15–29, where he takes the Pythagoreans to task for inventing another entity merely to permit summing the types of heavenly bodies to the holy number ten, thus force-feeding phenomena into their own theories.

8. Fully aware that the image is now discarded, music writer Jamie James nonetheless concludes his paean to past harmonies with a yearning for the "silence" of the spheres: "Yet now that science permits us actually to hear the soundtrack of the cosmos, in the form of random blips and howls picked up by

radio telescopes, how we long for silence." Jamie James, *The Music of the Spheres: Music, Science, and the Natural Order of the Universe* (New York: Copernicus / Springer Verlag, 1993), 241.

9. Boethius, *De institutione musica,* in Cattin, *Music of the Middle Ages I,* 166.

10. Cattin, *Music of the Middle Ages I,* 157–161. The phrase "theoretical music" was commonly contrasted with "practical music." See, for example, Magister Lambertus or Pseudo-Aristotle (fl. 1270), *Tractatus de musica,* selection in F. Alberto Gallo, *Music of the Middle Ages II,* trans. Karen Eales (Cambridge: Cambridge University Press, 1985), 109. As suggested earlier, the followers of Aristotle were not sympathetic to the Pythagorean-Platonic efforts to utilize the diatonic scale for explaining order in the heavens or the human soul. Following the reemergence of Aristotelian texts in the thirteenth century, many scholastics sought to incorporate the Philosopher's epistemology into their teachings on music, treating it as an ancillary and subordinate discipline in the scholastic curriculum. In so doing, they generally restricted the theoretical dimension of music to the study of consonance, the harmonic mixture of opposing sounds. Still, very little of this crept into teaching music at the practical level. In his detailed doctoral study of music education in the Middle Ages, Gilles Rico summarizes, "Practical music including polyphony was simply not part of the [Paris] University's educational agenda." Gilles Rico, "Music in the Arts Faculty of Paris in the Thirteenth and Early Fourteenth Centuries" (DPhil thesis, University of Oxford, 2005), 183–187, 296, 295–307, http://www.diamm.ac.uk/documents/42/RicoFull.pdf.

11. Among his many writings, Augustine of Hippo articulated the religious rationale for both liturgical singing and psalmody. In contrast to the reflective, contemplative nature of philosophical music (which he also endorsed), sung music provided a different, inexplicable kind of "understanding, . . . a kind of sound indicating that the heart is giving birth to what cannot be spoken." Therefore, "sing in jubilation," he urged, for the emotive, spiritual intensity captured in ephemeral sound is "fitting . . . for the ineffable God." If only for a brief moment, sung or instrumental music could transport one into the embrace of the divine. Augustine, "Commentary on Psalm 32," in Cattin, *Music of the Middle Ages I,* 162–163.

12. For accessible introductions to plainchant, see Cattin, *Music of the Middle Ages I,* 1–20 and throughout; Richard H. Hoppin, *Medieval Music* (New York: W. W. Norton, 1978), 30–91; Calvin M. Bower, "The Transition of Ancient Musical Theory into the Middle Ages," in *The Cambridge History of Western Music Theory,* ed. Thomas Christensen (Cambridge: Cambridge University Press, 2002), 136–167; and Jeremy Yudkin, *Music in Medieval Europe* (Upper Saddle River, NJ: Prentice Hall, 1989), 33–82. The name "plainchant" comes from the Latin *cantus planus* (whose etymology is uncertain), first appearing in the tenth century.

13. For example, Guido d'Arezzo, "Epistola de ignoto cantu," in Cattin, *Music of the Middle Ages I,* 176.

14. The term *neume* (sign) derives from a bowdlerized form of the Greek *pneuma* ("breath," "to breathe," from which our 'pneumatic'). Although not depicting the earliest, Western musical symbols, manuscripts of neumatic notation located in the monastery of Saint Gall in Switzerland and dating from the ninth and tenth centuries do provide the earliest examples of musical repertory written in neumes. See Richard Rastall, *The Notation of Western Music* (New York: St. Martin's, 1982), 15–26, and Carl Parrish, *The Notation of Medieval Music* (1957; New York: W. W. Norton, 1959), 3–40, plus facsimiles. Two other key documents from this period, "Musica enchiriadis" (Manual of music) and "Scholia enchiriadis" (Commentary on the manual), were the most widely copied manuscripts of music in the ninth and tenth centuries; they too stand among the early evidence of musical neumes and of musical theorizing associated with the rise of polyphony.

15. Hoppin, *Medieval Music,* 187. Echoing the significance of polyphony are Victor Zuckerkandl, *The Sense of Music* (Princeton, NJ: Princeton University Press, 1959), 137, and Parrish, *Notation of Medieval Music,* who writes: "The notation of rhythm in a clear and unequivocal manner is intimately linked to the rise of polyphony, and is one of the most absorbing episodes in the whole history of music" (193–194).

16. Vincenzo Galilei, "Il primo libro della prattica del contrapunto intorno all'uso delle consonanze" (1589), cited in Claude V. Palisca, *Music and Ideas in the Sixteenth and Seventeenth Centuries* (Chicago: University of Illinois Press, 2006), 42.

17. Hoppin, *Medieval Music,* 73.

18. Isidore of Seville, *Isidore of Seville's Etymologies: The Complete English Translation of "Isidori Hispalensis Episcopi Etymologiarum sive Originum," Libri 20,* vol. 1, bks. 1–10, trans. Priscilla Throop (Charlotte, VT: Medieval MS, 2005), bk. 3, ch. 15, sec. 2. Isidore also observed that "the use of letters was devised to remember things. . . . The vast variety of information cannot be learned by hearing, nor held in memory." See ibid., bk. 1, ch. 3, sec. 2.

19. "The written note is the signifier, the sound is the signified," observed Johannes de Muris (ca. 1290–ca. 1355), French mathematician and proponent of the *ars nova,* the "new art" of music. Johannes de Muris, *Notitia artis musice* (1321), selection in Gallo, *Music of the Middle Ages II,* 114. Somewhat later, Gioseffo Zarlino noted that "musicians, . . . realizing . . . sounds could in no way be written or otherwise depicted on paper . . . , devised certain signs or characters which they called figures or notes." Gioseffo Zarlino, *The Art of Counterpoint: Part Three of "Le Institutioni harmoniche," 1558,* trans. Guy A. Marco and Claude V. Palisca (1968; New York: W. W. Norton, 1976), 3.

20. Although musicians commonly use the terms interchangeably (as we shall here, except when otherwise indicated), 'pitch' and 'tone' are conceptually quite distinct. The former denotes the acoustical, physical occurrence of sound, caused

by vibrations and their frequencies; the latter refers to the sounds of music per se and their variously described dynamic qualities. Thus we hear rising and descending pitches (as in the whir of a jet engine) that have nothing to do with music (except, perhaps, to a pilot's ear), even though without pitch we hear no tones at all. Pitch is necessary, but not sufficient, for musical tone.

21. These ratios exhibit a rare case in which the practical and contemplative dimensions of early numeracy coincided thoroughly. Still, like ancient mathematics, music remained thoroughly subordinated to the classifying temper—falling in the categories of the quadrivium and being further subdivided into a variety of theoretical and practical genera, species, and subspecies. For one of many examples, see Isidore of Seville, *Etymologies,* bk. 3, chs. 15–23.

22. Besides their philosophical interest in music, the Greeks had developed an extensive and practical "musical science," much of which had been compiled and formulated by a student of Aristotle, Aristoxenus (fl. 335 BCE), whose output was reported as a prodigious, Isaac Asimov–rivaling 453 books. Of all these, only incomplete writings on harmonics and rhythm have survived. While the essential mathematical and theoretical dimensions of music in the Pythagorean-Platonic tradition were passed along to the Middle Ages through the works of Ptolemy, Boethius, and others, most works of practical Greek music were lost in the vanishing of much classical culture. A few manuscripts, among which Aristoxenus's *Harmonics,* were later recovered during the Renaissance. The development of plainsong and its notation into polyphony thus represents a fresh start in practical, Western music. For an entry into Greek music theory and practice, see Henry S. Macran, introduction to *The Harmonics of Aristoxenus,* ed. and trans. Henry S. Macran (Oxford: Oxford University Press, 1902), 4–15 and more generally 1–93; J. Murray Barbour, *Tuning and Temperament: A Historical Survey* (1951; Mineola, NY: Dover, 2004), 15–24; Claude V. Palisca, *Humanism in Italian Renaissance Musical Thought* (New Haven, CT: Yale University Press, 1985), 35–50; and Thomas J. Mathiesen, "Greek Music Theory," in Christensen, *Cambridge History of Western Music Theory,* 109–135.

23. Among others, Parisian music theorist Johannes de Grocheio (ca. 1255–ca. 1320) noted the similarities between the alphabet, arithmetic, and music: "And just as the grammarian is able to write down any word from a few letters through their connection and position, and the arithmetician from a few figures can construct any infinite number by placing them before and behind, so the musician can write down any measured song from three figures [long, breve, semibreve]." Quoted in Anna Maria Busse Berger, *Mensuration and Proportion Signs: Origins and Evolution* (Oxford: Oxford University Press, 1993), 5–6. In her fascinating, though technical study, Busse Berger suggests a "closely parallel development" between the history of "language-writing culminating in the invention of the Greek alphabet" and that of a "mensural system" in music during the fourteenth through sixteenth centuries (4–7 and throughout).

24. Hoppin, *Medieval Music,* 187. Only the barest outline of polyphony's early development is known; its documentation gathers momentum after the turn of the first millennium. Plainsong melodies were called *organum* in Latin (pl. *organa*), from the Greek *organon,* which meant "organ, instrument, or tool." Adding a parallel octave or fifth gave rise to the "parallel organum."

25. Guido of Arezzo, "Prologus antiphonarii sui," in Strunk, *Source Readings in Music History,* 118–119.

26. "O Saint John, in order that thy servants may be able to sing the praises of the marvels of thy deeds upon loosened harpstrings, cleanse those accused of the stain of sin." Over the centuries, *Ut* was replaced by *Do* in most countries, with the French still preferring the original. The *Si* of the scale was formed by combining the first two letters of Sancte *I*ohannes, now often indicated as *Ti.* Guido d'Arezzo, "Epistola de ignoto cantu," 178.

27. Ibid., 176, 177. "For just as there are seven days in a week, so there are seven notes in music. The others, which are added above the seven, are the same, . . . except that they double the sound higher up" (ibid., 179); Hoppin, *Medieval Music,* 60–64; Rastall, *Notation of Western Music,* 128–140.

28. Géza Szamosi, *The Twin Dimensions: Inventing Time and Space* (New York: McGraw-Hill, 1986), 104, and following. I have relied considerably on Szamosi's penetrating discussion of time and polyphony and their role in the Scientific Revolution.

29. Included among these innovators were two successive organists at Notre Dame, composers whose names for the first time in history were directly associated with their own compositions, Magisters Léonin (fl. 1150s–ca. 1200) and Pérotin (fl. early 1200s), the latter known as "Pérotin le Grand."

30. *Anonymous IV,* trans. and ed. Luther Dittmer, Musical Theorists in Translation (Brooklyn, NY: Institute of Mediaeval Music, 1959), 1:9–10.

31. The chief authority for this argument remains William G. Waite, *The Rhythm of Twelfth-Century Polyphony: Its Theory and Practice* (New Haven, CT: Yale University Press, 1954), 8, 27–40. See also Craig Wright, "Leoninus, Poet and Musician," *Journal of the American Musicological Society* 39, no. 1 (1986): 1–35.

32. *Anonymous IV,* 9–10, 12. The terms *perfecta* and *imperfecta* referred to subdivisions of notes, with perfect divisions being done in threes and imperfect in twos. The former's "perfection," often called ternary division, betokened a reference to the Holy Trinity. The latter, now called the 'duple' or binary system, has been the standard of division since the sixteenth century. Triplets, dotted notes (halves, quarters, and so on), musical ties, and time signatures (such as, 3 / 4, 6 / 8) are all devices for capturing ternary divisions in a duple system.

33. *Anonymous IV,* 47. Szamosi, *Twin Dimensions,* 105, remarks that in the thirteenth century one of the chief difficulties in moving toward a satisfactory account of motion lay with the inability of scholastics to separate time from motion, for they remained "mesmerized by the circular Aristotelian idea that bodily

motion measures time and time measures motion." Galileo would turn this perception inside out, but even he struggled to separate time and motion, as we shall see.

34. Victor Zuckerkandl, in *Sound and Symbol: Music and the External World,* trans. R. Trask (1956; Princeton, NJ: Princeton University Press, 1969), 171, 180, 199–203, and more generally 151–264, provides an incisive treatment of meter, rhythm, and the "musical concept of time." Zuckerkandl's writings reward careful, reflective reading, as he combines the perspectives of neo-Kantian philosophy and gestalt psychology in rich and probing analyses of our engagements with music. See also his *Man the Musician,* vol. 2 of *Sound and Symbol,* trans. Norbert Gutterman (Princeton, NJ: Princeton University Press, 1973), and the more accessible Zuckerkandl, *Sense of Music.*

35. Szamosi, *Twin Dimensions,* 107, 97. It bears digressing to summarize Szamosi's explanation of the unprecedented step from keeping track of time to measuring it (chs. 1–5). In the course of their evolution, plants and animals developed "biological clocks" as a means of keeping track of regular, temporal changes in the environment, such as the movement from day to night or season to season. Adaptations that produce behaviors governed by these clocks generate a process known as photoperiodism, which includes such widespread phenomena as mating, pelt growth, migration, nesting, and hibernation, among many others. With their language ability and capacity for creating symbolic meaning, humans far surpassed other mammals in this endeavor. Ancient cultures devised various technologies—calendars, sundials, sand glasses, burning candles, and the like—to supplement biological clocks in keeping track of the rhythms of the natural environment and to adjust or organize their lives in accordance with them. See, for example, Robert Hannah's account in *Time in Antiquity* (Loudon: Routledge, 2009). With the advent of polyphony, time became measured in regular units of duration, musical beats, and expressed in note symbols. The measuring apparatus was the human ear, the single most sensitive device for keeping the beat before the appearance of electronic instruments in the twentieth century. We shall see this again with Galileo, who used musical beat to measure time in his free-fall experiments.

36. Szamosi, *Twin Dimensions,* 107. To appreciate the difference between strict meter and rhythm, think of reciting a poem, such as William Shakespeare's Sonnet 73, which was written in iambic pentameter, a line of verse consisting of five rhythmic patterns (or feet) of short (unstressed) and long (stressed) syllables. For example, the expression "Ta DAH" is one iambic foot. (See Appendix A.3 for the correlation between poetic meter and rhythmic mode, exemplified in Sonnet 73.) When we deliver a line emphasizing the regular beats, the "metric structure" predominates: "Ta DAH Ta DAH Ta DAH Ta DAH Ta DAH." But the meaning of the verse often gets lost. If we stress the meaning, however, with the help of grammatical punctuation, we let the "true rhythm" of the verse prevail and the metric constraints control its temporal structure:

That time of year thou mayst in me behold
When yellow leaves, or none, or few, do hang
Upon those boughs which shake against the cold,
Bare ruin'd choirs, where late the sweet birds sang.

Thus do meter and rhythm make the meaning all the "more significant and beautiful." So too in music does meter allow true rhythm to enhance the meaning and beauty of melody. To borrow the distinction from philosophers, the creation of metric time established a necessary, though not sufficient, condition for musical rhythm. See Zuckerkandl, *Sound and Symbol,* 151–200.

37. Franco's principal notes, the core of the mensural system:

Duplex (double long) (▀┐), long (▘┐), breve (■), semibreve (◆). Later additions included the minim (♩), the semiminim (♩), and the fusa (♪), counterparts to our modern half, quarter, and eighth notes.

These notes are the forerunners to our modern system of musical notation. Franco of Cologne, *Ars cantus mensurabilis,* in Strunk, *Source Readings in Music History,* 139–159. Though his name is associated with it through his manuscript, the mensural system was by no means Franco's exclusive creation; after him many others contributed significantly to its clarification and development. See Willi Apel, *The Notation of Polyphonic Music, 900–1600,* 4th ed. (Cambridge, MA: Mediaeval Academy of America, 1949), 96–112, 188–195; Rastall, *Notation of Western Music,* 46–78; Parrish, *Notation of Medieval Music,* 108–141; Busse Berger, *Mensuration and Proportion Signs,* 4–5, 33–34, 49, 51–83.

38. Thus Johannes Grocheio defined the *tempus* as the "interval in which the smallest pitch or smallest note is fully presented or can be presented." Quoted in Alfred W. Crosby, *The Measure of Reality: Quantification and Western Society, 1250–1600* (Cambridge: Cambridge University Press, 1997), 152. The *minim* was the "elementary unit of value" in the theory of Philippe de Vitry (1291–1361), while for Giovanni Spataro (ca. 1458–1541), the *tempo* ('time') provided "that invariable *mensura* and point of departure." Quoted in Busse Berger, *Mensuration and Proportion Signs,* 60–61, 80–81. In the Renaissance various terms—*tempus, mensura, battuta,* and *tactus* (German)—all generally came to mean what we call musical beat.

39. Franco of Cologne, *Ars cantus mensurabilis,* 150. In the phrase of music theorist Zuckerkandl, the rest, "a void, a nothing, becomes the conveyor of motion." *Sound and Symbol,* 121.

40. Quoted in Busse Berger, *Mensuration and Proportion Signs,* 13–14. See n. 32 in the present chapter. Along with time signs, *modus* (mode) signs were also used in mensural notation, adding innumerable complexities essential for performers and musicologists, but ones we need not pursue.

41. Busse Berger, *Mensuration and Proportion Signs,* 34 and more generally 34–50. In general, the mensuration of a breve was technically termed a *tempus* and existed in two forms, perfect (ternary), whose symbol was the circle (O), and imperfect (binary or duple), whose symbol was the semicircle (C). The mensuration of the semibreve was called a *prolatio* ("prolation"), symbolized in its perfect form by (ʘ) and in its imperfect form by (Ͼ). See Apel, *Notation of Polyphonic Music,* 96–100, and his more general discussion, 96–125.

42. In Roman times, for example, on the abacus O designated an *uncia* (the origin of our 'ounce' and 'inch'). An *uncia* was one-twelfth of the largest monetary unit, the *as* or *libra,* a copper coin. (The modern equivalent of the *libra* until lately was the British pound, whose symbol, £, as well as that of the Italian lira, ₤, is based on an abbreviation of *libra,* or "pound.") The *uncia* was further divided into one-half, one-quarter, and one-third, each with its own symbol. The symbol for one-half was a C with a squiggly stroke below, in order to separate it from the C of one hundred. Because the Roman system was duodecimal (based on integer divisions of twelve), Roman fractions were all expressed in multiples of two and three; fractions involving five or seven were either rounded off or multiplied so as to be expressed in larger multiples. See Busse Berger, *Mensuration and Proportion Signs,* 34–50, for an extended comparison of Roman symbols for fractions with those of mensural notation.

43. Apel, *Notation of Polyphonic Music,* 146–147, 188–195, and more generally 145–195. "Noncumulative" meant that each new proportion was formed in relation to the initial time sign of the composition, not in relation to the immediately preceding section.

44. Technically, this example is in "perfect time, minor prolation." Busse Berger, *Mensuration and Proportion Signs,* 204–205.

45. Ibid., 26, 183–185, 205–207. Even for modern arithmephobes these calculations are grade-school easy, further indication of a Renaissance world marked by the general absence of mathematics. Several hundred years in acquisition, facility with basic counting and figuring in Hindu-Arabic numerals spread with excruciating slowness.

46. Thus Michael Masi writes that such "proportional architecture" was "not perceptible" to the Renaissance listener but rather was invoked to develop a "higher understanding derived from music." Michael Masi, "Boethian Number Theory and Music," in Boethius, *Boethian Number Theory: A Translation of the De Institutione Arithmetica,* trans. Michael Masi (Amsterdam: Editions Rodophi, 1983), 25–30. Curt Sachs, *Rhythm and Tempo: A Study in Music History* (New York: W. W. Norton, 1953), 205–217, sees these exuberances simply as a matter of "pen and ink." Composers could make new calculations, and they did.

47. For tuning systems and temperament in general, in addition to the previously cited Barbour (*Tuning and Temperament*), Stuart Isacoff, *Temperament: How Music Became a Battleground for the Great Minds of Western Civilization* (New

York: Vintage Books, 2001), and Ross W. Duffin, *How Equal Temperament Ruined Harmony (and Why You Should Care)* (New York: W. W. Norton, 2007), have both written informative, delightful, and engaging introductions to the subject. A good introduction to the mathematics of the Pythagorean system is Neil Bibby, "Tuning and Temperament: Closing the Spiral," in *Music and Mathematics: From Pythagoras to Fractals,* ed. John Fauvel, Raymond Flood, and Robin Wilson (Oxford: Oxford University Press, 2003), 13–27.

48. Claude V. Palisca, introduction to *Dialogue on Ancient and Modern Music,* by Vincenzo Galilei, trans. Claude V. Palisca (New Haven, CT: Yale University Press, 2003), xxx–xxxvii. In virtually all cultures instrumental music had existed primarily to accompany song, dance, and ritual. So too had it served antiquity and Europe of the Middle Ages. But no music was written specifically for instruments before the fifteenth century. The earliest keyboard score dates from slightly later, 1523, and thereafter the proliferation of instrumental scores multiplied the possibilities of musical experience dramatically. See also Apel, *Notation of Polyphonic Music,* 3–5.

49. For the dispute, see the authoritative account in Palisca, *Humanism,* 244–250, 265–279.

50. Like the famous theorem associated with his name, the Pythagorean scale most likely predates its namesake, perhaps dating to Babylonian times, circa 1800 BCE. Still, legend has it that he discovered the harmonious relationships of sound upon hearing dulcet tones ringing from a blacksmith's hammer and then experimenting to come up with the numerical ratios that underlay them, ratios keyed to the *tetractys.* The earliest account of the legend is that of Nicomachus, *The Manual of Harmonics of Nicomachus the Pythagorean,* trans. Flora R. Levin (Grand Rapids, MI: Phanes, 1994), 83–85. In his "Life of Pythagoreas," in *The Pythagorean Sourcebook and Library,* ed. David Fideler (Grand Rapids, MI: Phanes, 1987, 1988), 86–88, Iamblichus copied the legend pretty much in its entirety.

51. In modern times we interpret these ratios as intervals of pitch, whose frequencies are caused by vibrations from the sound source (measured by hertz, cycles per second, abbreviated Hz). Middle C on the piano, for instance, equals 261.63 Hz; doubling that to 523.26 Hz creates C', an octave higher, the 2:1 ratio. Set at 440 Hz, the note A above the piano's middle C has been adopted as a widespread standard in modern, equal temperament tuning. Before modern times the intervals were interpreted as differences in string lengths (assuming the same size and tension of strings) and the vibrations they produced. While the modern version yields a more thorough account of the acoustical science behind pitch, the numerical ratios operate the same, whether of pitch frequencies or string lengths. Still, some care is needed to understand the order of the numbers. A 2:1 ratio can be read as referring to a string length double that of the original. In this case the octave will be *lower* than the original tone. Or it can be read as a string length half the original, with the octave being *higher.* Halving the length means doubling the

vibrations, while doubling the length means halving them. The Greeks formed their scales in tetrachords (four notes) by lowering tones (say, A, G, F, E). Modern convention presents scales as rising tones (thus, E, F, G, A).

52. Claude V. Palisca, introduction to Zarlino, *Art of Counterpoint,* ix.

53. Zarlino, *Art of Counterpoint,* 264–265.

54. Thus Zarlino accepted the unheard harmony of the spheres and the consonance of the soul's parts, which he grouped under the category of "animistic music" (*musica animistica*). He also accepted the premise that sonorous number and fixed proportion produce all music, animistic as well as organic (*musica organistica*), the latter divided into harmonic (vocal) and natural (instrumental). Music remained for him the grand harmonizing agent of the universe. Palisca, introduction to Zarlino, *Art of Counterpoint,* xviii–xix.

55. Zarlino, *Art of Counterpoint,* 1–2.

56. Gioseffo Zarlino, *Le Institutioni harmoniche,* pt. 1, ch. 14, cited in Benito V. Rivera, "Theory Ruled by Practice: Zarlino's Reversal of the Classical System of Proportions," *Indiana Theory Review* 16 (1995): 151 and more generally 145–170. See also Robert W. Wienpahl, "Zarlino, the Senario, and Tonality," *Journal of the American Musicological Society* 12, no. 1 (1959): 27–41.

57. Vincenzo Galilei, *Discorso intorno alle opera di Gioseffo Zarlino* (1589), quoted in Rivera, "Theory Ruled by Practice," 145–146.

58. Galilei, *Dialogue on Ancient and Modern Music,* 81. Pianist, composer, and music writer Stuart Isacoff notes that today's barbershop quartets, singing in "tight formation" without benefit of accompaniment, do not rely on any given tuning system; rather, they employ all of them as needed in their "ever-shifting" pursuit of harmonious accord among their voices. Isacoff, *Temperament,* 140.

59. See Galilei's analyses in the *Dialogue on Ancient and Modern Music,* 33–74; also see Palisca, *Humanism,* 271. The Pythagorean ratio for the major third (the interval from C to E on the piano) was 81:64, while that of the minor third (E to G) was 32:27. In Zarlino's system, flattening the major third by a small amount called a comma (ratio of $\frac{81}{80}$) produced $\frac{81}{64} \div \frac{81}{80} = \frac{81}{64} \times \frac{80}{81} = \frac{6480}{5184} = \frac{5}{4}$. Similarly, sharpening the minor third by a comma produced $\frac{32}{27} \times \frac{81}{80} = \frac{2592}{2160} = \frac{6}{5}$. Comma placement here made these tones "just." However, at the same time, the five steps from D to A yielded a ratio of a bit less than 3:2, "flattening" this particular fifth, even while maintaining the fifth's "perfect" consonance from C to G.

60. Galilei, *Dialogue on Ancient and Modern Music,* 83–89.

61. Mathematically each of the twelve semitones in equalized temperament is the irrational result of a computation—$a = \sqrt[12]{2}$ (where a is a semitone, and 2 is the octave, whose ratio is 2:1). For convenience, modern musicologists use a system of "cents" to figure equal intervals within an octave. An octave equals 1,200 cents, with each semitone valued at 100. A fifth is thus 700 cents, whereas in pure, com-

putational form it equals 702 cents. Other notes have a larger deviation of cents from their pure ratios.

62. The tension experiment, among others, had been suggested to Galilei by his correspondent Girolamo Mei (1519–1594), one of the foremost Renaissance students of Greek music. Mei provided Galilei (who did not read Greek) a great deal of information about ancient tuning, especially the "intense" diatonic of Aristoxenus, which was, in effect, an equal temperament. See Palisca, *Music and Ideas,* 149–153, and Palisca, *Humanism,* 265–279, for a brief account of Galilei's experimental work. Stillman Drake, "Renaissance Music and Experimental Science," *Journal of the History of Ideas* 31, no. 4 (1970): 483–500, explores some of the wider implications of this work for the development of experimental science, especially in the work of Vincenzo Galilei's famous son, Galileo. Galileo recorded his father's findings in his own work *Two New Sciences,* which first appeared in 1638. Galileo Galilei, *Two New Sciences,* trans. Stillman Drake (Madison: University of Wisconsin Press, 1974), 99–102.

63. See Duffin, *How Equal Temperament Ruined Harmony.*

64. *Anonymous IV,* 53.

65. Vincenzo Galilei, "Discorso particolare intorno alla diversità delle forme del diapason," quoted in Palisca, *Humanism,* 277.

66. The quotations are from Michael Stifel, *Arithmetica integra* (1544), and found in Peter Pesic's excellent article "Hearing the Irrational Music and the Development of the Modern Concept of Number," *Isis* 101, no. 3 (2010): 509–510, and more generally 501–530.

67. For Cardano's remarks, see ibid., 510–511.

68. It does merit noting that many Renaissance natural philosophers and mathematicians were either musicians themselves or thoroughly conversant with current music theory and practice—to wit, Nicole Oresme, Johannes de Muris, Rabbi Levi ben Gershon (Latinized as Gersonides, 1288–1344), Philippe de Vitry, Franchino Gaffurio (1451–1522), Giovanni Battista Benedetti, Girolamo Cardano, and Leonardo da Vinci (1452–1519), among others.

7. Geometry

1. The gulf between continuity and discreteness is most profoundly manifest in the wave-particle duality of light, energy, and matter found in the microworld of quantum mechanics. Arthur Zajonc has included a captivating account of this duality in his study of light and its history: *Catching the Light: The Entwined History of Light and Mind* (New York: Oxford University Press, 1993), 225–329.

2. Boethius had started a Latin translation of the *Elements,* but it survived only in fragments in the early Middle Ages and contained no proofs. For an account of geometry in medieval schools, see Lon R. Shelby, "Geometry," in *The Seven Liberal*

Arts in the Middle Ages, ed. David L. Wagner (Bloomington: Indiana University Press, 1986), 196–217; Edward Grant, *The Foundations of Modern Science in the Middle Ages: Their Religious, Institutional, and Intellectual Contexts* (Cambridge: Cambridge University Press, 1996), 44–47; and Menso Folkerts, "The Importance of the Latin Middle Ages for the Development of Mathematics," in *Essays on Early Medieval Mathematics: The Latin Tradition* (Burlington, VT: Ashgate, 2003), sec. 1.

3. Hugh of Saint Victor, *Practical Geometry (Practica Geometriae),* trans. Frederick A. Homann (Milwaukee: Marquette University Press, 1991), 33–34. See too Shelby, "Geometry," 197–199, 202–203; Paul Rorem, *Hugh of St. Victor* (Oxford: Oxford University Press, 2009), 42–45; and Glen Van Brummelen, *The Mathematics of the Heavens and the Earth: The Early History of Trigonometry* (Princeton, NJ: Princeton University Press, 2009), 224–226.

4. Plato had even less use for visual art than he did for auditory music, seeing it as an imitation, "an inferior thing that consorts with another inferior thing to produce an inferior offspring." *Republic,* bk. 10, 603b, in *Plato: Complete Works,* ed. John M. Cooper (Indianapolis: Hackett, 1997). See the masterful treatment of medieval aesthetic themes in the work of Umberto Eco, *Art and Beauty in the Middle Ages,* trans. Hugh Bredin (New Haven, CT: Yale University Press, 1986), 1–42 and throughout.

5. Among others, the terms "proportion," "symmetry," "order," "disposition," and "eurhythmy" (harmonious movement) initially applied to architecture but became the lingua franca of more general discussions of beauty. Eco, *Art and Beauty,* 1–42. Vincent of Beauvais assembled and wrote the most famous and extensive of medieval encyclopedias, a work consisting of some 4.5 million words, comprising four main parts, eighty books, and 9,885 chapters—seven folio volumes of over five hundred pages each. For his achievement, see Ann M. Blair, *Too Much to Know: Managing Scholarly Information before the Modern Age* (New Haven, CT: Yale University Press, 2010), 41–46.

6. "Proportion is a correspondence among the measures of the members of an entire work, and of the whole to a certain part selected as standard . . . as in the case . . . of a well-shaped man. . . . From this result the principles of symmetry." Marcus Vitruvius Pollio, *The Ten Books on Architecture,* trans. Morris Hicky Morgan (1914; New York: Dover, 1960), 72. Macrobius (fl. fifth century) called the "universe a huge man and man a miniature universe." Macrobius, *Commentary on "The Dream of Scipio,"* trans. William Harris Stahl (1952; New York: Columbia University Press, 1990), 224. Also see Eco, *Art and Beauty,* 29, 35–42.

7. In geometry, the golden ratio occurs when a line AB is divided by point C so that the ratio of AB to AC equals that of AC to CB: A C B (AB : AC :: AC : CB). The same holds with many polygons and solid figures. More generally, two quantities are considered in the golden ratio if the ratio between them equals the ratio of their sum to the larger of the two quantities. For two quantities, *a* and *b*, with $a > b$, this is can be represented algebraically by the equation $\frac{a+b}{a} = \frac{a}{b} = \varphi$

and expressed in quadratic form as $\varphi^2 - \varphi - 1 = 0$. Its solution, $\varphi = \frac{1+\sqrt{5}}{2}$, yields the unending, irrational decimal fraction 1.6180339887. . . . Nature and history alike have supplied countless examples and applications of this ratio, whose first recorded definition was made by Euclid with his "extreme and mean ratio." Euclid, *The Thirteen Books of the Elements,* 2nd ed., 3 vols., trans. Thomas L. Heath (1926; New York: Dover, 1956), bk. 7, prop. 30, 2:267. Leonardo's description was written in his notebooks: Leonardo da Vinci, *Notebooks,* trans. Jean Paul Richter, sel. Irma A. Richter, ed. Thereza Wells (New York: Oxford University Press, 2008), 138–143. Many of its mathematical dimensions are discussed by Roger Herz-Fischler in *A Mathematical History of the Golden Number* (1987; Mineola, NY: Dover, 1998), throughout, but especially 134–159 for the medieval and Renaissance periods. Also, for the ratio's place in Renaissance drawing and painting, as well as its more general, aesthetic role, see Mario Livio, *The Golden Ratio: The Story of Phi, the World's Most Astonishing Number* (New York: Broadway Books, 2002), 124–142, and H. E. Huntley, *The Divine Proportion: A Study in Mathematical Beauty* (New York: Dover, 1970).

8. For the influence of Neoplatonism and its prescriptive mathematics, especially "harmonic proportion," on Renaissance architecture, see the earlier work of Rudolf Wittkower, *Architectural Principles in the Age of Humanism* (1949; London: Academy Editions, 1988), 104–137 and throughout. A more recent approach that explores the "Vitruvian tradition . . . as a catalyst for communication and exchange between learning and skill" is Pamela O. Long, *Artisan/Practitioners and the Rise of the New Sciences, 1400–1600* (Corvallis: Oregon State University Press, 2011), 93, 62–93, and throughout.

9. "Mathematics for the metaphysicians made no inroads into mathematics proper." Thus does J. V. Field succinctly sum up this state of affairs in *Piero Della Francesca: A Mathematician's Art* (New Haven, CT: Yale University Press, 2005), 95. Important advances in trigonometry were, however, made by Muslim mathematicians, slowly spreading to Europe beginning roughly in the eleventh century. See the informative account in Van Brummelen, *Mathematics of the Heavens,* 135–222.

10. Edward Grant, *A History of Natural Philosophy: From the Ancient World to the Nineteenth Century* (Cambridge: Cambridge University Press, 2007), 321–323. Music, too, was occasionally seen as a mixed science.

11. Hugh associated the "mechanical" arts with adultery, deriving the term from the Latin *moechus* (adulterer) rather than from the Greek *mekhane* (machine) via the Latin *mechanica.* For him the mechanical art was the work of an "artificer" who "imitates nature" in a "clever and . . . delicate way," one that lies "beyond detection." Thus did the mechanical arts also entail deceit, a quality later surfacing in linear perspective. *The Didascalicon of Hugh of Saint Victor: A Medieval Guide*

to the Arts, trans. Jerome Taylor (New York: Columbia University Press, 1961), 56, 191. For the relation between the mechanical arts, mixed sciences, and natural philosophy, see the Cambridge offerings: Katharine Park and Lorraine Daston, "Introduction: The Age of the New," in *The Cambridge History of Science,* vol. 3, *Early Modern Science,* ed. Katharine Park and Lorraine Daston (Cambridge: Cambridge University Press, 2006), 1–21; William A. Wallace, "Traditional Natural Philosophy," in *The Cambridge History of Renaissance Philosophy,* ed. Charles B. Schmitt et al. (Cambridge: Cambridge University Press, 1988), 201–206; and James A. Weisheipl, "The Interpretation of Aristotle's *Physics* and the Science of Motion," in *The Cambridge History of Later Medieval Philosophy: From the Rediscovery of Aristotle to the Disintegration of Scholasticism, 1100–1600,* ed. Norman Kretzmann, Anthony Kenny, and Jan Pinborg (Cambridge: Cambridge University Press, 1982), 525 and more generally 521–536. Following the term "mixed mathematics" into modernity are Gary I. Brown, "The Evolution of the Term 'Mixed Mathematics,'" *Journal of the History of Ideas* 52 (1991): 81–102; and Peter Dear, "Mixed Mathematics," in *Wrestling with Nature: From Omens to Science,* ed. Peter Harrison et al. (Chicago: University of Chicago Press, 2011), 149–172.

12. The "winged eye" was featured centrally on the personal medallion of Leon Battista Alberti, a key figure in the following pages.

13. Exemplars of exalted claims for perspective include those of Samuel Y. Edgerton Jr., who in *The Renaissance Rediscovery of Linear Perspective* (New York: Basic Books, 1975), 3, writes that perspective "was to change the modes, if not the course of Western history." In his recent work, *The Mirror, the Window, and the Telescope: How Linear Perspective Changed Our Vision of the Universe* (Ithaca, NY: Cornell University Press, 2009), Edgerton broadens his scope to the "vision of the universe" declared in his title. David Summers, in *Vision, Reflection, and Desire in Western Painting* (Chapel Hill: University of North Carolina Press, 2007), 160, claims that "perspective became a modern Western metaphor for human being-in-the world." Such comments derive from one of the founding fathers of modern art history and criticism, Erwin Panofsky, *Perspective as Symbolic Form,* trans. Christopher S. Wood (German original, 1924–1925; New York: Zone Books, 1991), 67–68, who wrote nearly a century ago that "the history of perspective . . . is as much a consolidation and systematization of the external world as an extension of the domain of the self." These are all very good books, well worth the reading. Still, sometimes such declarations can leave one a bit breathless. But agree or disagree, more than parallel lines and vanishing points stands at stake in these arguments—and perspectives.

14. For gazing examples, see Hans Belting, *Florence and Baghdad: Renaissance Art and Arab Science,* trans. Deborah Lucas Schneider (Cambridge, MA: Harvard University Press, 2011), 9, 151, and throughout, and Norman Bryson, *Vision and Painting: The Logic of the Gaze* (New Haven, CT: Yale University Press, 1983). In her review of Belting's book, "The Invention of Space," *New Republic,* December 13, 2011, Ingrid Rowland addresses concisely and incisively the dubious matter of the

"gaze," explaining its deviation and derivation from the French *regard* and its theory-bound origins in the *(outré)* "scopic regimes" of Jacques Lacan.

15. James R. Banker records over forty interpretations of Piero della Francesca's enigmatic *Flagellation of Christ* alone in *Piero della Francesca: Artist & Man* (Oxford: Oxford University Press, 2014), 135.

16. Belting, *Florence and Baghdad*, 212; Leon Battista Alberti, *On Painting*, trans. Cecil Grayson, introduction and notes by Martin Kemp (New York: Penguin Books, 1991). The term *costruzione legittima* was coined later by Pietro Accolti (ca. 1579–1642) to affirm the correctness of Alberti's method. Though "silly," it stuck, writes J. V. Field in *The Invention of Infinity: Mathematics and Art in the Renaissance* (Oxford: Oxford University Press, 1997), 25–30, 165.

17. Leon Battista Alberti, *On Sculpture*, trans. Jason Arkles (self-pub., lulu.com, 2013), 14, 21–25, 30. A sculptor of monumental and other forms of statuary who lives and teaches in Florence, Italy, Arkles deftly explains Alberti's pathbreaking account of "dimensions" and "terminal points," as well as other technical matters pertaining to scaling and sculpture. In restricting my comments mainly to perspective, proportionality, and geometric continuity, I shall be leaving aside many other technical and qualitative features of pictorial representation, such as texture, color, value (light and dark), and the like, including the "psychology of perception" and narrative content (what Alberti and others termed *historia*), all subjects more suited to the investigations of art history per se. My more limited aim here is to highlight widely used features of the visual arts that contributed to the emerging information technology of numeracy (though not so called at the time), which in turn enabled the transformation from thing-mathematics to relation-mathematics and the reverse engineering of modern analysis.

18. In its resolution of 1406, the city's governing signoria had charged all the major and middle guilds with commissioning statues of their patron saints to embellish the facades of the church. Site not only of earlier grain storage but also of numerous reported miracles, the palatial Orsanmichele commanded the city's symbolic center of nourishment, both body and soul, its pride of place among Florence's public buildings second only to the still-unfinished cathedral. The signoria viewed its embellishment as top priority among its civic projects, and expectations ran high.

19. The story comes to us from Giorgio Vasari, *The Lives of the Artists*, trans. Julia Conaway Bondanella and Peter Bondanella (Oxford: Oxford University Press, 1991), 150–151. While always charming, Vasari's lives are not always reliable, and this tale is most likely apocryphal, although it harbors a historical truth, Donatello's use of the new perspective in his pathbreaking statuary. For an enjoyable introduction to the Florentine art world of the early *quattrocentro*, see Paul Robert Walker, *The Feud That Sparked the Renaissance: How Brunelleschi and Ghiberti Changed the Art World* (New York: HarperCollins, 2002), esp. 32–35, 60–65, 75–76. Ross King delightfully relates the biography of Brunelleschi in *Brunelleschi's Dome: How a Renaissance Genius Reinvented Architecture* (New York: Bloomsbury, 2000),

while Michael Baxandall skillfully situates painting and artistic practices within their Renaissance social context in *Painting and Experience in Fifteenth Century Italy: A Primer in the Social History of Pictorial Style*, 2nd ed. (Oxford: Oxford University Press, 1972).

20. According to Brunelleschi's next-generation biographer, Antonio Manetti (1423–1497), Donatello and Brunelleschi had spent several years together in Rome in the early 1400s, digging among classical ruins, taking measurements of buildings (including the Pantheon), eyeing sculptures, making drawings, and exploring the remnants of antiquity. Defending his hero, Manetti wrote that Brunelleschi "told [Donatello] nothing of his ideas, either because he did not find Donatello apt or because he was not confident of prevailing." Still, it's hard to imagine that something of "perspective" did not emerge from these shared experiences for both men, close friends at the time (their falling-out would come much later). Antonio Manetti, *The Life of Brunelleschi*, trans. Catherine Enggass, introduction and notes by Howard Saalman (University Park: Pennsylvania State University Press, 1970), 52, more generally 50–54. Moreover, Donatello clearly used perspective lines in constructing his relief *St. George and the Dragon* (1415–1417) a mere two or three years later. See Field, *Invention of Infinity*, 40–41; and Walker, *Feud*, 90, 189–191.

21. The comment of Domenico da Prato, quoted in King, *Brunelleschi's Dome*, 34, was made in 1413, but whether the experiment had already been performed remains uncertain. Estimates of its actual date range from between 1401 and 1409, to before 1413, to the year 1425. See Michael Kubovy, *The Psychology of Perspective and Renaissance Art* (Cambridge: Cambridge University Press, 1986), 32. As well as being a secretive and somewhat prickly engineering genius, Brunelleschi was also a notorious prankster, and he likely designed something on the order of a trompe l'oeil painting to both amaze and amuse. See Antonio Manetti's yarn *The Fat Woodworker*, trans. Robert L. Martone and Valerie Martone (New York: Italica, 1991), for the comical depiction of an elaborate *beffe*, a hoax perpetrated by Brunelleschi and friends on an unsuspecting woodworker named Manetto (who was "actually a bit simple," wrote the storyteller). Through an elaborate array of circumstances, the pranksters convinced him that he did not exist as himself. Mary McCarthy used the tale to extol the genius of Brunelleschi, "which found the way to calculate the vanishing point, [and which] could make a bulky man vanish or seem to himself to vanish." Mary McCarthy, *The Stones of Florence* (New York: Harcourt, 1963), 116–118.

22. Brunelleschi likely chose the baptistery for its regular, octagonal shape and symmetrical construction, which would minimize the image reversal in the mirror. Edgerton claims (perhaps rightly) that Brunelleschi painted the panel in reverse from the mirror reflection itself (with his back to the baptistery) by means of a two-point construction, so that the demonstration would be in effect the reversal of a reversal and hence identical to the direct view through the sight line without the mirror. Edgerton, *Mirror*, 65–68.

23. Manetti, *Life of Brunelleschi*, 42, 44, 46. The term "artificial perspective" (*perspectiva artificialis*) was generally used in the Renaissance to distinguish the

geometrical, linear perspective of artists from *perspectiva naturalis,* or optics. Last seen in the possession of Lorenzo the Magnificent, Brunelleschi's actual painting was lost, probably during the French occupation of Florence in 1494. All versions of the experiment stem from Manetti, who claimed to have seen the small panel in Brunelleschi's later years. Beyond Manetti's few facts, virtually everything about the demonstration—its specific date, its layout and orientation, its antecedents, and its interpretation—remains shrouded in scholarly controversy. Edgerton argues, citing Archbishop Fra Antonio of Florence (Antonio Pierozzi, 1389–1459), that Brunelleschi's motive may have been an expression of "spiritual geometry . . . to measure temporal things . . . not as quantities, but as virtues within God." He also notes that Brunelleschi might have painted the baptistery to "see through" (*perspicere*) the famous baptistery doors of Lorenzo Ghiberti (1378–1455), thus showing up his rival, by whom he had been bested in the competition for sculpting the doors. See Edgerton, *Mirror,* 48–50, and Walker, *Feud,* 13–25. Whatever the motivation, a consensus remains: Brunelleschi's discovery made explicit the geometric rationalization of depicting visual space.

24. Dante Alighieri, *The Banquet,* trans. Elizabeth Price Sayer (1887; Hollywood: Aegypan, [2013?]), 64. The history of early optics lies beyond present concerns, but it is well recounted in David C. Lindberg, *Theories of Vision: From Al-Kindi to Kepler* (Chicago: University of Chicago Press, 1976), 1–17; A. Mark Smith, *Ptolemy and the Foundations of Ancient Mathematical Optics: A Source Based Guided Study* (Philadelphia: American Philosophical Society, 1999), 23–49; Edgerton, *Renaissance Rediscovery,* 64–78; and Zajonc, *Catching the Light,* 10–37.

25. Their conjectures ranged between two alternatives, depending on what one considered the motion's source. The atomists (Democritus, Epicurus, Lucretius) believed that all vision came to the eye via light, a "material effluence" conveyed from a visible object (the "intromission theory"). Others coalesced around those (including Pythagoras and Empedocles) who held that the eye radiated an "internal fire," which then intermingled with light from the sun to produce vision (the "extramission theory"). Both Plato and Aristotle drew elements from these theories, with the latter particularly stressing a middle ground that emphasized invisible light or "luminosity" as the medium linking the visible object and the observer. Other thinkers worked within the assumptions laid down by these early forays into optics, including Galen of Pergamon, the first to devise an extensive account of the eye as a visual instrument.

26. For Alhazen's optics, see Lindberg, *Theories of Vision,* 58–86. Lindberg has also translated and edited Pecham's treatise on optics: *John Pecham and the Science of Optics: Perspectiva communis,* ed. and trans. David C. Lindberg (Madison: University of Wisconsin Press, 1970). His introduction (3–58) provides a good synopsis of medieval optics, particularly for the expressed desire of many authors "to illustrate spiritual truths" through their understanding of vision and light (19). Other scholastic writers on optics included Robert Grosseteste (ca. 1175–1253), Roger Bacon (ca. 1214–1294), and Witelo (Erazmus Ciolek Witelo, ca. 1230–ca.

1280–1314). Alhazen's work was also translated into Italian in the fourteenth century, and Ghiberti, among others, had extracted and copied portions of it in his own commentaries.

27. Alberti, *On Painting,* 39–40.

28. Coordinates and reference points of various sorts had long been employed, as had geometric symbols, in use ever since the first, simple pictographs—for example, Δ for triangle. As well, many ancient manuscripts featured diagrams of geometrical problems, particularly in the ruler-and-compass constructions of the Greeks in which letter symbols also referred to lines. Occasionally, too, in our time period one can find cases of substituting the symbol of an ideogram for a word, as with ~ for 'similar,' while in the sixteenth century algebraic symbols began creeping into geometrical use, just as they did with early, rhetorical algebra. But of the systematic use of grid lines, one finds very little until Ptolemy. See the discussion later; also see Florian Cajori, *A History of Mathematical Notations: Two Volumes Bound as One,* vol. 1, *Notations in Elementary Mathematics;* vol. 2, *Notations Mainly in Higher Mathematics* (LaSalle, IL: Open Court, 1928, 1929; New York: Dover, 1993), 401–431.

29. One started with a center, "in a definite place," he noted, to which "lines should naturally correspond with due regard to the point of sight and the divergence of visual rays." The resulting "deception" would provide "a faithful representation of the appearance of buildings." Vitruvius, *Ten Books on Architecture,* 198. Beyond such barebones observations, however, Vitruvius did not venture, and his ambiguity of expression, along with the absence of any actual diagrams or grid-line construction, has prompted scholarly skepticism as to whether he understood much, if anything at all, about linear perspective. See, for example, Belting, *Florence and Baghdad,* 17; John White, *The Birth and Rebirth of Pictorial Space,* 3rd ed. (Cambridge, MA: Harvard University Press, 1987), 250–258; and Kirsti Anderson, *The Geometry of an Art: The History of the Mathematical Theory of Perspective from Alberti to Monge* (New York: Springer, 2007), 60.

30. Euclid's diagram of the visual cone was as follows:

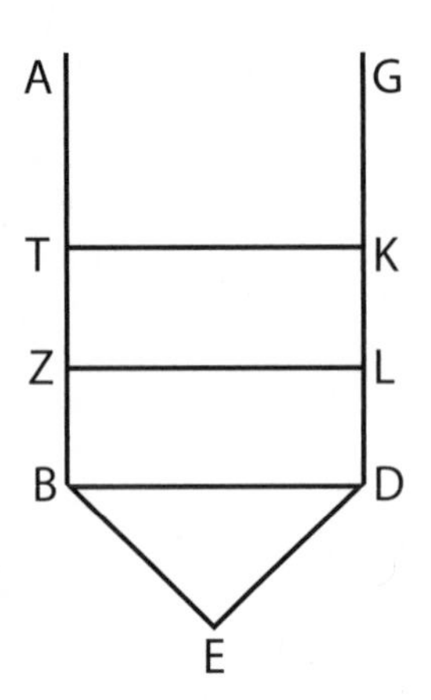

Closer objects, represented here by the line BD, appear larger because the angle of vision, BED, is larger than for objects of the same size seen at greater distance (lines ZL and TK). Further, the space between parallel lines AB and GD will appear to grow smaller as the lines BD, ZL, and TK move farther away from the eye, E. This occurs because the angles forming the visual cones (BED, ZEL, and TEK) grow smaller. Euclid, "The Optics of Euclid," trans. Harry Edwin Burton, *Journal of the Optical Society of America* 35, no. 5 (1945): 358 and more generally, 357–372.

31. Ibid., 372. Mark A. Peterson develops this observation in *Galileo's Muse: Renaissance Mathematics and the Arts* (Cambridge, MA: Harvard University Press, 2011), 103–106.

32. Claudius Ptolemy, *Optics,* bk. 2, 50, excerpted in Smith, *Ptolemy,* 53. Ptolemy also extended the study of geometrical optics to include the reflection of rays (catoptrics) and their refraction (dioptrics).

33. The rays near the center of the visual cone, for example, were closer to the eye, returning clearer and more precise impressions than those on the periphery, which helped explain convexity and concavity, critical to discerning the shapes of objects. Alberti would later single out the "centric ray" as most important, describing it as "keen and vigorous, . . . the leader and prince of rays." Claudius Ptolemy, *Optics,* bk. 2, 64–67, excerpted in Smith, *Ptolemy,* 63; Alberti, *On Painting,* 43–44.

34. Claudius Ptolemy, *Optics,* bk. 2, 76, excerpted in Smith, *Ptolemy,* 64. "Visual perception of the phenomena associated with locomotion depends primarily on the visual faculty itself," Ptolemy wrote (ibid.).He was keenly aware that vision alone could not determine whether celestial bodies circumnavigated a stationary earth or whether the earth also moved. Because it utilized haptic (tactile), as well as visual information, physics trumped astronomy, whose function was to save the appearances (discussed earlier).

35. In the 1390s a number of Florentine intellectuals and businessmen formed a study group for the purpose of learning Greek, in the course of which they hired a Byzantine scholar named Manuel Chrysoloras, a gifted and inspiring teacher. The group soon exhausted the local supply of Greek works, whereupon Chrysoloras and one of the group members, Jacopo d'Angiolo, journeyed to Constantinople in search of more manuscripts. Among their finds was a Greek copy of Ptolemy's *Geography* (previously unknown in the West), which they proceeded to translate into Latin, making the manuscript available from 1406 forward. See Edgerton, *Renaissance Rediscovery,* 91–123.

36. Calculating such lines traces back to Eratosthenes and Hipparchus, but Ptolemy became the first to apply them systematically to a map of the entire world. For an introduction to Ptolemy's geography, see David Buisseret, *The Mapmakers' Quest: Depicting New Worlds in Renaissance Europe* (Oxford: Oxford University Press, 2003), 10–24; and J. Lennart Berggren and Alexander Jones, *Ptolemy's "Geography": An Annotated Translation of the Theoretical Chapters* (Princeton, NJ:

Princeton University Press, 2000). Especially valuable is the latter's introduction, 3–54.

37. The first was a simple, conical projection with the parallels represented by concentric circles and the meridians by straight lines converging on a single point. Distortion entered here when the parallels south of the equator diminished while the arcs of the "cone" increased in length. In the second, "pseudo-conical" projection, Ptolemy curved the meridian lines, producing in effect the image of a hemisphere. The axis of the visual cone connected the eye and the center of the hemisphere, from which point the parallels and meridians were calculated. But distortions entered in these calculations, for a circular arc can connect any three noncollinear points but not four, making it impossible for Ptolemy to keep the distances measured along more than three of the parallels in exact proportionality with the distances along the central meridian. Ptolemy, *Geography,* bk. 1, 21–24, in Berggren and Jones, *Ptolemy's "Geography,"* 83–93. See too 35–40.

38. Ptolemy, *Geography,* bk. 7, 6–7, in Berggren and Jones, *Ptolemy's "Geography,"* 112–117. The editors (39) underscore the significance of Ptolemy's world map as "the unique example of a construction according to linear perspective surviving from antiquity."

39. Beyond the evidence from optics and cartography, occasional references to plots, grids, plans, and graphs can be found antedating the creation of linear perspective: the "centurians" of Roman surveyors, plats of huge, rectangular fields; plans of monasteries and abbeys, most notably a detailed drawing of Saint Gall in Switzerland, dating from circa 820; in physics, graphs of uniformly accelerating bodies (the mean speed theorem), created by Nicole Oresme and the *calculatores* of the Merton school; in music, as we have seen, the plotting of tone and time against a grid of horizontal lines, the musical staff.

40. Architecture per se was not yet a recognized occupation, but Manetti, in *Life of Brunelleschi,* 52, noted that Brunelleschi and Donatello "drew the elevations [of buildings] on strips of parchment graphs with numbers and symbols which Filippo alone understood," and Richard Krautheimer calls Brunelleschi's perspective method "fundamentally an architectural tool." For this argument, see his *Lorenzo Ghiberti,* vol. 1, with T. Krautheimer-Hess, excerpted in *Brunelleschi in Perspective,* ed. Isabelle Hyman (Englewood Cliffs, NJ: Prentice-Hall, 1974), 129–139.

41. For Jacob Burckhardt, the nineteenth-century historian who invented the term and the study of the "Renaissance" as a cultural era and entity, Alberti provided the model of the "many-sided," universal man: intellectual and athlete, philosopher and inventor, ambitious courtier and practitioner of the fine arts, a humanist who transformed quotidian activities into the art of living. Jacob Burckhardt, *The Civilization of the Renaissance in Italy* (1878; Mineola, NY: Dover, 2010), 85–87. Anthony Grafton, in *Leon Battista Alberti: Master Builder of the Italian Renaissance* (Cambridge, MA: Harvard University Press, 2000), has written the definitive contemporary account of Alberti's life and times, which places this

"Renaissance man" firmly in the context of fifteenth-century Italy, interweaving in particular the humanistic and practical dimensions of Alberti's remarkable career.

42. Alberti, *On Painting,* 34–35.

43. Ibid., 39–48. For "a skin stretched over the whole extent of the surface," the Italian reads, "come buccia sopra tutto il dosso della superficie." Alberti fleshed out his own metaphor in book 2, urging that when painting animals one sketches the bones first, then, in order, the sinew, muscle, flesh, and skin (72–73). Although beyond the present scope, light, shadow, and color also contributed significantly to the illusions of depth and solidity in Alberti's account.

44. A *braccio* (arm) served as a measuring unit during the Renaissance, just as other body parts, such as a foot or hand, were and are in use still today (for instance, we measure horse heights by 'hands'). The *braccio* varied from city to city; in Sansepolcro, it measured 56 modern centimeters, while in Florence it was 58.36. Generally it was somewhere between 56 and 69 centimeters, about two modern feet.

45. In their practices painters devised a number of ways to lay out a painting using perspective. "Snap lines" (like a modern carpenter's chalk line) were sometimes used to provide both orthogonals and transversals. For frescoes, the *spolvero* technique predated perspective, but employed it as well. A bag filled with dust (*spolvero*) would be gently tapped on a pricked drawing, leaving a charcoal outline on the fresco surface. Alberti also described using a "veil," thin horsehairs placed in a square grid over the model, then transposed in scale to a larger work. "Cartoons" (figure outlines) were also commonplace.

46. With "change of position, the properties inherent in a surface appear to be altered," wrote Alberti in *On Painting,* 39.

47. Ibid., 76–81.

48. Field, *Piero Della Francesca,* 272. Along similar lines, Descartes would later philosophically separate "indefinite" and "infinite," referring to matter as "indefinitely extended" to contrast with the positive infinity of God. René Descartes, *Philosophical Letters,* trans. and ed. Anthony Kenny (Oxford: Clarendon, 1970), 242. The classic treatment of the shift in science and philosophy from a closed, finite world to an infinite spatial universe is Alexandre Koyré, *From the Closed World to the Infinite Universe* (Baltimore: Johns Hopkins University Press, 1957). See esp. 1–57, 110–124.

49. Leonardo da Vinci, *Notebooks,* 120–122.

50. Thus Proclus, in *A Commentary of the First Book of Euclid's Elements,* trans. Glenn R. Morrow (Princeton, NJ: Princeton University Press, 1992), 30–31: "These exact sciences [mathematics, including arithmetic and geometry] exclude infinity from plurality and magnitude and concern themselves straightway with the Limited." Infinity entered Greek conceptions, if at all, only when they imagined the unending repetition of an individual act, such as expressed in Zeno's paradoxes—to

wit, the unending repetition of crossing a room halfway (and thus never reaching the other side). Continuity was likewise finite, such as seen with the diagonal of a square; it could not be cast as discrete and 'rational' because its fractional expression led to unending repetitions—as with, say, the square root of two. Later on, for Christians imbued with Greek thought, 'infinity' meant something in a Platonic vein, like 'absolute' ("that than which nothing greater can be conceived" was Saint Anselm's definition of God in his famous ontological argument). It was a positive quality, like Perfection and Being, like Goodness and Truth. As such, one couldn't "see" it, except in such glimpses as God granted to mystics or supreme intellects or a humble Saint Francis.

51. Euclid defined a line as "breadthless length," and in definition 23 ("parallel straight lines") and postulate 5 (the famous "parallel postulate"), he did speak of lines "produced indefinitely." But the thrust of the discussions concerning parallels, both before and after Euclid's time, centered not on their infinite extension but on the "equidistance" and the "direction" theories of their justification, and on whether the postulate itself is derived from more-primitive definitions or is "really indemonstrable," as Heath concludes in his commentary. See Euclid, *Elements,* bk. 1, defs. 2, 23, post. 5, 1:153–155, as well as Heath's extended remarks, 1:158–165, 190–194, 202–220.

52. The mathematical concept of infinity would not be explicitly articulated until the projective geometry of Desargues, although philosophically, speculatively, and artistically it gained currency in the preceding two centuries. Theologian-mathematician Nicolas of Cusa, for example, speculated on the plurality of worlds and the infinity of space. The connections between Cusanus and Alberti, infinity and perspective, are explored by Karsten Harries, who claims that "the conception of space as infinite . . . underlies Leon Battista Alberti's perspective construction." Karsten Harries, *Infinity and Perspective* (Cambridge, MA: MIT Press, 2001), 66, 42–77, and throughout. With horizons expanding by leaps and bounds, we should also mention Paolo Toscanelli (1397–1482), astrologer, cartographer, and one of the most renowned mathematicians of the fifteenth century. Friend of Brunelleschi, Alberti, and other Tuscans, Toscanelli generously shared his knowledge with numerous associates, although publishing little during his lifetime. Most notably, he produced and sent to Christopher Columbus the map of the globe that provided the impetus of the latter's maiden voyage of "discovery."

53. Other evidence—such as the "column paradox" discovered by Piero and Leonardo da Vinci—demonstrated the difficulty, impossibility even, of rendering illusions in precise perspective. Facing a row of columns, the ones on either side will appear wider than those in the center due to their extension away front the frontal plane of the elevation. See E. H. Gombrich's explanation of this and related matters of perspective, which leads to his claim that "perspective is merely a convention and does not represent the world as it looks." E. H. Gombrich, *Art and Illusion: A Study in the Psychology of Pictorial Representation* (Princeton, NJ:

Princeton University Press, 1960), 254, and more generally 254–257. See too the more extensive analysis of the "marginal distortions" of perspective in Kubovy, *Psychology of Perspective,* 104–126.

54. The story is told by Vasari, *Lives of the Artists,* 74–83.

55. See Wittkower, *Architectural Principles,* 41–59, 111–113.

56. Quoted in Grafton, *Leon Battista Alberti,* 287. Grafton remarks that Alberti's comments on architecture amount to a "magnificent jumble of contents that surprises, and sometimes disappoints, readers." *Leon Battista Alberti,* 266.

57. Scholarly practice, followed here, refers to della Francesca by his first name, Piero. More than anyone, the noted Italian art historian Roberto Longhi consolidated Piero's reputation in modern times. See Roberto Longhi, *Piero della Francesca,* trans. David Tabbat, introduction by Keith Christiansen (Italian ed., 1927; Riverdale-on-Hudson, NY: Stanley Moss-Sheep Meadow, 2002). Field, *Piero Della Francesca,* and Field, *Invention of Infinity,* have thoroughly and interestingly integrated Piero's mathematical and painterly careers, while in a recent biography (*Piero della Francesca*), lifelong Piero scholar James Banker has masterfully investigated and historically reconstructed this elusive figure's life from a paucity of documents. Although too heavy-handed with Piero's alleged "Platonism," Larry Witham offers another recent, worthwhile introduction to Piero's work and subsequent reputation: *Piero's Light: In Search of Piero della Francesca: A Renaissance Painter and the Revolution in Art, Science, and Religion* (New York: Pegasus Books, 2014).

58. His other mathematical works consisted of *Trattato d'Abaco* (*Treatise on the Abacus*) and *Libellus de quinque corporibus regularibus* (*The Little Book on the Five Regular Solids*). For these, see Banker, *Piero della Francesca,* 79–95, 198–205; and Field, *Piero Della Francesca,* 12–31, 119–128.

59. Piero della Francesca, *De prospectiva pingendi,* bk. 1, prop. 30, excerpted in Field, *Piero Della Francesca,* 162–163; see also Field, *Piero Della Francesca,* 303. Luca Pacioli later wrote of *De prospectiva pingendi,* "In this work of every ten words, nine of them translate into proportion." Quoted in Banker, *Piero della Francesca,* 217.

60. Field, *Piero Della Francesca,* 58–59.

61. "Traces," recall, refers to Aquinas: "Every effect in some degree represents its cause, . . . [with] such a representation . . . called a *trace.*" Aquinas, *Summa Theologica,* pt. 1, qu. 45, art. 7, in *Basic Writings,* 1:444 (italics in the original).

62. Alberti, *On Sculpture,* 15, 30, 33, and more generally 14–38. In Alberti's description, the "setting of points" meant "to draw and develop all the contours and determine angles, hollows, and projections" of a model, which were then transferred to the sculpting block. The device for doing this consisted of a disc (called the "horizon") fixed to the top of a model, a long boom, and plumb lines, which were attached to the boom with pegs. Both the disc and the boom were marked off in equal units: "feet," "degrees," and "minutes." By careful positioning, the sculptor

could situate any given point on the model, and then adjust the scales proportionately to transfer it to the marble or other stone designated for carving (30–32).

63. The parallels here with music are striking. Zarlino had prescribed the ideal relations from the first six numbers as the basis of harmony, while Galilei had used numbers only to describe the relations between sounds actually heard, numbers having no intrinsic and prescriptive, "sonorous" qualities. Similarly, Alberti's proportional measurements described body parts and their relations to one another, which carried no intrinsic or prescriptive, no philosophical connection to "ideal" figures. Peterson, in *Galileo's Muse,* 136–147, skillfully dissects the "popular mathematical mystery culture" of φ, which stemmed centrally from Pacioli's *Divine Proportion* (*De Divina Proportione*). Although Pacioli's work drew heavily on Piero's *De prospectiva pingendi* (Vasari claimed Pacioli plagiarized much from it), Piero himself had little to do with speculative interpretations of numbers. Banker summarizes: "There is no evidence . . . that these geometrical or numeric worlds possessed some underlying religious or philosophical meaning." Banker, *Piero della Francesca,* 218.

64. Grafton, *Leon Battista Alberti,* 125.

65. Alberti, *On Painting,* 53. René Descartes, "Rules for the Direction of the Mind," in *The Philosophical Works of Descartes in Two Volumes,* trans. and ed. Elizabeth S. Haldane and G. R. T. Ross (1911; Cambridge: Cambridge University Press, 1970), 1:15. Banker similarly concludes, "It is a long way from Piero in the fifteenth century to modern science, but it is significant that he had prepared for this numerical conceptualization of the basic structures of nature." Banker, *Piero della Francesca,* 216.

66. Gombrich, *Art and Illusion,* 87.

67. Galileo was certainly aware of these developments. His first job application (unsuccessful) was for the post of mathematician at a drawing school. He was friends with the painter Cigoli (Ludovico Cardi, 1559–1613), who had designed an instrument for aiding in drawing perspective. His friend and benefactor, Guidobaldo del Monte, had sent him a book on the mathematics of perspective in the 1590s, which had occasioned the exchange of opinions between the two. As we shall see, by then he was also working on his "geometric and military compass," a mechanical means of computing proportionalities, later versions of which were inscribed with scales for determining visual perspective. He also, of course, had an interest in optics, which he taught privately in 1601 and which became manifest in devising his own telescopes.

8. Astronomy

1. Thus Thomas Aquinas, in "Commentary," *I Sentences,* dist. 19, qu. 2, art. 2: "The *now* of time is not time, the *now* of eternity is really the same as eternity."

Thomas Aquinas, *Philosophical Texts,* sel. and trans. Thomas Gilby (New York: Oxford University Press, 1960), 84. For Whitehead's remark, see Alfred North Whitehead, *The Concept of Nature* (Cambridge: Cambridge University Press, 1920), 61.

2. The literature on time is virtually as vast as the skies themselves, with something on the order of 130,000 books published since 1900 carrying "time" in the title, according to Denis Feeney, *Caesar's Calendar: Ancient Time and the Beginnings of History* (Berkeley: University of California Press, 2007), 1. A worthy introduction to philosophical issues is Richard M. Gale, ed., *The Philosophy of Time: A Collection of Essays* (Garden City, NY: Anchor Books, 1967), while an invaluable general study is that of time scholar–scientist and founder of the International Society for the Study of Time (1966), J. T. Fraser, *Time, the Familiar Stranger* (Redmond, WA: Tempus Books of Microsoft Press, 1987). Among his other writings, see *The Genesis and Evolution of Time: A Critique of Interpretation in Physics* (Amherst: University of Massachusetts Press, 1982). Equally prominent are the works of G. J. Whitrow (all from Oxford University Press): *The Natural Philosophy of Time,* 2nd ed. (1980, 1st ed. 1961); *What Is Time?* (1972); and *Time in History: Views of Time from Prehistory to the Present Day* (1988). Studies addressing time as a dimension of modern historical consciousness are nearly as extensive as those of the scientific and philosophical traditions. Much of this literature is summed up and trumped by the insights of Zachary Sayre Schiffman in *The Birth of the Past* (Baltimore: Johns Hopkins University Press, 2011).

3. This, the "giant impact" theory, was developed in the wake of evidence brought back from the Apollo moon explorations. See the works of cosmologist William K. Hartmann, specifically, with Donald R. Davis, "Satellite-Sized Planetesimals and Lunar Origin," *Icarus* 24 (1975): 504–515, and, more generally, *Moons and Planets,* 5th ed. (Belmont, CA: Brooks / Cole-Thomson, 2005), 121–122. For an accessible account of time in contemporary cosmology, one can start with George Smoot and Keay Davidson, *Wrinkles in Time* (New York: Avon Books, 1993), the popular work of Stephen W. Hawking, *A Brief History of Time: From the Big Bang to Black Holes* (New York: Bantam Books, 1988), and Steven Weinberg, *The First Three Minutes: A Modern View of the Origin of the Universe* (New York: Basic Books, 1977).

4. Augustine of Hippo, *The Confessions of St. Augustine,* trans. J. G. Pilkington (New York: Boni and Liveright, 1927), bk. 11, 285–286: "What then is time? If no one ask of me I know; if I wish to explain it to him who asks, I know not. . . . I know that if nothing passed away, there would not be past time; and if nothing were coming, there would not be future time; and if nothing were, there would not be present time. Those two times, therefore, past and future, how are they, when even the past now is not, and the future is not as yet? But should the present always be present, and should it not pass into time past, time truly it could not be, but eternity. If then, time present—if it be time—only comes into existence because it passes into time past, how do we say that even this is, whose cause of being is that

it shall not be—namely, so that we cannot truly say that time is, unless because it tends not to be?" Augustine clearly captured the peculiarity of time, but not its information.

5. The phrase is from Arthur Koestler, *The Sleepwalkers: A History of Man's Changing Vision of the Universe* (1959; New York: Grosset and Dunlap, 1963), 191. See too Owen Gingerich's entertaining account, *The Book Nobody Read: Chasing the Revolutions of Nicolaus Copernicus* (New York: Walker, 2004).

6. In modern times, Claude Shannon, the chief progenitor of our computer-driven information age, distinguished between information's "channel" ("medium" in popular parlance) and its message. In Shannon's flow, a message was encoded, then transmitted, then decoded back into its initial form. The object of the entire process lay with lessening the "noise," the extraneous intrusions into the message itself. The contemporary means of achieving this lay with combining mathematical algorithms and electronic circuitry, a link Shannon had produced in his own master's thesis, where Boolean logic provided the heartbeat of the entire electronic system and the backbone of microinformation. Although the messages encoded and transmitted could be quite literally anything, in the human world they typically conveyed the "stuff" brought to the mind by the sensations, the macroinformation of quotidian life. See James Gleick, *The Information: A History, a Theory, a Flood* (New York: Vintage Books, 2011), 222–223. These same micro- and macrofunctions were performed in earlier ages by other information technologies: letters of the alphabet (with their syntax and semantics), Hindu-Arabic numerals (with their cardinal and ordinal dimensions), musical notes (with their structures of time and tone), perspective lines (with their mapping and geometrical rules).

7. Aristotle, *Physics,* bk. 4, 219a4–219b13. See too Aristotle, *On the Soul,* bk. 3, 425a14–425b12, for the ties between the perception of "common sensibles" and motion.

8. Aristotle, *Physics,* bk. 6, 235a12–237a6. Also see the discussion of early numeracy in Chapter 3. Throughout the centuries, numerous philosophers have criticized Aristotle's definitions of time and motion as circular or contradictory. See Whitrow, *Natural Philosophy of Time,* 1–2, 25–27, 48–49.

9. In his quite technical account, *Aristotle on Time: A Study of the Physics* (Cambridge: Cambridge University Press, 2011), Tony Roark has essayed Aristotle's rescue from critics by interpreting his theory of time as a "hylomorphic" compound of matter and form, with motion being time's matter and perception its form. In so doing, Roark reveals the still-current preoccupation of many philosophers, reminiscent of Aristotle himself, with seeking a "definition" that can "succeed in revealing the essential nature of time" (77). Roark's claim that 'before' and 'after' "can . . . be explicated independently of temporal notions" (80) gives pride of place to hair-splitting definitions that often beggar common sense, and even logic, yielding an interpretation that unwittingly exposes the deficiencies of defi-

nitions and conventional language in trying to bridge continuity and discreteness. (Roark does note that time for Aristotle "seems to be *both*" [111].) This is not to suggest that philosophical issues pertaining to time were or would be overcome by introducing new technological means of analytically separating it from motion, the theme of the present chapter. Nor could an information technology based on empty and abstract symbols resolve logical and linguistic conundrums regarding time, which have bedeviled the deepest meditations of philosophers from antiquity to the present. It is to suggest, however, that time remains philosophically indeterminable and a mystery, but technologically manageable as a set of problems. Friedrich Waismann, among others, voiced this idea: "To the question whether a word like 'time' *can* be defined . . . we are inclined to say it cannot." Friedrich Waismann, "Analytic-Synthetic," in Gale, *Philosophy of Time,* 57 and more generally 55–63. Writing in the wake of the later Wittgenstein, Waismann aptly characterized philosophy as "trying to catch the shadows cast by the opacities of speech." See Gale's introduction to "Part I: 'What, Then, Is Time?' " in *Philosophy of Time,* 7.

10. William Shakespeare, *All's Well That Ends Well,* 5.3.

11. Referring to the seventeenth century, astronomer and anthropologist Anthony Aveni remarks, "Once quite comfortable together, science and religion were on a collision course, and time was the issue." Anthony Aveni, *Empires of Time: Calendars, Clocks, and Cultures* (New York: Basic Books, 1989), 143.

12. Herodotus, to cite a somewhat random example, noted that the Battle of Salamis (480 BCE) occurred "during the archonship of Calliades at Athens." Herodotus, *The Histories,* trans. Robin Waterfield, intro. Carolyn Dewald (Oxford: Oxford University Press, 1998), 504. For a valuable introduction to ancient calendars and time reckoning, as well as ancient astronomy, see James Evans, *The History and Practice of Ancient Astronomy* (Oxford: Oxford University Press, 1998), 163–204, and throughout.

13. Feeney, *Caesar's Calendar,* 15–16. The Scythian monk Dionysius Exiguus (Dennis the Small, ca. 470–ca. 544) is credited with inventing half of the axis—the AD (*anno Domini*) linked to Jesus's birth. But this was in the service of calculating Easter, not of universal dating. Similarly, Bede and other medieval writers used AD in reference primarily to "divine time" and "ecclesiastical events" (such as the death of an archbishop) or astronomical events portending the divine (comets, eclipses), not a general chronology. Along with Feeney, see Donald J. Wilcox, *The Measure of Times Past: Pre-Newtonian Chronologies and the Rhetoric of Relative Time* (Chicago: University of Chicago Press, 1987), 7–9, 203–208.

14. For example, business was transacted on the *dies fasti* (allowed days), triumphs were celebrated on the *fasti triumphales,* and other days were similarly set aside for court appearances, assemblies, religious functions, and comparable civic events. The *fasti diurni* evolved into a yearbook of urban and rural activities, while the *fasti consulares* listed accomplishments of each consul, who held office for one

year. See Feeney, *Caesar's Calendar,* 138–142, 167–193; and Robert Hannah, *Time in Antiquity* (Loudon: Routledge, 2009), 32–42.

15. By the first century BCE, lunar cycles were falling as far as three months behind the solar year (celebrating the spring planting during harvest season). After defeating Pompey and having been declared dictator of Rome, according to the Roman historian Suetonius (ca. 69–122), Caesar turned his attention to the "reorganization of the state, reforming the calendar." During his sojourn in Egypt he had encountered a solar calendar, and on the advice of Sosigenes (first century BCE), a Greek astronomer from Alexandria, he introduced it to Rome by fiat. Gaius Suetonius Tranquillus, *Lives of the Caesars,* trans. Catharine Edwards (Oxford: Oxford University Press, 2000), 20. Plutarch claimed that Caesar had called in the "best philosophers and mathematicians of his time" to correct the calendar, a generally welcomed reform, which nevertheless gave offense to some. When asked whether "Lyra would rise" the next morning, Cicero grumpily replied, "Yes, in accordance with the edict." (Lyra is a small constellation whose brightest star is Vega, visible during the summer months in the northern latitudes.) Plutarch, *Lives of the Noble Romans,* trans. John Dryden, rev. Arthur Hugh Clough, ed. Edmund Fuller (New York: Dell, 1959), 231–232. A slight adjustment was made during the reign of Augustus to clear up a confusion over leap years, but after the year 8 CE, the calendar remained unaltered until the reforms of Gregory XIII in the sixteenth century.

16. Thus O. Neugebauer, in *The Exact Sciences in Antiquity,* 2nd ed. (Princeton, NJ: Princeton University Press, 1952; New York: Dover, 1969), 103: for the Babylonians, "the zodiac of 12 times 30 degrees [served] as a reference system for solar and planetary motion . . . [as did] the use of arithmetic progressions to describe periodically variable quantities."

17. As Feeney writes, "The diffusion of the [Julian] calendar was limited and culture-specific, and many other civil calendars remained quite happily in place for as long as the Empire lasted." *Caesar's Calendar,* 198. See too, 209–211, as well as the comments of Medieval scholar Arno Borst in *The Ordering of Time: From the Ancient Computus to the Modern Computer,* trans. Andrew Winnard (Chicago: University of Chicago Press, 1993), 18.

18. Bonnie Blackburn and Leofranc Holford-Strevens, *The Oxford Companion to the Year: An Exploration of Calendar Customs and Time-Reckoning* (Oxford: Oxford University Press, 1999), 675, 762–765.

19. Faith Wallis, introduction to *Bede: The Reckoning of Time,* trans. Faith Wallis (Liverpool: Liverpool University Press, 1999), xcvi–xcvii. Wallis's translation and commentary offer exemplary and incisive scholarship surrounding Bede and the myriad issues pertaining to the calendar during the Middle Ages. It bears mention that, although distinct, astronomy and astrology were often interwoven until the scientific revolution (and even after, in many minds).

20. Take, for example, the Metonic cycle, named for its Greek discoverer, Meton (fl. fifth century BCE). (Like other discoveries attributed to the Greeks, this too was handed down by the Babylonians.) Meton found that nineteen solar years very nearly equaled 235 synodic or lunar months, both of which add up to roughly 6,940 days, and the Metonic cycle was incorporated in early Christian conventions (discussed later). But by modern reckoning, nineteen tropical years actually total 6,939.6018 . . . days, while 235 synodic months equal 6,939.688 . . . days, and correlating the two could never be accomplished without remainders, producing a difference of nearly two hours in the cycle, or one full day every 228 years. Hannah, *Time in Antiquity,* 32–42.

21. See Aveni, *Empires of Time,* 111 and more generally, 85–118: "The story of the attempts to make months fit into the years constitutes a major chapter of our Western Calendar." Borst has formulated succinctly the broader conceptual issue: "Time can either be aligned with perceptible experiences, in which case it will not be consistent, or else incorporated into a logical system of thought, in which case it will not be accurate." Borst, *Ordering of Time,* 6.

22. At age seven, Bede entered the monastic school at Wearmouth-Jarrow in Britain's Northumbria, which was founded in 674 by Saint Benedict Biscop (628–689), its first abbot. Bede essentially remained there until his death in 735, sharing his "delight to learn or to teach or to write" with fellow monks. The monastery at Wearmouth-Jarrow was for over two centuries the most highly reputed Anglo-Saxon center of learning in northern England. Biscop himself made five journeys to Rome, during which he accumulated a large collection of books for the monastery's unrivaled library, of which Bede took full advantage as its most illustrious patron. Bede, *The Ecclesiastical History of the English People,* trans. Bertram Colgrave, ed. Judith McClure and Roger Collins (1969; Oxford: Oxford University Press, 1994), 293–295. For the little known about Bede's life, see Judith McClure and Roger Collins, introduction to ibid., ix–xxxii. Before Bede's lifetime, the word *computus* covered a large semantic field and included our terms 'reckoning,' 'counting,' 'computing,' and 'estimating,' but by the time of Bede's work and subsequently throughout the Middle Ages, it was largely restricted to calendrical computations. Borst, *Ordering of Time,* 16–32.

23. Bede, *Reckoning of Time,* 249. For example, the emperor Constantine (ca. 272–337), who called the Council of Nicaea (325), sought unity among Christians, decreeing that the Easter feast should be "observed by all according to one and the same order and certain rule." Quoted in Stephen C. McCluskey, *Astronomies and Cultures in Early Medieval Europe* (Cambridge: Cambridge University Press, 1998), 85. The details of these early disputes lie beyond present concerns, but in addition to McCluskey's fine account (77–96), they are well covered in the authoritative works of Borst, *Ordering of Time,* 24–41; Charles W. Jones, *Bede, the Schools and the Computus,* ed. Wesley M. Stevens (Hampshire, UK: Variorum, 1994), secs.

6–10; Wesley M. Stevens, *Cycles of Time and Scientific Learning in Medieval Europe* (Hampshire, UK: Variorum, 1995), sec. 1; Olaf Pedersen, "The Ecclesiastical Calendar and the Life of the Church," in *Gregorian Reform of the Calendar,* ed. G. V. Coyne et al. (Vatican City: Pontificia Academia Scientiarum, 1983), 17–74; Gustav Teres, "Time Computations and Dionysius Exiguus," *Journal for the History of Astronomy* 15 (1984): 177–188; Blackburn and Holford-Strevens, *Oxford Companion to the Year,* 772–776, 791–797; and Wallis, introduction to *Reckoning of Time,* xv–lxiii.

24. Even this was subject to different interpretations for Christians. In the Hebrew Bible or Old Testament, Jewish law ordained the slaughtering of the Paschal lamb on the day that began at nightfall on the fourteenth of Nisan (Lev. 23:5; Num. 28:16; Deut. 16:1). In the synoptic gospels of the New Testament, the Last Supper was a Passover meal (Matt. 26:17; Mark 14:12; Luke 22:8), but in John 13:1 and 19:14, it occurred on the day of preparation. Still, all four gospels agreed that the Resurrection happened on the first day of the week and on the third day (counting inclusively) after Jesus's execution or Passion.

25. In 325 the Council of Nicaea issued no specific guidelines regarding Easter, except for the anti-Semitic restriction that it could not be celebrated "with the Jews" on the fourteenth of Nisan, even if it fell on a Sunday. (A bit earlier, in 321, Constantine had assigned Sunday as the weekly day of worship and rest, thus displacing both the Roman "Saturn-day" and the Jewish Sabbath as the beginning of the week.) After Nicaea, the general rule for establishing Easter emerged from the bishops in Alexandria: it would be observed on the first Sunday after the first full moon (14 Nisan, or luna XIV) that fell on or after the vernal equinox. The latter was specified as March 21, following Ptolemy's astronomical tables, and supplanted March 25, the day most Romans had observed as the equinox until then. Wallis, introduction to *Reckoning of Time,* xxxvii–xxxix; Blackburn and Holford-Strevens, *Oxford Companion to the Year,* 791–794.

26. Bede, "Letter to Wicthed," in *Reckoning of Time,* 420. For Bede's sources, see Wallis, introduction to *Reckoning of Time,* lxxii–lxxxv. Bede also encouraged naming the time of the Christian resurrection after the pagan goddess Eostre, the light of spring arising in the east. Bede, *Reckoning of Time,* 54. Also see Borst, *Ordering of Time,* 40–41.

27. Otherwise termed a "synodic month" or "period," a 'lunation' refers to the average number of days (29.5) it takes the moon to circumnavigate the earth, measured from its alignment, or conjunction, with the sun, whereas a "sidereal" month is a period of 27.3 days and is measured from a distant star. The difference of slightly over two days represents the sun's advancement relative to the earth during the time of the moon's cycle. In contrast to *computus,* which dealt only with average periods, Babylonian and Greek astronomy sought to predict exact positions of planets and stars, thereby accounting for deviations from mean motion, which *computus* could not explain. McCluskey, *Astronomies and Cultures,* 80.

28. Bede, *Reckoning of Time,* 11, more generally 9–13.

29. In calendar terms, a 'cycle' refers to the patterned repetition of a number of days. Thus, the days from Sunday through Saturday form a pattern, after which it repeats, the weekly cycle. Similarly with astronomically reckoned lunar and solar cycles. Bede's full Paschal tables included cyclical correlations with the years, "concurrent" days of the week, lunar epacts, lunar cycles, full moons (luna XIV), and Easter Sundays, whose correlated repetitions required 532 years, as we shall see. Thus if Easter fell on a given day in year Y, it would fall on the same day in Y + 532, or Y + 1064, and so on. Bede, *Reckoning of Time,* 121–124, 155–156, 392–404; Blackburn and Holford-Strevens, *Oxford Companion to the Year,* 768–771, 812.

30. Bede, *Reckoning of Time,* 63, 71.

31. Bede drew heavily on the earlier work of Dionysius Exiguus, mentioned earlier. Bede, *Reckoning of Time,* 155–156.

32. Nineteen lunar years of 354 days each equals 6,726 days (19 × 354), to which were added the 210 days of seven intercalary months, plus another 4.75 days for the leap year insertions, totaling 6,940.75. The difference between the solar and lunar figures stems from the fact that an average lunation in modern calculation is approximately 29.530589 days, some forty-four minutes per lunation over the figure of 29.5 days used in earlier computations.

33. The details of computing Easter were somewhat more complicated than the foregoing summary indicates. Each year in the Metonic cycle was assigned a number, known as the "golden number" (from one to nineteen), and it was assumed that new moons would always fall on the same days in the years indicated by the same golden number. Bede measured the epacts from January 1, the beginning of the Julian year, which meant determining the moon's age on that date and using it to calculate the moons of the other eleven months. From this he computed Easter based on the epact of March 22, the day after the equinox. If on March 22, for example, the moon was eleven days old (luna XI), the Paschal moon would fall three days later, on March 25 (luna XIV), with Easter occurring the following Sunday. The insertions of embolismic months and bissextile years made these reckonings all the more difficult, but they were still quite manageable arithmetically. Bede, *Reckoning of Time,* 64–73, 130–132; Wallis, commentary to ibid., 293–296.

34. Bede, *Reckoning of Time,* 13–14; Borst, *Ordering of Time,* 38–39.

35. Borst, *Ordering of Time,* 39.

36. Bede, *Reckoning of Time,* 157–158, 247, and more generally 157–237, 246–249. See the commentary of Wallis, ibid., 353–366; and Charles W. Jones, "Bede as Early Medieval Historian," in *Bede,* sec. 3.

37. For Bede's figural history, see Schiffman, *Birth of the Past,* 126–134, as well as Bede, *Ecclesiastical History.*

38. Bede, *Reckoning of Time,* 9–13, 279. Wallis, in her introduction to ibid., lxxxvi, notes that 240 manuscripts of all or part of the work have survived; Bede

also wrote an earlier, shorter work on time and chronology, *On Time* (*De Temporibus*). See too Borst, *Ordering of Time,* 36–41, and Charles W. Jones, "Bede's Place in Medieval Schools," in *Bede,* sec. 5.

39. For these terms and their calculations, see John D. North, *The Universal Frame: Historical Essays in Astronomy, Natural Philosophy, and Scientific Method* (London: Hambledon, 1989), 40–44.

40. Although it lies beyond the scope of present concerns, one of the motivations behind medieval astronomy lay in determining the astrological significance of various heavenly bodies for medical practices, some of which depended on the waxing and waning of the moon. In general these motivations grew out of the association of the celestial macrocosm and the human microcosm, mentioned earlier, and became increasingly important with the revival and expansion of astrology throughout the High Middle Ages and Renaissance. For the latter topic, see John D. North, *The Norton History of Astronomy and Cosmology* (New York: W. W. Norton, 1995), 119–124, 259–271; Olaf Pedersen, *The Two Books: Historical Notes on Some Interactions between Natural Science and Theology,* ed. George V. Coyne and Tadeusz Sierotowicz (Vatican City: Vatican Observatory Foundation, 2007), 91–95, 165–169; and Jim Tester, *A History of Western Astrology* (Rochester, NY: Boydell, 1987), 98 and following.

41. Olaf Pedersen, "Astronomy," in *Science in the Middle Ages,* ed. David C. Lindberg (Chicago: University of Chicago Press, 1978), 312. Also see McCluskey, *Astronomies and Cultures,* 180–185, and Jonathan Lyons, *The House of Wisdom: How the Arabs Transformed Western Civilization* (New York: Bloomsbury Press, 2009), 125–127. The older epacts had also been based on the Paschal or full moon (luna XIV), but observationally these were slippery, for to the naked eye the moon appears full for some two or three nights in succession.

42. Pedersen, "Astronomy," 312, more generally, 303–337; Stevens, *Cycles of Time,* vii, sec. 1; Charles W. Jones, "Bedae Pseudepigrapha (with Indices)," in *Bede,* 108–140. See too North, *Norton History of Astronomy,* 177–223, and Bernard R. Goldstein, "Astronomy as a 'Neutral Zone': Interreligious Cooperation in Medieval Spain," *Medieval Encounters* 15 (2009): 159–174. A stunning and valuable medieval manuscript, *The Calendar and the Cloister,* Oxford, Saint John's College, MS 17, was created in the first decade of the twelfth century at Thorney Abbey, a Benedictine monastery in Cambridge shire. As a scholars' resource, it has been thoroughly digitized and made available at http://digital.library.mcgill.ca/ms-17 (accessed October 27, 2017).

43. In 1143 an anonymous monk in southern Germany introduced Hindu-Arabic numerals into a *computus,* which were used soon after in 1171 by the dean of Paderborn Cathedral, Reiner, whose own *computus* exploited the fractional possibilities of the numerals in correcting received values for solar and lunar years. Borst, *Ordering of Time,* 73–78.

44. Evans, *History and Practice,* 107–109; North, *Norton History of Astronomy,* 203–223.

45. A good summary treatment of the issues is Bernard R. Goldstein, "Historical Perspectives on Copernicus' Account of Precession," *Journal for the History of Astronomy* 25 (1994): 189–197.

46. Various theories were advanced to explain the discrepancies, including the so-called theory of trepidation or access and recess (long since discarded), which held that over a period of seven thousand years the equinoxes oscillated in their precessions, making the full precession a time span of some forty-nine thousand years. The explanations derived from interpretations of celestial movements on the eighth sphere, that of the fixed stars, and were also invoked to account for variations observed in the obliquity of the ecliptic, the changing angle between the plane of the celestial equator and that of the ecliptic. These topics are well covered in Evans, *History and Practice,* 54–55, 59–60, 274–280. For technical dimensions of the historical issues, in addition to Goldstein, "Historical Perspectives," see Emmanuel Poulle, "The Alfonsine Tables and Alfonso X of Castile," *Journal for the History of Astronomy* 19 (1988): 97–113; and Jerzy Dobrzycki, "Astronomical Aspects of the Calendar Reform," in Coyne et al., *Gregorian Reform,* 117–127. For a more general account, see North, *Norton History of Astronomy,* 203–223, and his more recently expanded *Cosmos: An Illustrated History of Astronomy and Cosmology* (Chicago: University of Chicago Press, 2008), 215–301.

47. By modern calculations, a sidereal year consists of 365 days, 6 hours, 9 minutes, and 9.5 seconds, while a tropical year measures 365 days, 5 hours, 48 minutes, and 45.2 seconds, a difference of 20 minutes and 24.3 seconds. Even without considering sidereal reckoning, the difference between the value taken for the Julian year (365.25 days) and that of the true tropical year (365.24219 days, by modern figuring) meant equinoctial drift could be observed. Yet more serious was the fact that the Paschal moon was slipping well ahead of its tabulated date, causing havoc in determining Easter. E. G. Richards, *Mapping Time: The Calendar and Its History* (Oxford: Oxford University Press, 1998), 239–256.

48. North, *Cosmos,* 286; John D. North, *God's Clockmaker: Richard of Wallingford and the Invention of Time* (New York: Hambledon and London, 2005), 60–64.

49. In its final report, the council said nothing explicitly about adjusting the calendar, but it did decree reform of the missal and breviary, which Pope Pius V (1504–1572) interpreted as implying calendar reform. His successor, Gregory, likewise appealed to the authority of the council in calling for reform. Aloysius Lilius (ca. 1510–1576), an Italian astronomer, philosopher, and chronologist, was largely responsible for the main outlines of the reform. Throughout these years, a key proponent, time reckoner, and adviser to Gregory was the noted Jesuit mathematician and astronomer Christopher Clavius (1538–1612), called by his contemporaries the "Euclid of the sixteenth century." In 1603, Clavius, whom we shall meet again,

explained the reforms in a detailed, 680-page tome (*Romani calendarii a Gregorio XIII P.M. restituti explicatio*). See August Ziggelaar, "The Papal Bull of 1582 Promulgating a Reform of the Calendar," in Coyne et al., *Gregorian Reform,* 201–239, for details surrounding reform of the calendar; see too Richards, *Mapping Time,* 239–256.

50. For Montaigne, see *The Complete Essays of Montaigne,* trans. Donald M. Frame (Stanford, CA: Stanford University Press, 1965), 773. Current projections have the Gregorian year falling behind the tropical year slightly over three days after the passing of some 10,000 years, which might have eased Montaigne's fretting, although continued small decreases in the tropical year and increases in the "natural day" mean that the discrepancy might be greater. In time, further reforms will be needed. More complicated but more accurate is the Iranian solar cycle of 2,820 years, which fall shy of 2,820 tropical years by only two minutes. For these and similar topics, including a discussion of the Gregorian lunar calendar and the reception of the Gregorian reforms, see Blackburn and Holford-Strevens, *Oxford Companion to the Year,* 692, 735–738, 817–828.

51. Wilcox, *Measure of Times Past,* 197. Scaliger almost single-handedly introduced a "radical change in chronological studies" in formulating the new science of chronology (*chronologia*); the name itself had been coined by sixteenth-century humanists. Anthony T. Grafton, "Joseph Scaliger and Historical Chronology: The Rise and Fall of a Discipline," *History and Theory* 14, no. 2 (1975): 157, and more generally 156–185. For Scaliger's life and work, see Grafton's definitive account, *Joseph Scaliger: A Study in the History of Classical Scholarship,* 2 vols. (New York: Oxford University Press, 1983, 1993). Volume 2 addresses Scaliger's historical chronology. See too the chapter "Scaliger's Chronology: Philology, Astronomy, World History," in Anthony T. Grafton, *Defenders of the Text: The Traditions of Scholarship in an Age of Science, 1450–1800* (Cambridge, MA: Harvard University Press, 1991), 104–144; as well as Borst, *Ordering of Time,* 104–106; and Wilcox, *Measure of Times Past,* 195–203. "Bottomless pit" quoted in Alfred W. Crosby, *The Measure of Reality: Quantification and Western Society, 1250–1600* (Cambridge: Cambridge University Press, 1997), 90.

52. Using the indiction cycle as a component of a Julian period would appear to be a departure from relying on the "natural" Metonic and solar cycles. But bear in mind that all time reckoning is a human endeavor and is thus arbitrary (as is the seven-day week, which has little to do with "nature" per se).

53. Joseph Scaliger, *De emendatione temporum,* quoted in Wilcox, *Measure of Times Past,* 200. Further problems arose in dating Egyptian and other dynasties reckoned to begin not only before the creation of the world but before Scaliger's Julian period itself. See Grafton, "Joseph Scaliger and Historical Chronology," for these consequences. Dating with the Julian period has proved extremely useful in coordinating ancient calendars and chronology, while Scaliger's "Julian" day continues in use among modern astronomers for dating celestial events.

54. Wilcox, *Measure of Times Past,* 206, and more generally 203–208. The practice of dating with the BC/AD grid became widespread only in the eighteenth century. As well, thinking in centuries has been around for only the past three and a half centuries, and thinking in decades for only the past seven or eight decades.

55. For instance, see the sixteenth-century chronologer Joannes Temporarius's *Chronologicarum Demonstrationum Libri Tres (1596),* facsimile ed. (Whitefish, MT: Kessinger, 2010). Many other examples may be found in Daniel Rosenberg and Anthony Grafton, *Cartographies of Time: A History of the Timeline* (New York: Princeton Architectural, 2010), 26–95 and throughout.

56. Geoffrey Chaucer, *The Treatise on the Astrolabe,* in *The Complete Works of Chaucer,* ed. Walter W. Skeat (Oxford: Clarendon, 1900), 3:175. Earlier scholarly disputes over whether Littel Lowis was actually Chaucer's son generated enough conflicting arguments to cancel one another; as a result we may take the dedication at face value.

57. According to an Islamic legend, Ptolemy discovered the essentials of the apparatus when he dropped a celestial globe he had been carrying and the donkey he was riding stepped on it. The flattened sphere awakened in him the notion of a stereographic projection. Closer to the truth, Hipparchus and Vitruvius had already described such a projection, with the latter using it to design an early clock. Marcus Vitruvius Pollio, *The Ten Books on Architecture,* trans. Morris Hicky Morgan (1914; New York: Dover, 1960), 270–277. Ptolemy applied such knowledge to his early maps, but the first astrolabes themselves came from the Islamic world of astronomers and craftsmen, who developed and perfected the instruments. Astrolabes became known in Europe beginning in the eleventh century, introduced along with Arabic-to-Latin translations of Greek and Muslim scientific works, and by Chaucer's lifetime were, if not commonplace, certainly not rarities. For his account Chaucer relied heavily on the description of Messahala (Masha'allah ibn Atharī, ca. 740–815), a leading Persian Jewish astronomer-astrologer of the eighth century whose work was readily available in Latin translation (*Compositio et Operatio Astrolabie*). The oldest surviving astrolabe was probably crafted in 927 or 928. For an introduction to the astrolabe's history and inner workings, see John D. North, "The Astrolabe," *Scientific American* 230, no. 1 (1974): 96–106; and Donald R. Hill, *A History of Engineering in Classical and Medieval Times* (1984; London: Routledge, 1996), 319–326. Engineer James E. Morrison has compiled a handsome encyclopedia of material concerning every aspect of the astrolabe: *The Astrolabe* (Rehoboth Beach, DE: Janus, 2007).

58. Derek de Solla Price, *Science since Babylon,* enlarged ed. (New Haven, CT: Yale University Press, 1975), 31.

59. For these developments, see especially Gerhard Dohrn-van Rossum, *History of the Hour: Clocks and Modern Temporal Orders,* trans. Thomas Dunlap (Chicago: University of Chicago Press, 1996), 1–15, 45–123, and throughout; David S. Landes, *Revolution in Time: Clocks and the Making of the Modern World* (Cambridge, MA:

Harvard University Press, 1983), 1–97; Carlo M. Cipolla, *Clocks and Culture, 1300–1700* (New York: W. W. Norton, 1978), 37–75; and Whitrow, *Time in History,* 99–114. Jo Ellen Barnett provides a clear description of escapement mechanisms in *Time's Pendulum: From Sundials to Atomic Clocks, the Fascinating History of Timekeeping and How Our Discoveries Changed the World* (New York: Harcourt, 1998), 64–85.

60. Early monastic organization remained quite freely tied to variable, natural time divisions, such as daylight and seasons. The offices themselves derived from Psalms 119:164 and 119:62: "Seven times daily I praise you" and "I get up at midnight to thank you." Dohrn-van Rossum, *History of the Hour,* 35–39; Landes, *Revolution in Time,* 58–66.

61. Not all in the cloister were equally enthusiastic about improving timekeeping. Aquinas, for example, noted that "the Church does not strive for restriction through clever study of time. One does not need . . . an astrolabe to know when it is time to eat." Quoted in Dohrn-van Rossum, *History of the Hour,* 79.

62. Time numbers were not understood as designating a strict sequence. The terce, sext, and none of monastic practice, for instance, translate literally into "third," "sixth," and "ninth," yet these were highly variable (for example, the none eventually slipped backward from the ninth hour of the day—roughly midafternoon—to our midday, 'noon'). More to the point, they were used primarily to indicate the quarterly divisions of daylight. Thus "terce" referred to the first quarter of the day and so on. We must dispel numerical precision from our thoughts when reading about premodern time.

63. One of the earliest mentions of a "mechanical" clock dates from a manuscript written in 1271 for students at Montpelier by one Robertus Anglicus (Robert the Englishman), an astronomer about whom virtually nothing is known save his commentary on the Sphere of Sacrobosco. In it Robert lamented that current clocks "could not follow . . . astronomy with . . . accuracy" and noted that clockmakers were "trying to make a wheel which will make one complete revolution" for each day but were unable to "perfect their work." A really accurate clock, he believed, would be worth much "more than an astrolabe or other astronomical instrument for reckoning the hours." Then he proceeded to describe how one could be built, starting with a "disk of uniform weight" powered by a lead weight hanging from its axle "so that it would complete a revolution from sunrise to sunset." This could be done more accurately if an astrolabe were constructed on whose rete an "entire equinoctial" circle was divided up, thus connecting a flow-measuring timepiece with an astronomical reckoning of a "natural day." Still, it was a flow he referred to, not an oscillating escapement. Robertus Anglicus, "Commentary on the Sphere of Sacrobosco," in *The Sphere of Sacrobosco and Its Commentators,* ed. and trans. Lynn Thorndike (Chicago: University of Chicago Press, 1949), 199–246, esp. 229–231.

64. Later, with the invention of the pendulum clock by Christiaan Huygens in 1656, the order would be reversed, with the pendulum mechanism driving an oscillating anchor escapement, whose pallets then moved a toothed "escape wheel" continuously forward.

65. Early variants of our word 'clock'—*clokke, Glocke, cloche,* and so on—referred to bells, while the generic *horologium* denoted any timekeeping means, mechanical or otherwise.

66. Dohrn-van Rossum, *History of the Hour,* 1–15 and throughout.

67. In 1579 a Swiss horologist named Jost Bürgi, the finest clockmaker of his age and well ahead of his contemporaries, produced a clock that could mark seconds as well as minutes, with a variation of less than a minute per day. Landes, *Revolution in Time,* 105.

68. See Samuel L. Macey, *Clocks and the Cosmos: Time in Western Thought and Life* (Hamden, CT: Archon Books, 1980), 123–132.

69. Bede, *Reckoning of Time,* 24–26.

70. Augustine of Hippo, *Confessions,* bk. 11, 286–291, 302–304; Charles Taylor, *A Secular Age* (Cambridge, MA: Harvard University Press, 2007), 56–59.

71. Walter Benjamin, *Illuminations,* trans. Harry Zohn, ed. Hannah Arendt (New York: Schocken Books, 1968), 261.

72. See Schiffman, *Birth of the Past,* 138–216.

73. For the "immanent frame," see Taylor, *Secular Age,* 539–556. For a good survey of prophecy and its connection with early science, see Charles Webster, *From Paracelsus to Newton: Magic and the Making of Modern Science* (Cambridge: Cambridge University Press, 1982), 42, 15–47, and throughout. There is a voluminous body of literature on all these topics, which lie well beyond our present, narrower point—namely, that religious and philosophical minds viewed the content of prophecy, eschatology, and religious meaning in general as falling within the framework of empty, not gathered, time. Here I have relied on Charles Taylor's pathway through much of this dense thought.

74. Francis Bacon, *Novum Organum,* trans. James Spedding, in *The Works of Francis Bacon,* 2 vols. (New York: Hurd and Houghton, 1878), 1:350.

75. Taylor, *A Secular Age,* 58.

The Moment of Modern Science

1. Galileo Galilei, *Operations of the Geometric and Military Compass,* trans. Stillman Drake (Washington, DC: Smithsonian Institution, 1978). About this time Galileo also published pseudonymously two brief, controversial books on the nova of 1604. Galileo Galilei, *Galileo against the Philosophers in his "Dialogue of Cecco di Ronchitti" (1605) and "Considerations of Alimberto Mauri" (1606),* trans. Stillman Drake (Los Angeles: Zeitlin and Ver Brugge, 1976).

2. The earliest, manuscript versions of the *Operations* appeared in 1596 with investigations into the mathematical possibilities of a drafting instrument, probably devised by his friend and patron, the Marquis Guidobaldo del Monte. The compass was made up of three components: (1) two legs, pivotally hinged, on which were engraved various scales, front and back; (2) the quadrant, also engraved with scales, which could be attached to the legs, thus fixing them at ninety degrees for use as a "sector" or military *squadra* to measure elevations; and (3) a clamp or cursor inserted into one of the legs to keep the instrument vertical and extend the leg. Before its publication Galileo had written several variations of the *Operations* to accompany each of the devices he had sold or given away, in order to foil its economic exploitation by others.

3. See Stillman Drake, "Galileo and the First Mechanical Computing Device" and "Tartaglia's Squadra and Galileo's Compasso," in *Essays on Galileo and the History and Philosophy of Science,* 3 vols., sel. N. M. Swerdlow and T. H. Levere (Toronto: University of Toronto Press, 1999), 3:5–14, 15–32; and Peter Machamer, "Galileo's Machines, His Mathematics, and His Experiments," in *The Cambridge Companion to Galileo,* ed. Peter Machamer (Cambridge: Cambridge University Press, 1998), 53–79.

4. The expression is from Domenico Bertoloni Meli, *Thinking with Objects: The Transformation of Mechanics in the Seventeenth Century* (Baltimore: Johns Hopkins University Press, 2006).

5. In recent years, scholars have focused a great deal on the close associations of theory and practice in the early formation of modern science. These have included, on the one hand, material developments, engineering practices, instrument design and production, and artisanal practices in general and, on the other hand, investigations into motion and mechanics, creation of mathematical techniques of description and analysis, and experimental testing of new ideas. A good introduction to these studies may be found in Meli, *Thinking with Objects,* 1–104. See too Pamela O. Long, *Artisan/Practitioners and the Rise of the New Sciences, 1400–1600* (Corvallis: Oregon State University Press, 2011); Alexander Marr, *Between Raphael and Galileo: Mutio Oddi and the Mathematical Culture of Late Renaissance Italy* (Chicago: University of Chicago Press, 2011); and Jürgen Renn, "Editor's Introduction: Galileo in Context: An Engineer-Scientist, Artist, and Courtier at the Origins of Classical Science," in *Galileo in Context,* ed. Jürgen Renn (Cambridge: Cambridge University Press, 2001), 1–8. Renn describes "the Galilean moment" in the history of science as "an explosion of technical and scientific knowledge." Renn, "Editor's Introduction," 3.

6. John Henry summarizes the consensus view, held by many historians, that although Galileo pursued his investigations with the same "modus operandi" as other *mathematici,* he was nonetheless the "greatest figure" in the "movement" of combining "mathematical analysis and experimental investigation" during the

early days of the Scientific Revolution. See John Henry, *The Scientific Revolution and the Origins of Modern Science* (New York: St. Martin's, 1997), 14–19.

7. A huge and sprawling scholarly industry awaits any serious student of the "Scientific Revolution," and the extensive bibliographies in any of the major studies of the past seventy-five years or so—Peter Dear, Margaret J. Osler, John Henry, Herbert Butterfield, Hugh Kearney, Steven Shapin, David Wootton, A. R. Hall, Marie Boas, Salomon Bochner, Alexandre Koyré, among others (there are far too many to list here)—can serve as initial, salutary guides. The previously cited work of H. Floris Cohen, *The Scientific Revolution: An Historiographical Inquiry* (Chicago: University of Chicago Press, 1994), offers an exhaustive discussion of this literature up to the end of the twentieth century. Many current scholars still question whether there even was such a series of events that can be sensibly rendered in a unified narrative. For instance, in his study *The Scientific Revolution* (Chicago: University of Chicago Press, 1996), 1, Steven Shapin sums up decades of this scholarship with his provocative opening salvo: "There was no such thing as the Scientific Revolution, and this is a book about it." But as mentioned earlier, David Wootton, in *The Invention of Science: A New History of the Scientific Revolution* (New York: HarperCollins, 2015), 3–54, harbors no such trepidation in offering a worthy analysis of current historiographical arguments surrounding the "idea of the scientific revolution" and in defending the positions that (a) it actually happened and that (b) it is amenable to a master narrative interpretation. Wootton dates the Scientific Revolution from the nova of 1572 to the publication of Newton's *Optics* in 1704, and emphasizes developments in astronomy. Much of Wootton's argument is quite persuasive, and in general I have followed the outlines of his most welcome account. Nonetheless, despite his claim that the "Scientific Revolution started as a revolution of the mathematicians" (*Invention of Science,* 445), his cursory treatment of "the mathematization of the world" (163–210) simply does not go far enough toward explaining the origins and developments of mathematics and physics, along with their consequences for the separation of scientific analysis and religious belief, the central narrative of this study.

8. Galileo, *Operations,* 92. According to Stillman Drake, the device was "capable of solving quickly and easily every practical mathematical problem that was likely to arise at the time." Stillman Drake, *Galileo at Work: His Scientific Biography* (New York: Dover, 1978), 45.

9. Galileo, *Operations,* 48–50. To envision its operation, imagine two similar isosceles triangles, one superimposed on the other, as depicted here, whose apex, E, lies at the "hinge" of the compass and whose sides, EA and EB, are its legs. Because the triangles EDC and EAB are similar, the ratios between the sides and bases will be proportional. Thus, with the arithmetic scales in place, which lines Galileo divided into 250 equal parts, adjusting the angle DEC or the length of a leg EA permits computing proportionalities mechanically. For example, since

ED:EA :: DC:AB, one can produce an equation that captures the rule of three: ED / EA = DC / AB, or, in algebraic notation, $a / b = c / x$. To solve for x, we set the compass for the known numerical values of ED and DC, then drop to the known A (length EA) and read the unknown value in the measurement from A to B. This yields an analog computation whose accuracy more than sufficed for sixteenth-century purposes. Galileo, *Operations,* 48–51. In his *Optics,* Euclid had described essentially the same method for determining a "given length" (AB), with the eye at E and its "rays" falling on EA and EB (Euclid, "The Optics of Euclid," trans. Harry Edwin Burton, *Journal of the Optical Society of America* 35, no. 5 [1945]: 361), but Galileo's hinged compass permitted far more efficient and speedy calculations of changing ratios.

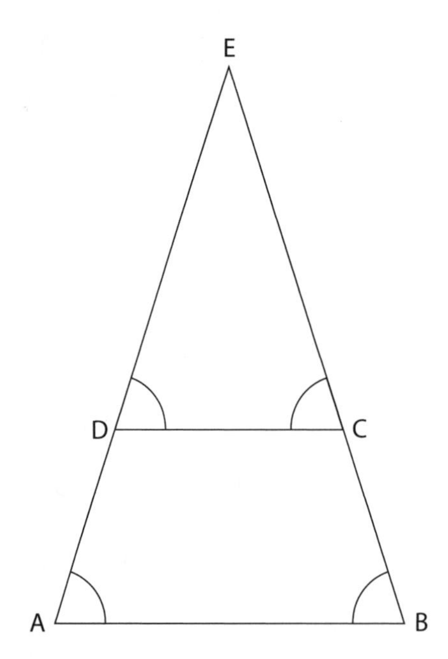

10. Galileo, *Operations,* 52–91.

11. David Wootton, *Galileo: Watcher of the Skies* (New Haven, CT: Yale University Press, 2010), 25.

12. Leonardo of Pisa, too, had devised geometric methods for handling arithmetic and algebraic computations without benefit of a symbolized algebra, which was developed only later. Others followed in his practical footsteps, even while the academic treatment of arithmetic and geometry remained buried in the separate categories of the quadrivium. For the earlier treatment of "geometrical algebra," see Thomas L. Heath, *A History of Greek Mathematics,* vol. 1, *From Thales to Eu-*

clid (1921; New York: Dover, 1981), 150–154, and O. Neugebauer, *The Exact Sciences in Antiquity,* 2nd ed. (Princeton, NJ: Princeton University Press, 1952; New York: Dover, 1969), 147–150. As mentioned previously, this interpretation of classical mathematics sources has been largely discarded.

13. Galileo, *Galileo against the Philosophers,* 37. Galileo's remark—"A pox on those goat-turds at Padua"—reveals both his antipathy toward, especially, his Aristotelian colleagues (he did not suffer fools and believed them foolish) and his willingness to adopt a mask, for the publication was anonymous. Early on in his career Galileo learned well how to dissimulate and particularly as regarded religion generally kept his most private opinions private. See Wootton, *Galileo,* 240–250, and J. L. Heilbron, *Galileo* (Oxford: Oxford University Press, 2010), 64–78.

14. See Paul Lawrence Rose, *The Italian Renaissance of Mathematics: Studies on Humanists and Mathematicians from Petrarch to Galileo* (Geneva: Librairie Droz, 1975), 280–291. A side note: in 1592 Giordano Bruno had traveled to Padua seeking the same position Galileo won. A second side note: William Harvey attended while Galileo taught there, although there is no record of their having met.

15. Michael H. Shank, "Setting the Stage: Galileo in Tuscany, the Veneto, and Rome," in *The Church and Galileo,* ed. Ernan McMullin (Notre Dame, IN: University of Notre Dame Press, 2005), 62–63.

16. Wolfgang Lefèvre, "Galileo Engineer: Art and Modern Science," in Renn, *Galileo in Context,* 11–27.

17. Karl Marx, "The Eighteenth Brumaire of Louis Bonaparte," in *Surveys from Exile,* trans. Ben Fowkes, ed. David Fernbach, Political Writings, vol. 2 (New York: Vintage Books, 1974), 146.

18. Confirmation of astrology's vacuity came in January of1609, when the Grand Duchess Christina wrote Galileo requesting that he cast a horoscope for her husband, Ferdinando, who had fallen seriously ill. Galileo complied, forecasting many happy years for the duke. Ferdinando died three weeks later. Atle Naess, *Galileo Galilei: When the World Stood Still,* trans. James Anderson (Berlin: Springer, 2005), 42; Heilbron, *Galileo,* 2–3, 91–94, 325–326.

19. Peter Dear, in *Revolutionizing the Sciences: European Knowledge and Its Ambitions, 1500–1700* (Princeton, NJ: Princeton University Press, 2001), 72–79, shows how Galileo's polemical use of the label "physico-mathematics" both rejected the disciplinary boundary between natural philosophy and mathematics and stressed the merging of "knowing, doing, and mathematics" that would supplant philosophy itself in the "true understanding of nature." The emphasis on "doing" (and experimenting) recalls what Amos Funkenstein, earlier mentioned, has termed the active, "ergetic ideal of knowing" of modern science ("knowing by doing or knowing by construction"), vis-à-vis the passive, "contemplative" ideal prevalent in antiquity and throughout the Middle Ages. Moreover, Funkenstein notes, the "formalization of mathematics was, in a sense, its mechanization," as

mathematics became the predominant tool of the new constructivist paradigm of knowledge. Amos Funkenstein, *Theology and the Scientific Imagination: From the Middle Ages to the Seventeenth Century* (Princeton, NJ: Princeton University Press, 1986), 312–327.

20. Galileo Galilei, *The Assayer,* in *Discoveries and Opinions of Galileo,* trans. Stillman Drake (New York: Anchor Books, 1957), 237–238.

21. Contrast this perception with that of John of Salisbury, who wrote that "nature is the mother of all the arts," and furthermore that "the rules of poetry clearly reflect the ways of nature, and require anyone who wishes to become a master in this art to follow nature as his guide." John of Salisbury, *The Metalogicon of John of Salisbury: A Twelfth-Century Defense of the Verbal and Logical Arts of the Trivium,* trans. Daniel D. McGarry (1955; Berkeley: University of California Press, 1962), bk. 1, ch. 11, 34, and bk. 1, ch. 17, 51. As a Renaissance humanist, Galileo's thought processes were also heavily influenced by concerns for rhetoric and the arts of persuasion. See, for example, the noteworthy account by Jean Dietz Moss, *Novelties in the Heavens: Rhetoric and Science in the Copernican Controversy* (Chicago: University of Chicago Press, 1993). But while rhetoric may have been critical in trying to persuade naysayers of the importance and veracity of demonstrable, scientific claims, Galileo was crystal clear that it had nothing to do whatsoever with the content of those claims, a point we shall be developing in the chapters that follow.

22. Galileo Galilei, *Two New Sciences,* trans. Stillman Drake (Madison: University of Wisconsin Press, 1974), 133.

23. Arthur Koestler, *The Sleepwalkers: A History of Man's Changing Vision of the Universe* (1959; New York: Grosset and Dunlap, 1963).

24. For the former, see Mario Biagioli, *Galileo, Courtier: The Practice of Science in the Culture of Absolutism* (Chicago: University of Chicago Press, 1993); for the latter, see Heilbron, *Galileo.*

9. The Birth of Analysis

1. William R. Shea and Mariano Artigas, *Galileo in Rome: The Rise and Fall of a Troublesome Genius* (Oxford: Oxford University Press, 2003), 5–6, 15–16; Stillman Drake, *Galileo at Work: His Scientific Biography* (New York: Dover, 1978), 12–14.

2. The new, Italian edition of Euclid's *Elements* had been published in 1543 by Niccolò Tartaglia, with whom Ricci probably studied. Later, Ricci himself taught in Italian and almost certainly used a 1565 reprint of this text in his tutorial sessions with Galileo, rather than the Latin versions found in the universities. Neither of the two available Latin editions of Euclid (one based on an Arabic text, another based on a Greek text) contained a correct account of Eudoxus's theory of proportion, which Euclid incorporated in book 5. (Eudoxus, whom we met earlier, was an astronomer and mathematician who studied briefly with Plato.) Tart-

aglia's Italian translation and commentary did and thus, through Ricci, placed in Galileo's hands this powerful, analytical tool. See Drake, *Galileo at Work,* 2–4; Stillman Drake, "Velocity and Eudoxian Proportion Theory," in *Essays on Galileo and the History and Philosophy of Science,* 3 vols., sel. N. M. Swerdlow and T. H. Levere (Toronto: University of Toronto Press, 1999), 2:264–280; Stillman Drake, "Euclid, Book V from Eudoxus to Dedekind," in *Essays on Galileo,* 3:61–75; and Euclid, *The Thirteen Books of the Elements,* 2nd ed., 3 vols., trans. Thomas L. Heath (1926; New York: Dover, 1956), 2:112–129.

3. See Galileo Galilei, *"On Motion" and "On Mechanics,"* comprising *De Motu* (ca. 1590), trans. I. E. Drabkin, and *Le Meccaniche* (ca. 1600), trans. Stillman Drake (Madison: University of Wisconsin Press, 1960). More on this later.

4. An exception is William R. Shea, *Galileo's Intellectual Revolution: Middle Period, 1610–1632* (New York: Science History, 1972), who situates Galileo's "Revolution" in the years following his return to Florence.

5. Thus Alexandre Koyré, *Metaphysics and Measurement: Essays in Scientific Revolution* (Cambridge, MA: Harvard University Press, 1968), 13: "For it is thought, pure unadulterated thought, and not experience or sense-perception . . . that gives the basis for the 'new science' of Galileo Galilei." Also see Thomas B. Settle, "An Experiment in the History of Science," *Science* 133 (1961): 19–23. Drake's main analyses are cited later.

6. Many of these "worksheet" folios were excluded from the standard, scholarly version of Galileo's works edited by Antonio Favaro and were brought to light in the 1970s by Stillman Drake. The entire collection, "Galileo Galilei's Notes on Motion: Ms. Gal. 72, Folios 33 to 196," is now sponsored on the Internet by the Biblioteca Nazionale Centrale (Florence), the Istituto e Museo di Storia della Scienza (Florence), and the Max Planck Institute for the History of Science (Berlin): http://www.imss.fi.it/ms72/INDEX.HTM.

7. The pioneering work of Drake remains the starting point for analyses of these folios. From them he himself actually developed several accounts of Galileo's free-fall discovery. See Stillman Drake, "Galileo's Discovery of the Law of Free Fall," *Scientific American* 228, no. 5 (1973): 84–92; Drake, *Galileo at Work,* 83–104; Stillman Drake, *History of Free Fall: Aristotle to Galileo* (Toronto: Wall and Thompson, 1989); and Stillman Drake, *Galileo: Pioneer Scientist* (Toronto: University of Toronto Press, 1990), 9–31. See too the extensive collection of Drake's essays, *Essays on Galileo,* and in particular "Galileo and Mathematical Physics" and "Galileo's Constant," 2:332–345, 357–369. Other scholars have challenged various technical details of Drake's analyses, especially regarding how Galileo arrived at his theory of projectile motion. Collectively, the work of all these scholar-experimenters is quite remarkable and, disputes notwithstanding, offers an intriguing look into historical detective work at its finest. See Ronald H. Naylor, "Galileo's Theory of Projectile Motion," *Isis* 71 (1980): 550–570; Ronald H. Naylor, "The Role of Experiment in Galileo's Early Work on the Law of Fall," *Annals of*

Science 37 (1980): 363–378; Ronald H. Naylor, "Galileo's Method of Analysis and Synthesis," *Isis* 81 (1990): 695–707; Winifred Lovell Wisan, "Galileo and the Process of Scientific Creation," *Isis* 75 (1984): 269–286; Winifred Lovell Wisan, "The New Science of Motion: A Study of Galileo's *De Motu Locali*," *Archive for History of Exact Sciences* 13 (1974): 103–306; David K. Hill, "Dissecting Trajectories: Galileo's Early Experiments on Projectile Motion and the Law of Fall," *Isis* 79 (1988): 646–668; David K. Hill, "Galileo's Work on 116v: A New Analysis," *Isis* 77 (1986): 283–291; P. Damerow et al., *Exploring the Limits of Preclassical Mechanics* (New York: Springer-Verlag, 1992), 135–273; and Jürgen Renn et al., "Hunting the White Elephant: When and How Did Galileo Discover the Law of Fall?," in *Galileo in Context*, ed. Jürgen Renn (Cambridge: Cambridge University Press, 2001), 29–149.

8. Thus Jürgen Renn et al. ("Hunting the White Elephant," 32–33) write, "The cornerstones of 1604 and 1609 define a scaffolding for more or less speculative stories about what really happened at those times." The letters are partially reproduced in Stillman Drake, "Galileo's 1604 Fragment on Falling Bodies," in *Essays on Galileo*, 2:188–189; and Stillman Drake, "Galileo's Experimental Confirmation of Horizontal Inertia," in *Essays on Galileo*, 2:156.

9. A recent Galileo biographer summarizes succinctly: "What seems essential to know is not who made what discoveries when, but why Galileo came to see them as fundamentally important, as the basis of a new science of motion." J. L. Heilbron, *Galileo* (Oxford: Oxford University Press, 2010), 128. In *The Invention of Science: A New History of the Scientific Revolution* (New York: HarperCollins, 2015), David Wootton highlights the invention of discovery for its critical role in launching the scientific revolution. As with trying to envision a premathematical age, so too does it tax our modern imagination to understand a world in which discovery was not only little noticed or acknowledged but lay beyond the common field of ordinary perception. In Wootton's estimation, Christopher Columbus changed all that. For more along these lines, see Anthony Grafton, *New Worlds, Ancient Texts: The Power of Tradition and the Shock of Discovery*, with April Shelford and Nancy G. Siraisi (Cambridge, MA: Harvard University Press, 1992). Still, even with discovery in the air, Galileo, writing in Latin, used the word *exploro* in referring to his discoveries of the moons of Jupiter. Only from late Latin (*discooperire*, meaning 'uncover') and medieval French (*descovrir*) does the word 'discover' gradually enter etymologically into western Europe, although initially it denoted not 'finding' something new but 'divulging' or 'revealing,' such as betraying or exposing someone the way an informant might do.

10. The imagery is Francis Bacon's, from "Aphorism LXXXIV," in *Novum Organum*, trans. James Spedding, in *The Works of Francis Bacon*, 2 vols. (New York: Hurd and Houghton, 1878), 1:117: "Time . . . is the author of authors, nay rather of authority. For rightly is truth called the daughter of time, not of authority."

11. This is the comment of particle physicist Lawrence M. Krauss, *Fear of Physics* (New York: Basic Books, 2007), 58 (italics in the original). Krauss begins his en-

thusiastic, charming, and incisive introduction to modern physics with the recognition that Galileo "literally *created* modern science by describing motion" (8).

12. Galileo's *On Motion* challenged Aristotelian positions in ways similar to those of his immediate predecessors, Tartaglia, Benedetti, and Guidobaldo del Monte, by analyzing moving bodies in a medium (hence the concern for hydrostatics and the Archimedean balance). For a more thorough explication of this context, see Domenico Bertoloni Meli, *Thinking with Objects: The Transformation of Mechanics in the Seventeenth Century* (Baltimore: Johns Hopkins University Press, 2006), 18–65; Renn et al., "Hunting the White Elephant," 46–79; and Stillman Drake, introduction to *Mechanics in Sixteenth-Century Italy,* ed. Stillman Drake and I. E. Drabkin, trans. Stillman Drake and I. E. Drabkin (Madison: University of Wisconsin Press, 1969), 3–60.

13. Galileo, *On Motion,* 30, 58. A bit later, in lecture notes to himself for a class on cosmography (1602), Galileo wrote of addressing the subject with "observations . . . hypotheses . . . geometrical demonstrations . . . [and] arithmetical operations, . . . leaving to the Natural Philosophers consideration of the qualities." Reproduced in Drake, *Galileo at Work,* 52.

14. Today we would express this relation in formulaic terms, such that $v \propto \frac{f}{r}$, with v standing for speed and f and r standing for force and resistance, respectively, while the symbol $\propto$ designates proportionality.

15. Galileo, *On Motion,* 30, 27. Galileo repeated several of these reductio arguments later in Galileo Galilei, *Two New Sciences,* trans. Stillman Drake (Madison: University of Wisconsin Press, 1974), 65–71, voiced by his representative figure, Salviati. Although Galileo did not invoke them in the opinions of the dialogue's Aristotelian spokesman, Simplicio, there were some decent, commonsense replies available. For instance, imagine that two people join their hands together while in free fall (say, skydiving). Neither fact nor Aristotle would double their speed at that moment. Because, for Aristotle, speed is a function also of resistance, as well as force, joining two people would double their resistance as well as doubling their weight; any change would be a wash, with both plummeting as before. Still, Galileo's attacks on Aristotle's qualitative definitions were based ultimately on analyses tied to quantification and the new numeracy, against which backing and filling with further refinements of the definitions could only go so far.

16. Cited in Drake, *Galileo at Work,* 15, 24.

17. Quoted by Heilbron, *Galileo,* 164. See too David Wootton, *Galileo: Watcher of the Skies* (New Haven, CT: Yale University Press, 2010), 46–51; Roberto Vegara Caffarelli, *Galileo Galilei and Motion: A Reconstruction of 50 Years of Experiments and Discoveries,* trans. Christine Valerie Pennison (Berlin: Springer, 2009), 119–125.

18. A recent analysis by Paolo Palmieri has concluded, for example, that with his inclined planes Galileo was investigating "the evanescent degrees of speed"

and that his discovery of the "times-squared law may just have been a fortunate incident." Paolo Palmieri, "Galileo's Construction of Idealized Fall in the Void," *History of Science* 43 (2005): 363. See too Meli, *Thinking with Objects,* 69–74.

19. Much later, Galileo described these investigations: "In order to make use of motions as slow as possible . . . I also thought of making moveables [moving bodies] descend along an inclined plane not much raised above the horizontal. On this, no less than in the vertical, one may observe what is done by bodies differing in weight. Going further, I wanted to be free of any hindrance that might arise from contact of these moveables with the said tilted plane. Ultimately, I took two balls, one of lead and one of cork, the former being at least a hundred times as heavy as the latter, and I attached them to equal thin strings four or five braccia long, tied high above." Galileo, *Two New Sciences,* 87.

20. Potentially important in determining the order of Galileo's experiments and of his thinking, the dating and sequencing of folios and some of the entries have been only partially achieved through analyses of the paper and watermarks, chemistry of the inks, handwriting (some of the notes were copied by Galileo's students, Mario Guiducci and Niccoló Arrighetti), and context. See Stillman Drake, "Galileo Gleanings—XXI: On the Probable Order of Galileo's Notes on Motion" and "Dating Unpublished Notes, Such as Galileo's on Motion," in *Essays on Galileo,* 1:171–184, 1:185–200. As regards this particular worksheet, folios 107r and 107v, the analysis of the inks performed by Renn and his colleagues shows little difference in the chemical composition of the numbers in all three columns, leading him to conclude that one could not prove them written at different times, a claim Drake maintains. However, Drake's reading of the column C numbers as measurements from an inclined plane, established by musical beat, remains largely unaffected by Renn's analysis, which poses its own difficulties. (Renn claims the numbers represent distances measured from a hanging chain or catenary, very close to a shallow parabola, which would have had to be over twenty-five yards in length and over seven feet above ground level—not very plausible.) Renn et al., "Hunting the White Elephant," 99–101.

21. Recent research has even challenged the consensus calculation (0.94 millimeters) of Galileo's unit of measure, the *punto.* See Roberto Vegara Caffarelli, "How Long Is Galileo's *Punctus?,*" accessed January 2016, https://independent.academia.edu/RobertoVergaraCaffarelli. Caffarelli's refiguring sets the standard at 2.29 millimeters, and from this basis he offers a further interpretation of the folio 107v numbers and calculations, fixing Galileo's "timing" on pendulum swings one through eight (column B in the table). See Caffarelli, *Galileo Galilei and Motion,* 135–153. But Drake and others derived the consensus standard in part from Galileo's own gradations on the arithmetic scale of his geometric compass, from a direct measurement on folio 166r, as well as other calculations from the worksheets, while Caffarelli ignores these and bases his unit for the *punto* (also named *picciolo*) on comparative tables from over a century after Galileo's death

and on informal renderings from later writings. A larger point looms here. Drake, Caffarelli, and Renn, among others, all introduce ad hoc adjustments to their analyses of folio 107v, adjustments they assert must have been made by Galileo, in order to ensure their correct interpretations of the numbers. In part this situation derives from the measurements in column C themselves. Because, in effect, they "map" onto the formula of a parabola or free fall, researchers are able to interpret them in a variety of ways. Thus, to reiterate, we don't know how Galileo got his experimental results with any certainty; we can surmise with probability, and Drake's account of folio 107v is at least as probable as, and probably more so than its rivals.

22. In his last work, Galileo bolstered the likelihood of this scenario when he noted that by imagining "time divided into moments, that is, into minimum equal tiny parts," one could correlate the "ratio of the numbers of vibrations [of strings] and impacts of air waves that go to strike our eardrum, which likewise vibrates according to the same measure of times." Galileo, *Two New Sciences,* 106, 104.

23. In the early fifteenth century (ca. 1400–1430) frets were added to Renaissance lutes. The frets were made of gut and tied around the neck of the instrument, thus making them movable. Galileo would have had only to work out a means of depressing the fret-gut into the groove on an inclined plane, which he could do simply by placing another piece of gut over the groove's entire length (lying on top of and perpendicular to the fret-guts) and pressing each of the frets to the bottom of the channel. (In my own reconstructions I used a piece of copper wire to achieve the same result.) As previously mentioned, Vincenzo Galilei not only was a leading lute player and composer but also experimented with the acoustics of lute strings, measuring the ratios between their variable tensions. In addition to Galileo's own musical interests and abilities with the lute, it is very likely that he was still living at home and present in 1588–1589 when Vincenzo was conducting his experiments and likewise that Galileo was thoroughly familiar with the dispute between his father and Zarlino over the tuning question. See Stillman Drake, "Renaissance Music and Experimental Science," *Journal of the History of Ideas* 31, no. 4 (1970): 483–500; Stillman Drake, "The Role of Music in Galileo's Experiments," in *Essays on Galileo,* 2:307–315; Mark A. Peterson, *Galileo's Muse: Renaissance Mathematics and the Arts* (Cambridge, MA: Harvard University Press, 2011), 149–173; and Heilbron, *Galileo,* 8–11.

24. Drake maintains that the time intervals matched a modern metronome tempo of 109 beats per minute, or 0.55 seconds for each beat and hence each time interval.

25. Stillman Drake, "Galileo's Work on Free Fall in 1604," in *Essays on Galileo,* 2:281–291. See too I. Bernard Cohen, "Galileo's Rejection of the Possibility of Velocity Changing Uniformly with Respect to Distance," *Isis* 47 (1956): 231–235; and I. Bernard Cohen, *The Birth of a New Physics,* rev. ed. (1960; New York: W. W. Norton, 1985), 85–106, 196–206. Galileo could have simply "normalized" his

numbers (as we term it now) by taking the first number, 33, as his base unit, then dividing it into the remaining entries in column C, rather than dividing the intervals between the entries. This would produce 33 ÷ 33 = 1; 130 ÷ 33 = 3.94; 298 ÷ 33 = 9.03; 526 ÷ 33 = 15.94; 824 ÷ 33 = 24.97; 1,192 ÷ 33 = 36.12; 1,620 ÷ 33 = 49.09; 2,104 ÷ 33 = 63.76. Rounding these results directly nets the "square" sequence of 1, 4, 9, 16, 25, 36, 49, and 64. It's possible, although not likely, given (a) how long it took him to correlate distance and time squared; (b) the odd number sequence on folio 107v, which suggests he worked with the intervals rather than with the entries themselves; and (c) the enormous stumbling block of making physical sense of time squared.

26. This famous story was told first by Vitruvius, whom we met earlier. Archimedes leaped from his bath and ran naked through the streets of Syracuse crying, "Eureka!" (I have found it!) when he realized that the volume of water displaced from his bath must equal the parts of his body submerged, and that displaced water could be used to measure the volumes of irregular objects. This gave him the means of weighing King Hiero's crown and determining whether it was pure gold or an alloy. Marcus Vitruvius Pollio, *The Ten Books on Architecture,* trans. Morris Hicky Morgan (1914; New York: Dover, 1960), 253–254.

27. Working with ratios and proportions, Galileo frequently used squares and square roots in his calculations, and had no difficulty manipulating them, even without algebra. Correlating "derived" magnitudes directly with physical phenomena, however, required a substantial reorientation in his thinking.

28. Recall the famous line from Albert Einstein, "How can it be that mathematics, being after all a product of human thought independent of experience, is so admirably adapted to the objects of reality?" Albert Einstein, "Geometry and Experience," lecture before the Prussian Academy of Sciences, January 27, 1921, in Albert Einstein, *Ideas and Opinions,* trans. Sonja Bargmann (1954; New York: Three Rivers Press, 1982), 233.

29. Galileo Galilei, "Letter to Guidobaldo del Monte," November 29, 1602, reproduced in Drake, *Galileo at Work,* 69–73.

30. Galileo, *Two New Sciences,* 170. Water "clocks" dated to antiquity, but the notion of using one as a water "stopwatch" for measuring time was yet another of Galileo's innovations.

31. As with other early forms of common measurement (such as foot, hand, arm—the last, *braccio* in Italian), grain weights began as available conventions for comparison of small masses and volumes. Grains used were generally barley, wheat, or carob seeds. Over time these became somewhat standardized, so that by Galileo's period there were 480 grains to a fluid ounce. Drake estimates that the flow rate of Galileo's water clock was approximately three ounces per second or 1,440 grains per second. In the cgs (centimeter-gram-second) system, 16 grains is almost exactly one gram. See Drake, *History of Free Fall,* 39, 43.

32. In my classes on Galileo's experimental science, my students and I have made measurements in this fashion. It requires a bit of practice, timing, and the right sort of anticipation, much like starting a dash in track. After a few attempts, a pair of "timers" can produce remarkably consistent readings for very small amounts of time, literally fractions of a second. Galileo's attention to experimental detail, though quite extraordinary in his day, has since become commonplace in scientific practice.

33. See Galileo, "Notes on Motion," fols. 154v, 151v, 90v, and 189v, and the analysis of Drake in *History of Free Fall,* 35–51. Also see Peter Machamer and Brian Hepburn, "Galileo and the Pendulum: Latching onto Time," *Science and Education* 13 (2004): 333–347.

34. Later he would posit the period of the pendulum itself as a kind of regular intervallic or periodic movement that could drive more accurate clocks when combined with a workable escapement mechanism. Galileo is thus sometimes credited with the invention of the pendulum clock, even though he never made one himself. His son Vincenzo was working on a prototype of it when he died in 1649. Soon after, the Dutch physicist and mathematician Christiaan Huygens (1629–1695) produced a very accurate and workable model in 1656 and is generally considered the pendulum clock's inventor.

35. These two time measurements correlated with pendulum lengths of 6,960 and 13,920 *punti,* respectively, as can be seen in columns A and B, rows 4 and 5, of the tabulated data shown in Appendix A.11. The mean proportional of 6,960 and 13,920 is 9,843.

36. Drake was further able to confirm Galileo's experimental operations with an intriguing discovery concerning the last number in the length column, as shown in the Appendix A.11 data. In 1609 Galileo had written a note regarding a separate topic on the blank side of a folio, which he cut and pasted onto folio 90. At Drake's request, the curators lifted the pasted-on portion from folio 90, and he then discovered hidden on the backside the number 27,834, along with sufficient words to identify it as a diameter. The number represents an adjustment Galileo made to 27,840, the last pendulum length on row 6 in Appendix A.11. The calculations were probably made on the discarded portion of folio 90^{b}v. Drake claims the number played a crucial role in the computations that allowed Galileo to link the time of a freely falling body through a distance equal to the length of a pendulum and take the final step to the law of free fall, which he did on folio 189v. Drake, *Galileo: Pioneer Scientist,* 17–31. Drake's analysis of folio 189v has been branded as "fancyful" by Renn et al., "Hunting the White Elephant," 34.

37. Drake, *Galileo: Pioneer Scientist,* 91. Early in his experiments Galileo thought that free falls and pendulums were the same, a conjecture he tied to the chord theorem, which he had found in 1602 and which states that the times of descent along all chords of a vertical circle to their lowest points are equal. Because the

vertical diameter is itself a chord to the lowest point, the time of a ball falling freely along that diameter of a circle should be the same as that of a bob descending along any arc of the pendulum. But his experiments showed otherwise. Indeed, when an arc is extremely wide, the pendulum takes a perceptibly longer time to reach the vertical, and it also undergoes further perturbations, which he realized. Still, the similarities between the time and length ratios of each could not be ignored, and in another deft move Galileo devised an additional ratio between the timings of free falls and those of pendulum drops to use as a basis for calculating free-fall distances in general. See Galileo, "Notes on Motion," fols. 151v and 154v, for these figures and calculations. Drake summarizes this calculation as "Galileo's Constant." He argues that Galileo had used the ratio between the timing of a pendulum of length 1,740 *punti,* which was 942 grains, and the timing of a free fall through the same 1,740 *punti,* whose time was 850 grains, as the basis for calculating further free falls. This ratio, $\frac{942}{850}$, equals 1.1082. Drake calculated its modern equivalent as 1.1107, a result that reveals the remarkable experimental accuracy of Galileo's times ratio. Drake, "Galileo's Constant," in *Essays on Galileo,* 2:357–369.

38. Galileo, *Two New Sciences,* 165. As we discussed in Chapter 4, the question of how much Nicole Oresme anticipated or influenced Galileo has been of central importance in the arguments over the "continuity thesis," the claim that there was no real rupture or break between medieval and modern science, but rather a long, slowly evolving series of developments. Beyond the specific references cited in Chapter 4, see Marshall Clagett, *The Science of Mechanics in the Middle Ages* (Madison: University of Wisconsin Press, 1959), 331–381 and throughout. In "Galileo and the Oxford *Calculatores:* Analytical Languages and the Mean-Speed Theorem for Accelerated Motion," in *Reinterpreting Galileo,* ed. William A. Wallace (Washington, DC: Catholic University of America Press, 1986), 53–108, Edith Dudley Sylla thoroughly documents her conclusion (107) that throughout the course of his investigations "it is unlikely that Galileo learned much detail from the Calculatory tradition" (including Oresme). Meli, *Thinking with Objects,* 75, agrees with this assessment.

39. It bears noting that the law of free fall also presupposes the idea of inertia. For Aristotle velocity was a function of the force impressed on a body; remove the force and the body would return to a state of rest. It followed that a freely falling body acquired its full downward velocity instantly at the moment of release. In Galileo's free fall, adding force (for example, the weight of the ball) at each interval of the ball's downward motion produces not constant velocity but uniform and continuous acceleration, as Newton's laws of motion would later make explicit. To assume the inertial movement of bodies rather than rest as their natural state marks one of the central, conceptual shifts away from the common sense of Aristotelian physics.

40. Niccolò Tartaglia, *Nova Scientia* (Venice, 1537), in Drake and Drabkin, *Mechanics in Sixteenth-Century Italy,* 80.

41. As with the account of Galileo's discovery of free fall, there is no scholarly consensus as to whether the parabolic trajectory depicted on these documents was a new discovery or the experimental and mathematical confirmation of an earlier, qualitative judgment that projectiles followed some sort of symmetrical curve—a parabola, hyperbola, or catenary. In both the cases of free fall and projectile trajectory, argues Renn, "the context of discovery is indistinguishably intermingled with the context of its justification." Renn et al., "Hunting the White Elephant," 126.

42. The main folios in question are 81r, 114v, 116v, and 117r, while the key scholars who have scrutinized them are the previously cited Drake, Wisan, Naylor, and Hill. As a result of in-class exercises in reproducing these experiments with my students over the years, I have come to rely on Hill's incisive analysis from his articles "Dissecting Trajectories" and "Galileo's Work on 116v."

43. The precise purpose of the experiment, what Galileo initially hoped to accomplish, remains uncertain. Drake, for example, suggests that Galileo devised it to test either (a) his restricted concept of inertia or (b) the "double distance rule" that, in a time equal to that of its uniform acceleration, a body would move twice as far as it previously traveled during the accelerating period. Drake, *Galileo at Work,* 127–133. Hill more persuasively argues that having confirmed the speed law ($v \propto \sqrt{d}$) in folio 114v, and having deduced it from the law of chords, Galileo then sought to corroborate it with horizontal projectiles. Hill, "Galileo's Work on 116v."

44. Galileo, *Two New Sciences,* 169.

45. Renn et al., "Hunting the White Elephant," 35–70.

46. See Hill, "Galileo's Work on 116v." Hill argues that Galileo was led to correlating the speed law with parabolic trajectory because he had earlier proved the chord theorem—that the times of descent along all chords of a vertical circle to their lowest points are equal—the mathematical equivalent of the speed law, $v^2 = d$.

47. Drake, *Galileo: Pioneer Scientist,* 197–198.

48. Galileo Galilei, *The Assayer,* in *Discoveries and Opinions of Galileo,* trans. Stillman Drake (New York: Anchor Books, 1957), 276–277. Thus, "if ears, tongues, and noses [that is, sensations] were removed, shapes and numbers and motions would remain, but not odors or tastes or sounds."

49. Krauss writes, "It has been an implicit part of physics since Galileo that irrelevant information must be discarded." *Fear of Physics,* 226. In our own day, physicists often use terms such as "coarse" or "fine graining" to focus on the level of detail or magnitude within which their investigations are conducted, likewise discarding the irrelevant features of questions at hand. Discovering the right irrelevancies in any given inquiry, then, stands as an integral feature of analytically unearthing natural laws, a counterpart to the process of abstraction.

50. Abraham Pais, *Subtle Is the Lord: The Science and Life of Albert Einstein* (Oxford: Oxford University Press, 1982), 253. Einstein greatly admired Galileo for many reasons, notably the "imaginative power" exhibited in his discoveries related to "acceleration" and moving bodies. See Albert Einstein, foreword (1952) to *Galileo: Dialogue concerning the Two Chief World Systems,* rev. ed., ed. and trans. Stillman Drake (1953; Berkeley: University of California Press, 1967), vi–xx. In *To Explain the World: The Discovery of Modern Science* (New York: HarperCollins, 2015), 248, physicist Steven Weinberg articulates succinctly the emotional experience of Galileo and Einstein: "We get intense pleasure when something has been successfully explained," he writes. "The scientific theories and methods that survive are those that provide such pleasure."

51. During the early phases of the development of modern science, the replicability of experiments was exceedingly spotty at best. Observers and scientists had to become acclimated and trained in what to look for and how to proceed. Even something as seemingly simple as seeing new wonders when looking through a telescope required enormous patience and practice to gain a stable image at the right place and time in the night skies. Building confidence was a slow process in these years.

10. Toward the Mathematization of Matter

1. The term 'capital,' for one among hundreds of examples, evolved from the Latin *caput,* meaning 'head,' which was the Roman term for counting livestock (we still count cattle herds by 'head') and therefore referred to wealth. Nowadays, despite its linguistic association and resonance with 'stocks' and 'stock markets,' the term refers to an entirely different sort of wealth, one determined by the context of our own times. Karl Marx, of course, wrote three large volumes describing 'capital' not as a thing but as a historical process. Still, historical processes, too, differ from the logical procedures of abstract mathematics.

2. Peter Harrison, *The Territories of Science and Religion* (Chicago: University of Chicago Press, 2015), 3 and throughout. Harrison's erudition is immense, his scholarship careful. But restricting himself to words and their referents leads to some odd conclusions. For example, on the "Platonic" mathematical "model," he notes, nature would consist of "idealized geometrical entities rather than divinely instituted symbols" (77), an assessment that smacks of what Amos Funkenstein termed the medieval "inventory of mathematical objects." Thus does Harrison sidestep entirely the abstract and functional characteristics of modern relation-mathematics, which reside at the core of Galilean and modern physics. For Wootton, see David Wootton, *The Invention of Science: A New History of the Scientific Revolution* (New York: HarperCollins, 2015). As mentioned in an earlier note, Wootton's otherwise salutary and very welcome study falls short in its explanation of "the mathematization of the world" (163–210). We cannot emphasize enough the

emerging bifurcation between words, the language of speech and writing, and number, "the language of science" (see Tobias Dantzig, *Number: The Language of Science: A Critical Survey Written for the Cultured Non-mathematician,* 4th ed. [1930; Garden City, NY: Doubleday, 1954]), a bifurcation that created and separated the territories of science and religion.

3. Friedrich Waismann, *Introduction to Mathematical Thinking: The Formation of Concepts in Modern Mathematics* (1951; New York: Harper and Row, 1959), 27 and throughout, esp. 25–48, "The Rigorous Construction of the Theory of Integers." Waismann (1896–1959), whom we met earlier, in Chapter 8, was a prominent, second-generation member of the Vienna Circle before relocating to England where he taught philosophy of science and mathematics at Cambridge and Oxford until his death.

4. As our own shorthand for these developments, one has merely to think here of the logically self-contradictory "instantaneous rate of change," later developed and applied to the operations of calculus. An instant has no time lapse; change requires it; thus the contradiction in terms. The divine bishop George Berkeley (1685–1753), himself no slouch when it came to intellectual rarefactions, would even look on such a linguistic oxymoron as actually serving religion: "He who can digest a second or third fluxion [Newton's term for the calculus] . . . need not, methinks, be squeamish about any point in Divinity." George Berkeley, *The Analyst: A Discourse Addressed to an Infidel Mathematician* (1734; Whitefish, MT: Kessinger Legacy Reprints, 2015), 4.

5. As with other dimensions of his investigations, Galileo was not alone in his tentative forays into early modern atomism. For instance, Julius Caesar Scaliger (1484–1558), an Aristotelian and father of Joseph Scaliger (who thus named the Julian cycle after both his father and Caesar), had devised a theory of "natural minima," drawn largely from Averroes. A half century later, Daniel Sennert (1572–1657), Sebastian Basso (ca. 1573–?), and David van Goorle (1591–1612) revived interest in Democritus, while the generations after Galileo saw René Descartes, Pierre Gassendi, and Robert Boyle (1626–1691) all develop atomism or "corpuscular philosophy." See Andrew G. van Melsen, *From Atomos to Atom: The History of the Concept Atom* (1952; New York: Dover, 2004), 73–77, 81–115.

6. Through numerous books—an entire scholarly career, in fact—William A. Wallace has sought to recover and promote the philosophy of Aristotle for the modern world and to show that, properly understood, from his youth forward Galileo was a good Aristotelian (so too Newton, in Wallace's view). In *The Modeling of Nature: Philosophy of Science and Philosophy of Nature in Synthesis* (Washington, DC: Catholic University of America Press, 1996) Wallace synthesizes much of this work and tenders a classificatory philosophy of carefully defined and explicated terms of which any scholastic thinker would have approved and been proud. But he too suppresses critical developments in mathematics—its origins in the new abstractions of an emerging information technology, its core of modern

numeracy, and the ensuing metamorphosis from early thing-mathematics into modern relation-mathematics. This is too stark a transition to ignore. The new mathematics resided at the heart both of Galileo's hostility to Aristotle and of his analytical approach to investigating nature. His friend and fellow Lincean Federico Cesi (discussed later in this chapter) put it bluntly: "What we are engaged in is the destruction of the principal doctrines of the philosophy which is currently dominant, the doctrine of 'the master of those who know' [Dante's cipher for Aristotle]." Quoted in David Wootton, *Galileo: Watcher of the Skies* (New Haven, CT: Yale University Press, 2010), 33.

7. Technically speaking, at Trent the terms 'essence' and 'accident' were not used in the exact definition of the sacrament. Rather, 'species' referred to the appearance of the bread and wine and 'substance' to its inner reality: "After the consecration of the bread and wine, our Lord Jesus Christ, true God and man, is truly, really, and substantially contained under the species of those sensible things." This "conversion is . . . suitably and properly called Transubstantiation." "The Council of Trent: The Thirteenth Session," in *The Canons and Decrees of the Sacred and Oecumenical Council of Trent,* ed. and trans. J. Waterworth (London: Dolman, 1848), 76, 78, and more generally 75–91, http://history.hanover.edu/texts/trent/ct13.html. In theology one should slight technicalities only with trepidation, but in this case the terminology is close enough as to be essentially the equivalent.

8. Recently, one scholar, Pietro Redondi, has attempted to show that Galileo's atomism lay at the basis of his trial and condemnation. See Pietro Redondi, *Galileo, Heretic,* trans. Raymond Rosenthal (Princeton, NJ: Princeton University Press, 1987). As regards the trial, Redondi outpaces his evidence by far, venturing into never-never-land, as we shall address later on; nonetheless, he is certainly correct in claiming that atomism posed a recognizable threat to Catholic doctrine and authority, as was even more apparent in the later work of Descartes and Gassendi.

9. In the preface to his *Great Instauration,* in *The Works of Francis Bacon,* 2 vols. (New York: Hurd and Houghton, 1878), 1:34, Francis Bacon claimed to have "established for ever a true and lawful marriage between the empirical and the rational faculty," a union born of "dwelling purely and constantly among the facts of nature." As suggested earlier, the impulses for gathering the "facts of nature" had many sources in the Renaissance and early modern centuries. And the new facts were gathered from far and wide—from the New World, from new collections of flora and fauna supplied by a growing interest in natural history, from information gleaned with new instruments, from simply paying closer attention to nature's behavior. Much of this new information was expressed in lists with words, or with classification systems (like Bacon's own organization of knowledge), or with various linguistic descriptions and compilations, such as travelogues and curiosity cabinets. Bacon too proffered a way of arranging new in-

formation, one among many: "For all nature rises to a point like a pyramid. Individuals, which lie at the base of nature, are infinite in number; these are collected into Species, which are themselves manifold; the Species rise again into Genera; which also by continual gradations are contracted into more universal generalities, so that at last nature seems to end as it were in unity." No paragon of modern science here. See Francis Bacon, *De Augmentis Scientiarum,* quoted in Robert McRae, "The Unity of the Sciences: Bacon, Descartes, and Leibniz," *Journal of the History of Ideas* 18, no. 1 (1957): 30–31. The issues were twofold: gathering more facts from nature; making sense of them. The Baconian, empirical tradition spurred fact gathering. But with the advance of years and decades, classifying facts, regardless of the system, would be relied on less and less to make sense of them. Analysis would. For Bacon on classification, see McRae, ibid., 27–48, and Sachiko Kusukawa, "Bacon's Classification of Knowledge," in *The Cambridge Companion to Bacon,* ed. Markku Peltonen (Cambridge: Cambridge University Press, 1996), 47–74. For the Baconian heritage, see Anonio Pérez-Ramos, "Bacon's Legacy," 311–334, in the same volume.

10. Galileo Galilei, *Sidereus Nuncius, or The Sidereal Messenger,* trans. Albert Van Helden (Chicago: University of Chicago Press, 1989), 1–24.

11. Galileo to Belisario Vinta (secretary of state for Florence), May 7, 1610, reproduced in Stillman Drake, *Galileo at Work: His Scientific Biography* (New York: Dover, 1978), 160.

12. Quoted in William R. Shea, *Galileo's Intellectual Revolution: Middle Period, 1610–1632* (New York: Science History, 1972), 14.

13. In September, Galileo attended a subsequent gathering on the topic, sponsored by Cosimo II. Present at this meeting were the Mantuan and Florentine cardinals Ferdinando Gonzaga and Maffeo Barberini (the future Urban VIII). Gonzaga sided with the Aristotelians while Barberini endorsed Galileo's position.

14. The details surrounding the debate may be found in Stillman Drake, "Galileo Gleanings VIII: The Origin of Galileo's Book on Floating Bodies and the Question of the Unknown Academician," *Isis* 51, no. 1 (1960): 56–63.

15. Stillman Drake and Galileo Galilei, *Cause, Experiment, and Science: A Galilean Dialogue Incorporating a New English Translation of Galileo's "Bodies That Stay atop Water or Move in It"* (Chicago: University of Chicago Press, 1981), 99, and more generally 29–34, 89–102, 129, 212. This can easily be imagined if one pictures a glass bowl, heavier than water, floating on the surface. The bowl sinks into the water, the water level rises up the side of the bowl, and inside it, below the water level, is air; the same occurs when the bowl is inverted so that the enclosed air pocket keeps it afloat. Now imagine making the bowl increasingly shallow until it is flat; the flat piece of glass (or ebony or wax or other material) still floats, but slightly below the surface, as Galileo observed. For Galileo's early use and development of the term *momento* in *On Mechanics* (*De Meccaniche*), see Drake, *Galileo at Work,* 56, 191.

16. Galileo, *Bodies That Stay,* 98–99. Galileo's drawing may be found in an early English edition of *Discourse on Floating Bodies,* trans. Thomas Salusbury (London, 1663; Project Gutenberg, 2011), 46, http://www.gutenberg.org/files/37729/37729-h/37729-h.htm.

17. Galileo Galilei, *Two New Sciences,* trans. Stillman Drake (Madison: University of Wisconsin Press, 1974), 19.

18. See Chapter 2. Also see Aristotle, *Physics,* bk. 3, 200b10–201b15, and bk. 4, 208a27–209a30.

19. Galileo, *Bodies That Stay,* 117; Shea, *Galileo's Intellectual Revolution,* 28.

20. Galileo, *Bodies That Stay,* 174–177.

21. In the early seventeenth century several roughly synonymous terms represented the smallest bits of matter: atom, particle, corpuscle. Stillman Drake translates Galileo's expression "moltitudine di corpicelli minimi" as "multitude of minute particles . . . [of] fire" and the term *ignicoli* as "fire-corpuscles." Galileo Galilei, *The Assayer,* in *The Controversy on the Comets of 1618,* trans. Stillman Drake and C. D. O'Malley (Philadelphia: University of Pennsylvania Press, 1960), 312. Shea renders *ignicoli* as "fire-atoms." Shea, *Galileo's Intellectual Revolution,* 28.

22. Van Melsen, *From Atomos to Atom,* 9–25.

23. L. E. Le Grand, "Galileo's Matter Theory," in *New Perspectives on Galileo: Papers Deriving from and Related to a Workshop on Galileo Held at Virginia Polytechnic Institute and State University, 1975,* ed. Robert E. Butts and Joseph C. Pitt (Dordrecht, Holland: D. Reidel, 1978), 197–208. See also A. Mark Smith, "Galileo's Theory of Indivisibles: Revolution or Compromise?," *Journal of the History of Ideas* 36, no. 4 (1976): 571–588.

24. Galileo, *Assayer,* 292.

25. The paradox is found in Aristotle's *Mechanics,* 855a29–856a40, a later work probably written by one of his followers, which addressed several issues of applied mathematics. Hence the author of the paradox is often referred to as Pseudo-Aristotle. Structurally, the problem of the wheel parallels the paradoxes of Zeno, especially the dichotomy and Achilles and the tortoise. All these cases depend on the presumed impossibility of making a one-to-one correspondence between an infinite set and one of its parts. See Israel E. Drabkin, "Aristotle's Wheel: Notes on the History of a Paradox," *Osiris* 9 (1950): 162–198. See too the more recent accounts of Zvi Biener, "Galileo's First New Science: The Science of Matter," *Perspectives on Science* 12, no. 3 (2004): 262–287; and Amir Alexander, *Infinitesimal: How a Dangerous Mathematical Theory Shaped the Modern World* (New York: Scientific American / Farrar, Straus and Giroux, 2014), 86–93. Alexander deftly and incisively situates the dispute over infinitesimals in the context of the struggle between the "Galileans" and the Jesuits. Alexander, *Infinitesimal,* 17–180.

26. Galileo, *Two New Sciences,* 28.

27. Aristotle, *Mechanics,* 855b30–35.

28. Galileo, *Two New Sciences,* 32–33, and more generally 28–34. Looking backward, Galileo's approach of considering ever-larger polygons as approaching a circle was reminiscent of the Greek "method of exhaustion," introduced by Antiphon the Sophist (ca. 430 BCE), made rigorous by Eudoxus, and effectively applied by Archimedes. The method involved finding the area of a shape, such as a circle, by inscribing inside it a sequence of polygons whose areas converge to (or "exhaust") the area of the containing shape. Looking forward, Galileo's approach prefigures such concepts as "asymptote" and "limit," central to the invention of analytical geometry and calculus. See Edna E. Kramer, *The Nature and Growth of Modern Mathematics* (Princeton, NJ: Princeton University Press, 1981), 173–174 and ch. 8.

29. Galileo, *Two New Sciences,* 41. Alexander, *Infinitesimal,* 80–93; Dantzig, *Number,* 172–180. It is generally conceded that Galileo anticipated the point of view of Georg Cantor, whose work launched set theory in modern mathematics, providing the technical means for resolving Zeno's paradoxes mathematically. See Kramer, *Nature and Growth,* 577–597. Cantor's contemporary Richard Dedekind (1831–1916) joined him in formulating the Dedekind-Cantor axiom, which states, "It is possible to assign to any point on a line a unique real number, and conversely, any real number can be represented in a unique manner by a point on a line." This axiom makes explicit the correlation between the realms of number and space, arithmetic (or algebra) and geometry, which many seventeenth-century mathematicians, including Galileo, employed intuitively.

30. The phrase is Dantzig's, *Number,* 63 and more generally 58–77.

31. Galileo, *Two New Sciences,* 40–41.

32. Ibid., 51.

33. Ibid., 46.

34. In our own age philosopher Max Black has challenged the Galilean mathematical resolution of Zeno's paradoxes by arguing that the expression 'infinite series of acts' is "self-contradictory." Responding to Achilles's dilemma, Black imaginatively creates an "infinity machine" to demonstrate the logical incompatibility between a "finite number of real things" Achilles must accomplish (winning the race) and the "infinite series of numbers" that describes his accomplishment. Black's argument thus returns us to the Aristotelian threshold of linguistic incommensurability between the finite and infinite. See Max Black, "Achilles and the Tortoise," in *Zeno's Paradoxes,* ed. Wesley C. Salmon (1970; Indianapolis: Hackett, 2001), 67–81. In a series of articles, also found in *Zeno's Paradoxes,* Adolf Grünbaum provides counterarguments that build on the aforementioned work of Cantor and his critical concept of "super-denumerable infinity." See Adolf Grünbaum, "Modern Science and Refutation of the Paradoxes of Zeno" (164–175), "Zeno's Metrical Paradox of Extension" (176–199), and "Modern Science and Zeno's Paradoxes of Motion" (200–250).

35. Galileo, *Two New Sciences,* 33.

36. Ibid., 19.

37. Bertrand Russell, *An Outline of Philosophy* (1927; London: Unwin Hyman, 1989), 124–125. It goes without saying that Russell's remarks apply, mutatis mutandis, to the world of subatomic particles—quarks, leptons, gluons, bosons, and the like.

38. Physicists now use the term 'singularity' to characterize, among other phenomena, the collapse of a star into itself, which creates a 'black hole' from which no light can emanate. As light (and matter) disappears into a black hole, important physical properties such the density of matter or temperature become infinitely large, along with the curvature of space time: "Matter and space time shrink to a mathematical point." Thus writes physicist Géza Szamosi, who then concludes that at such moments "not only does our visual imagination stop, not only do our words and grammar fail, but even mathematics breaks down completely." *The Twin Dimensions: Inventing Time and Space* (New York: McGraw-Hill, 1986), 179–180. Galileo took the first, halting steps along this mind-boggling pathway.

39. In this vein, modern computer abstractions are at once simpler and more robust than even those of relational numeracy. Algorithmic coding of zeros and ones allows us to process not only the information of mathematics but also of conventional language and myriad other audio and visual applications. See Michael E. Hobart and Zachary S. Schiffman, *Information Ages: Literacy, Numeracy, and the Computer Revolution* (Baltimore: Johns Hopkins University Press, 1998), 213–218.

40. The investigations compiled in this work had been virtually completed by 1609. See Stillman Drake, introduction to *Two New Sciences,* by Galileo, ix–xii.

41. Herbert Butterfield, *The Origins of Modern Science, 1300–1800,* rev. ed. (1949, 1957; New York: Free Press, 1965), 7. While scholars have challenged and amended many of Butterfield's particular claims as regards causes and consequences of the Scientific Revolution, his general assessment of the importance of the scientific thinking it sired remains for most quite unassailable.

42. Galileo Galilei, *Galileo on the World Systems: A New Abridged Translation and Guide,* ed. Maurice A. Finocchiaro (Berkeley: University of California Press, 1997), 101, 113–114. As used in ordinary logic, "extensive" and "intensive" mark different kinds of meaning attributed to general terms. "Extensive" terms (or "extension") are those whose meaning is determined by the objects to which the term may be applied, often called a "denotative" (or "synthetic" or "a posteriori") definition. "Intensive" terms (or "intension") are those whose meaning pertains to common properties possessed by all the objects falling within it, frequently called a "connotative" (or "analytic" or "a priori") definition. Thus the extensive definition of 'cow' would be all those animals one could point to as cows (Angus, Hereford, Holstein, crossbreeds, and so on), whereas the intensive definition would include all the properties of a cow (for example, ruminant, cloven hoof, and the

like). See Irving M. Copi, *Introduction to Logic* (New York: Macmillan, 1953), 107–112. For Galileo, "extensive" referred to the "intelligible things" that belonged to the terms 'human' and 'divine' as applied to their 'understanding'; thus between 'human' and 'divine' there could be no comparison extensively. By contrast, "intensive" for him meant the logical, analytical connections between intelligible "things," such as the objects and relations of mathematics, and here we could have the same intensive certainty and perfection in our reasoning as the divine itself.

11. Demonstrations and Narrations

1. If Vincenzo had bequeathed to his son an antiauthority attitude, it's probably fair to say that Galileo inherited certain "bullying and devious" tendencies from the formidable Giulia. Much remains speculative, for we know little of her and her influence on Galileo's formative years, certainly not enough for a book on "Galileo's Mother." Still, during the Padua years, from her home in Florence she encouraged Galileo's servants to spy on him; tried unsuccessfully to wrest money from him by stealing, then selling the lenses for his telescopes; meddled in his relationship with his mistress, Marina Gamba, and the rearing of their children; and was even reported as having given evidence against him in an earlier brush with the Inquisition, swearing not only that did Galileo not attend church (although he did not speak against religion) but also that in response to her deposition he had sworn at her and called her "a whore and an ugly old cow." David Wootton, *Galileo: Watcher of the Skies* (New Haven, CT: Yale University Press, 2010), 93–95. One can only imagine Galileo having been inured to conflict from an early age, and from this infer that there may well have been limits on his ability to be circumspect. He clearly liked a good fight, if not always choosing his battles wisely.

2. A recent summary of Galileo's youth, family, and livelihood may be found in J. L. Heilbron's biography *Galileo* (Oxford: Oxford University Press, 2010), 1–27, 83–88, and throughout, along with this summary judgment: "Galileo did not shirk the financial, only the emotional, responsibility of maintaining his children" (164). See, too, Wootton, *Galileo,* 9–13, 67–69, 93–95, and throughout. Both these fine, well-written works incorporate the most recent scholarship surrounding the life and career of Galileo and nit-picking aside, reward their readers with full and interesting biographies.

3. For contemporary reactions to the censure and trial, see Maurice A. Finocchiaro, *Retrying Galileo, 1633–1992* (Berkeley: University of California Press, 2005), 1–42, and throughout. For the aftermath of these events and issues pertaining to the creation and debunking of the Galilean myth, see Michael Segre, *In the Wake of Galileo* (New Brunswick, NJ: Rutgers University Press, 1991).

4. One such recent attempt has been that of Pietro Redondi, *Galileo, Heretic,* trans. Raymond Rosenthal (Princeton, NJ: Princeton University Press, 1987).

Redondi argues that Galileo was condemned for heresy not for his Copernicanism but because of his atomism and its subversive implications for the doctrine of the Eucharist. Redondi's interpretation has been largely discredited by critical challenges to his circumstantial and speculative inferences and by more-careful analysis of his documentary evidence. See Vincenzo Ferrone and Massimo Firpo, "From Inquisitors to Microhistorians: A Critique of Pietro Redondi's *Galileo eretico*," *Journal of Modern History* 58 (1986): 485–524. Having reexamined the same documents Redondi cited, William R. Shea and Mariano Artigas, in *Galileo Observed: Science and the Politics of Belief* (Sagamore Beach, MA: Science History, 2006), 179, present a reliable postmortem to Redondi's account: "Other documents may one day surface and enlighten us on the circumstances that led to Galileo's trial. We do not believe, however, that the facts about the Galileo Affair will be undermined. Galileo was condemned for failing to comply with a formal order not to teach that the earth moves." And finally, commenting on the 1998 opening to all scholars of the archive of the Congregation for the Doctrine of the Faith (including archives of the Holy Office and of the Congregation of the Index), Francesco Beretta concludes: "The systematic review of the records conducted by the team led by Ugo Baldini [does] not suggest that any revolutionary new discoveries will be forthcoming." See Francesco Beretta, "The Documents of Galileo's Trial," in *The Church and Galileo,* ed. Ernan McMullin (Notre Dame, IN: University of Notre Dame Press, 2005), 204 and more generally 191–212.

5. "Suspicion of heresy" was a technical crime in canon law, distinguished from "formal heresy," the more severe offense against the faith. "Formal heresy" referred to willful rejection of key Church doctrines, such as "decrees of Sacred Councils," "Holy Scripture," or articles of "Holy Faith" (for example, denial of the Trinity or the divinity of Christ), whereas "suspicion of heresy" denoted acts and words such as reading forbidden books, or professing religious orders while married, or uttering words that "occasionally . . . offend" listeners. Further, within "suspicion," two additional subcategories, "vehement" and "slight," were separated, "vehement" being the more serious for exhibiting a greater degree of willful intent. See "Sentence (22 June 1633)," in *The Galileo Affair: A Documentary History,* ed. Maurice A. Finocchiaro (Berkeley: University of California Press, 1989), 287–291, and Finocchiaro's introduction to the volume, 14–15. For an excellent and thoroughly detailed account of the canonical judicial system pertaining to heresy and Galileo's trial, see Jules Speller, *Galileo's Inquisition Trial Revisited* (Frankfurt am Main, Germany: Peter Lang, 2008), 19–50 and throughout.

6. The more generally known, modern title, *Dialogue on the Two Chief World Systems,* dates from 1744, when permission was granted to reprint the book with an explanatory preface added by a Catholic theologian, Giuseppe Toaldo of Padua. Finocchiaro, *Retrying Galileo,* 126–138. In correspondence Galileo referred to his manuscript in progress as "Dialogue on the Tides." Stillman Drake, in *Galileo:*

Pioneer Scientist (Toronto: University of Toronto Press, 1990), 186–187, 192–206, argues that he never intended the book to be merely a comparison of the two rival world systems (Copernican and Ptolemaic-Aristotelian); rather, it was to be an analysis of which system better explained the tides *ex suppositione* (hypothetically).

7. Richard J. Blackwell, *Galileo, Bellarmine, and the Bible* (Notre Dame, IN: University of Notre Dame Press, 1991), 131, 177.

8. "Decree of the Index (5 March 1616)," in Finocchiaro, *Galileo Affair,* 148–149. See, too, "Inquisition Minutes (25 February 1616)," in ibid., 147.

9. "Galileo's Letter to the Grand Duchess Christina (1615)," in Finocchiaro, *Galileo Affair,* 96. Galileo attributed the epigram to Cardinal Baronius (Cesare Baronio, 1538–1607), Oratorian, ecclesiastical historian, and Vatican librarian; its full text stands at the beginning of this chapter.

10. "Galileo to Castelli (21 December 1613)," in Finocchiaro, *Galileo Affair,* 52; "Galileo's Letter to the Grand Duchess," 93, 96. Later, in a letter dated January 15, 1633, and addressed to Elia Diodati (1576–1661), a Swiss lawyer living in Paris and a Galileo supporter, Galileo wrote, "Thus the world is the work and the Scriptures the word of the same God." Quoted in Olaf Pedersen, "Galileo's Religion," in *The Galileo Affair: A Meeting of Faith and Science,* ed. George V. Coyne et al. (Vatican City: Vatican Observatory, 1985), 90.

11. "Galileo's Letter to the Grand Duchess," 92–94, 102, 104–106, 113. As a "protection against error," Galileo noted, one should start from the "point of natural and mathematical arguments," with interpretations of scripture to follow. See "Galileo's Unpublished Notes (1615)," in Blackwell, *Galileo, Bellarmine,* 273, and more generally 269–276.

12. Jean Dietz Moss, *Novelties in the Heavens: Rhetoric and Science in the Copernican Controversy* (Chicago: University of Chicago Press, 1993), 241.

13. The question of Galileo's actual religious beliefs is quite complicated, turning on the extent to which one takes his many comments at face value. Although not a regular churchgoer, he always claimed to be a good Catholic and was mostly careful, cautious, and circumspect about religion. But he lived in a Counter-Reformation Catholic culture and world in which Inquisition watchdogs were widespread, and speaking otherwise would have prompted serious consequences. Even Ignatius Loyola (1491–1556), founder of the Jesuits, had been briefly imprisoned by the Inquisition. This meant likely paying much lip service to authorities and their official positions. Dissimulation was clearly part of Galileo's character, as was using various "masks" to shield his true beliefs and intentions. As his friend Paolo Sarpi put it, "I am compelled to wear a mask; perhaps there is no one who can survive in Italy without one." Quoted in Wootton, *Galileo,* 249. Further, Galileo's letters reveal little in the way of deeper emotions or religious commitments. All these various features have led to his being termed everything from a favorite son of the Church, to a "sincere believer" (according to Pope John Paul II), to a "run-of-the-mill"

Catholic, to a deist. In "Galileo's Religion,"Pedersen argues that Galileo was a pious Christian. Wootton's recent biography builds the most believable case for "Galileo's (un)belief," his not untypical blend of public belief and private skepticism, and concludes that "deist" fits as accurately as any other "label." Wootton, *Galileo,* 240–250.

14. "Galileo to Prince Leopold of Toscana," 1640, reproduced in Paul Feyerabend, *Against Method: Outline of an Anarchistic Theory of Knowledge* (1975; London: Verso, 1978), 69–70.

15. In *Galileo and the Art of Reasoning: Rhetorical Foundations of Logic and Scientific Method* (Dordrecht, Holland: D. Reidel, 1980), 3, Maurice A. Finocchiaro writes that Galileo sought "to persuade the appropriate Church officials of the truth of Copernicanism and thereby . . . repeal . . . the condemnation of 1616." With the *Dialogue,* Galileo's intentions were undoubtedly quite complicated, and his audience quite diverse. See Moss, *Novelties in the Heavens,* 278–300 and throughout.

16. Blackwell narrates a similar story of "how science and religion encountered each other as each emerged from the sixteenth century in its recognizably modern garb . . . a significant part of a large-scale turning point in Western culture." Focusing on theology and biblical exegesis, however, Blackwell omits entirely the changes in technology, especially those underlying the emergence of modern numeracy, and with it modern science and its parting from religion. See his *Galileo, Bellarmine,* 2 and throughout.

17. "Galileo to Castelli (21 December 1613)" in Finocchiaro, *Galileo Affair,* 49–54; "Galileo's Letter to the Grand Duchess Christina (1615)," in ibid., 87–118.

18. In a letter to Johannes Kepler in 1597, Galileo said he had "embraced" the "hypothesis" of Copernicus "already many years ago." Quoted in Annibale Fantoli, *The Case of Galileo: A Closed Question?,* trans. George V. Coyne (Notre Dame, IN: University of Notre Dame Press, 2012), 20. Only four people published anything in support of Copernicus within the first fifty years after publication of *De Revolutionibus* (1543). See Wootton, *Galileo,* 53. Between then and the book's censure by the Catholic Church in 1616, no more than a handful of astronomers actually identified themselves as followers of Copernicus, some quite cautiously. Professor Owen Gingerich began his recent search for extant first- and second-edition copies of *De Revolutionibus* (estimated to be between 900 and 1,050 total) by producing a list of prospective readers during the period who might have owned a copy. There were only nine or ten. Eventually, of course, there would be more, and by all accounts the book had an enormous impact, notwithstanding the paucity of early readership. See Owen Gingerich, *The Book Nobody Read: Chasing the Revolutions of Nicolaus Copernicus* (New York: Walker, 2004), 22 and throughout. Still, before Galileo's discoveries with the telescope and the censure a few years later, there was only a tiny amount of interest and controversy. The censorship

pushed Copernicus and his book, Galileo and his telescope, and a host of related matters dramatically into public awareness; Galileo's heresy trial even more so.

19. Galileo initially hoped his letter to Castelli would help forestall any adverse judgments from Church officials regarding Copernicanism. It was quite widely distributed and was seriously considered by Bellarmine and the authorities of the Holy Office and the Congregation of the Index in Rome, but to little of the effect Galileo had desired. In the letter to Christina he had expanded the themes of the separation of science and religion expressed in the Castelli letter. And he had clearly advocated Copernicanism as a "real" system, not just a mathematical "hypothesis." But Bellarmine's letter to Paolo Antonio Foscarini, which Galileo saw, raised a red flag regarding this position, leading Galileo and his friends to withhold the Christina letter from circulation. Behind-the-scene exchanges and actions by participants in both the Galileo and Church camps are clearly documented and trenchantly explained by Fantoli, *Case of Galileo,* 93 and, more generally, 33–120, and in his longer work, Annibale Fantoli, *Galileo: For Copernicanism and for the Church,* 2nd ed. (Notre Dame, IN: University of Notre Dame Press, 1996), 208, 253, and in general, 169–270. See "Castelli to Galileo (14 December 1613)," in Finocchiaro, *Galileo Affair,* 47–48, and "Galileo to Castelli (21 December 1613)," along with related correspondence from the period. See also Finocchiaro's introduction to *Galileo Affair,* 27–32; Moss, *Novelties in the Heavens,* 190–195; and Ernan McMullin, "Galileo on Science and Scripture," in *The Cambridge Companion to Galileo,* ed. Peter Machamer (Cambridge: Cambridge University Press, 1998), 271–347.

20. Throughout the twentieth century, Galileo's reliance on this terminology led philosophically minded scholars to debate the place of "reason and experience" in his "scientific method" as a form of Platonic, mathematical idealism versus an Aristotelian "empiricism." Scholars in both camps easily supplied examples from his works to support to their interpretations. But today such a framework seems retrograde, failing to capture the innovative, analytical dimensions of his scientific work. Despite such traditional philosophical resonances, Galileo was neither an idealist in the Platonic mold nor an empiricist in the lineage of Aristotle. Both cautious and confident, he crossed the threshold into analysis with his science. Similarly, his reflections on method reveal a comparable mix of hesitancy and self-assuredness. But within the context of traditional concepts, Galileo transformed them sufficiently so as to create a new approach to the investigation of natural phenomena. For a discussion of the older debates, see Dudley Shapere, *Galileo: A Philosophical Study* (Chicago: University of Chicago Press, 1974), 3–21, 126–145. For Galileo's Platonism, see Alexandre Koyré, "Galileo and Plato," in *Metaphysics and Measurement: Essays in Scientific Revolution* (Cambridge, MA: Harvard University Press, 1968), 16–43. For Galileo's Aristotelianism, see William A. Wallace, *The Modeling of Nature: Philosophy of Science and Philosophy of*

Nature in Synthesis (Washington, DC: Catholic University of America Press, 1996), 262–275 and throughout, and William A. Wallace, "Galileo's Pisan Studies in Science and Philosophy," in Machamer, *Cambridge Companion to Galileo,* 27–52. While discussing both Galileo's Platonism and his Aristotelian empiricism, and aligning him more with Plato, William Shea does note that Galileo did not really fit either category, even though he exhibited features of both. William R. Shea, *Galileo's Intellectual Revolution: Middle Period, 1610–1632* (New York: Science History, 1972), 150–163.

21. *Galileo on the World Systems: A New Abridged Translation and* Guide, trans. and ed. Maurice A. Finocchiaro (Berkeley: University of California Press, 1997), 97. Drake translates this passage as "by making use of analytical methods." *Galileo: Dialogue concerning the Two Chief World Systems,* rev. ed., ed. and trans. Stillman Drake (1953; Berkeley: University of California Press, 1967), 51.

22. Paolo Antonio Foscarini, "A Letter to Fr. Sebastiano Fantone, General of the Order [of Carmelites] concerning the Opinion of the Pythagoreans and Copernicus . . . ," reproduced in Blackwell, *Galileo, Bellarmine,* 217–251. See too Blackwell's account of Foscarini's interpretations, Blackwell, *Galileo, Bellarmine,* 87–110.

23. In 1615 Foscarini, perhaps in common strategy with Galileo, was trying to persuade Church officials that a heliocentric cosmology should be accepted as compatible with scripture. But in March 1616, along with the Church's censure of key Copernican ideas, his own tract was also placed on the Index of Forbidden Books.

24. A "scholastic theologian" and Jesuit, Bellarmine's "existence had been shaped by the struggle against heresy." Richard S. Westfall, *Essays on the Trial of Galileo* (Vatican City: Vatican Observatory, 1989), 4, 15–16, and more generally 1–30. Bellarmine held a "Biblical cosmology" that was quite non-Aristotelian but relied heavily on a literal reading of scripture, which included the earth's immobility. This central "fact" was the basis of his *demonstratio quia* (for the term, see Chapter 1, note 56) from which other deductions followed in syllogistic rigor. Their rationale stemmed directly from Aristotle's *Prior* and *Posterior Analytics,* still then the standard account of scientific and theological explanation, and the backbone of Jesuit training. Blackwell, *Galileo, Bellarmine,* 40–45. Ever since the earliest announcement of Galileo's telescopic findings, Bellarmine had also been quite concerned about their compatibility with scripture and soon after had asked Clavius and other Jesuit mathematicians at the Collegio Romano to inform him as to whether the new results were "well founded." By 1615, Federico Cesi was cautioning Galileo that "Bellarmine . . . considers it [the Copernican theory] heretical and . . . contrary to Scripture." "Cesi to Galileo, 12 January 1615," quoted in Westfall, *Essays,* 13. Also see Fantoli, *Case of Galileo,* 54–57, and George V. Coyne and Ugo Baldini, "The Young Bellarmine's Thoughts on World Systems," in Coyne et al., *Galileo Affair,* 103–109.

25. "Cardinal Bellarmine to Foscarini (12 April 1615)," in Finocchiaro, *Galileo Affair,* 67–69.

26. The Council of Trent had decreed that "in matters of faith and morals," the Church's interpretations would prevail over those of any individual. "Decrees of the Council of Trent Session IV (8 April 1546)," in Blackwell, *Galileo, Bellarmine,* 183. Bellarmine took "matters of faith" as entailing certain statements of fact. His hermeneutical position regarding biblical interpretation has been described as based on the "limitation principle," which states that certain factual claims found in scripture, including a geocentric cosmology, are necessary for salvation. This stood in opposition to Galileo's "independence principle," which holds that factual statements about natural phenomena, even when present in the Bible, have nothing to do with salvation. For modern arguments defending Bellarmine's stance, see Marcello Pera, "The God of Theologians and the God of Astronomers: An Apology of Bellarmine," in Machamer, *Cambridge Companion to Galileo,* 367–387.

27. Fantoli notes that Bellarmine's life in polemics with Protestant theologians led to a "progressive hardening of his positions" and an "instinctive suspicion" concerning any potential threat to Catholicism. Thus "the possibility of a future proof of Copernicanism [was] for all practical purposes excluded a priori." Fantoli, *Case of Galileo,* 81–83. The evidence from Galileo's telescope served mainly to disprove the Aristotelian-Ptolemaic picture of the cosmos but fell short of an absolute proof of the Copernican. Galileo refused to consider the alternative put forth by Tycho Brahe (1546–1601), that the earth remained at the center of the universe, that the moon and sun orbited the earth, and that the rest of the planets orbited the sun. For Tycho, his system had advantages of both the Ptolemaic and Copernican versions, but for Galileo it had no physical basis. By not considering the Tychonic option, Galileo presented his readers with a disjunction: either Aristotle-Ptolemy or Copernicus. This was a rhetorical strategy, as well as a scientific claim. With the disjunction, logically disproving the former meant confirming the latter. Many Jesuits favored the Tychonic alternative, and quite reasonably refused to accept the disjunction as presented by Galileo. "Galileo's Considerations on the Copernican Opinion (1615)," in Finocchiaro, *Galileo Affair,* 74–75.

28. Drake, *Galileo: Dialogue,* 50–51.

29. Finocchiaro, *Galileo on the World Systems,* 128. We are reminded of another well-known comment from Einstein, that "as far as the laws of mathematics refer to reality, they are not certain; and as far as they are certain, they do not refer to reality." Albert Einstein, "Geometry and Experience," lecture before the Prussian Academy of Sciences, January 27, 1921, in Albert Einstein, *Ideas and Opinions,* trans. Sonja Bargmann (1954; New York: Three Rivers Press, 1982), 233.

30. "Galileo's Letter to the Grand Duchess," 106.

31. As with the term 'analysis,' I am using 'functional dependency' here in an intuitive, historical, and general manner. Eventually this intellectual process

would be captured technically in the concept of a "function" articulated entirely in mathematical symbols. In calculus, for instance, the standard way of symbolizing the function of a variable, generally x, is $f(x)$, expressed rhetorically as the "function of x" or simply the "f of x." In the equation $y=f(x)$, a change in x produces a change in y, and if x is known, y can be determined. Thus x is known as the 'independent variable' and y as the 'dependent variable.'

32. Later, in *Two New Sciences,* trans. Stillman Drake (Madison: University of Wisconsin Press, 1974), 225, Galileo wrote, "No firm science can be given of such events of heaviness, speed, and shape, which are variable in infinitely many ways. Hence to deal with such matters scientifically, it is necessary to abstract from them."

33. Drake, *Galileo: Dialogue,* 207–208. Joseph C. Pitt makes the similar point that for Galileo there is a "one to one correspondence between the structures of geometry and the world." See Joseph C. Pitt, "Galileo: Causation and the Use of Geometry," in *New Perspectives on Galileo: Papers Deriving from and Related to a Workshop on Galileo Held at Virginia Polytechnic Institute and State University, 1975,* ed. Robert E. Butts and Joseph C. Pitt (Dordrecht, Holland: D. Reidel, 1978), 190–191 and more generally 181–195.

34. Edna E. Kramer, *Nature and Growth of Mathematics* (Princeton, NJ: Princeton University Press, 1981), 134 and more generally 134–166.

35. Modern mathematicians have devised operationally formal and technical definitions of functional dependency appropriate to their needs. For instance, a contemporary website devoted to mathematics defines 'function' as "a *relation* that uniquely associates members of one *set* with members of another *set.*" Christopher Stover and Eric W. Weisstein, "Function," MathWorld, last updated September 20, 2017, http://mathworld.wolfram.com/Function.html. Another site defines it as "a *relation* from a set of inputs to a set of possible outputs where each input is related to exactly one output." Duane Q. Nykamp, "Function Definition," Math Insight, http://mathinsight.org/definition/function, accessed November 2, 2017. The key intuitive components in these definitions are the cardinal correspondence between members of sets or between inputs and outputs, and the ordinal recurrence of counting, features that have pervaded mathematics from its earliest, symbolic expression until now.

36. Recognizing the functional dependency in Galileo's mathematical physics allows us to understand what Galileo perceived only hazily but couldn't explain about his method, and to address a puzzle posed by Galilean scholarship. On the one hand, he sought to separate himself from the traditional, definition-based approach to natural phenomena of the Aristotelians, substituting for it the direct investigation of nature in the language of mathematics. Yet, on the other hand, he wanted to be a natural philosopher and not just a mathematician, as he insisted when he moved to the Medici court. This meant addressing problems in the "real"

world, and thus searching for causalities in some fashion or another. With the feature of functional dependency becoming apparent in his abstract mathematics, Galileo could ease away from the essentialist language of definitions employed by Aristotelians and still investigate how different features of phenomena related to others—or caused them—in the real world.

37. As suggested earlier, some scholars following Wallace continue to claim that Galileo's "method" was derived predominantly from Aristotle's *Prior* and *Posterior Analytics,* citing not only Galileo's early notebooks and formative training but also the continued use of Aristotelian terminology throughout his mature scientific work. See William A. Wallace, "Galileo's Concept of Science: Recent Manuscript Evidence," in Coyne et al., *Galileo Affair,* 15–40. To repeat, this interpretation overlooks entirely the understanding that Galileo derived from the new technology and its relation-mathematics, an understanding that led, for instance, to his acceptance of early "infinitesimals" in analysis. See Amir Alexander, *Infinitesimal: How a Dangerous Mathematical Theory Shaped the Modern World* (New York: Scientific American / Farrar, Straus and Giroux, 2014), 80–93. These were quantities or factors Aristotle had defined out of existence by separating geometry and arithmetic. Mostly though, Galileo's awareness of what we are terming functional dependency signaled an enormous departure from the Aristotelian definition cage in the approach to correlating mathematics and physical phenomena. See Drake's nuanced treatment of just how, probably around 1600 but certainly in the Padua years, "Galileo turned his back on philosophy in favor of the direct investigation of nature." Of course Galileo used Aristotelian terminology; it was science's lingua franca of the times. But he moved steadily away from it, and by the end of his last book, in 1638, was ridiculing philosophy's "causal enquiries." Stillman Drake, "Galileo and the Career of Philosophy," in *Essays on Galileo and the History and Philosophy of Science,* 3 vols., sel. N. M. Swerdlow and T. H. Levere (Toronto: University of Toronto Press, 1999), 1:257–272.

38. Galileo variously called these "rigorous" or "palpable" or "natural" demonstrations, or "manifestly demonstrated conclusions." See Shea, *Galileo's Intellectual Revolution,* 54; Galileo Galilei and Christoph Scheiner, *On Sunspots,* ed. and trans. Eileen Reeves and Albert Van Helden (Chicago: University of Chicago Press, 2010), 89, 256; and "Galileo's Unpublished Notes," 274.

39. Drake, *Galileo: Dialogue,* 55; Galileo and Scheiner, *On Sunspots,* 281–282. In the letter to Christina, Galileo wrote, "I feel I have conclusively demonstrated in my *Sunspot Letters*" that the sun completes "an entire rotation in about one month." "Galileo's Letter to the Grand Duchess," 116.

40. This was a tiresome quarrel, for sunspots had been seen by the naked eye in antiquity; the earliest telescopic observation was made by Thomas Harriot in December of 1610, although Galileo may have seen them a bit before then. The earliest printed reference to them occurred in the summer of 1611. The

only interesting matter of priority concern lay with who figured them out. That kudos went to Galileo. See Eileen Reeves and Albert Van Helden, introduction to *On Sunspots,* by Galilei and Scheiner, 1–6.

41. For a good summary of the dispute, see Shea, *Galileo's Intellectual Revolution,* 49–74.

42. In 1612 Galileo first began attaching a grid of lines to a panel fixed to the side of his telescope, by means of which he could plot more accurately the moons of Jupiter and other celestial phenomena. He then used his 'plot map' to predict their positions. This was a crude micrometer, a forerunner of ones nowadays attached to the inside of telescopes. See Wootton, *Galileo,* 129.

43. Galileo and Scheiner, *On Sunspots,* 111.

44. Diagram source: Galileo Galilei and Marcus Welser, *Istoria e dimostrazioni intorno alle macchie solari e loro accidenti comprese in tre lettere scritte all'illvstrissimo signor Marco Velseri,* ed. Angelo De Filiis (Rome: G. Mascadi, 1613), 127, Lessing J. Rosenwald Collection, Library of Congress, https://lccn.loc.gov/65059245.

45. Galileo and Scheiner, "Galileo's Third Letter," in *On Sunspots,* 276–278. See Reeves and Van Helden, "Appendix 5," in Galileo and Scheiner, *On Sunspots,* 366–369, for a modern mathematical rendition of Galileo's calculations, which shows the ratio between the spot on the edge and the one in the middle as 1:5.95, very close to Galileo's 1:6. And even though Galileo erred at one point in comparing the two illustrations, somewhat uncharacteristically (for him) "looking up the wrong cosine," his overall calculations proved conclusively that the spot rested on the surface.

46. "Galileo's Letter to the Grand Duchess," 101.

47. A committed Aristotelian, Scheiner continued to advance fierce polemics against Galileo, but even he eventually accepted Galileo's analysis, and its implication that Aristotle's theory of the incorruptible heavens was incorrect. Fantoli, *Case of Galileo,* 143–145.

48. Post-Copernican astronomers, including Tycho, Kepler, and Galileo, all looked for stellar parallax to prove the earth's orbit of the sun. Stellar parallax is the change in the angle of an observer's viewpoint relative to a point of reference, in this case a distant star. But their instruments were not refined enough to measure the angles by the arc seconds needed. Such a direct, observable proof would not be forthcoming until a German astronomer, Friedrich Bessel, made the first successful measure of stellar parallax in 1838. For Galileo on parallax, see Drake, *Galileo: Dialogue,* 372–373, 383–387.

49. "Galileo's Letter to the Grand Duchess," 104–105; "Galileo to Dini (May 1615)," quoted in Fantoli, *Case of Galileo,* 83–84.

50. See Westfall, *Essays,* 21–23. The tag of "formally heretical" came from the "Consultant's Report on Copernicanism (24 February 1616)," but it was not in-

cluded in the decree itself, which stated only that the doctrine is "altogether contrary to the Holy Scripture." The wording's ambiguity created problems for the Holy Office later on during the trial. See the "Report" and the "Decree," in Finocchiaro, *Galileo Affair,* 146–147 and 148–149 respectively. Cardinal Maffeo Barberini (1568–1644), who would become Pope Urban VIII, likely had a major hand in keeping the Congregation of the Index from declaring the doctrine heretical. See Speller, *Galileo's Inquisition Trial Revisited,* 60–61, 71, 77–79.

51. In the letter to Foscarini, Bellarmine had written that in scripture, "the Holy Spirit [speaks] through the mouth of the prophets and the apostles." Therefore, if not "as regards the topic," then "as regards the speaker," the earth's immobility is certainly a matter of faith and as such "it would be heretical" to deny it. "Cardinal Bellarmine to Foscarini (12 April 1615)," in Finocchiaro, *Galileo Affair,* 68.

52. "Galileo to Dini (16 February 1615)," in Blackwell, *Galileo, Bellarmine,* 204.

53. Thus, Galileo wrote, science "tends to subvert all natural philosophy." Drake, *Galileo: Dialogue,* 37. Still, we should remember that at the time there remained serious objections to Galileo's science per se.

54. The corrections were to include statements to the effect that Copernicus's work was done suppositionally, only as a mathematical device to save the phenomena, not as a claim about the real constitution of the cosmos. Copernicus had clearly claimed otherwise, as Galileo remarked in "Galileo's Considerations on the Copernican Opinion," 70–71.

55. "Cardinal Bellarmine's Certificate (26 May 1616)," in Finocchiaro, *Galileo Affair,* 153.

56. Minutes of an earlier meeting had resulted in a "special injunction" issued by Bellarmine. Although Galileo never signed them, they did record that he was "warned" not to "hold, teach, or defend *in any way whatever,* either orally or in writing," the "erroneous opinion" (italics added). Were he to do so, the "Holy Office would start proceedings against him." Galileo "acquiesced" and "promised to obey." "Special Injunction (26 February 1616)," in Finocchiaro, *Galileo Affair,* 146–148. As indicated by the italicized words, the discrepancy between the wording of these minutes and Bellarmine's certificate, where they were absent, would become critical at the trial.

57. Galileo's career as a practicing scientist basically had ended with his book on sunspots. By 1612, "every important discovery that he was going to make he had now made," writes Wootton in *Galileo,* 134. Henceforth his attention would be governed by gathering together his materials at hand, refining and presenting them to the public. In so doing, he often proceeded indirectly, engaging in issues tangential to Copernicanism per se (for example, the dispute over comets), or by using a surrogate (Guiducci) to express some of his opinions, or by speaking more generally about the methods of natural philosophy rather than its specific content (*The Assayer*).

58. For elaboration of this theme, see Drake's introduction to *The Controversy on the Comets of 1618,* by Galileo Galilei et al., trans. Stillman Drake and C. D. O'Malley (Philadelphia: University of Pennsylvania Press, 1960), vii–xxv.

59. "Galileo to Castelli (21 December 1613)," 50. In "Galileo's Letter to the Grand Duchess," 93, Galileo expanded the thought: "Nature is inexorable and immutable, never violates the terms of the laws imposed upon her, and does not care whether or not her recondite reasons and ways of operating are disclosed to human understanding."

60. Drake, *Galileo: Dialogue,* 53.

61. See Chapter 2 and Drake, *Galileo: Dialogue,* 367.

62. Drake, *Galileo: Dialogue,* 103, 237, 367–368. Finocchiaro interprets this last passage as Galileo's criticism of "teleological anthropocentrism." Finocchiaro, *Galileo on the World Systems,* 259.

63. Drake, *Galileo: Dialogue,* 367. Again, the expression was, plausibly, ironic; it's hard to see how Galileo could have said otherwise given the Counter-Reformation Catholic culture that surrounded him. This is among the phrases that make us wonder about his "lip-service" in the name of belief. See note 13 on Galileo's religion. The Holy Office clearly looked askance at his lip-service denials of being a Copernican, given the overall weight and tenor of his rhetoric throughout the *Dialogue.* As Galileo intended, reading his "play" left open many questions and avenues of interpretation.

64. "Galileo's Letter to the Grand Duchess," 103; Drake, *Galileo: Dialogue,* 5–7, 38. Galileo would well have appreciated the later incisive remark of novelist George Eliot that "the very breath of science is a contest with mistake." George Eliot, *Middlemarch* (1872; New York: Barnes and Noble Classics, 2003), 703.

65. "The Fourth Day," in Drake, *Galileo: Dialogue,* 418. The third motion of the system was the now-termed "precession of the equinoxes," described in Chapter 8. See, too, Appendix A.8.

66. "The Fourth Day," in Drake, *Galileo: Dialogue,* 416–465; for the quotes, see 418, 424–425.

67. Philosophically, this is known as the problem of "induction," whose modern formulation dates from David Hume in the eighteenth century. One can describe it in the context of modern, propositional logic, where inference depends on the direction of argument, from antecedent to consequent. The standard "modus ponens" inference reads, "If P, then Q. | P, therefore Q." If (P) 'Copernican theory is true,' then (Q) 'The earth's mobility explains tides' is also true. But the argument doesn't work the other way. "If P, then Q. | Q, therefore P" is not a valid logical inference but rather a fallacy, sometimes called "affirming the consequent"—that is, (Q) 'The earth's mobility explains tides' does not imply (P) 'Copernican theory is true.' (By contrast, in "modus tollens" inferences, negative consequents logically *disprove* an antecedent: "If P, then Q. | Not Q, therefore not P.") These considerations have led to fleshing out the modus ponens inferences with background and

supplementary claims, the accumulation of which can corroborate the antecedent proposition, though never prove it logically. Thus, if a preponderance of inferences drawn from the assumption of heliocentricity led to the discovery of numerous local truths, then the probability of the truth of Copernicanism would increase, although never to the level of logical certainty. This was Galileo's goal in providing numerous local demonstrations of the Copernican hypothesis.

68. "Galileo's Letter to the Grand Duchess," 100.

69. In the context of courtly politesse and exalted expression, Barberini often signed his letters to Galileo simply with *come fratello* (like a brother); while not fast friends, their relationship generally exceeded mere cordiality.

70. "Francesco Niccolini to Andrea Cioli (Tuscan Secretary of State)—5 September 1632," in Finocchiaro, *Galileo Affair,* 229.

71. Finocchiaro, *Galileo on the World Systems,* 307. Published on April 1, 1631, *Inscrutabilis* was primarily directed against astrologers for arrogating to themselves a divine-like status in predicting events, including on several occasions the imminent demise of Urban himself. The words, however, and the sentiments underlying them reveal the fullness of papal power behind the theme of the limitations of human knowledge: "The inscrutable height of the judgments of God does not suffer that the human intellect, enchained by the dark prison of the body, raising itself above the stars, should with abominable curiosity presume not only to explore the arcana hidden in the divine mind and unknown to the most blessed spirits themselves, but also, by an arrogant and dangerous example, to peddle them as [already] explored in contempt of God, to the disturbance of the commonwealth and danger of princes." Interpreting divinity lay within the collective wisdom of the Church, the living body of Christ on earth. Quoted in Michael H. Shank, "Setting the Stage: Galileo in Tuscany, the Veneto, and Rome," in McMullin, *Church and Galileo,* 77.

72. "Niccolini to Cioli—13 March 1633," in Finocchiaro, *Galileo Affair,* 247.

73. "Special Commission Report on the *Dialogue* (September 1632)," in Finocchiaro, *Galileo Affair,* 221; Finocchiaro, *Galileo on the World Systems,* 307.

74. The theological argument is more complicated than the foregoing suggests. Urban's skepticism was basically twofold: Copernicanism could never be directly proved because one cannot get far enough away to perceive the earth in motion; therefore, the Copernican view could only be proved indirectly with chains of human reason, and these were necessarily fallible. In contrast to frail and fallible human reason, divine omniscience and omnipotence must carry the day. See Speller, *Galileo's Inquisition Trial Revisited,* 375–396; Ernan McMullin, "The Church's Ban on Copernicanism, 1616," in McMullin, *Church and Galileo,* 150–190; and Finocchiaro, *Galileo on the World Systems,* 42, 77–82, 112–116, 306–308. In his letter to Christina, Galileo laid out the counterargument: "I do not think one has to believe that the same God who has given us senses, language, and intellect would want to set aside the use of these and give us by other means the information we can acquire with them." "Galileo's Letter to the Grand Duchess," 94.

75. Some of these descriptions came from the "Special Commission Report," which confirmed that Galileo "spoke absolutely" in holding and defending the Copernican theory. For a detailed account of the commission and its findings, see Speller, *Galileo's Inquisition Trial Revisited,* 152–201, 392.

76. Scholars continue to debate the details of the behind-the-scenes maneuvering between Galileo and the Inquisition. The main documents are reproduced in Finocchiaro's collection, *Galileo Affair.* I have followed Speller's account, *Galileo's Inquisition Trial Revisited,* which is the most detailed reconstruction available. Other worthwhile interpretations include Fantoli, *Case of Galileo* and *Galileo;* Shea and Artigas, *Galileo Observed;* and Blackwell, *Galileo, Bellarmine.* See too Richard J. Blackwell's fine *Behind the Scenes at Galileo's Trial: Including the First English Translation of Melchior Inchofer's "Tractatus syllepticus"* (Notre Dame, IN: University of Notre Dame Press, 2006).

77. As Drake has written, the popular version of Galileo's utterance at the time of his adjuration was clearly preposterous; the consequences would have been dire, as he well knew. Still, a modicum of plausibility perhaps lies behind the story, with the utterance occurring at a different time and place. In 1911 a Belgian family was having cleaned a painting they owned, which came from the Spanish painter Bartolomé Esteban Murillo, or someone of his school, and which was dated 1643 (or 1645). The painting depicted an old Galileo in the dungeons of the Inquisition, pointing a finger at the dungeon wall just where the canvas was folded under. When unfolded during cleaning it revealed the hidden words "eppur si muove." It would have been in character for Galileo, upon leaving the home of his good friend Archbishop Ascanio Piccolomini, to have stamped the ground, winked, and uttered the words. (After the trial, the first location of his house arrest was Piccolomini's estate; he moved from there to his own residence in Arcetri at the end of 1633.) Ascanio's brother, Ottavio, was a professional soldier, at the time stationed in Madrid. It's quite possible, even plausible, that the story lived on orally among the Piccolomini family and close friends in quiet circulation, out of concern for Galileo's safety, and was picked up in Madrid by the artist to be included, but concealed in the painting. Stillman Drake, *Galileo at Work: His Scientific Biography* (New York: Dover, 1978), 356–357. In any event, it makes for a nice story.

78. "Galileo's Letter to the Grand Duchess," 93, 104. See too "Galileo to Castelli (21 December 1613)," 50–51.

Epilogue

1. Francis Bacon, *The Great Instauration,* in *The Works of Francis Bacon,* 2 vols. (New York: Hurd and Houghton, 1878), 1:67.

2. Philosopher C. D. Broad charmingly and trenchantly styled "Inductive Reasoning" as the "glory of Science" but the "scandal of Philosophy." C. D. Broad, *The Philosophy of Francis Bacon: An Address Delivered at Cambridge on the Occasion*

of the Bacon Tercentenary, 5 October 1926 (Cambridge: Cambridge University Press, 1926), http://www.ditext.com/broad/bacon.html.

3. Eugene Wigner, "The Unreasonable Effectiveness of Mathematics in the Natural Sciences," *Communications in Pure and Applied Mathematics* 13, no. 1 (1960): 2 and more generally 1–14.

4. The phrase comes from linguistics and combines "fixed and invariant principles" with a "kind of switch box of parameters" to help explain the hierarchical "architecture" of language. As presented in this study, information technologies can be seen as combining a similar structure of basic rules and the parameters of their application. See Robert C. Berwick and Noam Chomsky, in *Why Only Us: Language and Evolution* (Boston: MIT Press, 2015), 7, 68–69.

5. As mentioned earlier, these comments refer primarily to the "Western" heritage of object-oriented languages, leaving aside other linguistic traditions. Still, we need to bear in mind that modern science began in western Europe, and that it did so growing out of the new information technology of modern numeracy, whose backstory and context was provided by alphabetic literacy.

6. Alfred North Whitehead, *Science and the Modern World* (1925; New York: New American Library / Mentor Books, 1948), 33–34.

7. Thus, according to David Wootton, Galileo "preferred the abstraction, the perfection, and the theoretical nature of mathematics." *Galileo: Watcher of the Skies* (New Haven, CT: Yale University Press, 2010), 232.

8. See Richard Dawkins, *The Blind Watchmaker: Why the Evidence of Evolution Reveals a Universe without Design* (New York: W. W. Norton, 1986), 13, where he also introduces the summary expression "hierarchical reductionism." Strong reductionists, Dawkins and E. O. Wilson for example, appeal to reductionism as a way of explaining virtually all topics in the natural and social sciences. Others challenge this position by showing its inadequacy to account for emergent phenomena, or complexity, or feedback loops between levels, and comparable topics. In biology, the works of Stuart Kauffman, including *At Home in the Universe: The Search for Laws of Self-Organization and Complexity* (Oxford: Oxford University Press, 1995) and others, present a strong case for antireductionist, more holistic methods of scientific investigation.

9. A host of terms fills the current lexicon for sorting out levels of appropriate scientific explanation and arenas of analysis—'coarse' and 'fine graining,' 'explanation framing' or 'explanation space,' 'boundary conditions,' 'system analysis,' 'micro' versus 'macro,' to name but a few. Many of these serve holistic, as well as reductionist, modes of mathematical and physical description.

10. See Michael E. Hobart and Zachary S. Schiffman, *Information Ages: Literacy, Numeracy, and the Computer Revolution* (Baltimore: Johns Hopkins University Press, 1998), 201–259.

11. Foundational works in these developments include the symbolic logic of George Boole (1815–1864) and the set theory of the earlier-mentioned Georg Cantor. Cantor's pathbreaking work was refined by Ernst Zermelo (1871–1953),

Abraham Fraenkel (1891–1965), and other twentieth-century mathematical logicians who created modern, axiomatic set theory. Among his other innovations Cantor redefined the notion of cardinality to show how infinite, countable sets can have different cardinalities, the first of which, in his notation, was an $\aleph_0$ (aleph zero or aleph null), the infinite, countable set of natural numbers (which could, in turn, be followed by $\aleph_1, \aleph_2 \ldots$, and other "transfinite" numbers). For a brief and manageable introduction to this involved and far-reaching topic, see Edna E. Kramer, *The Nature and Growth of Modern Mathematics* (Princeton, NJ: Princeton University Press, 1981), 577–597. Of particular note for us, Kramer shows how Cantor's set theory built on the intuitive anticipations of Galileo regarding infinity (introduced in his considerations of matter) and finally resolved Zeno's paradoxes.

12. The expression is from Steven Weinberg, *Dreams of a Final Theory: The Scientist's Search for the Ultimate Laws of Nature* (New York: Vintage Books, 1992).

13. See David Wootton, *The Invention of Science: A New History of the Scientific Revolution* (New York: HarperCollins, 2015), 251, and more generally 251–309, for an account of "killer facts: facts which required the abandonment of well-established theories" and for the development of the "fact" as the "very foundation of all knowledge."

14. Richard P. Feynman, *Six Easy Pieces* (New York: Basic Books, 1963), 15. Feynman used the metaphor of a chess game to describe scientific understanding. Nature is "like a great chess game," he noted. "We are the observers," and if we watch long enough we may "catch on to a few of the rules. The *rules of the game* are what we mean by *fundamental physics*" (24, italics in original). Such metaphors may or may not help us understand what scientists are doing, but they are not the science per se. This was the distinction that Galileo made and exploited.

15. Alexander Pope, *An Essay on Man*, ed. Henry Morely (London: Cassell, 1891; Project Gutenberg, 2007), epistle 1, stanza 9, https://www.gutenberg.org/files/2428/2428-h/2428-h.htm.

16. David Hume, *An Enquiry concerning Human Understanding* (La Salle, IL: Open Court, 1956), 145, and in general the chapter "On Miracles," 120–145.

17. "Jefferson to Mr. Charles Thompson," in Thomas Jefferson, *The Jefferson Bible: The Life and Morals of Jesus of Nazareth* (1902; Mineola, NY: Dover, 2006), 12.

18. William Wordsworth, "The Tables Turned," line 28 (1798), in *The Complete Poetical Works,* intro. John Morley (London: Macmillan, 1888). The eighteenth and nineteenth centuries also witnessed numerous "Great Awakenings" of religious "enthusiasm," revival movements that sought to recapture religious mystery largely through greater attention to human emotion, rather than reason.

19. John F. W. Herschel, *A Preliminary Discourse on the Study of Natural Philosophy* (Chicago: University of Chicago Press, 1987), 5, 13, 16.

20. Michael Faraday, "Observations on Mental Education," in *Experimental Researches in Chemistry and Physics* (London: R. Taylor and W. Francis, 1859), 464, 471, and more generally 463–491.

21. Charles Darwin, *On the Origin of the Species,* facsimile of 1st ed., ed. Ernst Mayr (Cambridge, MA: Harvard University Press, 1964), 485.

22. A similar bifurcation has expressed itself in the recent culture wars between the sciences and humanities, the manifestation of what C. P. Snow, over a half century ago, advanced as a general description of an intellectual world divided into the "two cultures" of our own age. Although certainly not identical, the science-religion and science-humanities rifts of recent years overlap at many points, for both separations rest on the deeper division between the information technologies of literacy and numeracy. See the collection of lectures of C. P. Snow, *The Two Cultures and the Scientific Revolution* (Cambridge: Cambridge University Press, 1961).

23. The expression comes from the long-enduring, classic work on the Enlightenment by Carl L. Becker, *The Heavenly City of the Eighteenth-Century Philosophers* (New Haven, CT: Yale University Press, 1932).

24. Charles Taylor, *The Language Animal: The Full Shape of the Human Linguistic Capacity* (Cambridge, MA: Harvard University Press, 2016), 319 and more generally 291–319.

Appendixes

1. To add or combine the intervals, we actually multiply the ratios that correspond to them; to subtract them, we divide. For example, if we start with a given string and take another, two-thirds its length $(\frac{2}{3} \times 1)$, we go up in pitch a musical fifth (for example, from C to G on the piano keyboard). Taking three-quarters of the new string, we go up another fourth (from G to the piano's next-higher C′). Combining the two operations, $\frac{3}{4} \times \frac{2}{3} = \frac{1}{2}$. This corresponds to half the original string, the octave higher we have traveled on the piano. Thus a fourth plus a fifth equals an octave, and we derive that from multiplying their respective ratios.

2. A noteworthy introduction to the various interpretations may be found in William R. Shea and Mariano Artigas, *Galileo in Rome: The Rise and Fall of a Troublesome Genius* (Oxford: Oxford University Press, 2003). For the *warfare thesis,* see Andrew Dickson White, *A History of the Warfare of Science with Theology in Christendom,* 2 vols. (1896; New York: Dover, 1960), esp. vol. 1, ch. 3, and the sections devoted to the "war upon Galileo."

3. The *hubris hypothesis* derives from no less an authority than Robert Bellarmine himself, who counseled "prudence" in scientific matters: "Cardinal Bellarmine to Foscarini (12 April 1615)," in *The Galileo Affair: A Documentary History,* ed. Maurice A. Finocchiaro (Berkeley: University of California Press, 1989), 67. For a recent version, see Wade Rowland, *Galileo's Mistake: A New Look at the Epic Confrontation between Galileo and the Church* (New York: Arcade, 2001).

4. The *godfather scenario* emerges from Giorgio de Santillana's enjoyable and still valuable *Crime of Galileo* (Chicago: University of Chicago Press, 1955), xiii and throughout.

5. For the *courtier's demise*, see Mario Biagioli, *Galileo, Courtier: The Practice of Science in the Culture of Absolutism* (Chicago: University of Chicago Press, 1993), and Annibale Fantoli, *Galileo: For Copernicanism and for the Church,* 2nd ed. (Notre Dame, IN: University of Notre Dame Press, 1996), 456.

6. The *geopolitical web* is deftly spun by Michael H. Shank in "Setting the Stage: Galileo in Tuscany, the Veneto, and Rome," in *The Church and Galileo,* ed. Ernan McMullin (Notre Dame, IN: University of Notre Dame Press, 2005), 69–81.

7. For the *scholastic shuffle*, see Charles E. Hummel, *The Galileo Connection: Resolving Conflicts between Science and the Bible* (Downers Grove, IL: InterVarsity, 1986), ch. 5.

8. The *cast of the die* concludes with the remark of Ernan McMullin in "The Church's Ban on Copernicanism, 1616," in McMullin, *Church and Galileo,* 173.

Index